ELECTROMAGNETIC
METAMATERIALS

ELECTROMAGNETIC METAMATERIALS: TRANSMISSION LINE THEORY AND MICROWAVE APPLICATIONS

The Engineering Approach

CHRISTOPHE CALOZ
École Polytechnique de Montréal

TATSUO ITOH
University of California at Los Angeles

WILEY-
INTERSCIENCE

A JOHN WILEY & SONS, INC., PUBLICATION

For general information on our other products and services or for technical support, please contact our Customer Care Department within the United States at (800) 762-2974, outside the United States at (317) 572-3993 or fax (317) 572-4002.

Wiley also publishes its books in a variety of electronic formats. Some content that appears in print may not be available in electronic formats. For more information about Wiley products, visit our web site at www.wiley.com.

Library of Congress Cataloging-in-Publication Data:

Caloz, Christophe, 1969-
  Electromagnetic metamaterials : transmission line theory and microwave applications : the engineering approach / Christophe Caloz, Tatsuo Itoh.
       p.cm.
  "Wiley-Interscience publication."
  Includes bibliographical references and index.
  ISBN-10: 0-471-66985-7 (alk.paper)
  ISBN-13: 978-0-471-66985-2 (alk.paper)
    1. Magnetic materials. 2. Nanostructured materials. 3. Microwave transmission lines. I. Itoh, Tatsuo. II. Title.

TK454.4.M3C36 2006
620.1′18—dc22

                                                                              2005048976

10 9 8 7 6 5 4 3

To Dominique, Maxime, and Raphaël
Christophe

CONTENTS

3 TL Theory of MTMs   59

4 Two-Dimensional MTMs   133

PREFACE

This book is essentially the fruit of a research work carried out at University of California, Los Angeles (UCLA), from 2002 to 2004 in the context of a Multidisciplinary University Research Initiative (MURI) program. The main participants in this program, in addition to the authors, were John Pendry (Imperial College), David Smith (formerly University of California, San Diego (UCSD), now Duke University), Sheldon Schultz (UCSD), Xiang Zhang (formerly UCLA, now University of California, Berkeley), Gang Chen (formerly UCLA, now Massachusetts Institute of Technology (MIT)), John Joannopoulos (MIT) and Eli Yablonovitch (UCLA).

During these years of infancy for metamaterials (MTMs), which emerged from the first experimental demonstration of a left-handed (LH) structure in 2000, the vast majority of the groups involved in research on MTMs had been focusing on investigating from a physics point of view the fundamental properties of LH media predicted by Veselago in 1967. Not following this trend, the authors adopted an engineering approach, based on a generalized transmission line (TL) theory, with systematic emphasis on developing practical applications, exhibiting unprecedented features in terms of performances or functionalities. This effort resulted in the elaboration of the powerful composite right/left-handed (CRLH) MTM concept, which led to a suite of novel guided-wave, radiated-wave, and refracted-wave devices and structures.

This book presents electromagnetic MTMs and their applications in the general framework of CRLH TL MTMs. Chapter 1 introduces MTMs from a historical perspective and points out their novelty compared with conventional periodic backward-wave media. Chapter 2 exposes the fundamentals of ideal[1]

---

[1]"Ideal" designates here perfectly homogeneous and isotropic volumic (left-handedness along the three directions of space) materials.

LH MTMs, including the antiparallelism existing between phase/group velocities, the frequency dispersion required from entropy conditions, the modified boundary conditions, the modified Fresnel coefficients, the reversal of classical phenomena [Doppler effect, Vavilov-Čerenkov radiation, Snell's law (negative refractive index, NRI), Goos-Hänchen effect, lenses convergence/divergence], and focusing by a flat "lens" and subwavelength diffraction. Chapter 3. establishes the foundations of CRLH structures, in three progressive steps: ideal TL, LC network, and real distributed structure. Chapter 4 extends theses foundations to the two-dimensional (2D) case, develops a transmission matrix method (TMM) and a transmission line method (TLM) techniques to address the problem of finite-size 2D MTMs excited by arbitrary sources, illustrates by TLM simulations a few NRI effects occurring in 2D MTMs (backward-wave propagation, negative refraction by a RH-LH interface, a LH slab, a LH prism, inner reflection of Gaussian beams in a LH prism, and a LH "perfectly absorbing reflector"), and describes real distributed implementations of 2D MTMs. Chapter 5 presents guided-wave applications of CRLH TL MTMs, including dual-band components, enhanced-bandwidth components, a super-compact multilayer "vertical" structure, tight edge-coupled directional couplers, and negative/zeroth-order resonators. Chapter 6 presents radiated-wave applications of CRLH TL MTMs, including 1D or 2D frequency-scanned and electronically scanned leaky-wave antennas and reflecto-directive systems. Finally, future challenges and prospects of MTMs, including homogenized MTMs, quasi-optical NRI lenses, 3D isotropic LH structures, optical MTMs, magnetless magnetic MTMs, surface plasmonic MTMs, MTMs antenna radomes and frequency selective surfaces, nonlinear, and active MTMs, are discussed in Chapter 7.

It is hoped that the engineering approach of MTMs presented in this book will pave the way for a novel generation of microwave and photonic devices and structures.

C. CALOZ AND T. ITOH

Montéal, Québec, and Los Angeles, California

ACKNOWLEDGMENTS

Many researchers have contributed to the realization of this book.

The authors thank all the master's and Ph.D. students of UCLA who have been involved in this work: Anthony Lai, Sungjoon Lim, I-Hsiang Lin, Catherine Allen, Lei Liu, Simon Otto, Chen-Jung Lee, and Marc de Vincentis.

In addition, the authors express their gratitude to several visiting scholars of UCLA, who had various degrees of participation in this research, including Hiroshi Okabe (Hitachi Central Research Laboratory), Yasushi Horii (Kansai University), Taisuke Iwai (Fujitsu's Research Labs), and Dal Ahn (Soonchunhyang University). Special thanks are addressed to Atsushi Sanada (Yamagushi University), who played a major role in the development of several concepts and applications presented in the book.

Finally, very special thanks are addressed to Wolfgang Hoefer, University of Victoria, British Columbia, who kindly accepted to contribute to this work by authoring two sections in Chapter 4 on the transmission line method (TLM) analysis of 2D metamaterials.

This work was part of the MURI program "Scalable and Reconfigurable Electromagnetic Metamaterials and Devices." It was supported by the Department of Defense (program N00014-01-1-0803) and monitored by the U.S. Office of Naval Research.

C. C. AND T. I.

ACRONYMS

ATR	Attenuated total reflection
BCs	Boundary conditions
BE	Backfire-to-endfire
BWO	Backward-wave oscillator
BZ	Brillouin zone
CRLH	Composite right/left-handed
CLC	Coupled-line coupler
DB	Dual-band
DPDS	Dual-passband dual-stopband
FDTD	Finite-difference time domain
FEM	Finite-element method
FSS	Frequency selective surface
GVD	Group velocity dispersion
IC	Impedance coupling or coupler
LH	Left-handed
LTCC	Low-temperature cofired ceramics
LW	Leaky-wave
MIM	Metal-insulator-metal
MTM	Metamaterial
MLV	Multilayered vertical
MoM	Method of moments
MRI	Magnetic resonance imaging
MSBVW	Magnetostatic backward volume wave
NRI	Negative refractive index
OPL	Optical path length

PC	Phase coupling or coupler
PCB	Printed circuit board
PBCs	Periodic boundary conditions
PBG	Photonic band-gap
PPWG	Parallel-plate waveguide
PLH	Purely left-handed
PRH	Purely right-handed
QH	Quadrature hybrid
RH	Right-handed
RCS	Radar cross section
RDS	Reflecto-directive system
RRC	Rat-race coupler
SHPM	Subharmonically pumped mixer
SP	Surface plasmon
SPM	Self-phase modulation
SRR	Split-ring resonator
TE	Transverse electric
TEM	Transverse electric-magnetic
TH	Thin-wire
TL	Transmission line
TLM	Transmission line method
TM	Transverse magnetic
TMA	Transfer matrix algorithm
TMM	Transmission matrix method
TWT	Traveling-Wave Tube
UWB	Ultra wideband
ZOR	Zeroth-order resonator

1

INTRODUCTION

Chapter 1 introduces electromagnetic metamaterials (MTMs) and left-handed (LH) MTMs from a general prospect. Section 1.1 defines them. Section 1.2 presents the theoretical speculation by Viktor Veselago on the existence of "substances with simultaneously negative ε and μ" in 1967, which is at the origin of all research on LH MTMs. Section 1.3 describes the first experimental demonstration of left-handedness in 2000 by Smith et al., and Section 1.4 lists a number of further numerical, theoretical, and experimental confirmations of the fundamental properties of LH structures. The difference between "conventional" backward waves, known for many decades, and LH waves, as well as the essential novelty of LH MTMs, are explained in Section 1.5. Section 1.6 discusses the different terminologies used in the literature to designate LH MTMs. Next, Section 1.7 introduces the transmission line (TL) approach of LH MTMs, which has constituted a decisive step toward microwave applications, and Section 1.8 presents, in a concise manner, the generalized composite right/left-handed (CRLH) concept upon which the whole book is based. Finally, Section 1.9 points out the essential difference existing between photonic crystals or photonic band-gap (PBG) structures, more conventionally simply called "periodic structures", and MTMs.

1.1 DEFINITION OF METAMATERIALS (MTMs) AND LEFT-HANDED (LH) MTMs

Electromagnetic metamaterials (MTMs) are broadly defined as *artificial effectively homogeneous electromagnetic structures with unusual properties not readily*

Electromagnetic Metamaterials: Transmission Line Theory and Microwave Applications,
By Christophe Caloz and Tatsuo Itoh
Copyright © 2006 John Wiley & Sons, Inc.

available in nature. An *effectively homogeneous* structure is a structure whose *structural average cell size p* is *much smaller than the guided wavelength* λ_g. Therefore, this average cell size should be at least smaller than a quarter of wavelength, $p < \lambda_g/4$. We will refer to the condition $p = \lambda_g/4$ as the *effective-homogeneity limit* or *effective-homogeneity condition*[1], to ensure that *refractive phenomena* will dominate over *scattering/diffraction* phenomena when a wave propagates inside the MTM medium. If the condition of effective-homogeneity is satisfied, the structure behaves as a real material in the sense that electromagnetic waves are essentially *myopic to the lattice* and only probe the average, or effective, macroscopic and well-defined *constitutive parameters, which depend on the nature of the unit cell*; the structure is thus *electromagnetically uniform* along the direction of propagation. The constitutive parameters are the permittivity ε and the permeability μ, which are related to the refractive index n by

$$n = \pm\sqrt{\varepsilon_r\mu_r}, \qquad (1.1)$$

where ε_r and μ_r are the relative permittivity and permeability related to the free space permittivity and permeability by $\varepsilon_0 = \varepsilon/\varepsilon_r = 8.854 \cdot 10^{-12}$ and $\mu_0 = \mu/\mu_r = 4\pi \cdot 10^{-7}$, respectively. In (1.1), sign $\pm$ for the double-valued square root function has been a priori admitted for generality.

The four possible sign combinations in the pair (ε, μ) are $(+, +)$, $(+, -)$, $(-, +)$, and $(-, -)$, as illustrated in the $\varepsilon - \mu$ diagram of Fig 1.1. Whereas the first three combinations are well known in conventional materials, the last one $[(-, -)]$, with *simultaneously negative permittivity and permeability*, corresponds to the new class of left-handed (LH) materials.[2] LH materials, as a consequence of their double negative parameters, are characterized by antiparallel phase and group velocities, or negative refractive index (NRI) [negative sign in Eq. (1.1)], as will be shown in Chapter 2.

LH structures are clearly MTMs, according to the definition given above, since they are artificial (fabricated by human hands), effectively homogeneous $(p < \lambda_g/4)$, and exhibit highly unusual properties $(\varepsilon_r, \mu_r < 0)$. It should be noted that, although the term MTM has been used most often in reference to LH structures in the literature, MTMs may encompass a much broader range of

---

[1] This limit corresponds to a rule of thumb effectiveness condition. Microwave engineers often use the limit $p = \lambda_g/4$, where p is the size of the component considered, to distinguish lumped components $(p < \lambda_g/4)$ from quasi-lumped components $(\lambda_g/4 < p < \lambda_g/2)$ and distributed components $(p > \lambda_g/2)$ [1]. In the lumped case, the phase variation of the signal from the input to the output of the component is negligible, and the component may therefore be considered as a localized (sizeless) element. In contrast, in the distributed case, phase variation along the component cannot be ignored, and the component must consequently be considered as a transmission line section. MTMs are thus "distributed structures constituted of lumped elements."

[2] It should be noted that thin ferrimagnetic films biased in the plane of the film support magnetostatic backward volume wave (MSBVW) in the direction of bias, which may be seen as a LH phenomenon (Section 2.1) in a *real* material. However, this phenomenon is strictly dependent on the thin slab shape of the film and could not exist per se in a bulk ferrite and is therefore not an isotropic LH effect.

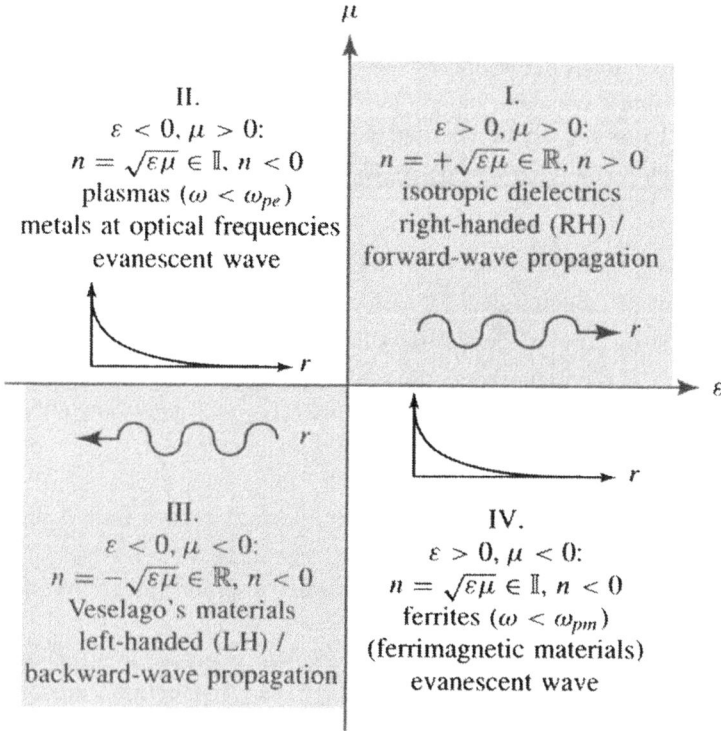

Fig. 1.1. Permittivity-permeability ($\varepsilon - \mu$) and refractive index (n) diagram. The time dependence $e^{+j\omega t}$, associated with the *outgoing* wave function $e^{-j\beta r}$ or *incoming* wave function $e^{+j\beta r}$, where β is the propagation constant, $\beta = nk_0$ ($k_0 = \omega/c$: free space wavenumber, ω: angular frequency, c: speed of light), is assumed. The angular frequencies ω_{pe} and ω_{pm} represent the electric and magnetic plasma frequencies, respectively. $\mathbb{R}$, purely real. $\mathbb{I}$, purely imaginary.

structures.[3] However, LH structures have been by far the most popular of the MTMs, due to their exceptional property of negative refractive index (Chapter 2). This book mainly deals with this class of MTMs and, more precisely, with composite right/left-handed (CRLH) MTMs (Section 1.8), which constitute a generalization of LH MTMs.

1.2 THEORETICAL SPECULATION BY VIKTOR VESELAGO

The history of MTMs started in 1967 with the visionary speculation on the existence of *"substances with simultaneously negative values of ε and μ"* (fourth

---

[3]Examples are MTMs with only one of the two constitutive parameters negative, anisotropic MTMs, or any type of functional effective engineered structure. In addition, many existing materials obtained by novel nanotechnology and chemistry processes may be regarded as MTMs.

quadrant in Fig 1.1) by the Russian physicist Viktor Veselago [3].[4] In his paper, Veselago called these "substances" LH to express the fact that they would allow the propagation of electromagnetic waves with the electric field, the magnetic field, and the phase constant vectors building a left-handed triad, compared with conventional materials where this triad is known to be *right-handed* (Section 2.1).

Several fundamental phenomena occurring in or in association with LH media were predicted by Veselago:

1. Necessary frequency dispersion of the constitutive parameters (Section 2.2).
2. Reversal of Doppler effect (Section 2.4).
3. Reversal of Vavilov-Cerenkov radiation (Section 2.5).
4. Reversal of the boundary conditions relating the normal components of the electric and magnetic fields at the interface between a conventional/right-handed (RH) medium and a LH medium (Section 2.3).
5. Reversal of Snell's law (Section 2.6).
6. Subsequent negative refraction at the interface between a RH medium and a LH medium (Section 2.6).
7. Transformation of a point source into a point image by a LH slab (Section 2.6).
8. Interchange of convergence and divergence effects in convex and concave lenses, respectively, when the lens is made LH (Section 2.6).
9. Plasmonic expressions of the constitutive parameters in resonant-type LH media (Section 1.3).

Veselago concluded his paper by discussing potential *real* (natural) "substances" that could exhibit left-handedness. He suggested that "gyrotropic substances possessing plasma and magnetic properties" ("pure ferromagnetic metals or semiconductors"), "in which both ε and μ are tensors" (anisotropic structures), could possibly be LH.[5] However, he recognized, "Unfortunately, . . . , we do not know of even a single substance which could be isotropic and have $\mu < 0$," thereby pointing out how difficult it seemed to realize a practical LH structure. No LH material was indeed discovered at that time.

1.3   EXPERIMENTAL DEMONSTRATION OF LEFT-HANDEDNESS

After Veselago's paper, more than 30 years elapsed until the first LH material was conceived and demonstrated experimentally. This LH material was not a natural substance, as expected by Veselago, but an *artificial* effectively homogeneous

---

[4]This paper was originally published in Russian in 1967 and then translated into English in 1968. In this translation, the date of 1964 is indicated by mistake as the publication year of the original Russian version.

[5]This is indeed the case, in a restricted sense, of ferrite films supporting MSBVWs, as pointed out in note 2.

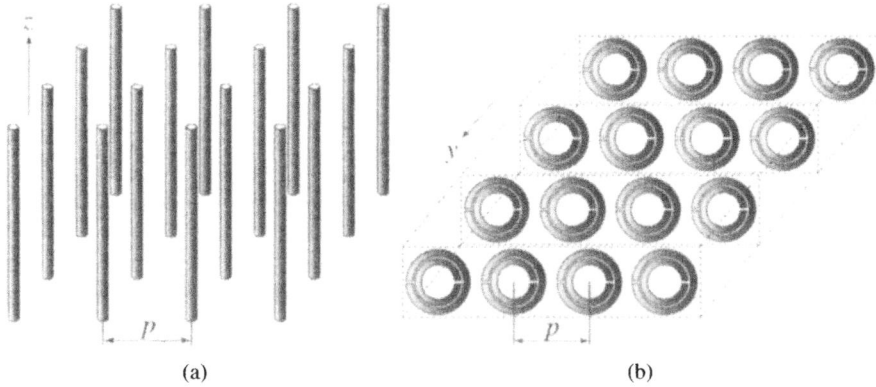

(a)   (b)

Fig. 1.2. First negative-ε/positive-μ and positive-ε/negative-μ MTM ($p \ll \lambda_g$), constituted only by standard metals and dielectrics, proposed by Pendry. (a) Thin-wire (TW) structure exhibiting negative-ε/positive-μ if $\mathbf{E}\|z$ [6]. (b) Split-ring resonator (SRR) structure exhibiting positive-ε/negative-μ if $\mathbf{H}\perp y$ [7].

structure (i.e., a MTM), which was proposed by Smith and colleagues at University of California, San Diego (UCSD) [4].[6] This structure was inspired by the pioneering works of Pendry at Imperial College, London. Pendry introduced the *plasmonic-type* negative-ε/positive-μ and positive-ε/negative-μ structures shown in Fig 1.2, which can be designed to have their *plasmonic frequency in the microwave range*.[7] Both of these structures have an average cell size p much smaller than the guided wavelength λ_g ($p \ll \lambda_g$) and are therefore effectively homogeneous structures, or MTMs.

The negative-ε/positive-μ MTM is the *metal thin-wire (TW) structure* shown in Fig 1.2(a). If the excitation electric field $\mathbf{E}$ is parallel to the axis of the wires ($\mathbf{E}\|z$), so as to induce a current along them and generate equivalent electric dipole moments,[8] this MTM exhibits a plasmonic-type permittivity frequency function of the form [5, 6]

$$\varepsilon_r(\omega) = 1 - \frac{\omega_{pe}^2}{\omega^2 + j\omega\zeta} = 1 - \frac{\omega_{pe}^2}{\omega^2 + \zeta^2} + j\frac{\zeta\,\omega_{pe}^2}{\omega(\omega^2 + \zeta^2)}, \qquad (1.2)$$

where $\omega_{pe} = \sqrt{2\pi c^2/[p^2\ln(p/a)]}$ (c: speed of light, a: radius of the wires) is the *electric plasma frequency*, tunable in the GHz range, and $\zeta = \varepsilon_0(p\omega_{pe}/a)^2/\pi\sigma$

[6]The fact that this first LH MTM was artificial, as all the other LH MTMs known to date!, does not at all exclude that natural substances exhibiting left-handedness could be discovered some day. Because left-handedness is allowed by fundamental physics laws in artificial structures, there is no reason why it would not be also possible in natural substances.

[7]This is distinct from conventional gas and metal plasmas, which have their plasma frequency far above the microwave range, preventing negative ε at microwave frequencies.

[8]If the electric field $\mathbf{E}$ is perfectly parallel to the axis of the wires (z), a maximum of effect is obtained. If it is exactly perpendicular to the wires, we have a situation of cross polarization, where no effect is produced. In the intermediate situation where the electric field is oblique with respect to the wires, a reduced effect occurs, decreasing as the angle with the wires increases.

(σ: conductivity of the metal) is a damping factor due to metal losses. It clearly appears in this formula that

$$\text{Re}(\varepsilon_r) < 0, \quad \text{for} \quad \omega^2 < \omega_{pe}^2 - \zeta^2, \tag{1.3a}$$

which reduces if $\zeta = 0$ to

$$\varepsilon_r < 0, \quad \text{for} \quad \omega < \omega_{pe}. \tag{1.3b}$$

On the other hand, permeability is simply $\mu = \mu_0$, since no magnetic material is present and no magnetic dipole moment is generated. It should be noted that the wires are assumed to be much longer than wavelength (theoretically infinite), which means that the wires are excited at frequencies situated far below their first resonance.

The positive-ε/negative-μ MTM is the *metal split-ring resonator (SRR)* structure[9] shown in Fig 1.2(b). If the excitation magnetic field **H** is perpendicular to the plane of the rings (**H**$\perp y$), so as to induce resonating currents in the loop and generate equivalent *magnetic dipole moments*,[10] this MTM exhibits a plasmonic-type permeability frequency function of the form [7]

$$
\begin{aligned}
\mu_r(\omega) &= 1 - \frac{F\omega^2}{\omega^2 - \omega_{0m}^2 + j\omega\zeta} \\
&= 1 - \frac{F\omega^2(\omega^2 - \omega_{0m}^2)}{(\omega^2 - \omega_{0m}^2)^2 + (\omega\zeta)^2} + j\frac{F\omega^2\zeta}{(\omega^2 - \omega_{0m}^2)^2 + (\omega\zeta)^2},
\end{aligned}
\tag{1.4}
$$

where $F = \pi(a/p)^2$ (a: inner radius of the smaller ring), $\omega_{0m} = c\sqrt{\frac{3p}{\pi \ln(2wa^3/\delta)}}$ (w: width of the rings, δ: radial spacing between the rings) is a magnetic resonance frequency, tunable in the GHz range, and $\zeta = 2pR'/a\mu_0$ (R': metal resistance per unit length) is the damping factor due to metal losses. It should be noted that the SRR structure has a magnetic response despite the fact that it does not include magnetic conducting materials due to the presence of artificial magnetic dipole moments provided by the ring resonators. Eq. (1.4) reveals that a frequency range can exist in which $\text{Re}(\mu_r) < 0$ in general ($\zeta \neq 0$). In the loss-less case ($\zeta \neq 0$), it appears that

$$\mu_r < 0, \quad \text{for} \quad \omega_{0m} < \omega < \frac{\omega_{0m}}{\sqrt{1 - F}} = \omega_{pm}, \tag{1.5}$$

where ω_{pm} is called the *magnetic plasma frequency*. An essential difference between the plasmonic expressions of ε and μ is that the latter is of *resonant*

---

[9]One single ring in the unit cell produces qualitatively identical effects, but the magnetic activity, effective permeability and bandwidth, is enhanced by the presence of a second ring due to larger overall current and slightly different overlapping resonances.

[10]A remark analogous to that of note 8 holds here under the substitutions $\mathbf{E} \to \mathbf{H}$ and $z \to y$.

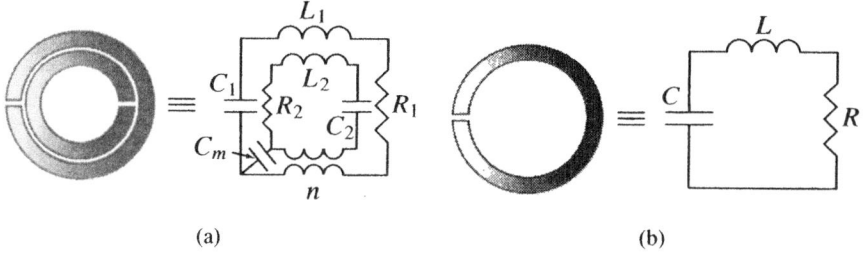

Fig. 1.3. Equivalent circuit model of SRRs. (a) Double SRR configuration (e.g., [7]). (b) Single SRR configuration.

nature $[\mu(\omega = \omega_{0m}) = \infty]$, whereas the former is a nonresonant expression. The resonance of the structure is due to the resonance of its SRRs, given in [7] by $\omega_{0m}^2 = 3pc^2/[\pi \ln(2w/\delta)a^3]$ (p: period, $c = 1/\sqrt{\varepsilon_0\mu_0}$: speed of light, w: width of the rings, δ: spacing between the rings).

The equivalent circuit of a SRR is shown in Fig 1.3 [8]. In the double ring configuration [Fig 1.3(a)], capacitive coupling and inductive coupling between the larger and smaller rings are modeled by a coupling capacitance (C_m) and by a transformer (transforming ratio n), respectively. In the single ring configuration [Fig 1.3(b)], the circuit model is that of the simplest RLC resonator with resonant frequency $\omega_0 = 1/\sqrt{LC}$. The double SRR is essentially equivalent to the single SRR if mutual coupling is weak, because the dimensions of the two rings are very close to each other, so that $L_1 \approx L_2 \approx L$ and $C_1 \approx C_2 \approx C$, resulting in a combined resonance frequency close to that of the single SRR with same dimensions but with a larger magnetic moment due to higher current density.

In [4], Smith et al. combined the TW and SRR structures of Pendry into the composite structure shown in Fig 1.4(a), which represented the first experimental LH MTM prototype. The arguments in [4] consisted of the following: 1) designing a TW structure and a SRR structure with overlapping frequency ranges of negative permittivity and permeability; 2) combining the two structures into a composite TW-SRR structure, which is shown in Fig 1.4(a); and 3) launching an electromagnetic wave $e^{-j\beta r}$ through the structure and concluding from a fact that a passband (or maximum transmission coefficient, experimentally) appears in the frequency range of interest proves that the constitutive parameters are simultaneously negative in this range on the basis of the fact that $\beta = nk_0 = \pm\sqrt{\varepsilon_r\mu_r}$ has to be real in a passband.

Although the arguments of [4] were questionable, because it ignored the fact that coupling interactions between the two constituent structures could yield properties totally different from the superposition of the properties of each structure taken separately,[11] a vivid experimental demonstration of the LH nature of the TW-SRR was provided in [10]. In this paper, the TW-SRR structure of Fig 1.4(b)

---

[11] In fact, this structure was really "working" only due to the judicious position of the wires in the symmetry center of the rings, which canceled interactions between the two structures due to induced

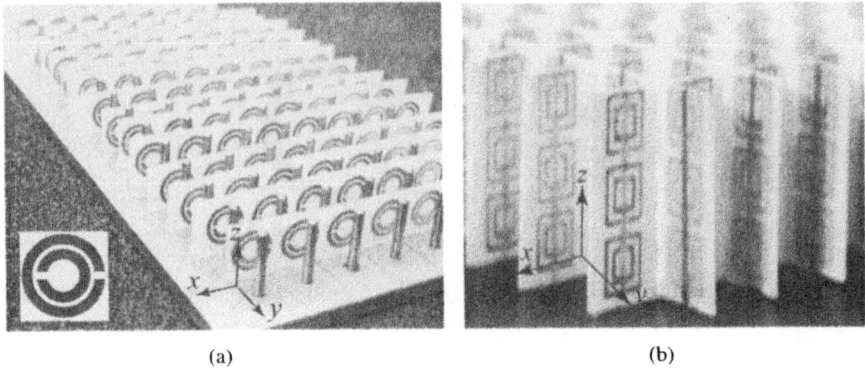

(a)   (b)

Fig. 1.4. First experimental LH structures, constituted of TWs and SRRs, introduced by the team of UCSD. (a) Monodimensionally LH structure of [4]. (b) Bidimensionally LH structure of [10].

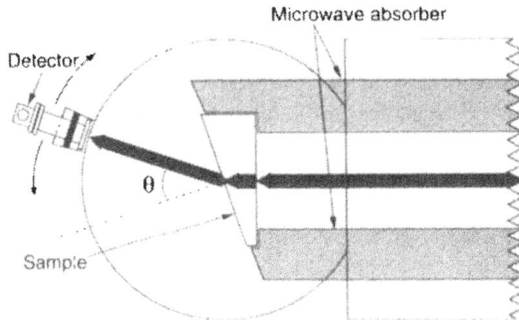

Fig. 1.5. Experimental setup used in [10] for the demonstration of left-handedness of the TW-SRR structure of Fig 1.4(b) at around 5 GHz. "The sample and the microwave absorber were placed between top and bottom parallel, circular aluminum plates spaced 1.2 cm apart. The radius of the circular plates was 15 cm. The black arrows represent the microwave beam as would be refracted by a positive index sample. The detector was rotated around the circumference of the circle in 1.5° steps, and the transmitted power spectrum was measured as a function of angle, θ, from the interface normal. The detector was a waveguide to coaxial adapter attached to a standard X-band waveguide, whose opening was 2.3 cm in the plane of the circular plates. θ as shown is positive in this figure." From [10]. © Science; reprinted with permission.

was cut into a wedge-shaped piece of MTM and inserted into the experimental apparatus depicted in Fig 1.5. Left-handedness of the TW-SRR structure was evidenced by the fact that a maximum of the transmission coefficient was measured in the negative angle (below the normal in the figure) with respect to the

currents with opposite signs in the SRRs and TWs [9]. Note that there an other possible symmetry location for the TWs is in the midplanes between the SRRs.

interface of the wedge, whereas a maximum in the positive angle (above the normal) was measured, as expected, when the wedge was replaced by a regular piece of teflon with identical shape. The result reported was in qualitative and quantitative agreement with *Snell's law*, which reads

$$k_1 \sin \theta_1 = k_2 \sin \theta_2, \tag{1.6a}$$

or, if the two media are isotropic (so that $k_1 = n_1 k_0$ and $k_2 = n_2 k_0$),

$$n_1 \sin \theta_1 = n_2 \sin \theta_2, \tag{1.6b}$$

where k_i, n_i, and θ_i represent the wavenumber, refractive index, and angle of the ray from the normal to the interface, respectively, in each of the two media considered ($i = 1, \ldots, 2$).

The MTMs described here are *anisotropic*[12] and characterized by uniaxial permittivity and permeability tensors

$$[\varepsilon] = \begin{bmatrix} \varepsilon_{xx} & 0 & 0 \\ 0 & \varepsilon_{yy} & 0 \\ 0 & 0 & \varepsilon_{zz} \end{bmatrix}, \tag{1.7a}$$

$$[\mu] = \begin{bmatrix} \mu_{xx} & 0 & 0 \\ 0 & \mu_{yy} & 0 \\ 0 & 0 & \mu_{zz} \end{bmatrix}. \tag{1.7b}$$

The structure shown in Fig 1.4(a) is monodimensionally LH, since only one direction is allowed for the doublet (**E**, **H**); we have $\varepsilon_{xx}(\omega < \omega_{pe}) < 0$ and $\varepsilon_{yy} = \varepsilon_{zz} > 0$, $\mu_{xx}(\omega_{0m} < \omega < \omega_{pm}) < 0$ and $\mu_{yy} = \mu_{zz} > 0$. The structure shown in Fig 1.4(b) is bidimensionally LH because, although **E** has to be directed along the axis of the wires, two directions are possible for **H**; then $[\varepsilon]$ is unchanged, but $\mu_{xx}, \mu_{yy} < 0$ for $\omega_{0m} < \omega < \omega_{pm}$ and $\mu_{zz} > 0$.

1.4 FURTHER NUMERICAL AND EXPERIMENTAL CONFIRMATIONS

In the few years after the first experimental demonstration of a LH structure by Smith et al., a large number of both theoretical and experimental reports confirmed the existence and main properties of LH materials predicted by Veselago.[13]

Characterization of the *TW-SRR LH structures* in terms of constitutive parameter functions similar to Eqs. (1.2) and (1.4) was provided by Smith et al. in [12], [13], and [14] and was then abundantly rediscussed and refined by various

---

[12]It is worth noting that various spatial filtering effects may be obtained from this anisotropic MTM [11].

[13]In 2002, some controversies temporarily cast doubt on these novel concepts [15, 16, 17]. However, these controversies were quickly shown to be based on physics misconceptions (e.g., [35]).

**TABLE 1.1. Verification of Main Properties and
Fundamental Physics of LH MTMs by Different
Approaches and Different Groups.**

Type of Investigation	References
Numerical FDTD	Ziolkowski et al. [27, 28]
Numerical FEM	Caloz et al. [29]
Numerical TMA	Markŏs et al. [20, 21]
Numerical TLM	So and Hoefer [30, 31]
Theoretical EM	Lindell et al. [32], Kong et al. [33, 34], Smith et al. [35], McCall et al. [36]
Experimental (bulk) TW-SRR	Shelby et al. [10], Greegor et al. [22], Ziolkowski [26], Marqués et al. [25]
Theoretical TL	Eleftheriades et al. [37, 38], Caloz and Itoh [39, 40], Oliner [41, 42]
Experimental (planar) TL	Iyer et al. [43], Caloz and Itoh [40], Sanada et al. [44]

groups, all confirming the LH nature of these structures (e.g., [18, 19, 20, 21, 22, 23, 24, 25, 26]).

Table 1.1 presents a list of key reports verifying, by different approaches, the main properties and fundamental physics of LH MTMs. The main *numerical methods* used are finite-difference time-domain (FDTD), finite-elements method (FEM), transfer matrix algorithm (TAM), and transmission line method (TLM). The *theoretical* verifications are subdivided into fundamental electromagnetic (EM) theory and transmission line (TL) theory approaches. *Experimental* demonstrations are provided both in TW-SRR bulk structures and planar TL-type structures.

1.5 "CONVENTIONAL" BACKWARD WAVES AND NOVELTY OF LH MTMs

Propagation of waves with antiparallel phase and group velocities is not a new phenomenon. It has been known for many years in the physics and microwave communities in various contexts. Back in the late 1940s, Brillouin and Pierce utilized the series-capacitance/shunt-inductance equivalent circuit model shown in Fig 1.6, which constitutes the starting point of the CRLH TL theory and applications presented in this book, to describe such antiparallel phase/group velocities propagation, and used the term *backward waves* to designate the corresponding waves [45, 46].[14] For Brillouin, backward waves referred to negative space

---

[14] We may also add Malyuzhinets, who also used the same LC model, after Brillouin and Pierce, in the description of the radiation condition in backward-wave media [47].

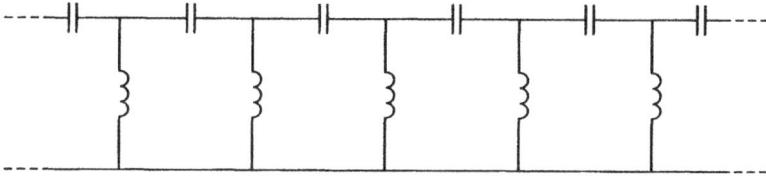

Fig. 1.6.   Equivalent circuit model for a line supporting a backward wave.

harmonics contributing in the Fourier series expansion of fields in periodic structures, while for Pierce and colleagues backward waves were related to the phenomenon of backward amplification in traveling wave tubes (TWT).[15] Several classical textbooks describing backward waves are available [48, 49].

However, in both Brillouin's and Pierce's cases, and it seems in all other known occurrences of backward-wave phenomena in the literature, perhaps with the exception of magnetostatic backward volume waves (MSBVWs) [2], backward waves are associated with either *space harmonics* or *higher-order modes* of the structures. Therefore, these structures have a period typically of the order of one-half the guided wavelength ($p \approx \lambda_g/2$), or multiple of half the guided wavelength for higher space harmonics or higher modes ($p \approx n\lambda_g/2$). Consequently, propagation along these conventional backward-wave structures are dominated by diffraction/scattering phenomena, so that they cannot be characterized by constitutive parameters ε and μ.[16] Extended to 2D or 3D, if possible, they would not exhibit the electromagnetic behavior of real materials, as for instance refractive properties obeying to simple laws such as Snell's law.

In contrast, LH MTMs are operating in their *fundamental mode* (effectively homogeneous), where $p \ll \lambda_g$, so that effective macroscopic ε and μ can be rigorously defined. They are thus behaving as "real" materials. In fact, as long as $p < \lambda_g/4$, the difference between a LH MTM and a natural dielectric material (such as glass or Teflon) is only *quantitative*. In the case of the natural dielectric material, the structural units inducing a given permittivity value are the constitutive molecules, which are of the order of the Angström, while in the case of current LH materials, the structural units are of the order of the centimeter in the microwave range; thus, in the microwave $L - X$ bands, the electrical size of the structural unit in the former case is of the order of $p/\lambda_g \approx 10^{-9}$ while it is of the order of $p/\lambda_g \approx 10^{-1} - 10^{-2}$ in the latter case. But *qualitatively*, similar refractive phenomena occur in LH MTMs as in conventional dielectrics. This is the point at which the fundamental originality of LH-MTMs lies: LH-MTMs represent artificial structures behaving in the same manner as conventional bulk dielectrics, hence the term "material" in meta-materials, but with negative constitutive parameters. The subsequent and fundamental originality of LH MTMs

---

[15]Backward-wave oscillators (BWOs), which are realized by coupling some of the output power of a TWT back into its input, also exhibit backward-wave properties reminiscent to those of LH structures.

[16]In this regard, they can be termed "structures" but not artificial "materials" or MTMs.

is that, due to their effective homogeneity, they can be extended to 2D and potentially 3D isotropic MTMs.

1.6 TERMINOLOGY

To date, there is no consensus regarding the terminology designating MTMs with simultaneously negative permittivity and permeability. Several equivalent terminologies have been used, each of which presenting advantages and disadvantages:

- *Left-handed (LH)* (e.g., [3, 12, 19, 20, 34, 40])—This is the terminology originally suggested by Veselago in [3]. It has the advantage of being related to the most fundamental property of these structures, which is antiparallel phase and group velocities (Section 2.1), and to be universal (applicable to 1D, 2D, and 3D structures). Some authors feel that the term "left-handed" could be confused with left-handed chiral materials, which may be considered a weakness of this term.
- *Double-negative (DNG)* (e.g., [28, 53])—In this case, the aforementioned potential confusion is not possible. However, this term does not tell what is negative and may be therefore seen as not self-consistent.
- *Negative-refractive-index (NRI)* (e.g., [38, 41])—NRI certainly represents a meaningful term for 2D or 3D structures and also has the advantage of emphasizing the effectively homogeneous nature of LH MTMs. However, it looses its sense in 1D LH structures, where propagation angles are not involved.
- *Backward-wave (BW)* (e.g., [32, 54])—The asset of this term may be that it points out the existence of antiparallel phase/group velocities, well known in conventional structures but fails to inform on the material (effectively homogeneous) nature of LH MTMs (Section 1.5).
- *Veselago medium* (e.g., [50, 51])—This choice pays tribute to the visionary idea of Veselago but does not provide any information on the properties of the medium.
- *Negative phase velocity medium (NPV)* (e.g., [52])—In our opinion, this term may be the best of all from a semantic point of view, but it has been so little used that it may be premature to adopt it definitely, unless some sort of consensus can be made between the main protagonists of MTM research.

In this book, we have made the choice of the terminology *left-handed (LH)* MTM.

1.7 TRANSMISSION LINE (TL) APPROACH

Although very exciting from a physics point of view, the initial TW-SRR MTMs (Fig 1.4) seem of little practical interest for engineering applications because these structures are *resonant*, and consequently exhibit *high loss* and *narrow bandwidth*.[17] A structure made of resonating elements generally does not constitute a good transmission medium for a modulated signal because of the quality factor intrinsically associated with each resonator [57]. In a resonator, the loaded quality factor Q_l is related to the unloaded quality factor Q_u and external quality factor Q_e by the relation

$$\frac{1}{Q_l} = \frac{1}{Q_u} + \frac{1}{Q_e},$$ (1.8)

which expresses the fact that the total transmission[18] loss ($\propto 1/Q_l$) through a resonator is equal to the sum of the dielectric/ohmic losses in the resonator ($\propto 1/Q_u$) and the coupling losses in the transitions with the external (source/load) circuits ($\propto 1/Q_e$). The loaded quality factor, which is the quantity actually measured and eventually relevant in terms of transmission, is also obtained from the magnitude of the transmission parameter (S_{21}) as

$$Q_l = \frac{f_r}{B},$$ (1.9)

where f_r is the resonance frequency and B is the -3-dB bandwidth (in Hz), while the unloaded quality factor is defined as

$$Q_u = \omega \frac{\text{average energy stored in resonator}}{\text{power dissipated in resonator}}.$$ (1.10)

These textbook formulas show that, for given dielectric (dielectric loss $\propto \tan \delta$) and metal (ohmic loss $\propto 1/\sigma$, σ: conductivity) materials, there is an unavoidable *trade-off between bandwidth and transmission level*. Minimum transmission loss, or equivalently maximum Q_l, is achieved at the resonance frequency f_r by minimizing the bandwidth B, according to Eq. (1.9), because in this case very little power is dissipated in the cavity since its bandwidth is extremely narrow so that Q_u is maximized, according to Eq. (1.10)[19]. So, in this case, good transmission characteristics can be obtained. But bandwidth is so restricted that a modulated signal, even with a modest bandwidth, cannot be transmitted without distortion

---

[17]The insertion loss reported in [13] for the first TW-SRR structure was higher than 30 dB! Optimization of coupling mechanisms later improved the transmission characteristics of this structure. For instance, minimum transmission loss of around 1 dB over a 3-dB fractional bandwidth of 6.5% for a 3-cell structure was reported in [22]. Unfortunately, such performances are still not acceptable for microwave engineering applications.

[18]We are considering here a two-port TL-type (series) resonator system because we are interested in transmission characteristics through the structure. In such a system, a maximum in the magnitude of the transmission parameter (S_{21}) occurs at the resonance [57]. A perfectly loss-less resonating two-port system would correspond to $Q_l = \infty$, implying $Q_u = \infty$ and $Q_e = \infty$.

[19]This consideration is based on the assumption, generally reasonable, that $Q_e \gg Q_u$, so that Q_l is mostly dependent on Q_u, from (1.8).

through the resonating structure. Bandwidth can naturally be increased, but this immediately results in a decrease of Q_l according to Eq. (1.9), and therefore in an increase of transmission loss. In conclusion, a modulated signal cannot be transmitted efficiently through a resonating propagation medium. Although this statement is a very basic fact of resonators theory, it might not have been sufficiently appreciated in the physics community investigating MTMs.

Due to the weaknesses of resonant-type LH structures, there was a need for alternative architectures. Therefore, recognizing the analogy between LH waves and conventional backward waves (Section 1.5), three groups introduced, almost simultaneously in June 2002, a transmission line (TL) approach of metamaterials: Eleftheriades et. al [37, 55], Oliner [41] and Caloz et al. [39, 56].

In fact, hypothetical "backward-wave" uniform TLs, without any suggestion for a practical implementation!, had been briefly described in a few textbooks, such as [48]. The incremental circuit model for such a TL, which is essentially the *dual*[20] of a conventional right-handed (RH) series-L/shunt-C TL, is shown in Fig 1.7.[21,22]

The fundamental characteristics of the TL of Fig 1.7 are straightforwardly derived by elementary TL theory. Let us consider here for simplicity the loss-less case, the lossy case being rigorously treated in Chapter 3. The complex

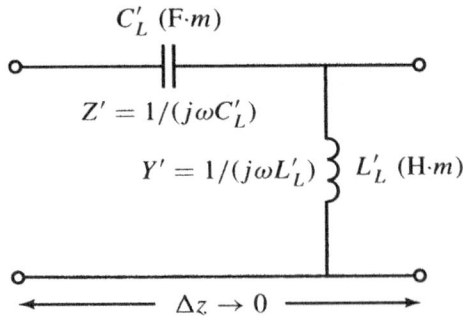

$$C'_L \ (\mathrm{F \cdot m})$$

$$Z' = 1/(j\omega C'_L)$$

$$Y' = 1/(j\omega L'_L) \qquad L'_L \ (\mathrm{H \cdot m})$$

$$\Delta z \to 0$$

Fig. 1.7. Incremental circuit model for a hypothetical uniform LH TL.

[20] In this book, the term "dual" is considered in a broad sense, although more specific definitions are preferred by some electromagnetic theoreticians. The series-C/shunt-L prototype is called the dual of the series-L/shunt-C prototype, since it represents the fundamental "complement" to it, as corroborated by the fact is associated with negative refractive index as a complement of positive refractive index.

[21] In the present book, primes are used to represent reactances related to *per-unit-length* immittances, i.e., per-unit-length impedances Z' in Ω/m and admittances Y' in S/m.

[22] Note the units of L'_L $(\mathrm{H \cdot m})$ and C'_L $(\mathrm{F \cdot m})$ ensuring that Z' and Y' correspond to per-unit-length immittances in Ω/m and S/m, respectively. We have $Z(\Omega) = 1/[j\omega C_L(\mathrm{F})] \to Z'(\Omega/m) = 1/[j\omega C_L(\mathrm{F})\Delta z] = 1/[j\omega C'_L(\mathrm{F \cdot m})]$ where $C'_L = C_L \cdot \Delta z$; similarly, $Y(S) = 1/[j\omega L_L(\mathrm{H})] \to Y'(S/m) = 1/[j\omega L_L(\mathrm{H})\Delta z] = 1/[j\omega L'_L(\mathrm{H \cdot m})]$ where $L'_L = L_L \cdot \Delta z$.

propagation constant γ, the propagation constant β, the characteristic impedance Z_c, the phase velocity v_p, and the group velocity v_g of the TL are given by

$$\gamma = j\beta = \sqrt{Z'Y'} = \frac{1}{j\omega\sqrt{L_L'C_L'}} = -j\frac{1}{\omega\sqrt{L_L'C_L'}}, \tag{1.11a}$$

$$\beta = -\frac{1}{\omega\sqrt{L_L'C_L'}} < 0, \tag{1.11b}$$

$$Z_c = \sqrt{\frac{Z'}{Y'}} = +\sqrt{\frac{L_L'}{C_L'}} > 0, \tag{1.11c}$$

$$v_p = \frac{\omega}{\beta} = -\omega^2\sqrt{L_L'C_L'} < 0, \tag{1.11d}$$

$$v_g = \left(\frac{\partial\beta}{\partial\omega}\right)^{-1} = +\omega^2\sqrt{L_L'C_L'} > 0. \tag{1.11e}$$

The last two equations immediately and unambiguously show that phase and group velocities in such a TL would be antiparallel, $v_p - \|v_g$. The phase velocity v_g, associated with the direction of phase propagation or wave vector β, is negative, whereas the group velocity v_g, associated with the direction of power flow or Poynting vector **S**, is positive (Section 2.1). Thus, the TL of Fig 1.7 is LH, according to the definition given in Section 1.2.

These considerations revealed that a LH TL medium could be engineered in a structural manner by conceiving LC structures corresponding to the model of Fig 1.6 *in condition that the average cell size p would be much smaller than the guided wavelength* λ_g ($p \ll \lambda_g$). If this could be accomplished, a *nonresonant* type, in the frequency range of interest, LH medium would be obtained. Such a design was shown to be perfectly realizable. The first practical LH TL was the microstrip structure shown in Fig 1.8(a) [39], with a 2D version of it in Fig 1.8(b).

Because of their nonresonant nature, TL MTMs can be designed to exhibit simultaneously *low loss* and *broad bandwidth*. Low loss is achieved by a "balanced" design (Section 1.8) of the structure and good matching to the excitation ports, whereas broad-bandwidth is a direct consequence of the TL nature of the structure and can be controlled by its LC parameters, which determine the cutoff frequency of the resulting high-pass structure. Another advantage of TL MTM structures is that they can be engineered in *planar configurations*, compatible with modern microwave integrated circuits (MICs). Finally, TL MTM structures can benefit from the efficient and well-established *TL theory* for the efficient design of microwave applications. Over 25 novel devices, with unique features or superior performances compared with conventional devices, have already been demonstrated [58, 59] and are described in the Chapters 3, 5 and 7.

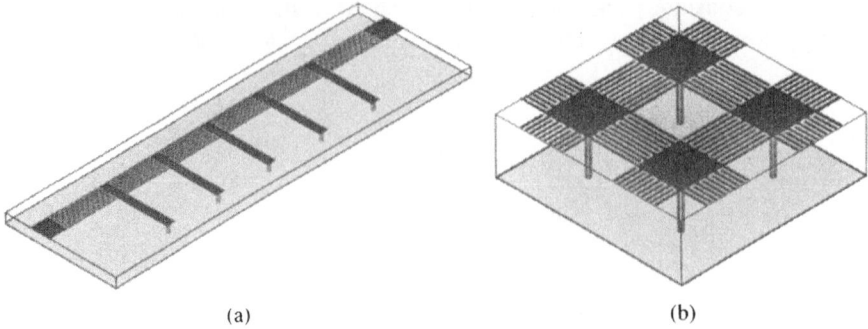

(a)   (b)

Fig. 1.8. Planar transmission line (TL) LH structures in microstrip technology constituted of series interdigital capacitors and shunt stub inductors. The gray areas represent the ground planes, and the black surfaces represent metal. (a) 1D (e.g., [39, 40]). (b) 2D (similar to [44]).

1.8   COMPOSITE RIGHT/LEFT-HANDED (CRLH) MTMs

The concept of composite right/left-handed (CRLH) MTM, introduced by Caloz et al. in [58], is the cornerstone of the theory and applications presented in this book and is therefore extensively developed in the next chapters. However, this section already introduces this concept as a broad-band generalization of LH MTMs and provides a quick explanation of its essence.

The TL structures shown in Fig 1.8 are constituted of series (interdigital) capacitors C_L and shunt (stub) inductors L_L, intended to provide left-handedness from the explanations of the previous section. However, as a wave propagates along the structures, the associated currents and voltages induce other natural effects. As currents flow along C_L, magnetic fluxes are induced and therefore a series inductance L_R is also present; in addition, voltage gradients exist between the upper conductors and the ground plane, which corresponds to a shunt capacitance C_R. As a consequence, *a purely LH (PLH) structure does not exist*, even in a restricted frequency range, since a real LH structure necessarily includes (L_R, C_R) contributions in addition to the (L_L, C_L) reactances. This was the motivation for the introduction of the term "composite right/left-handed" (CRLH), allowing to account for the exact natural of practical LH media [58, 60].

The essential characteristics of a CRLH TL MTM can be inferred from analysis of the equivalent circuit of Fig 1.9(a). At low frequencies, L_R and C_R tend to be short and open, respectively, so that the equivalent circuit is essentially reduced to the series-C_L/shunt-L_L circuit, which is LH since it has antiparallel phase and group velocities; this LH circuit is of *highpass* nature; therefore, below a certain cutoff, a LH stopband is present. At high frequencies, C_L and L_L tend to be short and open, respectively, so that the equivalent circuit is essentially reduced to the series-L_R/shunt-C_R circuit, which is RH since it has parallel phase and group velocities; this LH circuit is of *lowpass* nature; therefore, above a certain cutoff, a RH stopband is present. In general, the series resonance ω_{se} and shunt

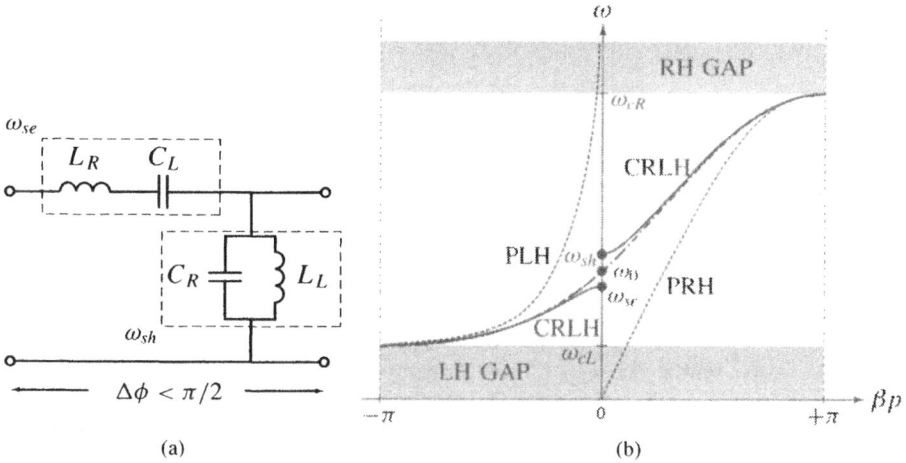

Fig. 1.9. Fundamentals of composite right/left-handed (CRLH) MTMs. (a) Unit cell TL prototype. (b) Dispersion diagram. The curves for a purely LH (PLH) structure $(L_R = C_R = 0)$ and for a purely RH (PRH) structure $(L_L = C_L = \infty)$ are also shown for comparison. Here, we have represented the case $\omega_{se} < \omega_{sh}$, but $\omega_{se} > \omega_{sh}$ is naturally also possible, depending on the LC parameters (Section 3.1).

resonance ω_{sh} are different, so that a gap exists between the LH and the RH ranges. However, if these resonances are made equal, or are "balanced," this gap disappears, and an *infinite-wavelength* $(\lambda_g = 2\pi/|\beta|)$ propagation is achieved at the transition frequency ω_0. Despite its filter nature, the CRLH structure is never operated at the edges of the Brillouin zone, where $p \approx \lambda_g/2$, but only in the vicinity of the transition frequency, where effective-homogeneity is ensured $(p < \lambda_g/4)$. Note that, although a CRLH structure has both a LH range and a RH range, the dispersion curve in each of these ranges significantly differs from that of the PLH and PRH structures, respectively, because of the combined effects of LH *and* RH contributions at all frequencies.

1.9 MTMs AND PHOTONIC BAND-GAP (PBG) STRUCTURES

The essential difference between MTM and photonic band gap (PBG) (or photonic crystals) structures has already been implicitly mentioned in Section 1.5 in connection with periodic structures. This section provides a more detailed explanation of this difference.

Photonic crystals [61] and PBG [62] structures are usually operated at frequencies where the lattice period p is of the order of a multiple of half a guided wavelength, $p \approx n\lambda_g/2$. In this case, as illustrated in Fig 1.10(a), the waves scattered by adjacent layers of the lattice interfere constructively for some specific angles of incidence. Therefore, net rejection of the incoming energy, corresponding to zero group velocity, occurs at these angles. This phenomenon is similar to

Bragg diffraction in X-ray optics [63] and is sometimes referred to as "Bragg-like" diffraction. The Bragg condition for maximum diffraction is given by

$$2p \sin \theta = m\lambda_g, \quad m = 0, 1, 2, \ldots, \tag{1.12}$$

where it is clearly seen that the Bragg angles are function of frequency (via λ_g). This condition, with $\beta = 2\pi/\lambda_g$, is equivalent to

$$\beta(\omega) = \frac{m\pi}{p \sin \theta(\omega)}, \quad m = 0, 1, 2, \ldots, \tag{1.13}$$

where the function $\beta(\omega)$ [or, more commonly, its inverse $\omega(\beta)$] is the *dispersion relation*, from which the *dispersion diagram* is computed. The points of the dispersion curves $\omega(\beta)$ where the Bragg condition is met have a zero slope (or tangent), since the slope of $\omega(\beta)$ corresponds to the group velocity, $v_g = d\omega/d\beta$, which is zero there. This means that Bragg points delimit the stop bands or *band gaps* in the dispersion diagram. The CRLH TL structure, whose dispersion diagram is shown in Fig 1.9(b), has two Bragg points: one is the highpass LH gap cutoff ω_{cL} and the other one is the lowpass RH gap cutoff ω_{cR}. At each of these two points, we have $|\beta|p = \pi$; i.e., $p = \pi/|\beta|$, which corresponds to (1.13) with $m = 1$.[23] These points correspond to resonating and noneffectively homogeneous frequencies, and are therefore *not in the range of MTM operation* of the CRLH structure.[24]

Whereas in a PBG structure period is of the order of wavelength [Fig 1.10(a)], period (or average period) in a MTM is much smaller than wavelength ($p/\lambda_g \ll 1$), as shown in Fig 1.10(b). Therefore, interference effects cannot take place in a MTM because the phase difference between adjacent cells is negligible. Instead, the wave simply travels through the material in a straight line, only probing average or effective constitutive parameters. This is what happens in the vicinity of the axis $\beta = 0$ (Γ spectral point) in a CRLH TL structure [Fig 1.9(b)]. In the balanced case, we have perfectly homogeneous transmission ($p/\lambda_g = 0$). In the unbalanced case, we have the two resonances ω_{se} and ω_{sh}. However, these resonances are *not Bragg* resonances but zeroth-order resonances (Section 5.5), in the sense that they are *infinite wavelength* ($\beta = 0$, fundamental mode) resonances.

Thus, a photonic crystal or PBG structure is operated in the Bragg regime, where periodicity plays a crucial role and the structure is strongly inhomogeneous, whereas a MTM is operated in the *long wavelength regime*, where the lattice does not need to be periodic and the structure is homogeneous. Consequently, (diffraction/scattering) *properties of photonic crystals are essentially determined by the lattice*,[25] while the *(refractive) properties of MTM are determined by the nature of the unit cell*, as detailed Section 3.1 [Eqs. (3.23) and (3.25)].

---

[23]In fact, $m = -1$ for the LH resonance, as it will be explained in Section 5.5.

[24]At these frequencies, the structure *is* a PBG but not a MTM.

[25]For instance, change the lattice of a 2D photonic crystal from square to hexagonal, without changing the unit cell or *basis*, results in dramatic changes in the PBG structure (dispersion diagram), whereas a change of the basis only hardly affects the PBG structure [61].

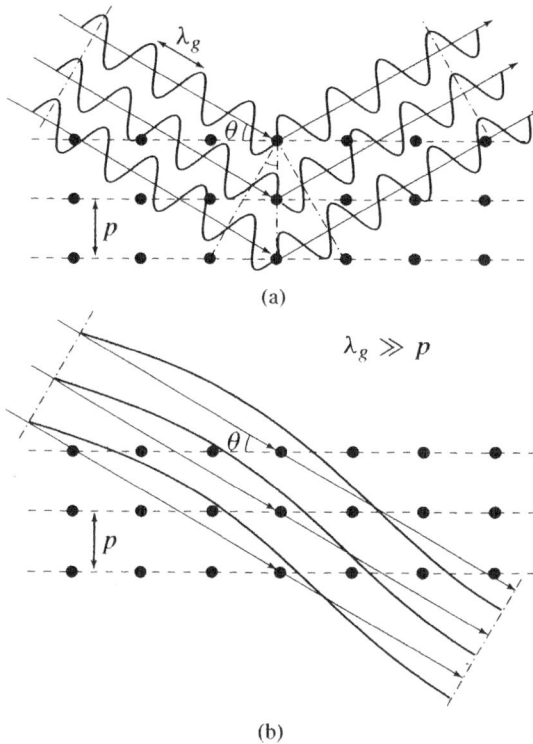

Fig. 1.10. Difference between the Bragg-scattering and long wavelength regimes. (a) Bragg regime, $p \approx \lambda_g/2$, usually prevailing in photonic crystals or PBG structures. (b) Long wavelength regime, $p \ll \lambda_g$, prevailing in MTMs.

Despite the fundamental differences pointed out above, several reports mentioning "negative refraction" in photonic crystals have been presented [64, 65, 66, 67].[26] The important distinction to make between these "refraction-like"crystals and LH MTMs is the following. A photonic crystal exhibits many higher-order modes (or bands) $\omega_m(\overline{\beta})$[27] with negative gradients $\nabla_{\overline{\beta}_m} \omega$ (m: mode index). For this reason, the corresponding m^{th} Fourier spectral contribution to the total field [Subsection 3.2.6, Eq. (3.107)] "propagates" with a phase in the opposite direction than that of power, which can result in NRI-like fields *outside* the photonic crystal (e.g., in a photonic crystal prism [67]). However, this "wave" $\overline{\beta}_m$ is only one Fourier component in the total physical field and is not this field itself. The physical field experiences scattering, not refraction, within the medium, as clearly shown in various reported simulations. No negative focusing would be possible

---

[26]It may be noted that the more careful term "refraction-like behavior" is used in [65].

[27]Note that $\overline{\beta}$ is a vector (hence the overline) if the dimension is superior than one. In the general 3D case, we have $\overline{\beta} = \beta_x \hat{x} + \beta_y \hat{y} + \beta_z \hat{z}$, where $\hat{x}, \hat{y}, \hat{z}$ denote unit vectors along the direction x, y, z, respectively.

inside a photonic crystal. In terms of engineering applications, this approach to obtain LH-like effects does not seem judicious because 1) such effects are more difficult to obtain in Bragg structures than in effectively homogeneous structures; 2) Bragg regime does not allow refractive effects inside the structure; and 3) Bragg structures necessarily result in large-size components since the unit cell itself is already very large, $p \sim m\lambda_g/2$.

1.10 HISTORICAL "GERMS" OF MTMs

In the previous sections, we reported the most significant milestones in the emergence of the new field of MTMs and more particularly of LH or CRLH MTMs. However, MTMs are related to such fundamental aspects of electromagnetic science and engineering that they necessarily had a few more or less diffuse "germs" in previous history. For instance in [68], Tretyakov presented an interesting "historical perspective," which uncovers some of these germs.[28] We give now a historical perspective partly inspired by this paper.

The earliest known speculation on negative refraction was made by Mandelshtam back in 1944 [69]. In this paper, Mandelshtam noticed that, given two media (labeled 1 and 2), for a given incidence angle θ_1 of the wave at the interface, Snell's law [Eq. (1.6a)] admits, mathematically, two solutions: not only the conventional solution θ_2 but also the "unusual" solution $\pi - \theta_2$[29]. Regarding this second solution, Mandelshtam wrote:

> "Demanding [as before] that the energy in the second medium propagates *from* the boundary, we arrive to the conclusion that the phase must propagate *towards* the boundary and, consequently, the propagation direction of the refracted wave will make with the normal the angle $\pi - \theta_2$. This derivation appears to be unusual, but of course there is no wonder because the phase velocity still tells us nothing about the direction of the energy transfer."

Although we note that Mandelshtam made no reference to *negative refraction* and did note carry out an in-depth study of this phenomenon, it is manifest that his argument refers to the same phenomenon. This fact reduces in no ways the pioneering merit of Veselago, who first addressed in 1967 the problem of LH materials in a systematic manner; Veselago may be considered as the father of LH MTMs with his decisive contribution of establishing their fundamental properties and predicting their unusual effects [3] (Section 1.2).

The year after his initial speculation (1945), with reference to Lamb (1904) who "gave examples of *fictitious* one-dimensional media with negative group velocity," Mandelshtam presented physical examples of structures supporting waves with "negative group velocity," such as structures with periodically varying

---

[28]The perspective of [68], as mentioned in the paper, is not systematic and therefore not necessarily exhaustive. It is likely that even former "germs" of MTMs have existed under various forms.

[29]This solution first seems to correspond to a ray in the half-plane of the reflection, but this solution is necessarily in medium 2, as Mandelshtam explained in the statement given here.

effective permittivity [70]. However, these structures were not MTMs in the sense defined in Section 1.1 but merely periodic structures, similar to those presented for example in [45, 46] at the same period, based on space harmonics excitation.

Maybe the first report of a hypothetical medium with simultaneously negative permittivity and permeability is the paper by Sivukhin in 1957 [71]. In this paper, Sivukhin noticed that a media with double negative parameters would be backward media, but this observation was still far from LH MTMs, as he pointed out "media with $\varepsilon < 0$ and $\mu < 0$ are not known. The question on the possibility of their existence has not been clarified."

As far as the TL approach of MTMs is concerned (Section 1.7), it has already been mentioned in Section 1.5 that LH TL circuit models with series capacitance and shunt inductances were used by several authors, such as for instance Brillouin and Pierce [45, 46], back in the 1940s in the context of periodic structures and microwave tubes. Malyuzhinets, in 1951, even showed a TL model similar to the CRLH TL model (Section 1.8) [47]. However, none of these approaches pertained to MTMs, as they were systematically used to describe space harmonics of periodic structures, as opposed to a fundamental propagation mode, and its seems that there was no clue at that time regarding how to design effective LH and CRLH TL structures. Moreover, most of the MTMs and CRLH MTM concepts presented in this book were apparently unknown. It should be noted that some authors, such as for instance Ramo et al., showed the PLH TL in their textbooks as a hypothetical effectively homogeneous TL, but without any suggestion on the existence or design of such TLs [48].

There is no doubt that the periodic slow-wave, backward-wave structures studied and developed in the 1940s and 1950s included important germs of MTMs and LH MTMs [45, 46, 47, 72]. It is worth noting that Silin discussed in 1959 the phenomenon of "negative refraction" in connection with periodic slow-wave structures (although the definition of a refractive index in a periodic structure operated in the Bragg regime is questionable, as pointed out in Section 1.9) [73]. Nowadays, there is still considerable effort in conceiving MTMs based on Bragg-regime periodic structures.

From the above reference with Mandelshtam, we see that Veselago was not the first to postulate the existence of LH media in [3], but it is an established fact that he was the first to conduct a systematic study of these media and to predict their most fundamental properties.

After Veselago, there has been made really no attempt to design a LH MTM before Smith et al. [4, 10]. However, resonant-type single ε negative and single μ negative structures, which are the constitutive elements of the SRR/TW LH MTMs (Section 1.3), have been reported under various forms long before the systematic studies of Pendry in [5] ($\varepsilon < 0$) and [6] ($\mu < 0$), who first developed a unified approach of these structure as effective media or MTMs. It seems that a wire medium, similar to the TW medium, was first described in the 1950s as a material for microwave lenses [74, 75]. Variants of the SRR were presented in the textbook by Schelkunoff and Friis published in 1952 [76], although these SRRs were not considered there as the artificial particles of a magnetic medium (as introduced by Pendry in [6]). Media with loops of various shapes (e.g., omega, helix, spiral) were abundantly reported in the 1980s and 1990s in the investigation of artificial *chiral and bi-isotropic materials* for microwave applications [77]. These chiral materials are effective media and are therefore MTMs according to our definition (Section 1.1); negative permeability was identified theoretically and experimentally and described in several papers about chiral media. It is worth noting that a chiral medium with double-ring structure implants very similar to the SRRs was presented [78]. To conclude these remarks on relations between chiral and LH MTMs let us also note that backward waves have been recently shown to be supported by materials with positive parameters if chirality is present [79, 80].

REFERENCES

1. J.-S. G. Hong, M. J. Lancaster. *Microstrip Filters for RF/Microwave Applications,* Wiley-Interscience, 2001.

2. D. D. Stancil. *Theory of Magnetostatic Waves,* Springer-Verlag, 1993.

3. V. Veselago. "The electrodynamics of substances with simultaneously negative values of ε and μ," *Soviet Physics Uspekhi,* vol. 10, no. 4, pp. 509–514, Jan., Feb. 1968.

4. D. R. Smith, W. J. Padilla, D. C. Vier, S. C. Nemat-Nasser, and S. Schultz. "Composite medium with simultaneously negative permeability and permittivity," *Phys. Rev. Lett.,* vol. 84, no. 18, pp. 4184–4187, May 2000.

5. J. B. Pendry, A. J. Holden, W. J. Stewart, and I. Youngs. "Extremely low frequency plasmons in metallic mesostructure," *Phys. Rev. Lett.,* vol. 76, no. 25, pp. 4773–4776, June 1996.

6. J. B. Pendry, A. J. Holden, D. J. Robbins, and W. J. Stewart. "Low frequency plasmons in thin-wire structures," *J. Phys. Condens. Matter,* vol. 10, pp. 4785–4809, 1998.

7. J. B. Pendry, A. J. Holden, D. J. Robbins, and W. J. Stewart. "Magnetism from conductors and enhanced nonlinear phenomena," *IEEE Trans. Micr. Theory. Tech.,* vol. 47, no. 11, pp. 2075–1084, Nov. 1999.

8. M. Makimoto and S. Yamashita. *Microwave Resonators and Filters Communications: Theory, Design and Applications,* Springer, 2000.

9. D. R. Smith, D. C. Vier, N. Kroll, and S. Schultz. "Direct calculation of permeability and permittivity for a left-handed metamaterial," *App. Phys. Lett.,* vol. 77, no. 14, pp. 2246–2248, Oct. 2000.

10. R. A. Shelby, D. R. Smith, and S. Schultz. "Experimental verification of a negative index of refraction," *Science,* vol. 292, pp. 77–79, April 2001.

11. D. R. Smith and D. Schurig. "Electromagnetic wave propagation in media with indefinite permittivity and permeability tensors," *Phys. Rev. Lett.,* vol. 90, pp. 077405:1–4, Feb. 2003.

12. D. R. Smith, D. C. Vier, N. Kroll, and S. Schultz. "Direct calculation of permeability and permittivity for a left-handed metamaterial," *App. Phys. Lett.,* vol. 77, no. 14, pp. 2246–2248, Oct. 2000.

13. D. R. Smith and N. Kroll. "Negative refractive index in left-handed materials," *Phys. Rev. Lett.,* vol. 85, no. 14, pp. 2933–2936, Oct. 2000.

14. R. A. Shelby, D. R. Smith, S. C. Nemat-Nasser, and S. Schultz. "Microwave transmission through a two-dimensional, isotropic, left-handed metamaterial," *App. Phys. Lett.,* vol. 78, no. 4, pp. 489–491, Jan. 2001.

15. M. Garcia, and M. Nieto-Vesperinas. "Left-handed materials do not make a perfect lens," *Phys. Rev. Lett.,* vol. 88, no. 20, pp. 207403:1–4, Jan. 2002.

16. M. Garcia, and M. Nieto-Vesperinas. "Is there an experimental verification of a negative index of refraction yet?," *Optics Lett.,* vol. 27, no. 11, pp. 885–887, June 2002.

17. P. M. Valanju, R. M. Walser, and A. P. Valanju. "Wave refraction in negative-index media: always positive and very inhomogeneous," *Phys. Rev. Lett.,* vol. 88, no. 18, pp. 187401:1–4, 2002.

18. P. Gay-Balmaz and O. J. F. Martin. "Electromagnetic resonances in individual and coupled split-ring resonators," *J. App. Phys.,* vol. 92, no. 5, pp. 2929–2936, Sept. 2002.

19. R. Marqués, F. Median, and R. Rafii-El-Idrissi. "Role of bianisotropy in negative permeability and left-handed metamaterials," *Phys. Rev. B,* vol. 65, pp. 144440:1–6, 2002.

20. P. Markŏs and C. M. Soukoulis. "Numerical studies of left-handed materials and arrays of split ring resonators," *Phys. Rev. E,* vol. 65, pp. 036622:1–8, 2002.

21. P. Markŏs and C. M. Soukoulis. "Transmission studies of left-handed materials," *Phys. Rev. B,* vol. 65, pp. 033401:1–4, 2002.

22. R. B. Greegor, C. G. Parazzoli, K. Li, B. E. C. Koltenbah, and M. Tanielian. "Experimental determination and numerical simulation of the properties of negative index of refraction materials," *Optics Express,* vol. 11, no. 7, pp. 688–695, April 2003.

23. E. Ozbay, K. Aydin, E. Cubukcu, and M. Bayindir. "Transmission and reflection properties of composite double negative metamaterials in free space," *IEEE Trans. Antennas Propagat.,* vol. 51, no. 10, pp. 2592–2595, Oct. 2003.

24. C. R. Simovski, P. A. Belov, and S. He. "Backward wave region and negative material parameters of a structure formed by lattices of wires and split-ring resonators," *IEEE Trans. Antennas Propagat.,* vol. 51, no. 10, pp. 2582–2591, Oct. 2003.

25. R. Marqués, F. Mesa, J. Martel, and F. Median. "Comparative analysis of edge- and broadside-coupled split ring resonators for metamaterial: design, theory and experiments," *IEEE Trans. Antennas Propagat.,* vol. 51, no. 10, pp. 2572–2581, Oct. 2003.

26. R. W. Ziolkowski. "Design, fabrication, and testing of double negative metamaterials," *IEEE Trans. Antennas Propagat.,* vol. 51, no. 7, pp. 1516–1529, July 2003.

27. R. W. Ziolkowski and E. Heyman. "Wave propagation in media having negative permittivity and permeability" *Phys. Rev. E,* vol. 64, pp. 056625:1–15, 2001.

28. R. W. Ziolkowski. "Pulsed and CW gaussian beam interactions with double negative metamaterial slabs," *Optics Express,* vol. 11, no. 7, pp. 662–681, April 2003.

29. C. Caloz, C. C. Chang, and T. Itoh. "Full-wave verification of the fundamental properties of left-handed materials (LHMs) in waveguide configurations" *J. App. Phys.,* vol. 90, no. 11, pp. 5483–5486, Dec. 2001.

30. P. P. M. So and W. J. R. Hoefer. "Time domain TLM modeling of metamaterials with negative refractive index," *IEEE-MTT Int'l Symp.,* pp. 1779–1782, Fort Worth, TX, June 2004.

31. P. P. M. So, H. Du, and W. J. R. Hoefer. "Modeling of metamaterials with negative refractive index using 2D-shunt and 3D-SCN TLM networks," *IEEE Trans. Microwave Theory Tech.*, pp. 1496–1505 vol. 53, no. 4, April. 2005.

32. I. V. Lindell, S. A. Tretyakov, K. I. Nikoskinen, and S. Ilvonen. "BW media –media with negative parameters, capable of supporting backward waves," *Micr. Opt. Technol. Lett.*, vol. 31, no. 2, pp. 129–133, Oct. 2001.

33. J. A. Kong, B.-I. Wu, and Y. Zhang. "A unique lateral displacement of a Gaussian beam transmitted through a slab with negative permittivity and permeability," *Micr. Opt. Technol. Lett.*, vol. 33, no. 2, pp. 136–139, April 2002.

34. J. Pacheco, T. M. Grzegorzcyk, B.-I. Wu, Y. Zhang, and J. A. Kong. "Wave propagation in homogeneous isotropic frequency-dispersive left-handed media," *Phys. Rev. Lett.*, vol. 89, no. 25, pp. 257401:1–4, Dec. 2002.

35. D. R. Smith, D. Schurig, and J. B. Pendry. "Negative refraction of modulated electromagnetic waves," *App. Phys. Lett.*, vol. 81, no. 15, pp. 2713–2715, Oct. 2002.

36. M. W. McCall, A. Lakhtakia, and W. S. Weiglhofer. "The negative index of refraction demystified," *Eur. J. Phys.*, vol. 23, pp. 353–359, 2002.

37. A. K. Iyer and G. V. Eleftheriades. "Negative refractive index metamaterials supporting 2-D waves," in *IEEE-MTT Int'l Symp.*, vol. 2, Seattle, WA, pp. 412–415, June 2002.

38. G. V. Eleftheriades, A. K. Iyer, and P. C. Kremer. "Planar negative refractive index media using periodically L-C loaded transmission lines," *IEEE Trans. Microwave Theory Tech.*, vol. 50, no. 12, pp. 2702–2712, Dec. 2002.

39. C. Caloz and T. Itoh. "Application of the transmission line theory of left-handed (LH) materials to the realization of a microstrip LH transmission line," *in Proc. IEEE-AP-S USNC/URSI National Radio Science Meeting*, vol. 2, San Antonio, TX, pp. 412–415, June 2002.

40. C. Caloz and T. Itoh. "Transmission line approach of left-handed (LH) structures and microstrip realization of a low-loss broadband LH filter," *IEEE Trans. Antennas Propagat.*, vol. 52, no. 5, pp. 1159–1166, May 2004.

41. A. A. Oliner. "A periodic-structure negative-refractive-index medium without resonant elements," in *URSI Digest, IEEE-AP-S USNC/URSI National Radio Science Meeting*, San Antonio, TX, pp. 41, June 2002.

42. A. A. Oliner. "A planar negative-refractive-index medium without resonant elements," in *IEEE-MTT Int'l Symp.*, Philadelphia, PA, pp. 191–194, June 2003.

43. A. K. Iyer, P. C. Kremer, and G. V. Eleftheriades. "Experimental and theoretical verification of focusing in a large, periodically loaded transmission line negative refractive index metamaterial," *Optics Express*, vol. 11, no. 7, pp. 696–708, April 2003.

44. A. Sanada, C. Caloz, and T. Itoh. "Planar distributed structures with negative refractive index," *IEEE Trans. Microwave Theory Tech.*, vol. 52, no. 4, pp. 1252–1263, April 2004.

45. L. Brillouin. *Wave Propagation in Periodic Structures*, McGraw-Hill, 1946.

46. J. R. Pierce. *Traveling-Wave Tubes*, D. Van Nostrand, 1950.

47. G. D. Malyuzhinets. "A note on the radiation principle," *Zhurnal Technicheskoi Fiziki*, vol. 21, pp. 940–942, 1951 [in Russian].

48. S. Ramo, J. R. Whinnery and T. Van Duzer. *Fields and Waves in Communication Electronics*, Third Edition, John Wiley & Sons, 1994.

49. J. A. Kong. *Electromagnetic Wave Theory,* Second Edition, EMW Pub., 2000.

50. P. A. Belov. "Backward waves and negative refraction in uniaxial dielectrics with negative dielectric permittivity along the anisotropy axis," *Microwave Opt. Technol. Lett.,* vol. 37, no. 4, pp. 259–263, March 2003.

51. D. Felbacq and A. Moreau. "Direct evidence of negative refraction media with negative ε and μ," *J. Opt. A,* vol. 5, pp. L9–L11, 2003.

52. A. Lakhtakia. "Positive and negative Goos-H anchen shifts and negative phase-velocity mediums (alias left-handed materials)," *Int. J. Electron. Commun.,* vol. 58, no. 3, pp. 229–231, 2004.

53. A. Alù and N. Engheta. "Guided modes in a waveguide filled with a pair of single-negative (SNG), double-negative (DNG) and/or double-positive (DPS) layers," *IEEE Trans. Microwave Theory Tech.,* vol. 52, no. 1, pp. 192–210, Jan. 2004.

54. S. A. Tretyakov, S. I. Maslovski, I. S. Nefedov, M. K. Kärkkäinen. "Evanescent modes stored in cavity resonators with backward-wave slabs," *Microwave Opt. Technol. Lett.,* vol. 38, no. 2, pp. 153–157, May 2003.

55. A. Grbic and G. V. Eleftheriades. "A backward-wave antenna based on negative refractive index L-C networks," in *Proc. IEEE-AP-S USNC/URSI National Radio Science Meeting,* vol. 4, San Antonio, TX, pp. 340–343, June 2002.

56. C. Caloz, H. Okabe, T. Iwai, and T. Itoh. "Anisotropic PBG surface and its transmission line model," in *URSI Digest, IEEE-AP-S USNC/URSI National Radio Science Meeting,* San Antonio, TX, pp. 224, June 2002.

57. R. E. Collin. *Foundations for Microwave Engineering,* Second Edition, McGraw-Hill, 1992.

58. C. Caloz and T. Itoh. "Novel microwave devices and structures based on the transmission line approach of meta-materials," in *IEEE-MTT Int'l Symp.,* vol. 1, Philadelphia, PA, pp. 195–198, June 2003.

59. A. Lai, C. Caloz and T. Itoh. "Transmission line based metamaterials and their microwave applications," *Microwave Mag.,* vol. 5, no. 3, pp. 34–50, Sept. 2004.

60. A. Sanada, C. Caloz and T. Itoh. "Characteristics of the composite right/left-handed transmission lines," *IEEE Microwave Wireless Compon. Lett.,* vol. 14, no. 2, pp. 68–70, Feb. 2004.

61. J. D. Joannopoulos, R. D. Meade, J. N. Winn. *Photonic Crystals,* Princeton University Press, 1995.

62. "Special issue on photonic band-gap structures," *Trans. Microwave Theory Tech.,* vol. 47, no. 11, Nov. 1999.

63. C. Kittel. *Introduction to Solid State Physics,* Seventh Edition, Wiley, 1995.

64. B. Gralak, S. Enoch, and G. Tayeb. "Anomalous refractive properties of photonic crystals," *J. Opt. Soc. Am.,* vol. 17, no. 6, pp. 1012–1020, June 2000.

65. M. Notomi. "Theory of light propagation in strongly modulated photonic crystals: refraction-like behavior in the vicinity of the photonic band gap," *Phys. Rev. B,* vol. 62, no. 16, pp. 10696–10705, Oct. 2000.

66. C. Luo, S. G. Johnson, J. D. Joannopoulos, and J. B. Pendry. "All-angle negative refraction without negative effective index," *Phys. Rev. B,* vol. 65, pp. 201104:1–4, 2002.

67. P. V. Parimi, W. T. Lu, P. Vodo, J. Sokoloff, J. S. Devov, and S. Sridhar. "Negative refraction and left-handed electromagnetism in microwave photonic crystals," *Phys. Rev. Lett.,* vol. 92, no. 12, pp. 127401:1–4, June 2004.

68. S. A. Tretyakov. "Research on negative refraction and backward-wave media: a historical perspective," in *Proc. Latsis Symposium,* Lausanne, Switzerland, pp. 30–35, March 2005.

69. L. I. Mandelshtam. "Lecture on some problems of the theory of oscillations," in *Complete Collection of Works,* vol. 5, Moscow: Academy of Sciences, pp. 428–467, 1944 [in Russian].

70. L. I. Mandelshtam. "Group velocity in a crystal lattice," in *Zhurnal Eksperimentalnoi i Teoreticheskoi Fiziki,* vol. 15, n. 9, pp. 476–478, 1945 [in Russian].

71. D. V. Sivukhin. "The energy of electromagnetic waves in dispersive media," in *Opt. Spektrosk.,* vol. 3, pp. 308–312, 1957.

72. R. G. E. Hutter. *Beam and Wave Electronics in Microwave Tubes,* Van Nostrand, 1960.

73. R. A. Silin. "Waveguiding properties of two-dimensional periodical slow-wave systems," in *Voprosy Radioelectroniki, Electronika,* vol. 4, pp. 11–33, 1959.

74. J. Brown. "Artificial dielectrics," in *Progress in Dielectrics,* vol. 2, pp. 195–225, 1960.

75. W. Rotman. "Plasma simulation by artificial and parallel plate media," in *IRE Trans. Ant. Propagat.,* vol. 10, pp. 82–95, 1962.

76. S. A. Schelkunoff and H. T. Friis. *Antennas: Theory and Practice,* John Wiley & Sons, 1952.

77. A. H. Sihvola, A. J. Viitanen, I. V. Lindell, and S. A. Tretyakov. *Electromagnetic Waves in Chiral and Bi-Isotropic Media,* Artech House, 1994.

78. M. V. Kostin and V. V. Shevchenko. "Artificial magnetics based on double circular elements," in Proc. *Bioanisotropics '94,* Périgueux, France, pp. 46–56, 1994.

79. S. Tretyakov, I. Nefedov, A. Sihvola, S. Maslovski, and C. Simovski. "Waves and energy in chiral nihility," *Journal of Electromagnetic Waves and Applications,* vol. 17, no. 7, pp. 695–706, 2003.

80. J. Pendry. "A chiral route to negative refraction," *Science,* vol. 306, pp. 1353–1955, 2004.

2

FUNDAMENTALS OF LH MTMs

Chapter 2 presents a concise overview on the fundamental theory of left-handed (LH) metamaterials (MTMs). Whereas this theory is of paramount importance in the study of any MTM or MTM-based structure, most of the MTM properties discussed here have not been directly exploited in practical applications yet.[1] However, it is useful still to describe them, for the sake of both completeness and illustration of the richness and wealth of potential applications, probably extending far beyond those presented in this book.

Section 2.1 shows how the antiparallelism existing between the phase and group velocities in a LH medium is inferred from Maxwell's equations and leads to negative refractive index (NRI). Section 2.2 demonstrates that a LH medium is necessarily frequency-dispersive and establishes the subsequent necessary entropy conditions on the LH constitutive parameters. Section 2.3 describes the modified boundary conditions at the interface between a right-handed (RH) medium and a LH medium compared with those at the boundary between two RH media. The next sections describe the reversals of the Doppler effect (Section 2.4) and the Vavilov-Čerenkov radiation (Section 2.5) occurring in LH media. The most important phenomenon, which is the reversal of Snell's law associated with NRI,

---

[1]We are referring essentially to refractive properties of 2D (or possibly 3D) MTM (Section 7.3), which are still mostly restricted to concepts at the time of this writing. This is in contrast to the properties exploited in the majority of the applications presented in this book, which are generally based on the CRLH TL concept (Section 1.8), not necessarily requiring refractive properties (at least not in 1D structures, where angles are not involved).

Electromagnetic Metamaterials: Transmission Line Theory and Microwave Applications,
By Christophe Caloz and Tatsuo Itoh
Copyright © 2006 John Wiley & Sons, Inc.

is presented in Section 2.6, which is followed by a section describing the LH "flat lens" (Section 2.7). Fresnel coefficients are given and discussed in Section 2.8. The next two sections describe the reversal of the Goos-Hänchen effect (Section 2.9) at a RH-LH interface and the reversal of the convergence/divergence effects induced by (curved) LH convex/concave lenses (Section 2.10). Finally, the concept of subwavelength diffraction is described in Section 2.11.

2.1 LEFT-HANDEDNESS FROM MAXWELL'S EQUATIONS

A LH material is an electromagnetic medium with *simultaneously negative permittivity ε and permeability μ*. We will show that the double negative nature of the constitutive parameters, ε and μ, results in the propagation of electromagnetic waves exhibiting *antiparallel phase and group velocities*[2] or LH waves. For this purpose, we start by writing Maxwell's equations[3]

$$\nabla \times \overline{\mathcal{E}} = -\frac{\partial \overline{B}}{\partial t} - \overline{\mathcal{M}}_s, \quad \text{(Faraday's law)} \tag{2.1a}$$

$$\nabla \times \overline{\mathcal{H}} = \frac{\partial \overline{D}}{\partial t} + \overline{\mathcal{J}}_s, \quad \text{(Ampere's law)} \tag{2.1b}$$

$$\nabla \cdot \overline{D} = \varrho_e, \quad \text{(electric Gauss' law)} \tag{2.1c}$$

$$\nabla \cdot \overline{B} = \varrho_m, \quad \text{(magnetic Gauss' law)} \tag{2.1d}$$

where $\overline{\mathcal{E}}$ (V/m) is the electric field intensity, $\overline{\mathcal{H}}$ (A/m) is the magnetic field intensity, $\overline{D}$ (C/m^2) is the electric flux density, $\overline{B}$ (W/m^2) is the magnetic flux density, $\overline{\mathcal{M}}_s$ (V/m^2) is the (fictitious) magnetic current density, $\overline{\mathcal{J}}_s$ (A/m^2) is the electric current density, ϱ_e (C/m^3) is the electric charge density, and ϱ_m (C/m^3) is the (fictitious) magnetic charge density. In addition, if the medium is linear (ε, μ not depending on $\overline{E}$ or $\overline{H}$) and nondispersive[4] (ε, μ not depending on ω), the vectors in the pairs [$\overline{D}, \overline{\mathcal{E}}$] and [$\overline{B}, \overline{\mathcal{H}}$] are related by the constitutive equations

$$\overline{D} = \varepsilon_0 \overline{\mathcal{E}} + \overline{\mathcal{P}} = \varepsilon_0 (1 + \chi_e) \overline{\mathcal{E}} = \varepsilon_0 \varepsilon_r \overline{\mathcal{E}} = \varepsilon \overline{\mathcal{E}}, \tag{2.2a}$$

$$\overline{B} = \mu_0 \overline{\mathcal{H}} + \overline{\mathcal{M}} = \mu_0 (1 + \chi_m) \overline{\mathcal{H}} = \mu_0 \mu_r \overline{\mathcal{H}} = \mu \overline{\mathcal{H}}, \tag{2.2b}$$

---

[2]Antiparallel vectors are collinear vectors with opposite signs, i.e., opposite directions.

[3]For the sake of generality, we consider here generalized symmetrical Maxwell's equations, including fictitious magnetic current density and magnetic charge density. Fictitious magnetic conductivity will also be admitted in the constitutive relations.

[4]In fact, LH media are necessarily dispersive, as shown in Section 2.2, where the corresponding constitutive relations and subsequent entropy conditions will be derived. However, Eqs. (2.2) are approximately true in a weakly dispersive media. (In practice, many dispersive media can be considered weakly dispersive.) In this section, we will assume weakly nondispersive LH media before presenting a rigorous discussion including dispersion in Section 2.2.

where $\overline{\mathcal{P}} = \varepsilon_0 \chi_e$ and $\overline{\mathcal{M}} = \mu_0 \chi_m$ are the electric and magnetic polarizations, respectively, χ_e and χ_m are the electric and magnetic susceptibilities, respectively, $\varepsilon_0 = 8.854 \cdot 10^{-12}$ (F/m) and $\mu_0 = 4\pi \cdot 10^{-7}$ (H/m) are the permittivity and permeability of free space, respectively, and $\varepsilon_r = 1 + \chi_e$ and $\mu_r = 1 + \chi_m$ are the permittivity and permeability, respectively, of the material considered. The latter can be written in the form

$$\varepsilon_r = \varepsilon' - j\varepsilon'' = \varepsilon' \left(1 - j \tan \delta_e\right), \quad \tan \delta_e = \frac{\omega \varepsilon'' + \sigma_e}{\omega \varepsilon'}, \tag{2.3a}$$

$$\mu_r = \mu' - j\mu'' = \mu' \left(1 - j \tan \delta_m\right), \quad \tan \delta_m = \frac{\omega \mu'' + \sigma_m}{\omega \mu'}. \tag{2.3b}$$

In the last two equations, the imaginary parts of ε and μ account for losses: $\omega \varepsilon''$ represents loss due to dielectric damping, σ_e loss due to finite electric conductivity, $\omega \mu''$ loss due to magnetic damping, and σ_m loss due to finite (fictitious) magnetic conductivity. Assuming harmonic fields with the time dependence[5] $e^{+j\omega t}$ and defining the corresponding generic phasor $\overline{F}(\overline{r})$ as

$$\overline{\mathcal{F}}(\overline{r}, t) = \text{Re}\left[\overline{F}(\overline{r})e^{+j\omega t}\right], \tag{2.4}$$

where $\overline{\mathcal{F}}$ represents any of the physical quantities in Eq. (2.1), Maxwell's equations and the constitutive equations can be written as

$$\nabla \times \overline{E} = -j\omega\mu\overline{H} - \overline{M}_s, \tag{2.5a}$$

$$\nabla \times \overline{H} = j\omega\varepsilon\overline{E} + \overline{J}_s, \tag{2.5b}$$

$$\nabla \cdot \overline{D} = \rho_e, \tag{2.5c}$$

$$\nabla \cdot \overline{B} = \rho_m, \tag{2.5d}$$

and

$$\overline{D} = \varepsilon\overline{E}, \tag{2.6a}$$

$$\overline{B} = \mu\overline{H}, \tag{2.6b}$$

respectively. Let us consider now the *plane wave*

$$\overline{E} = \overline{E}_0 e^{-j\overline{\beta}\cdot\overline{r}}, \tag{2.7a}$$

$$\overline{H} = \frac{\overline{E}_0}{\eta} e^{-j\overline{\beta}\cdot\overline{r}}, \tag{2.7b}$$

[5]The dependence $e^{+j\omega t}$ is the convention of engineers, whereas physicists generally use the phasor time dependence $e^{-i\omega t}$. Conversion from one formulation to the other can be accomplished by the imaginary variable substitution $i \rightarrow -j$. Particular care must be taken to this divergence of conventions in the study of LH media, where sign often plays a crucial role.

where $\eta = |\overline{E}|/|\overline{H}|$ denotes the wave impedance. Because any physical quantity can be expressed as a superposition of plane waves by virtue of the Fourier transform, consideration of an isolated plane wave will provide information on the fundamental response of the medium. This information is straightforwardly obtained by substituting the plane wave expressions (2.7) into the first two Maxwell's equations [Eq. (2.5)].

For simplicity, let us consider a loss-less medium[6] ($\varepsilon'' = \mu'' = 0$) in regions without sources ($\overline{M}_s = \overline{J}_s = 0$). In the case of a RH medium, $\varepsilon, \mu > 0$, and therefore

$$\overline{\beta} \times \overline{E} = +\omega\mu\overline{H}, \tag{2.8a}$$

$$\overline{\beta} \times \overline{H} = -\omega\varepsilon\overline{E}, \tag{2.8b}$$

which builds the familiar *right-handed* triad $(\overline{E}, \overline{H}, \overline{\beta})$ shown in Fig 2.1(a). In contrast, in the case of a LH medium, $\varepsilon, \mu < 0$, and therefore, since $|\varepsilon| = -\varepsilon > 0$ and $|\mu| = -\mu > 0$,

$$\overline{\beta} \times \overline{E} = -\omega|\mu|\overline{H}, \tag{2.9a}$$

$$\overline{\beta} \times \overline{H} = +\omega|\varepsilon|\overline{E}, \tag{2.9b}$$

which builds the unusual *left-handed* triad $(\overline{E}, \overline{H}, \overline{\beta})$ shown in Fig 2.1(b).

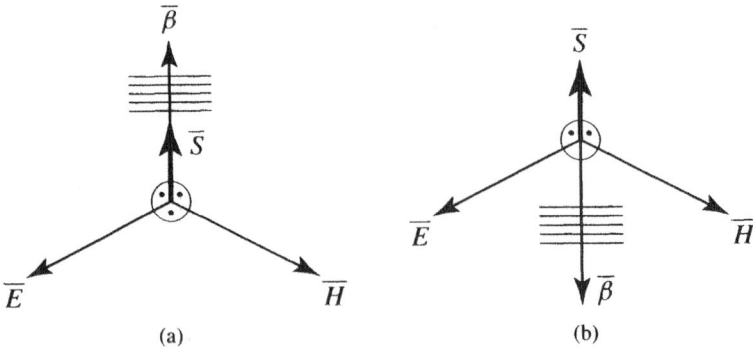

(a)   (b)

Fig. 2.1. Electric field–magnetic field–wave vector triad $(\overline{E}, \overline{H}, \overline{\beta})$ and Poynting vector $\overline{S}$ for an electromagnetic wave, as prescribed by Maxwell's equations [Eq. (2.1)] and Poynting's theorem [Eq. (2.19)], respectively. (a) Conventional, right-handed (RH) medium, where $\varepsilon, \mu > 0$. (b) Left-handed (LH) medium, where $\varepsilon, \mu < 0$.

---

[6]The following argument holds in the lossy case, because ε'' and μ'' represent an attenuation factor in Eq. (2.7), which is related to the evanescent nature of the wave and therefore does not affect phase in any way.

Thus, frequency being always a positive quantity, the phase velocity

$$\overline{v}_p = \frac{\omega}{\beta}\hat{\beta}, \quad \left(\hat{\beta} = \overline{\beta}/|\overline{\beta}|\right) \tag{2.10}$$

in a LH medium [Eq. (2.8)] is opposite to the phase velocity in a RH medium [Eq. (2.9)]. Moreover, because the wave number β is known to be positive in a RH medium (outward propagation from the source), it is negative in a LH medium (inward propagation to the source):

$$\text{RH medium: } \beta > 0 \quad (v_p > 0), \tag{2.11a}$$

$$\text{LH medium: } \beta < 0 \quad (v_p < 0). \tag{2.11b}$$

For generality, Eqs. (2.8) and (2.9) can be compacted into the single relation

$$\overline{\beta} \times \overline{E} = s\omega|\mu|\overline{H}, \tag{2.12a}$$

$$\overline{\beta} \times \overline{H} = -s\omega|\varepsilon|\overline{E}, \tag{2.12b}$$

where s is a "handedness" sign function defined as

$$s = \begin{cases} +1 & \text{if the medium is RH,} \\ -1 & \text{if the medium is LH.} \end{cases} \tag{2.13}$$

It follows from Eq. (2.9) [or Eq. (2.12) with $s = -1$] and Fig 2.1(b) that, in LH medium phase, which is related to phase velocity v_p, propagates backward to the source in the opposite direction than that of power, related to group velocity v_g. With the adopted time dependence $e^{+j\omega t}$ and assuming that power propagates in the direction of positive values of the space variable r, this backward-wave propagation implies that the fields have a time-space dependence

$$\overline{E}, \overline{H} \approx e^{+j(\omega t + |n|k_0 r)}. \tag{2.14}$$

In this expression, transverse electromagnetic (TEM) propagation, prevailing in a homogeneous and isotropic medium, has been implicitly assumed, so that the propagation constant has only one component that is equal to the wave number k_n in the medium

$$\beta = k_n = nk_0 = n\frac{\omega}{c}, \tag{2.15}$$

where

$$n = \pm\sqrt{\varepsilon_r \mu_r} \tag{2.16}$$

is the refractive index [Eq. (1.1)]. In a LH medium, since $\beta < 0$, we must have from Eqs. (2.15) and (2.16) a negative index of refraction (NRI), $n < 0$ (so that

$|n| = -n$). This demonstrates that *the index of refraction is negative in a medium with negative permittivity and permeability.* The index of refraction can thus be written in general as

$$n = s\sqrt{\varepsilon_r \mu_r}, \tag{2.17}$$

where Eq. (2.13) was used.

To better understand the implications of the above observations, let us consider also the Poynting theorem, which reads for the volume shown in Fig 2.2 [1]

$$P_s = P_0 + P_l + 2j\omega(W_e - W_m), \tag{2.18a}$$

where

$$P_s = -\frac{1}{2} \int_V \left(\overline{E} \cdot \overline{J}_s^* + \overline{H}^* \cdot \overline{M}_s \right) dv, \tag{2.18b}$$

$$P_0 = \frac{1}{2} \oint_S \overline{E} \times \overline{H}^* \cdot d\overline{s}, \tag{2.18c}$$

$$P_l = \frac{\sigma_e}{2} \int_V |\overline{E}|^2 dv + \frac{\sigma_m}{2} \int_V |\overline{H}|^2 dv + \frac{\omega}{2} \int_V \left(\varepsilon'' |\overline{E}|^2 + \mu'' |\overline{H}|^2 \right) dv, \tag{2.18d}$$

$$W_e = \frac{1}{4}\mathrm{Re} \int_V \overline{E} \cdot \overline{D}^* dv, \tag{2.18e}$$

$$W_m = \frac{1}{4}\mathrm{Re} \int_V \overline{H} \cdot \overline{B}^* dv. \tag{2.18f}$$

This theorem expresses the complex power P_s in a volume V delivered by the sources $\overline{M}_s$ and $\overline{J}_s$ in terms of the complex power flow P_0 out of the closed surface S delimiting V, the time-average dissipated power P_l in V due to electric/magnetic conductivities, dielectric and magnetic losses, and the time average

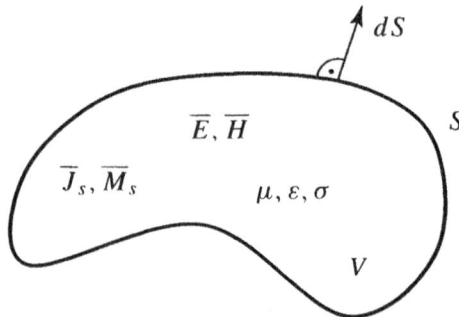

Fig. 2.2. Volume V formed by the closed surface S and containing the field intensities $\overline{E}, \overline{H}$ and the sources $\overline{J}_s, \overline{M}_s$.

electric and magnetic energy stored in V, W_e, and W_m, respectively. The power flow P_0 is related to the Poynting vector $\overline{S}$, defined as[7]

$$\overline{S} = \overline{E} \times \overline{H}^*, \qquad (2.19)$$

The Poynting vector $\overline{S}$, associated with the power flow P_0 in Eq. (2.18c), is oriented toward the direction of propagation of energy over time and is therefore *parallel* to the group velocity

$$\overline{v}_g = \nabla_{\overline{\beta}}\omega, \qquad (2.20)$$

which may be broadly defined as the velocity of a modulated signal in a distortionless medium. Thus, in contrast to the wave vector $\overline{\beta}$, the Poynting vector is *not* dependent on the constitutive parameters of the medium (ε and μ), but only on $\overline{E}$ and $\overline{H}$. This conclusion completes the characterization of the RH and LH media (Fig 2.1), which may be summarized as follows, considering positive the direction of the power flow

$$\text{RH medium:} \quad v_p > 0 \ (\beta > 0) \quad \text{and} \quad v_g > 0, \qquad (2.21a)$$

$$\text{LH medium:} \quad v_p < 0 \ (\beta < 0) \quad \text{and} \quad v_g > 0. \qquad (2.21b)$$

The fact that phase velocity is negative ($v_p < 0$) might seem troubling at first glance. However, this appears more acceptable if one remembers that phase velocity simply corresponds to the propagation of a *perturbation* and not of energy [2]. In contrast, a negative group velocity ($v_g < 0$) would violate causality, as it would correspond to transfer of energy toward the source! If often negative-gradient dispersion curves $\omega(\overline{\beta})$ associated with positive values of $\overline{\beta}$ (i.e., apparently negative group velocity modes) are observed in standard representations of dispersion diagrams, these curves should be interpreted as the eigen-solutions of the wave equation corresponding to the situation where the source placed at the other end of the medium (energy transfer from positive to negative values of the space variable).

2.2 ENTROPY CONDITIONS IN DISPERSIVE MEDIA

In this section, we will show that a LH medium, in contrast to a RH medium, is necessarily frequency dispersive, and we will derive a corresponding frequency-domain dispersion condition, called the entropy condition, which must be satisfied by the LH constitutive parameters ε and μ.

A *frequency-dispersive* or simply *dispersive* medium is a medium in which the propagation constant (β) is a nonlinear function of frequency, which results in

---

[7]This expression also holds in a LH (dispersive) medium, as shown in Section 2.2.

frequency-dependent group velocity and leads to distortion of modulated signals.[8] This implies from Eqs. (2.15) and (2.16) that either ϵ_r or μ_r (or both) have to be functions of frequency. An equivalent, more fundamental, definition is the following: dispersive medium is a medium in which the relation between $\overline{D}/\overline{B}$ and $\overline{\mathcal{E}}/\overline{\mathcal{H}}$[9] is a dynamic relation with "memory" rather than an instantaneous relation. The field intensities $\overline{\mathcal{E}}/\overline{\mathcal{H}}$ "create" the flux densities $\overline{D}/\overline{B}$ by inducing oscillation of the bound electrons in the atoms of the medium. These electrons respond to the excitation $(\overline{\mathcal{E}}/\overline{\mathcal{H}})$ with various delays to produce the collective response $(\overline{D}/\overline{B})$ of the system. As a consequence, a time delay exists between the cause and the effect, and arbitrary field intensities $\overline{\mathcal{E}}(t)/\overline{\mathcal{H}}(t)$ induce flux densities $\overline{D}(t)/\overline{B}(t)$ that are superpositions (assuming a linear medium) of the effects of $\overline{\mathcal{E}}(t')/\overline{\mathcal{H}}(t')$ at all times $t' \le t$. This corresponds to the convolutions[10]

$$\overline{D}(\overline{r}, t) = \varepsilon(t) * \overline{\mathcal{E}}(\overline{r}, t) = \int_{-\infty}^{t} \varepsilon(t - t')\overline{\mathcal{E}}(\overline{r}, t')dt', \tag{2.22a}$$

$$\overline{B}(\overline{r}, t) = \mu(t) * \overline{\mathcal{H}}(\overline{r}, t) = \int_{-\infty}^{t} \mu(t - t')\overline{\mathcal{H}}(\overline{r}, t')dt'. \tag{2.22b}$$

Using the Fourier transform pair

$$f(t) = \frac{1}{2\pi} \int_{-\infty}^{+\infty} \widetilde{f}(\omega)e^{+j\omega t} d\omega, \tag{2.23a}$$

$$\widetilde{f}(\omega) = \int_{-\infty}^{+\infty} f(t)e^{-j\omega t} dt, \tag{2.23b}$$

we thus have

$$\widetilde{\overline{D}}(\overline{r}, \omega) = \widetilde{\varepsilon}(\omega) \cdot \widetilde{\overline{\mathcal{E}}}(\overline{r}, \omega), \tag{2.24a}$$

$$\widetilde{\overline{B}}(\overline{r}, \omega) = \widetilde{\mu}(\omega) \cdot \widetilde{\overline{\mathcal{H}}}(\overline{r}, \omega). \tag{2.24b}$$

[8]Frequency dispersion in a LH TL medium is clearly apparent in the fundamental relations of Eq. (1.11) and in Fig 1.9(b).
[9]In this section, the notation x/y is a shorthand for "x, respectively, y," and does not mean x divided by y.
[10]We could make the following analogy with road traffic. The flux $\overline{\mathcal{F}}(r, t)$ of vehicles at different points r of a freeway's network as a function of time t depends on the traffic intensity at earlier times $\overline{\mathcal{I}}(r, t')$ ($t' < t$) weighted by the capacity $c(t)$ of the network to absorb traffic in time: $\overline{\mathcal{F}}(r, t') = \int_{-\infty}^{t} c(t - t')\overline{\mathcal{I}}(r, t')dt'$. If the network had the capacity to absorb traffic instantaneously, the flux of cars at a given time t would depend only on the traffic intensity at the same time, i.e., $c(t) = c\delta(t - t')$, and therefore we would have $\overline{\mathcal{F}}(r, t) = c\overline{\mathcal{I}}(r, t)$. If the network is capable of absorbing traffic very fast, the flux of vehicles is at any time fairly proportional to traffic intensity, $\overline{\mathcal{F}}(r, t) \approx c\overline{\mathcal{I}}(r, t)$; the response of the network is almost instantaneous, and the system is therefore said to be weakly dispersive. If in contrast the network cannot "follow" the traffic intensity, which generally results in a jam, the flux of vehicles strongly depends on the "history" of traffic along the network; we then have a strongly dispersive system, where $c(t')$ is a function exhibiting a slowly decreasing tail for $t' < t$ toward smaller values of t' [with $c(t' > t) = 0$].

Eqs. (2.22) show that, even for time harmonic excitations $\overline{\mathcal{E}}$ and $\overline{\mathcal{H}}$, $\partial/\partial t \neq j\omega$ in a dispersive medium. Therefore, the full time dependence of the different quantities must be retained. The Poynting vector [Eq. (2.19)], expressed in terms of the real fields $\overline{\mathcal{E}}/\overline{\mathcal{H}}$, can then be shown to read as

$$\overline{\mathcal{S}}(\overline{r}, t) = \overline{\mathcal{E}}(\overline{r}, t) \times \overline{\mathcal{H}}(\overline{r}, t). \tag{2.25}$$

This relation is formally identical to (2.19), except for the absence of the complex conjugate "*", which has become meaningless because all fields are now real quantities. The fact that this relation still holds in a dispersive medium can be demonstrated in the following manner. The continuity of the *tangential* components of $\overline{\mathcal{E}}/\overline{\mathcal{H}}$ at the boundary between two media induces the continuity of the *normal* component of $\overline{\mathcal{S}}$ at this interface. Because one of the two media considered can be nondispersive while the other one is assumed dispersive, the normal component of $\overline{\mathcal{S}}$, by continuity, is not affected by dispersion in the dispersive medium. It follows that the tangential components of $\overline{\mathcal{S}}$, although not continuous, are also unaffected by dispersion.[11] Therefore, the expression (2.19) established for a nondispersive medium still holds in the more general case of a dispersive medium, where it is more generally (also for nonharmonic fields) written as Eq. (2.25).

The rate of change of the energy per volume in a body V (see Fig 2.2) is $\nabla \cdot \overline{\mathcal{S}}$. With the use of Eq. (2.25), the sourceless Maxwell's equations (fields considered at locations where no source is present) (2.1a) ($\overline{\mathcal{M}}_s = 0$) and (2.1b) ($\overline{\mathcal{J}}_s = 0$) and the identity $\nabla \cdot (\overline{A} \times \overline{B}) = (\nabla \times \overline{A}) \cdot \overline{B} - (\nabla \times \overline{B}) \cdot \overline{A}$, this quantity becomes

$$\nabla \cdot \overline{\mathcal{S}} = -\left[\overline{\mathcal{E}} \cdot \frac{\partial \overline{\mathcal{D}}}{\partial t} + \overline{\mathcal{H}} \cdot \frac{\partial \overline{\mathcal{B}}}{\partial t}\right]. \tag{2.26}$$

At this point, some kind of assumption on the nature of the fields intensities has to be made in order to derive a useful expression in the frequency domain. Let us therefore consider quasi-harmonic fields with mean frequency ω_0. We can write such fields $\overline{\mathcal{F}}$ (standing for $\overline{\mathcal{E}}, \overline{\mathcal{H}}, \overline{\mathcal{D}}$ or $\overline{\mathcal{B}}$) in terms of the phasor $\overline{F}$ (standing for $\overline{E}, \overline{H}, \overline{D}$ or $\overline{B}$) [Eq. (2.4)]

$$\overline{\mathcal{F}} = \text{Re}\left[\overline{F}(t)e^{+j\omega_0 t}\right] = \text{Re}\left[\overline{\mathbf{F}}\right] = \frac{\overline{\mathbf{F}} + \overline{\mathbf{F}}^*}{2}, \quad \left[\overline{\mathbf{F}} = \overline{F}(t)e^{+j\omega_0 t}\right] \tag{2.27}$$

where here $\overline{F}(t)$ is function slowly varying in time in comparison with $e^{+j\omega_0 t}$, so that the field $\overline{\mathcal{F}}$ has a narrow spectrum centered around the mean value of ω_0, and $\overline{\mathbf{F}}$ (standing for $\overline{\mathbf{E}}, \overline{\mathbf{H}}, \overline{\mathbf{D}}$ or $\overline{\mathbf{B}}$) is the product of the phasor $\overline{F}$ by the harmonic function $e^{+j\omega_0 t}$. Inserting this expression for $\overline{\mathcal{E}}, \overline{\mathcal{H}}, \overline{\mathcal{D}}$, and $\overline{\mathcal{B}}$ into Eq. (2.26)

---

[11]Dispersion characteristics are naturally common to all the components of the fields. For instance, in the case of a plane wave, the same dependence $e^{-j\beta(\omega)r}$ is present in all the components.

and noting that the products $\overline{\mathcal{E}} \cdot \partial \overline{D}/\partial t$, $\overline{\mathcal{E}}^* \cdot \partial \overline{D}^*/\partial t$, $\overline{\mathcal{H}} \cdot \partial \overline{B}/\partial t$ and $\overline{\mathcal{H}}^* \cdot \partial \overline{B}^*/\partial t$ vanish when averaged over time, we obtain

$$\nabla \cdot \overline{S} = -\frac{1}{4} \left[\overline{E} \cdot \frac{\partial \overline{D}^*}{\partial t} + \overline{E}^* \cdot \frac{\partial \overline{D}}{\partial t} + \overline{H} \cdot \frac{\partial \overline{B}^*}{\partial t} + \overline{H}^* \cdot \frac{\partial \overline{B}}{\partial t} \right]. \qquad (2.28)$$

To develop the derivatives $\partial \overline{D}^*/\partial t$, $\partial \overline{D}/\partial t$, and subsequently $\partial \overline{B}^*/\partial t$, $\partial \overline{B}/\partial t$, let us expand $\overline{D}$ in its Fourier series

$$\begin{aligned}
\overline{D} = \overline{D}(t)e^{+j\omega_0 t} &= \left[\sum_{\omega'} \overline{D}_{\omega'} e^{+j\omega' t} \right] e^{+j\omega_0 t} = \sum_{\omega'} \overline{D}_{\omega'} e^{+j(\omega_0 + \omega')t} \\
&= \sum_{\omega'} \varepsilon(\omega_0 + \omega') \overline{E}_{\omega'} e^{+j(\omega_0 + \omega')t},
\end{aligned} \qquad (2.29)$$

where we remember that only terms with $\omega' \ll \omega_0$ will subsist in the series due to the assumption of slow time variation of the phasor $\overline{D}$. The derivative of the above expression reads

$$\frac{\partial \overline{D}}{\partial t} = \sum_{\omega'} f(\omega_0 + \omega') \overline{E}_{\omega'} e^{+j(\omega_0 + \omega')t}, \quad \text{with} \quad f(\omega) = j\omega\varepsilon(\omega), \qquad (2.30)$$

where $f(\omega)$ may be approximated by its Taylor expansion around $\omega = \omega_0$, since $\omega' \ll \omega_0$: $f(\omega) = f(\omega_0) + [df(\omega)/d\omega]_{\omega=\omega_0} (\omega - \omega_0)$ or $f(\omega_0 + \omega') = f(\omega_0) + \dot{f}(\omega_0)\omega'$. We then have

$$\begin{aligned}
\frac{\partial \overline{D}}{\partial t} &= \sum_{\omega'} \left[f(\omega_0) \overline{E}_{\omega'} e^{+j(\omega_0 + \omega')t} + \dot{f}(\omega_0)\omega' \overline{E}_{\omega'} e^{+j(\omega_0 + \omega')t} \right] \\
&= f(\omega_0) \left[\sum_{\omega'} \overline{E}_{\omega'} e^{+j\omega' t} \right] e^{+j\omega_0 t} + \dot{f}(\omega_0) \left[\sum_{\omega'} \omega' \overline{E}_{\omega'} e^{+j\omega' t} \right] e^{+j\omega_0 t} \\
&= f(\omega_0) \overline{E} e^{+j\omega_0 t} - j\dot{f}(\omega_0) \frac{\partial \overline{E}}{\partial t} e^{+j\omega_0 t} \\
&= f(\omega_0) \overline{E} - j\dot{f}(\omega_0) \frac{\partial \overline{E}}{\partial t} e^{+j\omega_0 t} \\
&= j\omega_0 \varepsilon(\omega_0) \overline{E} + \left[\frac{d(\omega\varepsilon)}{d\omega} \right]_{\omega=\omega_0} \frac{\partial \overline{E}}{\partial t} e^{+j\omega_0 t}.
\end{aligned} \qquad (2.31)$$

Omitting henceforward the subscript 0, we have thus for $\partial \overline{D}/\partial t$

$$\frac{\partial \overline{D}}{\partial t} = j\omega\varepsilon(\omega) \overline{E} + \frac{d(\omega\varepsilon)}{d\omega} \frac{\partial \overline{E}}{\partial t} e^{+j\omega t}, \qquad (2.32)$$

and similarly for $\partial \overline{B}/\partial t$

$$\frac{\partial \overline{B}}{\partial t} = j\omega\mu(\omega)\overline{H} + \frac{d(\omega\mu)}{d\omega}\frac{\partial \overline{H}}{\partial t}e^{+j\omega t}. \qquad (2.33)$$

Substituting these expressions in Eq. (2.28) and neglecting the imaginary parts of $\varepsilon(\omega)$ and $\mu(\omega)$ yields then

$$\nabla \cdot \overline{S} = -\frac{1}{4}\left\{\frac{d(\omega\varepsilon)}{d\omega}\left[\overline{E}^* \cdot \frac{\partial \overline{E}}{\partial t} + \overline{E} \cdot \frac{\partial \overline{E}^*}{\partial t}\right] + \frac{d(\omega\mu)}{d\omega}\left[\overline{H}^* \cdot \frac{\partial \overline{H}}{\partial t} + \overline{H} \cdot \frac{\partial \overline{H}^*}{\partial t}\right]\right\}$$

$$= -\frac{1}{4}\left\{\frac{d(\omega\varepsilon)}{d\omega}\frac{\partial}{\partial t}\left[\overline{E} \cdot \overline{E}^*\right] + \frac{d(\omega\mu)}{d\omega}\frac{\partial}{\partial t}\left[\overline{H} \cdot \overline{H}^*\right]\right\} \qquad (2.34)$$

since $\overline{E} \cdot \overline{E}^* = \overline{E} \cdot \overline{E}^*$ and $\overline{H} \cdot \overline{H}^* = \overline{H} \cdot \overline{H}^*$. Rewriting this expression in terms of the real fields $\mathcal{E}/\mathcal{H}$, we obtain the mean value of the electromagnetic part of the internal energy per unit volume of the medium, $\overline{W} = -\int \nabla \cdot \overline{S} \, dt$, as[12]

$$\overline{W} = \frac{1}{4}\left[\frac{d(\omega\varepsilon)}{d\omega}\overline{\mathcal{E}^2} + \frac{d(\omega\mu)}{d\omega}\overline{\mathcal{H}^2}\right]. \qquad (2.35)$$

If there is no dispersion, ε and μ are constants, and Eq. (2.35) reduces to

$$\overline{W} = \frac{\varepsilon\overline{E^2} + \mu\overline{H^2}}{4}, \qquad (2.36)$$

which corresponds, as expected, to the sum $\overline{W} = \overline{W}_e + \overline{W}_m$ of the time average electric and magnetic energies stored in the volume V, where $\overline{W}_e = \varepsilon\overline{\mathcal{E}^2}/4$ and $\overline{W}_m = \mu\overline{\mathcal{H}^2}/4$ according to Eqs. (2.18e) and (2.18f), respectively.

If the external supply of electromagnetic energy to the body is cut off, absorption ultimately converts the energy $\overline{W}$ entirely into heat. By the *law of entropy*, specifying that the entropy of a system is an ever increasing quantity, there must be evolution and not absorption of heat [3]. We therefore must have

$$\overline{W} > 0, \qquad (2.37)$$

which leads in the general (dispersive) case to the inequalities

$$\frac{d(\omega\varepsilon)}{d\omega} > 0, \qquad (2.38a)$$

$$\frac{d(\omega\mu)}{d\omega} > 0, \qquad (2.38b)$$

---

[12]It should be kept in mind that this relation is valid only for fields whose amplitude varies sufficiently slowly with time in comparison with $e^{+j\omega_0 t}$.

since $\overline{\mathcal{E}^2} > 0$ and $\overline{\mathcal{H}^2} > 0$. Eq. (2.37) is called the *entropy condition*, which is seen to be different in the general case of a dispersive medium [Eq. (2.35)] than in the particular case of a nondispersive medium [Eq. (2.36)].[13] Eqs. (2.38) are the *general entropy conditions for the constitutive parameters*. These entropy conditions show that *simultaneously negative ε and μ are physically impossible in a nondispersive medium* since they would violate the law of entropy. They also show that, in contrast *in a dispersive medium, simultaneously negative ε and μ are allowed*, as long as the frequency dependent ε and μ satisfy the conditions of Eqs. (2.38). For this to be achieved, ε and μ must be positive in some parts of their spectrum, to compensate for the negative parts, which shows that a LH medium is necessarily dispersive.

2.3 BOUNDARY CONDITIONS

The boundary conditions (BCs) at the interface between two media, which are directly derived from Maxwell equations and therefore clearly hold also in the case of LH media, read

$$\hat{n} \cdot (\overline{D}_2 - \overline{D}_1) = \rho_{es}, \qquad (2.39a)$$

$$\hat{n} \cdot (\overline{B}_2 - \overline{B}_1) = \rho_{ms}, \qquad (2.39b)$$

$$\hat{n} \times (\overline{E}_2 - \overline{E}_1) = -\overline{M}_s, \qquad (2.39c)$$

$$\hat{n} \times (\overline{H}_2 - \overline{H}_1) = \overline{J}_s, \qquad (2.39d)$$

where ρ_{se} is the electric surface charge density on the interface, ρ_{sm} is the magnetic (fictitious) surface charge density on the interface, and $\hat{n}$ is a unit vector normal to the interface pointing from medium 1 to medium 2. The first two equations of (2.39) state that, in the absence of charges ($\rho_{es} = \rho_{em} = 0$) at the interface, the normal components of $\overline{D}$ and $\overline{B}$ are continuous, whereas the last two questions of (2.39) state that, in the absence of sources ($\overline{J}_s = \overline{M}_s = 0$) at the interface, the normal tangential components of $\overline{E}$ and $\overline{H}$ are continuous, i.e.,

$$D_{1n} = D_{2n}, \qquad (2.40a)$$

$$B_{1n} = B_{2n}, \qquad (2.40b)$$

$$E_{1t} = E_{2t}, \qquad (2.40c)$$

$$H_{1t} = H_{2t}, \qquad (2.40d)$$

where the indexes n and t stand for normal and tangential, respectively.

Let us consider now the specific case of an interface between a RH and a LH media, illustrated in Fig 2.3 (medium 1: RH, medium 2: LH). The relations of

---

[13]In a nondispersive medium, the entropy conditions simply read $\varepsilon > 0$ and $\mu > 0$.

Medium 2: LH, ε_2, μ_2

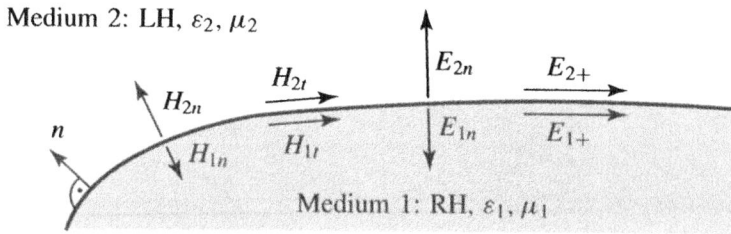

Fig. 2.3. Boundary conditions at the interface between a RH medium and a LH medium.

Eqs. (2.40) reveal the following. The BCs on the tangential components of $\overline{E}/\overline{H}$ are unaffected at this interface, since the relations on the tangential components do not depend on ε, μ. In contrast, the BCs on the normal components are necessarily changed, since they involve ε, μ with changes in signs. By assuming that the LH medium is weakly dispersive and that therefore the constitutive relations (2.2) or (2.6) are approximately valid, we obtain the following BCs at a RH/LH interface:

$$E_{1n} = -\frac{\varepsilon_2}{|\varepsilon_1|} E_{2n}, \tag{2.41a}$$

$$H_{1n} = -\frac{\mu_2}{|\mu_1|} H_{2n}, \tag{2.41b}$$

$$E_{1t} = E_{2t}, \tag{2.41c}$$

$$H_{1t} = H_{2t}. \tag{2.41d}$$

Thus *the tangential components of $\overline{E}/\overline{H}$ remain continuous while their normal components become antiparallel* at the interface between a RH medium and a LH medium. The BCs on the normal components of $\overline{E}/\overline{H}$ can be written in general, for and interface between arbitrary RH-LH media, as follows:

$$E_{1n} = s_1 s_2 \frac{|\varepsilon_2|}{|\varepsilon_1|} E_{2n}, \tag{2.42a}$$

$$H_{1n} = s_1 s_2 \frac{|\mu_2|}{|\mu_1|} H_{2n}, \tag{2.42b}$$

where s_i ($i = 1, 2$) represents the sign of handedness of the medium i, as defined in Eq. (2.13).

2.4   REVERSAL OF DOPPLER EFFECT

Consider a source S in motion along a direction z and radiating omnidirectionally an electromagnetic wave with angular frequency ω, as illustrated in Fig 2.4. In

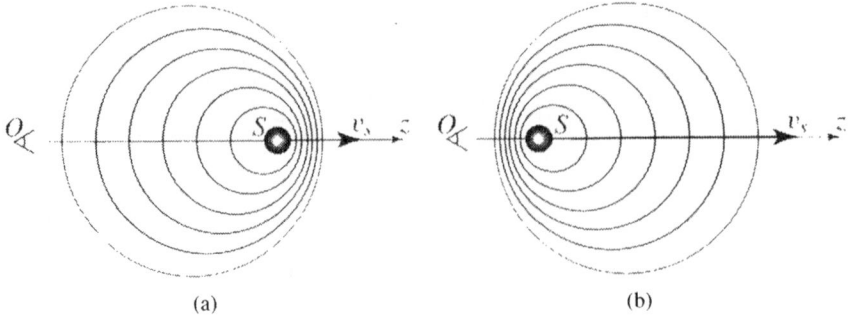

(a)   (b)

Fig. 2.4. Doppler effect. (a) Conventional, in a RH medium ($\Delta\omega > 0$). (b) Reversed, in a LH medium ($\Delta\omega < 0$).

the far-field of the source,[14] the radiated fields have the form [4]

$$\overline{E}(z, t), \overline{H}(z, t) \propto \frac{e^{j\varphi(\omega \cdot t)}}{r}, \quad \text{with} \quad \varphi(\omega, t) = \omega t - \beta r, \quad (2.43)$$

where β represents the wave number in the medium in which S moves and radiates and r the standard radial variable of the spherical coordinates system.

Let us consider what happens to the radiated wave along the direction of the motion of the source, i.e., for $r = z$ (in $\theta = 0$). If the sources moves toward positive values of z with a velocity $v_s = z/t$, its position as a function of time is $z = v_s t$. Consequently, the phase seen by an observer O located at the left-hand side of S (i.e., looking at S toward positive values of z) may be developed as follows along the z axis:

$$\varphi = \omega t - \beta v_s t = \omega \left(1 - \frac{\beta}{\omega} v_s\right) t = \omega \left(1 - \frac{v_s}{v_p}\right) t = \omega \left(1 - s \frac{v_s}{|v_p|}\right) t, \quad (2.44)$$

since $\omega/\beta = v_p$ according to Eq. (2.10). The coefficient of t is the Doppler frequency ω_{Doppler}, which is the difference of the frequency ω of the motionless ($v_s = 0$) source and the Doppler frequency shift $\Delta\omega$,

$$\omega_{\text{Doppler}} = \omega - \Delta\omega, \quad \text{with} \quad \Delta\omega = s \frac{v_s}{|v_p|}, \quad (2.45)$$

where s is the handedness sign function defined in Eq. (2.13). In a RH medium, $\Delta\omega > 0$ since $s = +1$, and therefore the frequency measured by the observer looking at the receding source is shifted downward or "red-shifted," as illustrated in Fig 2.4(a); on the other hand, an observer located on the right-hand side of the source, i.e., seeing a proceeding source, would measure a frequency shifted upward or "blue-shifted," because the sign of v_s would be changed for that observer [6]. In a LH medium, because $s = -1$, the whole phenomenon is

[14]The far-field region is defined as $r > 2D^2/\lambda$, where D is the largest size of the source (antenna) [4].

reversed, as shown in Fig 2.4(b): The Doppler frequency of a receding source is blue-shifted, whereas that of a proceeding source is red-shifted. This phenomenon of reversal of Doppler effect in a LH medium was pointed out by Veselago back in 1967 as an immediate consequence of left-handedness [5] (Section 1.2).

2.5 REVERSAL OF VAVILOV-ČERENKOV RADIATION

In his paper [5], Veselago also pointed out that a LH medium would induce reversal of Vavilov-Čerenkov radiation (Section 1.2). Vavilov-Čerenkov radiation is the visible electromagnetic radiation emitted by liquids and solids when bombarded by fast-moving electron beams with high velocity. For simplicity, let us consider a single electron $e-$ of charge q moving at the velocity v_e along a direction z, as illustrated in Fig 2.5. The current density for this charge is

$$\overline{J}(\overline{r}, t) = \hat{z} q v_e \delta(x)\delta(y)\delta(z - v_e t), \tag{2.46}$$

where $\hat{z}$ is a unit vector directed along the direction z. By solving the wave equation for this source in the cylindrical coordinate system, one obtains the electric field intensity solution [7]

$$\overline{E}(\overline{r}) = -\frac{q}{8\pi\omega\varepsilon}\left[\hat{z}k^2 - j\frac{\omega}{v_e}\nabla\right] H_0^{(1)}(k_\rho\rho)e^{-j\omega z/v_e}$$

$$\stackrel{k_\rho\rho \gg 1}{=} -\frac{q}{8\pi\omega\varepsilon}\sqrt{\frac{2jk_\rho}{\pi\rho}}\left[\hat{\rho}\frac{\omega}{v_e} - \hat{z}k_\rho\right]e^{-j(k_\rho\rho + \omega z/v_e)}, \tag{2.47a}$$

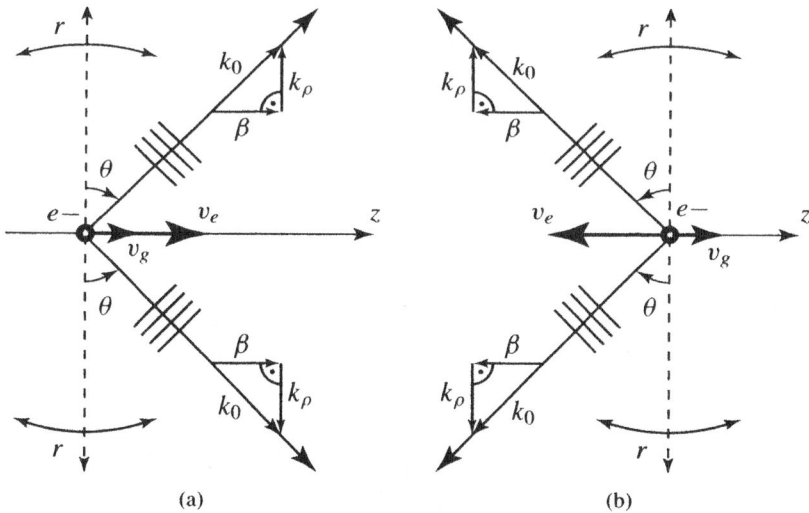

Fig. 2.5. Vavilov-Čerenkov radiation. v_g would represent the group velocity of the beam including the electron. (a) Conventional, in a RH medium ($\theta > 0$). (b) Reversed, in a LH medium ($\theta < 0$).

where

$$k_\rho = \sqrt{k^2 - \beta^2}, \quad \text{with} \quad \beta = \frac{\omega}{v_e}. \tag{2.47b}$$

Due to its wave nature, "the isolated electron considered *is* the wave" and therefore the phase velocity $v_p = \omega/\beta$ of the wave along z is the same thing as the velocity of the electron, $v_p = v_e$. This is the reason why we have $\beta = \omega/v_e$ in the above relations.[15]

It is worth noting that the phenomenon of Vavilov-Čerenkov radiation is fundamentally similar to the phenomenon of leakage radiation in traveling-wave antennas (Section 6.1). The link between the two descriptions may be understood by considering the fact that a beam of electrons in motion is equivalent to an electric current ($i = dq/dt$). From this point of view, a transmission line supporting a fast-wave ($v_p > c$) (leaky-wave structure) is equivalent to a transmission medium supporting a beam of fast-wave electrons ($v_e > c$). This analogy can be further extended to an array of antennas [4] where the phase difference $\varphi = -\beta p$ between the elements (separated by a distance p) may be considered equivalent to a wave propagating along the array with the velocity $v_p = \omega/\beta$ [8].

Examination of Eq. (2.47) explains the phenomena observed by Čerenkov in his experiments [7]:

- Since radiation requires $k_\rho \in \mathbb{R}$ (propagation along the direction perpendicular to the direction of motion of the beam), radiation can occur only in the frequency range where $k > \beta = \omega/v_e$, i.e., $v_e > \omega/k = c$. Thus, radiation occurs only at frequencies where the velocity of the electrons is larger than the speed of light. If $v_e < c$, k_ρ is imaginary, and therefore exponential decay of the fields occurs in the direction perpendicularly to the beam. In the leaky-wave approach, the dispersion or $\omega - \beta$ diagram is plotted along with the air lines $\omega = \pm k_0 c$, delimiting the "radiation cone" (Section 6.1). When the structure is excited at a point (ω, β) of the dispersion lying inside this cone, the wave in the direction of the line is called a *fast wave*, because $v_p > c$, and leakage radiation occurs; when the structure is excited at a point lying outside this region, the wave is called a *slow wave*, because $v_p < c$, and the wave is purely guided along the line, without radiation.

- Vavilov-Čerenkov radiation is directive, with a radiation angle depending on the velocity of the electrons. This radiation angl can be straightforwardly

---

[15]This fact is the most essential distinction between the Doppler "event" (Section 2.4) and the Vavilov-Čerenkov "event." In the case of the Doppler effect, the source in motion considered is not a relativistic particle, such as an electron, but macroscopic object (e.g., an antenna), the velocity (of motion) of which, v_s, has no relation with the phase velocity v_p of the wave it radiates ($v_p \neq v_e$). Otherwise, the Doppler effect and the Vavilov-Čerenkov radiation would be the same phenomenon! Another fundamental difference is that, in the case of Vavilov-Čerenkov, radiation occurs only if the source velocity is larger than the velocity of light, $v_e = v_p > c$ (fast-wave), whereas in the case of Doppler, radiation is a mechanism totally independent from the motion of the object, which can therefore occur for any source velocity v_s.

determined with the help of Fig 2.5(a)

$$\theta = \sin^{-1}\left(\frac{\beta}{k_0}\right) = \sin^{-1}\left(\frac{c}{v_e}\right). \tag{2.48}$$

It is clear in this relation that, if $v_e < c$, the angle θ is imaginary, consistent with the fact that there is no radiation in this case. This relation is exactly identical to that used in leaky-wave antennas (Section 6.1, Eq. 6.3).

- With regard to the polarization, $\overline{E}$ lies in the plane determined by $\overline{k}$ and $\hat{z}$. In addition, we have $\overline{E} \perp \overline{k}_0$ since $\overline{k}_0 \cdot \overline{E} = 0$. The following intuitive explanation is provided in [7] to explain this polarization. When the charged particle is at rest, the electric field points radially, whereas when the particle moves the field lines are bent at an angle proportional to the velocity of the particle, eventually breaking away from the charge to produce radiation when $v_e > c$.

From the above explanations and in particular Eq. (2.48), it is obvious that, in a RH medium, Vavilov-Čerenkov radiation occurs at positive θ angles (in the half-plane in the direction of motion of the particle), since $\beta > 0$, as shown in Fig 2.5(a). If in contrast the medium is LH, $\beta < 0$, radiation occurs at negative θ angles (in fact also the half-plane in the direction of motion of the particle but opposite to the group velocity of the beam including it). To better understand the meaning of this reversal of Vavilov-Čerenkov radiation, a source generating a modulated beam of electrons should be considered. In this case, if the source emits electrons toward positive values of z, power (v_g) would naturally propagate toward positive z, whereas the electron wave would propagate backward in the LH medium.

A theoretical study of Vavilov-Čerenkov radiation in a LH medium is presented in [9]. The conclusions of this paper correspond point by point to those obtained for the CRLH backfire-to-endfire leaky-wave antenna, which will be presented in Section 6.2.

2.6 REVERSAL OF SNELL'S LAW: NEGATIVE REFRACTION

One of the most remarkable property of LH media is their negative refractive index (NRI), already demonstrated in Section 2.1. In this section, we will show the consequences of NRI when the LH medium is interfaced with a RH medium.

Let us consider the classical problem of a plane wave incident upon the boundary between two homogeneous media, illustrated in Fig 2.6. In general, from the incident wave, $e^{-j\overline{k}_i \cdot \overline{r}}$, in medium 1, a reflected wave, $e^{-j\overline{k}_r \cdot \overline{r}}$, in medium 1, and a transmitted (or refracted) wave in medium 2, $e^{-j\overline{k}_t \cdot \overline{r}}$, are generated. The boundary condition [Eq. (2.41)] requires that the tangential components of $\overline{E}$ and $\overline{H}$ be continuous at $z = 0$ for all x and y. Calling the magnitudes of the tangential incident, reflected, and transmitted electric field[16] $E_{i,tan}$, $E_{r,tan}$ and $E_{t,tan}$, respectively, we must have (at $z = 0$) in all possible cases

$$E_{i,tan}e^{-j(k_{ix}x+k_{iy}y)} + E_{r,tan}e^{-j(k_{rx}x+k_{ry}y)} = E_{t,tan}e^{-j(k_{tx}x+k_{ty}y)}, \qquad (2.49)$$

since the total field is the sum of the incident and reflected field in medium 1 and the transmitted field in medium 2. The only way that this equation can be satisfied for all x and y (of the interface) is to have $E_{i,tan} + E_{r,tan} = E_{t,tan}$, and therefore

$$k_{ix} = k_{rx} = k_{tx} = k_x, \qquad (2.50a)$$

$$k_{iy} = k_{ry} = k_{ty} = k_y, \qquad (2.50b)$$

which show that the tangential component of the wave number $\overline{k}_{tan} = k_x\hat{x} + k_y\hat{y}$ is continuous at the interface between two media

$$\overline{k}_{1,tan} = \overline{k}_{2,tan}. \qquad (2.51)$$

This relation is the *phase matching-condition*, which is a direct consequence of the continuity of the tangential components of $\overline{E}$ and $\overline{H}$ and which also holds at the interface between a RH medium and a LH medium since the continuity of $\overline{E}$ and $\overline{H}$ is conserved in this case (Section 2.3). Another information provided by the phase matching condition is that the incident, reflected, and transmitted wave vectors must all lie in the *plane of incidence*, which is the plane determined by $\overline{k}_i$ and the normal to the interface (z axis).

The tangential components of the wave numbers can be expressed as a function of their corresponding angle with the help of Fig 2.6. For instance, for the x components we have

$$k_{ix} = k_i \sin\theta_i, \quad k_{rx} = k_r \sin\theta_r, \quad k_{tx} = k_t \sin\theta_t, \qquad (2.52)$$

where the wave numbers are

$$k_i = \frac{\omega\sqrt{\varepsilon_{r1}\mu_{r1}}}{c} = \frac{\omega n_1}{c} = k_r = k_1, \quad k_t = \frac{\omega\sqrt{\varepsilon_{r2}\mu_{r2}}}{c} = \frac{\omega n_2}{c} = k_2. \qquad (2.53)$$

Eqs. (2.51) and (2.52) applied to the incident and reflected waves lead to the relation $k_i \sin\theta_i = k_r \sin\theta_r$, which, with Eq. (2.53), yields *Snell's law of reflection*

$$\theta_r = \theta_i. \qquad (2.54)$$

This law is unchanged at the interface between a RH medium and a LH medium, because it relates fields in the same medium. Eqs. (2.51) and (2.52)

---

[16]Of course, the magnetic field can also be considered instead of the electric field.

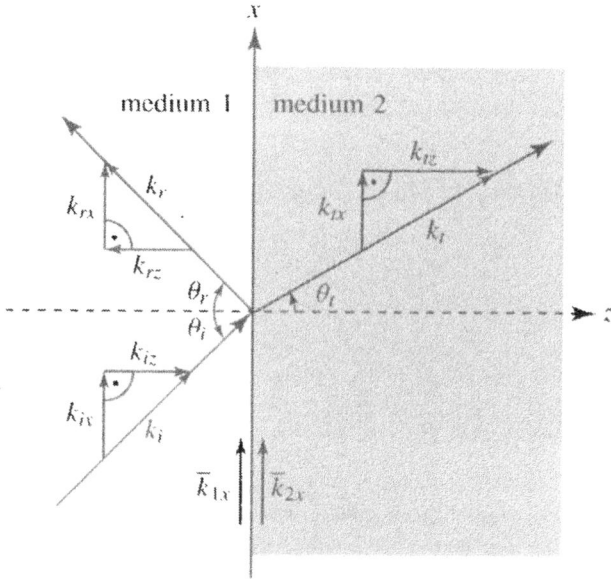

Fig. 2.6. Phase-matching at the boundary between two media. The angles θ_i, θ_r, and θ_t are the incident, reflected, and transmitted (or refracted) angles, respectively.

applied to the incident and transmitted waves lead to the relation $k_i \sin \theta_i = k_t \sin \theta_t$, which, with Eq. (2.53), yields *Snell's law of refraction*

$$n_1 \sin \theta_1 = n_2 \sin \theta_2. \tag{2.55}$$

This law is modified at the interface between a RH medium and a LH medium, due to the apparition of a negative sign in the refractive index of the LH medium (Section 2.1). Snell's law of refraction may be written in the more general form

$$s_1 |n_1| \sin \theta_1 = s_2 |n_2| \sin \theta_2, \tag{2.56}$$

where it also appears that if the two media are LH, Snell's law is unchanged due to mutual cancelation of the two minus signs of the two refractive indexes. Form (2.56), the reversal of Snell's law at the interface between a RH medium and a LH medium, illustrated in Fig 2.7, is manifest. A wave incident upon the interface between two media with same handedness (e.g., RH) experiences conventional *positive refraction*, characterized by a positive[17] refraction angle [Fig 2.7(a)], whereas a ray at the interface between two media of different handednesses (RH and LH) undergoes *negative refraction*, corresponding to negative refraction angle or NRI [Fig 2.7(b)].

---

[17]This is based on the convention that positive angles correspond to the counter-clockwise direction or rotation.

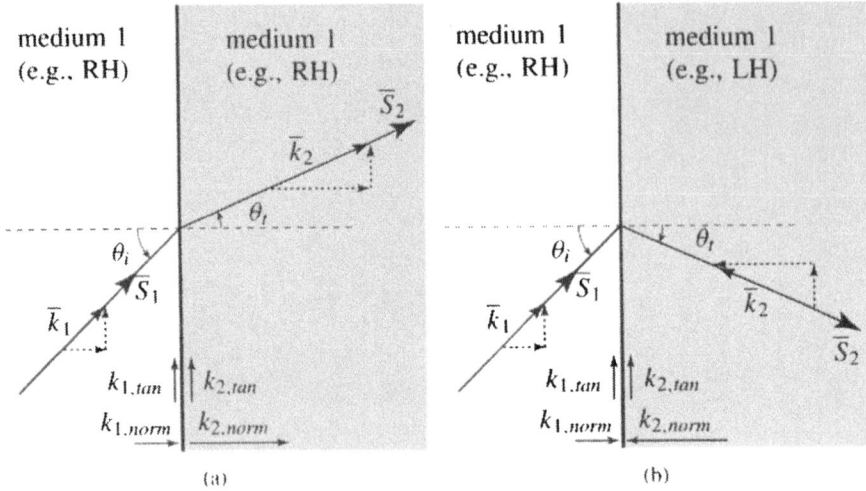

Fig. 2.7. Refraction of an electromagnetic wave at the interface between two different media. (a) Case of two media of same handedness (either RH or LH): positive refraction. (b) Case of two media of different handedness (one RH and the other one LH): negative refraction.

As a consequence of phase matching [Eq. (2.51)] and of the sign of the refractive index [Eq. (2.17)] in each of the two media, it is immediately apparent in Fig 2.7, that although the normal components of the wave vector $\bar{k}_{norm}$ are parallel at a RH/RH or LH/LH interface they are antiparallel at a RH/LH interface. The relation between the normal components of the wave vectors at the interface may be written in the general form

$$s_1|n_1| \cdot |\bar{k}_{2,norm}| = s_2|n_2| \cdot |\bar{k}_{1,norm}|. \tag{2.57}$$

Fig 2.7 also shows the Poynting vectors $\bar{S}_i$ which, according to Section 2.1, are parallel and antiparallel to the wave vector in RH medium and LH medium, respectively.

Various NRI effects will be illustrated by full-wave simulated fields distributions in Section 4.4, including the focusing effect induced by a flat lens, which is presented in the following section.

2.7 FOCUSING BY A "FLAT LH LENS"

By applying Snell's law [Eq. (2.56)] twice to a LH slab sandwiched between two RH media, also called a LH "lens,"[18] we obtain the *double focusing effect*

---

[18]This structure, although producing focusing, is not stricto senso a lens, considering that in principle a lens transforms a spherical wave into a plane wave, or vice versa, and not into another spherical wave, as it is the case for the LH slab.

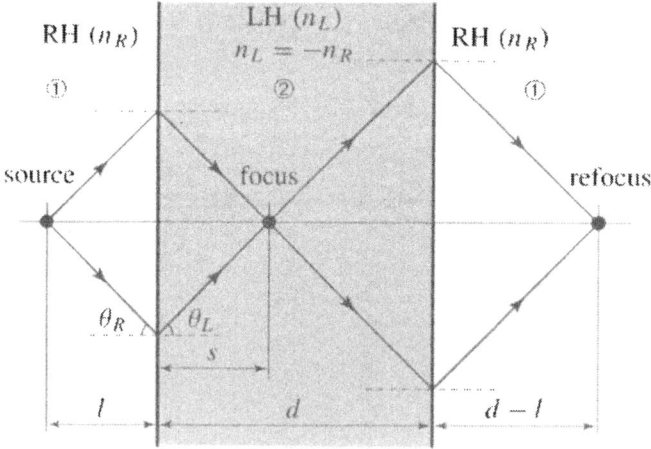

Fig. 2.8. Double focusing effect in a "flat lens," which is a LH slab of thickness d and refractive index n_L sandwiched between two RH media of refractive index n_R with $n_L = -n_R$.

depicted in Fig 2.8[19] [5]: two radiated rays with equal symmetric angles from a source at the distance l from the first interface are negatively refracted under an angle of same magnitude to meet at a distance s in the slab; then they focus again after a second negative refraction in the second RH medium at the distance $d - l$ from the second interface, where s is obtained by simple trigonometric considerations as

$$s = l \frac{\tan \theta_R}{\tan |\theta_L|}, \tag{2.58}$$

where the angle θ_R is the incidence angle and θ_L is obtained by Snell's law [Eq. (2.56)], $\theta_L = -\sin^{-1}\left[(n_R/n_L)\sin\theta_R\right]$. Formula (2.58) shows that if the two media have the same *electromagnetic density*, that is, refractive indexes of same magnitude ($n_L = -n_R$), focus is obtained at the mirror image of the source, $s = l$, since $|\theta_L| = \theta_R$ from Snell's law.

Let us now, instead of two isolated symmetric plane waves (Fig 2.8), consider a collection of plane waves (or rays), or more generally a cylindrical (2D problem) or spherical (3D problem) electromagnetic wave, as typically radiated by a point source. It is then necessary to have the two media with the same electromagnetic density to achieve good focusing. In this case, each pair l of symmetric rays with incidence angle $\theta_{R,l}$ focus at the same point, because $\theta_{L,l} = \theta_{R,l}$, $\forall l$, so that

---

[19]The motivation for using a slab with double interface, as opposed to single interface, for focusing is that it is assumed that the electromagnetic waves cannot be detected and manipulated *inside* the LH material but only in the air regions, as in the case of a conventional optical lens. However, this assumption is not necessarily pertinent in microwaves, as one could envision to detect and utilize a wave focused inside the LH structure.

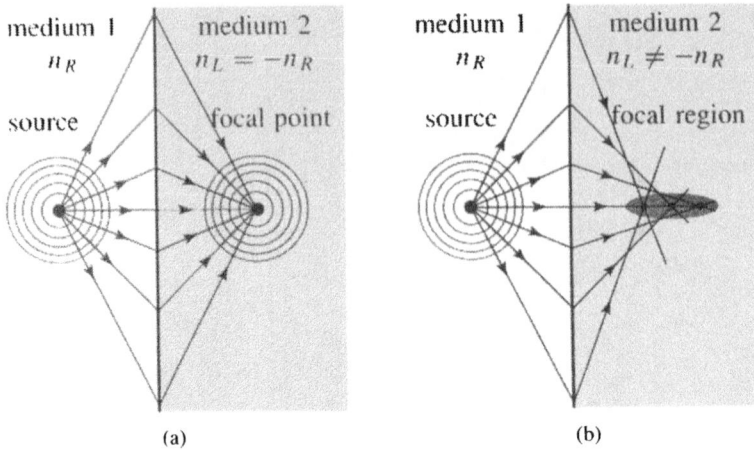

Fig. 2.9. Problem of spherical aberration. (a) When the two media have the same electromagnetic density $n_L = -n_R$, a pure focal point is formed. (b) When the two media have different electromagnetic densities $n_L \neq -n_R$, spherical aberration occurs.

the focal distances are the same for all pairs of rays, $s_l = s$ $\forall m$, according to Eq. (2.58), as illustrated in Fig 2.9(a). If the electromagnetic densities of the two media are different ($|n_L| \neq n_R$), then rays with different incidence angles refract to different focal points because the ratios $\tan \theta_{R,l} / \tan |\theta_{L,l}|$ of the different pairs of rays are all different, leading to different focal distances s_l. In that case, *spherical aberration* [11], illustrated in Fig 2.9(a), occurs, and the focal point degenerates into a diffuse focal spot with area increasing with the contrast of refractive indexes, $|n_L|/n_R$

2.8   FRESNEL COEFFICIENTS

In this section, we will see how Fresnel coefficients (reflection coefficient R and transmission coefficient T) at an interface between two arbitrary media are affected when one of the two media is LH. For simplicity, we restrict ourselves to the case of loss-less media [i.e., with zero imaginary parts of their constitutive parameters, according to Eq. (2.3)]. The reflection and transmission coefficients for parallel polarization ($\overline{E}$ parallel to the plane of incidence, also called E, TM$_z$, p) and perpendicular polarization ($\overline{E}$ perpendicular to the plane of incidence, also called H, TE$_z$, s) are given, with reference to Fig 2.6, by [10]

$$R_\parallel = \frac{\varepsilon_{r1}k_{2z} - \varepsilon_{r2}k_{1z}}{\varepsilon_{r1}k_{2z} + \varepsilon_{r2}k_{1z}} = \frac{\eta_2 \cos\theta_2 - \eta_1 \cos\theta_1}{\eta_2 \cos\theta_2 + \eta_1 \cos\theta_1}, \tag{2.59a}$$

$$T_\parallel = \frac{2(\varepsilon_{r1}\varepsilon_{r2}\mu_{r2}/\mu_{r1})k_{1z}}{\varepsilon_{r1}k_{2z} + \varepsilon_{r2}k_{1z}} = \frac{2\eta_2 \cos\theta_1}{\eta_2 \cos\theta_2 + \eta_1 \cos\theta_1}, \tag{2.59b}$$

$$R_\perp = \frac{\mu_{r2}k_{1z} - \mu_{r1}k_{2z}}{\mu_{r2}k_{1z} + \mu_{r1}k_{2z}} = \frac{\eta_2 \cos\theta_1 - \eta_1 \cos\theta_2}{\eta_2 \cos\theta_1 + \eta_1 \cos\theta_2}, \tag{2.59c}$$

$$T_\perp = \frac{2\mu_{r2}k_{1z}}{\mu_{r2}k_{1z} + \mu_{r1}k_{2z}} = \frac{2\eta_2 \cos\theta_1}{\eta_2 \cos\theta_1 + \eta_1 \cos\theta_2}, \tag{2.59d}$$

where η_i $(i = 1, 2)$ is the intrinsic impedance of medium i $(i = 1, 2)$

$$\eta_i = \sqrt{\frac{\mu_i}{\varepsilon_i}}. \tag{2.60}$$

These formulas, by setting $k_{z2} = -|k_{z2}|$ [Fig 2.7(b)], $\varepsilon_{r2} = -|\varepsilon_{r2}|$, and $\mu_{r2} = -|\mu_{r2}|$, show that *the Fresnel coefficients at a RH/LH interface are equal in magnitude to the Fresnel coefficients at a RH/RH interface.* The reason for conservation of the *magnitude* of the Fresnel coefficients is that these coefficients depend only on the tangential components of the fields, which are unchanged from a RH/RH interface to a RH/LH interface, as shown in Eqs. (2.41c) and (2.41d). In addition, *the Fresnel coefficients at a RH/LH interface are also equal in phase to the Fresnel coefficients at a RH/RH interface, except for the transmission coefficient for parallel polarization $T_\parallel$, the phase of which is reversed.*[20]

Whereas the indexes of refraction (2.17), related to the product of the permittivity and the permeability, determine the angle of transmission by Snell's law, the intrinsic impedances (2.60), related to the ratio of the permeability to the permittivity, determine the amount of reflection and transmission. Fresnel coefficients are therefore matching parameters, which are of particular importance in microwave and millimeter-wave structures. This observation emphasizes the following fact for the LH slab of Section 2.7 (Fig 2.8). In order to have "perfect" focusing, we must use a LH slab not only with the same refractive index as the index of the sandwiching RH media, to avoid spherical aberration, but also with the same intrinsic impedance, to obtain perfect transmission without reflection, as seen in Eq. (2.59) (where $\theta_1 = \theta_2$ for all pairs of symmetric rays if $|n_2| = n_1$). This means that we must have both same products and same ratios of permittivity and permeability and we therefore must have

$$\varepsilon_1 = \varepsilon_2, \tag{2.61a}$$

and

$$\mu_1 = \mu_2, \tag{2.61b}$$

which obviously represents a much more important design constraint than the isolated condition $n_2 = -n_1$. This is one of the reasons why focusing by a flat lens (Section 2.7) is relatively difficult to achieve.

---

[20]Note that it would be the phase of the transmission coefficient for *perpendicular* polarization $T_\perp$ that would be reversed in the complementary configuration of a RH slab in a LH environment.

2.9   REVERSAL OF GOOS-HÄNCHEN EFFECT

The Goos-Hänchen effect is the displacement d of a beam of light, or more generally of an electromagnetic wave, impinging on a planar interface with an optically rarer dielectric material under an angle where total reflection occurs. This phenomenon was first conjectured by Newton [14] and later demonstrated experimentally by Goos and Hänchen [15].

Let us rewrite the reflection coefficients given in Section 2.8 in the form

$$R_\| = \frac{1 - (\varepsilon_{r2}/\varepsilon_{r1})(k_{1z}/k_{2z})}{1 + (\varepsilon_{r2}/\varepsilon_{r1})(k_{1z}/k_{2z})}, \tag{2.62a}$$

$$R_\perp = -\frac{1 - (\mu_{r2}/\mu_{r1})(k_{1z}/k_{2z})}{1 + (\mu_{r2}/\mu_{r1})(k_{1z}/k_{2z})}, \tag{2.62b}$$

which is more tractable for the argument of this section. In the Goos-Hänchen effect, the incidence medium (medium 1) is rarer (has smaller electromagnetic density or magnitude of refractive index) than the medium with which it is interfaced (medium 2) (i.e., $n_1 > n_2$) and the incidence angle is larger than the critical angle ($\theta_i > \theta_c > \sin^{-1}|n_2/n_1|$), so that *total internal reflection* occurs. In this case, the normal component of the wave vector is real in the incidence medium and imaginary in the other medium[21]

$$k_{1z} = p_1 = \sqrt{n_1 k_0^2 - k_x^2}, \tag{2.63a}$$

$$k_{2z} = -jq_2 = -j\sqrt{k_x^2 - n_2 k_0^2}, \tag{2.63b}$$

where x is a variable along the plane of the interface, $p_1 \in \mathbb{R}$, $p_1 > 0$, assuming that medium 1 is RH, to ensure outgoing propagation away from the source, and $q_2 \in \mathbb{R}$, $q_2 > 0$ to ensure exponential decay away form the source. With these relations, the reflection coefficients (2.62) may be written in polar form as [16]

$$R_\| = |R_\||e^{j\varphi_\|} \quad \text{with} \quad \varphi_\| = -2\tan^{-1}\left(\frac{\varepsilon_{r2}}{\varepsilon_{r1}}\frac{p_1}{q_2}\right) = -2\tan^{-1}(\xi_\|), \tag{2.64a}$$

$$R_\perp = |R_\perp|e^{j\varphi_\perp} \quad \text{with} \quad \varphi_\perp = \pi - 2\tan^{-1}\left(\frac{\mu_{r2}}{\mu_{r1}}\frac{p_1}{q_2}\right) = \pi - 2\tan^{-1}(\xi_\perp), \tag{2.64b}$$

where p_1, q_2 in $\varphi_\|$ and $\varphi_\perp$ are functions of k_x via Eq. (2.63). The Goos-Hänchen shifts can be computed from these relations as

$$d_\| = -\frac{\partial\varphi_\|}{\partial k_x} = -\frac{2}{1 + \xi_\|^2}\frac{\varepsilon_{r2}}{\varepsilon_{r1}}\frac{k_x}{p_1 q_2}\left[1 - \left(\frac{p_1}{q_2}\right)^2\right], \tag{2.65a}$$

---

[21]The two media are assumed here to be loss-less for simplicity.

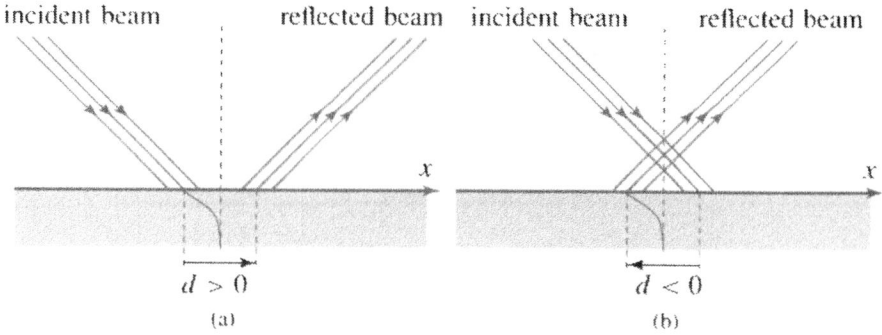

Fig. 2.10. Goos-Hänchen effect. (a) Conventional case, at the interface between two RH media. (b) Reversed effect, at the interface between a RH medium and a LH medium.

$$d_\perp = -\frac{\partial \varphi_\perp}{\partial k_x} = -\frac{2}{1 + \xi_\perp^2} \frac{\mu_{r2}}{\mu_{r1}} \frac{k_x}{p_1 q_2} \left[1 - \left(\frac{p_1}{q_2} \right)^2 \right], \qquad (2.65b)$$

where, according to (2.63),

$$\left(\frac{p_1}{q_2} \right)^2 = \frac{n_1 k_0^2 - k_x^2}{k_x^2 - n_2 k_0^2} > 0, \qquad (2.66)$$

since $p_1, q_2 \in \mathbb{R}$ and $p_1, q_2 > 0$. These relations with Eq. (2.64) reveal the following phenomenon. If both media are RH, $p_1/q_2 > 1$ since $n_1, n_2 > 0$ and $n_1 > n_2$, so that $d_\parallel$ and $d_\perp$ are positive quantities (i.e., with the same sign as k_x), according to Eq. (2.65). This corresponds to the positive shift of the beam observed by Goos and Hänchen, which is depicted in Fig 2.10(a).

If in contrast, medium 2 is replaced by a LH medium, the signs of ε_{r2} and μ_{r2} are reversed. Consequently, for parameters still satisfying the inequality of Eq. (2.66) after n_2 has been replaced by $-|n_2|$, negative phase shifts are obtained for $d_\parallel$ and $d_\perp$, as illustrated in Fig 2.10(b). Thus the Goos-Hänchen effect has been reversed [16, 17]. It should be noted that for parallel polarization medium 2 does not need to be LH but may only have negative permittivity (and positive permeability) and that, similarly, for parallel polarization medium 2 does not need to be LH but may only have negative permeability (and positive permittivity).

Goos-Hänchen shifts occurring in *layered structures* with negative refraction were analyzed by Shadrivov et al. in [18].

2.10 REVERSAL OF CONVERGENCE AND DIVERGENCE IN CONVEX AND CONCAVE LENSES

Due to their effectively homogeneous nature, in the frequency range of interest, MTMs represent a new paradigm to perform "conventional optics in non conventional materials at microwave frequencies," where in principle simple *ray*

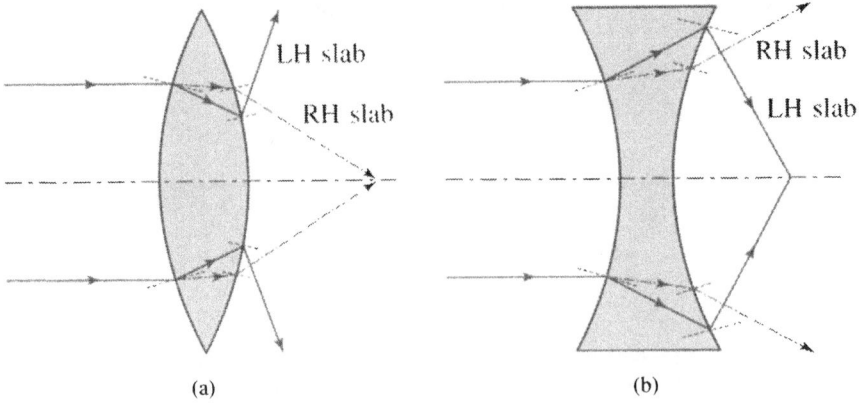

Fig. 2.11. Convergence and divergence induced by RH and LH (curved) lenses in air. (a) A LH convex lens is diverging, whereas a RH convex lens is converging. (b) A LH concave lens is converging, whereas a RH concave lens is diverging.

optics laws directly apply [11], as already shown in Sections 2.6, 2.7, 2.8, and 2.9. In practice, a typical limitation could be the too large electrical size of the unit cell p/λ_g at frequencies too far from the phase origin, where $p/\lambda_g = 0^{22}$ (Section 1.8). At those frequencies, some amount of diffraction or scattering is produced, which alters the purity of refraction.

All conventional ray optics may be revisited under the light of the unusual and exotic effects achieved with LH lenses [12, 13]. For instance, Fig 2.11 depicts the reversal of the convergence and divergence effects of convex (parabolic) and concave (hyperbolic) lenses, respectively, when the conventional RH lens is replaced by a LH lens [5]. The diverging effect of the convex lens and the converging effect of the converging lens are readily understood from simply ray optics based on Snell's law [Eq. (2.56)]. In practice, curved interface (e.g., parabolic or hyperbolic) with a LH MTM may be obtained only in a staircase fashion due to the structural nature of the MTM. However, if $p/\lambda_g \ll 1$, the scattering effects of this interface "quantization" are negligible.

Among the potential benefits of replacing conventional (RH) lenses by LH lenses, we may mention the following example [20]. The focal length in a thin lens is given by [11]

$$f = \frac{R}{|n - 1|}, \tag{2.67}$$

where R is the radius of curvature of the surface of the lens. From this formula, it appears that, for a given R, a lens with an index of $n = -1$ would have the

---

[22] As pointed out in Section 1.5, the rule of thumb $p < \lambda_g/4$ ensures predominance of refracted effects over diffraction/scattering effects, but at $p \approx \lambda_g/4$ there is nevertheless some amount of spurious scattering. Scattering may be considered negligible at frequencies where $p < \lambda_g/10, \lambda_g/15$.

same focal length as a lens of index $n = +3$. By the same reasoning, we see, by comparing a RH lens and a LH lens with same index magnitude, that the LH lens would have a smaller focal length and would therefore be more compact. Another observation from the focal length formula is that a RH lens of $n = +1$ would not focus electromagnetic waves (since $f = \infty$), whereas a LH lens with the same density, $n = -1$, would, with a focal length of $f = R/2$.

2.11 SUBWAVELENGTH DIFFRACTION

In [19], Pendry suggested that the subwavelength diffraction limit encountered in conventional lenses could be overcome by Veselago's LH "flat lens" (Fig 2.8) with $\varepsilon_r = \mu_r = -1$, and therefore $n = -1$. In this section, we reproduce the argument of [19], explain this phenomenon in the perspective of surface plasmons, and also point out the practical limitations of such a subwavelength lens.

Consider an infinitesimal dipole of frequency ω in front of a lens with axis z, as illustrated in Fig 2.12. The electric field produced by this source may be expressed in term of the 2D Fourier expansion

$$\overline{E}(r, t) = \overline{E}(\overline{r})e^{j\omega t} = \sum_m \sum_{k_x, k_y} \widetilde{\overline{E}}_m(k_x, k_y)e^{-j(k_x x + k_y y + k_z z - \omega t)}, \qquad (2.68)$$

with, from the wave equation,

$$k_{iz} = p_i - jq_i = \begin{cases} \sqrt{k_i^2 - (k_x^2 + k_y^2)} & \text{if } k_x^2 + k_y^2 = k_\rho^2 < k_i^2, \\ -j\sqrt{(k_x^2 + k_y^2) - k_i^2} & \text{if } k_x^2 + k_y^2 = k_\rho^2 > k_i^2, \end{cases} \qquad (2.69a)$$

where k_i is the wave number in the medium i ($i = 1, 2$; 1: air, 2: lens)

$$k_i = (\omega/c)\sqrt{\varepsilon_{ri}\mu_{ri}}. \qquad (2.70)$$

All of the information on the image (included in a $x - y$ plane) to be focused by the lens are included in their spectral components k_x and k_y, also referred to as *spatial frequencies*. By Fourier transform between the spatial (x, y) to the spectral (k_x, k_y) domains of the image, the meaning of the spatial frequencies k_ρ, associated with the image wavelength $\lambda_\rho = 2\pi/k_\rho$, is readily understood. Small k_ρs correspond to bulk features of the image, while large k_ρs correspond to detailed features of the image. The information on the image is carried to the lens in the air region by a wave propagating along z. Considering the z dependence of this wave, $e^{-jk_z z}$ ($k_z = k_{1z}$, air region), we see from Eq. (2.69) that, if $k_\rho < k_0$ or equivalently $\lambda_\rho > \lambda_0$, then k_z is real; therefore, the wave really propagates along z and the contents of the image are transmitted to the lens, which accomplishes focusing. This is the situation illustrated in Fig 2.12(a). In contrast,

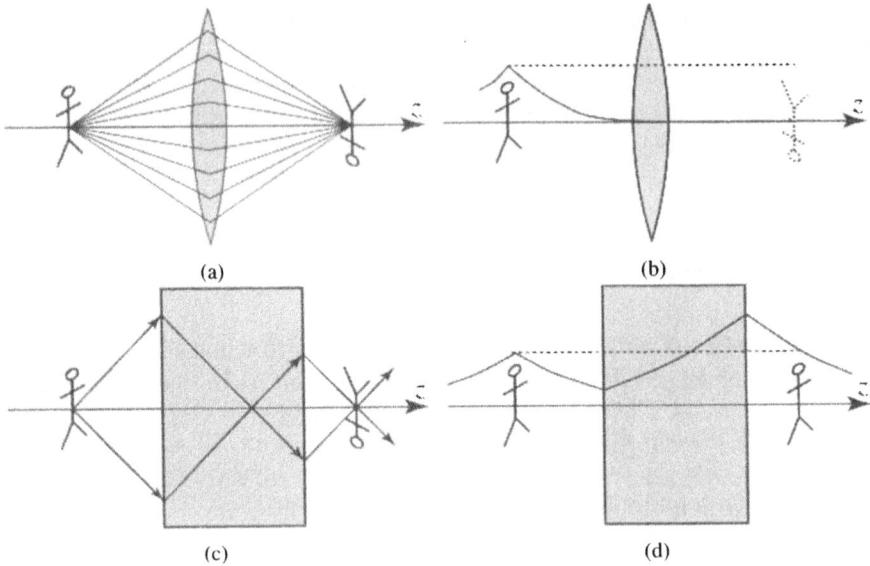

(a)   (b)

(c)   (d)

Fig. 2.12. Imaging subwavelength limit and subwavelength diffraction (the lenses are assumed to be in air). Reproduced from [20]. (a) Conventional (RH) lens for incident propagating waves, corresponding to low-spatial frequency components (bulk features) of the source $k_\rho < k_0$: focusing occurs. (b) Conventional (RH) focusing lens for incident evanescent waves, corresponding to high-spatial frequency components (details) of the source $k_\rho > k_0$: source information does not reach the source. (c) LH slab with $\varepsilon_r = \mu_r = -1$ for incident propagating waves, corresponding to low-spatial frequency components: focusing occurs. (d) LH slab with $\varepsilon_r = \mu_r = -1$ for incident evanescent waves, corresponding to high-spatial frequency components: information reaches the source due to energy enhancement related to surface plasmon resonance at each of the two interfaces.

if $k_\rho > k_0$ or $\lambda_\rho < \lambda_0$, k_z is imaginary; therefore, the wave is evanescent along z and the contents of the image either do not reach the lens or are strongly attenuated when reaching the location of focus. This is the situation illustrated in Fig 2.12(b). These considerations show that a conventional lens operates lowpass filter in terms of spatial frequencies of the image at the input of the system: bulk features of the image, transported by propagating components of the wave, are well transmitted, whereas details, included in the evanescent components of the wave, are lost in the transmission process. The spatial frequency limit is given as $k_\rho = k_0$, or equivalently $\lambda_\rho = \lambda_0$. This is the fundamental *diffraction limit* of conventional optics, which may be formulated as follows: the resolution Δ of an optical system is limited by the wavelength of the wave transporting the image

$$\Delta \approx \lambda_{\rho,min} = \frac{2\pi}{k_{\rho,max}} = \frac{2\pi}{k_0} = \frac{2\pi c}{\omega} = \lambda_0. \qquad (2.71)$$

Let us now show how Pendry demonstrated that this diffraction limit could, theoretically, be overcome in a LH slab with ε_r and $\mu_r = -1$. In this case,

$|k_{1z}| = |k_{2z}|$, and we have from Eq. (2.69),[23] where the subscripts 1 and 2 refer to air and to the LH medium of the slab, respectively,

$$k_{2z} = p_2 = -k_{1z} = -p_1 \quad \text{if} \quad k_{1z}, k_{2z} \in \mathbb{R} \quad \text{(propagating wave)}, \quad (2.72a)$$

$$k_{2z} = p_2 = +k_{1z} = p_1 \quad \text{if} \quad k_{1z}, k_{2z} \in \mathbb{I} \quad \text{(evanescent wave)}, \quad (2.72b)$$

where the first relation accounts for NRI of the wave when it is propagating and the second relation ensures exponential decay of the wave when it is evanescent ($\mathbb{R}$ stands for purely real and $\mathbb{I}$ for purely imaginary).

The Fresnel coefficients from medium 1 to medium 2 and from medium 2 to medium 1 are given by Eq. (2.59). Let us for instance consider the case of perpendicular polarization. We have

$$R_\parallel^{1\to2} = \frac{\mu_{r2}k_{2z} - \mu_{r1}k_{1z}}{\mu_{r2}k_{2z} + \mu_{r1}k_{1z}} = r, \quad T_\parallel^{1\to2} = \frac{2\mu_{r2}k_{1z}}{\mu_{r2}k_{2z} + \mu_{r1}k_{1z}} = t, \quad (2.73a)$$

$$R_\parallel^{2\to1} = \frac{\mu_{r1}k_{1z} - \mu_{r2}k_{2z}}{\mu_{r1}k_{1z} + \mu_{r2}k_{2z}} = r', \quad T_\parallel^{2\to1} = \frac{2\mu_{r1}k_{2z}}{\mu_{r1}k_{1z} + \mu_{r2}k_{2z}} = t'. \quad (2.73b)$$

It is then possible to compute the total transmission and reflection coefficients by summing the *multiple reflections* events

$$T_\parallel = tt'e^{-jk_{2z}d} + tt'r'^2e^{-j3k_{2z}d} + tt'r'^4e^{-j5k_{2z}d} + \ldots = \frac{tt'e^{-jk_{2z}d}}{1 - r'^2e^{-j2k_{2z}d}}, \quad (2.74a)$$

$$R_\parallel = r + tt'r'e^{-j2k_{2z}d} + tt'r'^3e^{-j4k_{2z}d} + \ldots = r + \frac{tt'r'e^{-j2k_{2z}d}}{1 - r'^2e^{-j2k_{2z}d}}. \quad (2.74b)$$

Let us now consider the overall transmission coefficient. By inserting Eq. (2.73) into Eq. (2.74a), we obtain with Eq. (2.72b) in the case of an *evanescent wave*

$$T_\parallel^{\text{evan.(LH)}} = \lim_{\substack{\mu_{r1}\to+1 \\ \mu_{r2}\to-1}} \frac{tt'e^{-jk_{2z}d}}{1 - r'^2e^{-j2k_{2z}d}}$$

$$= \lim_{\substack{\mu_{r1}\to+1 \\ \mu_{r2}\to-1}} \frac{2\mu_{r2}k_{1z}}{\mu_{r2}k_{2z} + \mu_{r1}k_{1z}} \frac{2\mu_{r1}k_{2z}}{\mu_{r1}k_{1z} + \mu_{r2}k_{2z}} \frac{e^{-jk_{2z}d}}{1 - \left(\frac{\mu_{r1}k_{1z}-\mu_{r2}k_{2z}}{\mu_{r1}k_{1z}+\mu_{r2}k_{2z}}\right)^2 e^{-j2k_{2z}d}}$$

$$= \frac{-2k_{1z}}{-k_{2z} + k_{1z}} \frac{2k_{2z}}{k_{1z} - k_{2z}} \frac{e^{-jk_{2z}d}}{1 - \left(\frac{k_{1z}+k_{2z}}{k_{1z}-k_{2z}}\right)^2 e^{-j2k_{2z}d}}$$

---

[23] Again, we consider a loss-less slab for simplicity.

$$= \frac{-4k_{1z}k_{2z}}{(k_{1z}-k_{2z})^2} \frac{e^{-jk_{2z}d}}{1 - \left(\frac{k_{1z}+k_{2z}}{k_{1z}-k_{2z}}\right)^2 e^{-j2k_{2z}d}}$$

$$\overset{(k_{2z}=k_{1z}=k_z)}{=\!=\!=} e^{+jk_zd} \overset{(k_z=-ja)}{=\!=\!=} e^{+qd} \quad (q \in \mathbb{R}, q > 0). \tag{2.75}$$

This result shows that *an evanescent wave is enhanced by a LH slab*, as illustrated in Fig 2.12(d).[24] This enhancement phenomenon is simply due to the excitation of a resonance mode of each of the interface, called a *surface plasmon*.[25] Thus, subwavelength details ($\lambda_\rho < \Delta \approx \lambda_0$) of the image can be focused by the LH slab. Bulk features of the image are clearly transmitted, as already shown in Section 2.7 and illustrated in Fig 2.12(c).

For the sake of completeness, let us consider what happens if the evanescent wave would be incident upon RH slab (for which we have $k_{2z} = k_{1z} = -jq$ for exponential decay at infinity)

$$T_\parallel^{\text{evan.(RH)}} = \lim_{\substack{\mu_{r1} \to +1 \\ \mu_{r2} \to +1}} \frac{tt'e^{-jk_{2z}d}}{1 - r'^2 e^{-j2k_{2z}d}}$$

$$= \lim_{\substack{\mu_{r1} \to +1 \\ \mu_{r2} \to +1}} \frac{2\mu_{r2}k_{1z}}{\mu_{r2}k_{2z} + \mu_{r1}k_{1z}} \frac{2\mu_{r1}k_{2z}}{\mu_{r1}k_{1z} + \mu_{r2}k_{2z}} \frac{e^{-jk_{2z}d}}{1 - \left(\frac{\mu_{r1}k_{1z}-\mu_{r2}k_{2z}}{\mu_{r1}k_{1z}+\mu_{r2}k_{2z}}\right)^2 e^{-j2k_{2z}d}}$$

$$= \frac{2k_{1z}}{k_{2z} + k_{1z}} \frac{2k_{2z}}{k_{1z} + k_{2z}} \frac{e^{-jk_{2z}d}}{1 - \left(\frac{k_{1z}-k_{2z}}{k_{1z}+k_{2z}}\right)^2 e^{-j2k_{2z}d}}$$

$$= \frac{4k_{1z}k_{2z}}{(k_{1z} + k_{2z})^2} \frac{e^{-jk_{2z}d}}{1 - \left(\frac{k_{1z}-k_{2z}}{k_{1z}+k_{2z}}\right)^2 e^{-j2k_{2z}d}}$$

$$\overset{(k_{2z}=k_{1z}=k_z)}{=\!=\!=} e^{-jk_zd} \overset{(k_z=-jq)}{=\!=\!=} e^{-qd} \quad (q \in \mathbb{R}, q > 0) \tag{2.76}$$

We see that, in this case, the typically expected exponential decay is observed.

Subwavelength diffraction with a $\varepsilon_r = \mu_r = -1$ LH slab suggests the theoretical idea of infinite resolution focusing. In practice, the severe constraint that the lens must have exactly $\varepsilon_r = \mu_r = -1$ limits this phenomenon. If this condition is not satisfied, spherical aberration (Section 2.7) and mismatch losses (Section 2.8)

[24]The same conclusion is straightforwardly obtained in the case of perpendicular polarization, where the fact that $\mu_{r1} \to +1$ $\varepsilon_{r2} \to -1$ is used in the derivations.
[25]MTM surface plasmons are illustrated by full-wave simulated fields in Section 4.4.4 and discussed in the perspective of potential novel applications in Section 4.5.1.

occur and ruin the theoretical interest of the lens. Such limitations are discussed in several papers (e.g., [21, 22]). Subwavelength focusing was demonstrated experimentally in the SRR-TW structure by Lagarkov et al. in [22] and in a 2D planar transmission line structure by Grbic et al. in [23]. In [24], regeneration of evanescent waves from a silver ($\varepsilon_r \approx -1$) superlens was demonstrated experimentally.[26] Note that performance is predicted from alternating parallel sided layers of $n = -1$ and $n = +1$ material, which also acts as a "lens" [25, 26].

REFERENCES

1. M. D. Pozar. *Microwave Engineering,* Third Edition, John Wiley & Sons, 2004.

2. L. Brillouin. *Wave Propagation and Group Velocity,* Academic Press, 1960.

3. E. M. Lifshitz, L. D. Landau, and L. P. Pitaevskii. *Electrodynamics of Continuous Media: Volume 8,* Second Edition, Butterworth-Heinemann, 1984.

4. J. D. Kraus and R. J. Marhefka. *Antennas,* Third Edition, McGraw Hill, 2001.

5. V. Veselago. "The electrodynamics of substances with simultaneously negative values of ε and μ," *Soviet Physics Uspekhi,* vol. 10, no. 4, pp. 509–514, Jan.-Feb. 1968.

6. R. J. Doviak, D. S. Zrnic, and R. J. Doviak. *Doppler Radar and Weather Observations,* Second Edition, Academic Press, 1993.

7. J. A. Kong. *Electromagnetic Wave Theory,* Second Edition, John Wiley & Sons, 1990.

8. C. Caloz and T. Itoh. "Array factor approach of leaky-wave antennas and application to 1D/2D composite right/left-handed (CRLH) structures," *IEEE Microwave Wireless Compon. Lett.,* vol. 14, no. 6, pp. 274–276, June 2004.

9. J. L, T. M. Grzegorczyk, Y. Zhang, J. Pacheo, B.-I. Wu, J. A. Kong, and M. Chen. "Čerenkov radiation in materials with negative permittivity and permeability," *Optics Express,* vol. 11, no. 7, pp. 723–734, April 2003.

10. A. Ishimaru. *Electromagnetic Wave Propagation, Radiation, and Scattering,* Prentice Hall, 1991.

11. M. Born and E. Wolf. *Principles of Optics,* Seventh Edition, Cambridge University Press, 2002.

12. D. Schurig and D. R. Smith. "Universal description of spherical aberration free lenses composed of positive or negative index media," *Phys. Rev. E,* eprint physics/0307088, 2003.

13. D. Schurig and D. R. Smith. "Negative index lens aberrations," *Phys. Rev. E,* eprint physics/0403147 v1, 2003.

14. H. K. V. Lotsch. "Beam displacement at total reflection: the Goos-Hänchen effect," *I. Optik,* vol. 32, pp. 116–137, 1970.

15. F. Goos and H. Hänchen. "Ein neuer und fundamentaler Versuch zur Totalreflexion," *Ann. Phys. Lpz,* vol. 1, pp. 333–346, 1947.

---

[26] A slab with only $\varepsilon_r = -1$ represents a simplification of the "perfect lens" (*p*-polarized fields), which is possible when all length scales are much smaller than the wavelength of light (quasistatic limit) [19]. Subwavelength resolution is then possible in principle for example with thin silver films, which exhibit a permittivity equal to -1 in the visible region of the spectrum, although losses still represent a severe limitations in this approach.

16. A. Lakhtakia. "Positive and negative Goos-Hänchen shifts and negative phase-velocity mediums (alias left-handed materials)," *Int. J. Electron. Commun.,* vol. 58, no. 3, pp. 229–231, 2004.

17. P. R. Berman. "Goos-Hänchen shift in negatively refractive media," *Phys. Rev. E,* vol. 66, pp. 067603:1–3, 2002.

18. I. V. Shadrivov, R. W. Ziolkowski, A. A. Zharov, and Y. S. Kivshar. "Excitation of guided waves in layered structures with negative refraction," *Optics Express,* vol. 13, no. 2, pp. 481–492, Jan. 2003.

19. J. B. Pendry. "Negative refraction makes a perfect lens," *Phys. Rev. Lett.,* vol. 85, no. 18, pp. 3966–3969, Oct. 2000.

20. D. R. Smith and J. B. Pendry. "Reversing light: negative refraction," *Physics Today,* pp. 1-8, Dec. 2003.

21. D. R. Smith, D. Schurig, M. Rosenbluth, and S. Schultz. "Limitations on subdiffraction imaging with a negative refractive index slab," *App. Phys. Lett.,* vol. 82, no. 10, pp. 1506–1508, March 2003.

22. A. N. Lagarkov and V. N. Kissel. "Near-perfect imaging in a focusing system based ona left-handed-material plate," *Phys. Rev. Lett.,* vol. 92, no. 7, pp. 077401:1–4, Feb. 2004.

23. A. Grbic and G. Eleftheriades. "Overcoming the diffraction limit with a planar left-handed transmission-line lens," *Phys. Rev. Lett.,* vol. 92, no. 11, pp. 117403:1–4, March 2004.

24. N. Fang, Z. Liu, T.-J. Yen, and X. Zhang. "Regenerating evanescent waves from a silver superlens," *Optics Express,* vol. 11, no. 7, pp. 682–687, March 2004.

25. E. Shamonina, V. A. Kalinin, K. H. Ringhofer, and L. Solymar. "Imaging compression and Poynting vector streamlines for negative permittivity materials," *Electron. Lett.,* vol. 37, no. 20, pp. 1243–1244, Sept. 2001.

26. S. .A. Ramakrishna, J. B. Pendry, M. C. K. Wiltshire, and W. J. Stewart. "Imaging the near field," *J. Mod. Opt.,* vol. 50, pp. 1419–1430, 2003.

3

TL THEORY OF MTMs

Chapter 3 presents the fundamental TL theory of MTMs. Because MTMs are effectively homogeneous structures (Section 1.1), they can be essentially modeled by one-dimensional (1D) transmission lines (TLs), whose propagation direction represents any direction in the material. Therefore, this chapter deals exclusively with 1D MTM, while a complementary theory for 2D MTMs is provided in Chapter 4.

As for the case of conventional RH materials [4], TL theory (Section 1.7) provides an insightful and powerful tool for the analysis and design of LH or, more generally, CRLH MTMs (Section 1.8) [1, 2, 3]. The TL approach of MTMs, along with the CRLH MTM concept, represents the foundation for most of the applications presented in this book.

The TL theory of 1D MTMs is presented here in three progressive steps, starting with the ideal TL model (Section 3.1), continuing on to the ideal LC network implementation (Section 3.2), and culminating with the real dispersive LC material (Section 3.3).

3.1 IDEAL HOMOGENEOUS CRLH TLs

A stricto senso *homogeneous*[1] TL is perfectly *uniform*, i.e., a TL that has invariant cross section along the direction of propagation. The TL is in addition called *ideal*

---

[1] Here "stricto senso homogeneous" is opposed to "only effectively homogeneous but structurally inhomogeneous" (constituted of cells much smaller than guided wavelength).

Electromagnetic Metamaterials: Transmission Line Theory and Microwave Applications,
By Christophe Caloz and Tatsuo Itoh
Copyright © 2006 John Wiley & Sons, Inc.

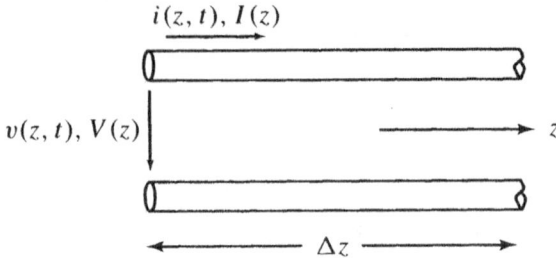

Fig. 3.1. Representation of an ideal (transmitting from $\omega = 0$ to $\omega \to \infty$) homogeneous (perfectly uniform) TL in the form of its incremental model (length $\Delta z \to 0$) with time-domain or steady-state voltage/currents, $v(z,t)/i(z,t)$ and $V(z)/I(z)$, respectively.

if it can transmit signals at all frequencies, from zero to infinity. Fig 3.1 shows a representation of an ideal homogeneous TL.

Although RH homogeneous TLs are commonly used (e.g., coaxial or microstrip lines), LH or CRLH homogeneous TLs do not seem to be possible, due to the unavailability of real homogeneous LH or CRLH materials. However, it is possible to construct *effectively homogeneous* artificial LC TL structures, or MTM TLs, that perfectly mimic ideal TLs in a restricted range of frequencies (Section 3.2). Despite the nonexistence of ideal homogeneous CRLH TLs, the analysis of such TLs is extremely useful because it provides insight regarding the fundamental aspects of CRLH MTMs with very simple relations (e.g., Section 1.7) and, more importantly, because it really describes the fundamental characteristics of MTMs due to their effective-homogeneity. The difference between a perfectly homogeneous TL and an effectively homogeneous TL is that in the former case we have an incremental length $\Delta z \to 0$, whereas in the latter case we must consider the restriction

$$\Delta z \ll \lambda_g \quad \left(\text{at least } \Delta z < \frac{\lambda_g}{4}\right), \tag{3.1}$$

where λ_g represents the guided wavelength and Δz is typically equal the average unit cell size p.[2]

In this section, we present the fundamentals of ideal CRLH TLs. Details of TL principles are available in classical textbooks such as [4, 6, 7, 8]. For the sake of simplicity, we will first consider the case of a loss-less CRLH TL.

3.1.1 Fundamental TL Characteristics

The model for a loss-less CRLH TL is shown in Fig 3.2. It consists of a per-unit length impedance Z' (Ω/m) constituted by a RH per-unit-length inductance L'_R

---

[2]The average unit cell size is called the *lattice constant* for crystals in solid state physics [5]. However, crystals are generally periodic (hence p is the period in crystals) structures, whereas MTMs may be nonperiodic (Fig 3.28).

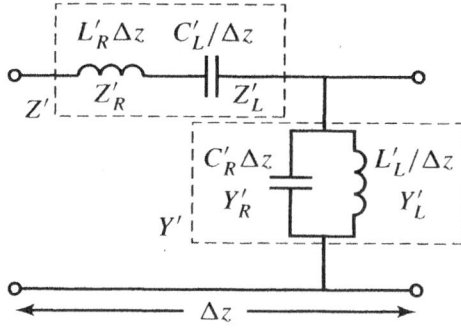

Fig. 3.2. Equivalent circuit model for the ideal CRLH TL.

(H/m) in series with a LH times-unit-length capacitance C'_L (F·m) and a per-unit-length admittance Y' (S/m) constituted by a RH per-unit-length capacitance C'_R (F/m) in parallel with a LH times-unit-length inductance L'_L (H·m)

$$Z' = j\left(\omega L'_R - \frac{1}{\omega C'_L}\right), \tag{3.2a}$$

$$Y' = j\left(\omega C'_R - \frac{1}{\omega L'_L}\right). \tag{3.2b}$$

If the LH immittances are zero, $Z'_L = -j/(\omega C'_L) = 0$ (or $C'_L = \infty$) and $Y'_L = -j/(\omega L'_L) = 0$ (or $L'_L = \infty$), only the RH immittances $Z'_R = j\omega L'_R$ and $Y'_R = j\omega C'_R$ remain and the model of Fig 3.2 reduces to the conventional RH TL model. In contrast, if it is the RH immittances, which are zero ($L'_R = 0$ and $C'_R = 0$), the model of Fig 3.2 includes only the elements C'_L and L'_L and becomes the *dual* of the RH TL model. For convenience, we call the TL including only L'_R and C'_R a purely RH (PRH) TL, and we call the TL including only L'_L and C'_L a purely left-handed (LRH) TL. If all four immittances Z'_R, Z'_L, Y'_R, Y'_L are nonzero, all of the elements L'_R, C'_L, C'_R, L'_L contribute to the transmission characteristic and the line is called *CRLH*.

As already pointed out in Section 1.8, a purely LH TL *cannot exist physically* because, even if we intentionally provide only series capacitance and shunt inductance, parasitic series inductance and shunt capacitance effects, increasing with increasing frequency, will unavoidably occur due to currents flowing in the metallizations and voltage gradients developing between the metal patterns of the trace and the ground plane. Thus, the CRLH model represents the most general MTM structure possible.

The behavior of the CRLH TL can be anticipated by the following asymptotic considerations. At low frequencies ($\omega \to 0$), $Z'_R \to 0$, $Y'_R \to 0$, so that the CRLH TL becomes equivalent to a PLH TL ($L'_R = C'_R = 0$). At high frequencies ($\omega \to \infty$), $Z'_L \to 0$, $Y'_L \to 0$, so that the CRLH TL becomes equivalent to A PRH TL

$(L'_L = C'_L = 0)$. At all other frequencies, the transmission characteristics depend on the combination of LH and RH contributions.

In order to carry out rigorous analysis of the CRLH TL, we start with the generalized telegraphist's equations for steady-state sinusoidal waves based on cosine phasors

$$\frac{dV}{dz} = -Z'I = -j\omega\left(L'_R - \frac{1}{\omega^2 C'_L}\right)I, \tag{3.3a}$$

$$\frac{dI}{dz} = -Y'V = -j\omega\left(C'_R - \frac{1}{\omega^2 L'_L}\right)V, \tag{3.3b}$$

where V and I are the position-dependent voltage and currents [$V = V(z)$ and $I = I(z)$] along the line, respectively. By solving simultaneously Eqs. (3.3a) and (3.3b), we obtain the wave equations for V and I

$$\frac{d^2V}{dz^2} - \gamma^2 = 0, \tag{3.4a}$$

$$\frac{d^2I}{dz^2} - \gamma^2 = 0, \tag{3.4b}$$

where γ (1/m) is the complex propagation constant. γ is expressed in terms of the per-unit-length immittances Z' and Y' as

$$\gamma = \alpha + j\beta = \sqrt{Z'Y'}, \tag{3.5}$$

and is associated with the $|z/ - z$-propagating $(e^{-\gamma z}/e^{+\gamma z})$ traveling wave solutions

$$V(z) = V_0^+ e^{-\gamma z} + V_0^- e^{+\gamma z}, \tag{3.6a}$$

$$I(z) = I_0^+ e^{-\gamma z} + I_0^- e^{+\gamma z} = \frac{\gamma}{Z'}\left(V_0^+ e^{-\gamma z} - V_0^- e^{+\gamma z}\right). \tag{3.6b}$$

The second equality in Eq. (3.6b) was obtained by taking the derivative of Eq. (3.6a) and equating the resulting expression with Eq. (3.3a). The characteristic impedance Z_c (Ω), relating the voltage and current on the line as $V_0^+/I_0^+ = Z_c = -V_0^-/I_0^-$, is then obtained in terms of Z' and Y' by comparing the two expressions of Eq. (3.6b) and using Eq. (3.5)

$$Z_c = R_c + jX_c = \frac{Z'}{\gamma} = \sqrt{\frac{Z'}{Y'}} = Z_c(\omega), \tag{3.7}$$

For convenience, we now introduce the variables

$$\omega'_R = \frac{1}{\sqrt{L'_R C'_R}} \quad \text{(rad·m)/s,} \tag{3.8a}$$

$$\omega_L' = \frac{1}{\sqrt{L_L' C_L'}} \quad \text{rad/(m·s)}, \tag{3.8b}$$

$$\kappa = L_R' C_L' + L_L' C_R' \quad \text{(s/rad)}^2, \tag{3.8c}$$

and the series and shunt resonance frequencies

$$\omega_{se} = \frac{1}{\sqrt{L_R' C_L'}} \quad \text{rad/s}, \tag{3.9a}$$

$$\omega_{sh} = \frac{1}{\sqrt{L_L' C_R'}} \quad \text{rad/s}, \tag{3.9b}$$

respectively. By inserting Eq. (3.2) into Eq. (3.5) and using Eq. (3.8), we obtain the following explicit expression for the complex propagation constant

$$\gamma = \alpha + j\beta = js(\omega)\sqrt{\left(\frac{\omega}{\omega_R'}\right)^2 + \left(\frac{\omega_L'}{\omega}\right)^2 - \kappa\omega_L'^2}, \tag{3.10}$$

where $s(\omega)$ is the following sign function,

$$s(\omega) = \begin{cases} -1 & \text{if} \quad \omega < \min(\omega_{se}, \omega_{sh}) \quad \text{LH range,} \\ +1 & \text{if} \quad \omega > \max(\omega_{se}, \omega_{sh}) \quad \text{RH range,} \end{cases} \tag{3.11}$$

which will be shortly further explained. Note that, because of the presence of the negative sign in the radicand of Eq. (3.10), the propagation constant γ is not necessarily purely imaginary $\gamma = j\beta$ (pass band); it can be purely real $\gamma = \alpha$ (stop band) in some frequency ranges despite the fact that the line is loss-less.

The CRLH dispersion/attenuation relation (3.10) is plotted in Fig 3.3. Fig 3.3(a) shows the CRLH dispersion and attenuation curves for energy propagation along both the positive and negative z directions, whereas Fig 3.3(b) shows these curves in comparison with the dispersion curves of the corresponding PRH and PLH TLs. This second graph shows how the CRLH dispersion curves tends to the PLH and PRH dispersion curves at lower and higher frequencies, respectively. We also note the presence of the predicted CRLH gap. This gap is due to the different series and shunt resonances (ω_{se}, ω_{sh}); when it occurs, the CRLH TL is said to be *unbalanced*. Section 3.1.3 will show that, when these two frequencies are equal, in which case the line will be called *balanced*, the CRLH gap closes up with nonzero group velocity.

The sign function of Eq. (3.11) is easily understood by considering Fig 3.3(a). If $\omega < \min(\omega_{se}, \omega_{sh})$, the phase velocity (slope of the line segment from origin to curve) and group velocity (slope of the curve) have opposite signs (i.e., they are antiparallel, $v_p - \|v_g$), meaning that the TL is LH and that β is therefore negative. In contrast, in the RH range $\omega > \max(\omega_{se}, \omega_{sh})$, the phase and group

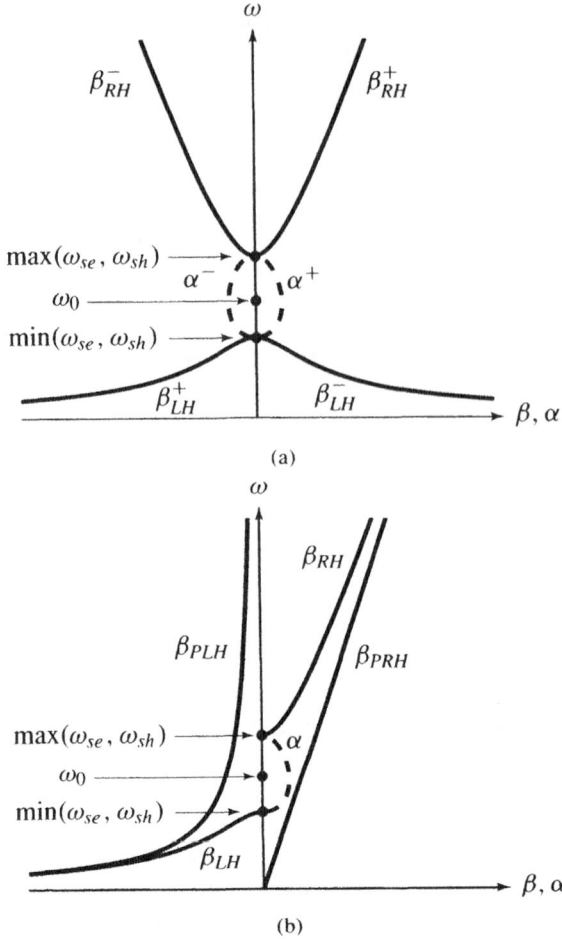

Fig. 3.3. Dispersion/attenuation diagrams computed by Eq. (3.10) for a CRLH TL. The labels "RH" and "LH" indicate the RH and LH frequency branches, respectively. (a) Energy propagation along $+z$ and $-z$ directions. (b) Comparison of the CRLH, PLH (β_{PLH}) and PRH (β_{PRH}) TLs for energy propagation along the $+z$ direction ($v_g > 0$).

velocities have the same sign ($v_p \| v_g$), meaning that the TL is RH and that β is therefore positive. Thus, the real dispersion diagram for propagation of energy (v_g) in the $+z$ direction is the one shown in Fig 3.3(b), which is flipped around the ω axis if propagation of energy occurs in the $-z$ direction.

The frequency of maximum attenuation ω_0 can be derived as the root of the derivative of the complex propagation constant (3.10)

$$\frac{d\gamma}{d\omega} = js(\omega) \frac{\omega/\omega_R'^2 - \omega_L'^2/\omega^3}{\sqrt{\left(\omega/\omega_R'\right)^2 + \left(\omega_L'/\omega\right)^2 - \kappa\omega_L'^2}} = 0, \tag{3.12}$$

yielding

$$\omega_0 = \sqrt{\omega'_R \omega'_L} = \frac{1}{\sqrt[4]{L'_R C'_R L'_L C'_L}}. \tag{3.13}$$

The CRLH characteristic impedance is obtained explicitly by inserting Eq. (3.2) into Eq. (3.7) with Eq. (3.9)

$$Z_c = Z_L \sqrt{\frac{(\omega/\omega_{se})^2 - 1}{(\omega/\omega_{sh})^2 - 1}}, \tag{3.14}$$

where Z_L is the PLH impedance

$$Z_L = \sqrt{\frac{L'_L}{C'_L}}, \tag{3.15}$$

and we also introduce for completeness and later convenience the PRH impedance

$$Z_R = \sqrt{\frac{L'_R}{C'_R}}. \tag{3.16}$$

It should be noted in Eq. (3.14) that the characteristic impedance has a zero and a pole at $\omega = \omega_{se}$ and $\omega = \omega_{sh}$, respectively,

$$Z_c(\omega = \omega_{se}) = 0, \tag{3.17a}$$

$$Z_c(\omega = \omega_{sh}) = \infty, \tag{3.17b}$$

corresponding to the series and shunt resonances, respectively. The CRLH impedance is plotted in Fig 3.4. As seen in Eq. (3.14), this impedance is purely imaginary in the gap, which extends from $\min(\omega_{se}, \omega_{sh})$ to $\max(\omega_{se}, \omega_{sh})$.[3] Because the characteristic impedance is a function of frequency, the CRLH TL can only be matched in a restricted frequency band, in general.

The other fundamental transmission quantities (defined only in the passband), guided wavelength λ_g, phase velocity v_p, and group velocity v_g, of the CRLH TL are readily derived from Eq. (3.10)

$$\lambda_g = \frac{2\pi}{|\beta|} = \frac{2\pi}{\sqrt{(\omega/\omega'_R)^2 + (\omega'_L/\omega)^2 - \kappa \omega'^2_L}}, \tag{3.18}$$

[3]It is interesting to observe that the resonances ω_{se} and ω_{sh} cannot be discriminated in the expression (3.10) for the propagation constant γ, but that they can be discriminated in the expression (3.14) for the characteristic impedance Z_c. In other words, by considering only the dispersion diagram (Fig 3.3), we cannot tell which of the two resonances is ω_{se} or ω_{sh}, whereas in the characteristic impedance graph (Fig 3.4) we can immediately identify the zero of Z_c as ω_{se} and its pole as ω_{sh}.

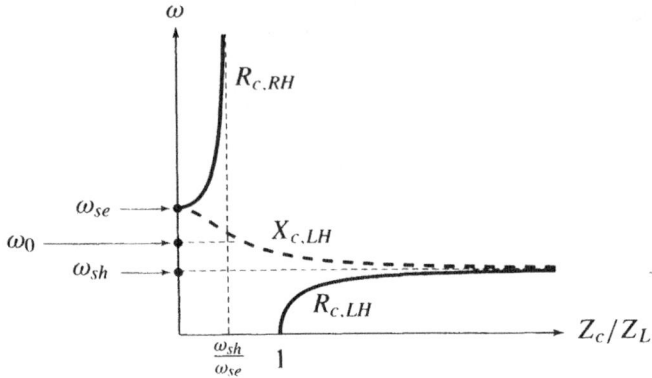

Fig. 3.4. Characteristic impedance [Eq. (3.14)] of the CRLH TL in the case $\omega_{sh} < \omega_{se}$. The labels "RH" and "LH" indicate the RH and LH frequency branches, respectively.

$$v_p = \frac{\omega}{\beta} = s(\omega)\frac{\omega}{\sqrt{(\omega/\omega_R')^2 + (\omega_L'/\omega)^2 - \kappa\omega_L'^2}}, \qquad (3.19)$$

$$v_g = \left(\frac{d\beta}{d\omega}\right)^{-1} = \frac{\left|\omega\omega_R'^{-2} - \omega^{-3}\omega_L'^2\right|}{\sqrt{(\omega/\omega_R')^2 + (\omega_L'/\omega)^2 - \kappa\omega_L'^2}}. \qquad (3.20)$$

Fig 3.5 shows the phase and group velocities for the PLH and CRLH TLs.

In the PRH TL, these velocities are well known to be constant and equal, $v_p^{PRH} = \omega_R'$ and $v_g^{PRH} = \omega_R'$. In the PLH TL, we find from Eqs. (3.19) and (3.20), with $L_R' = C_R' = 0$, that $v_p^{PLH} = -\omega^2/\omega_L'$ and $v_g^{PLH} = \omega^2/\omega_L'$, showing that phase velocity is negative and that the phase and group velocities are therefore antiparallel [Fig 3.5(a)], as expected. It appears that these PLH velocities are unbounded at high frequencies [$v_p, v_g(\omega \to \infty) \to \infty$]. The fact that v_g is

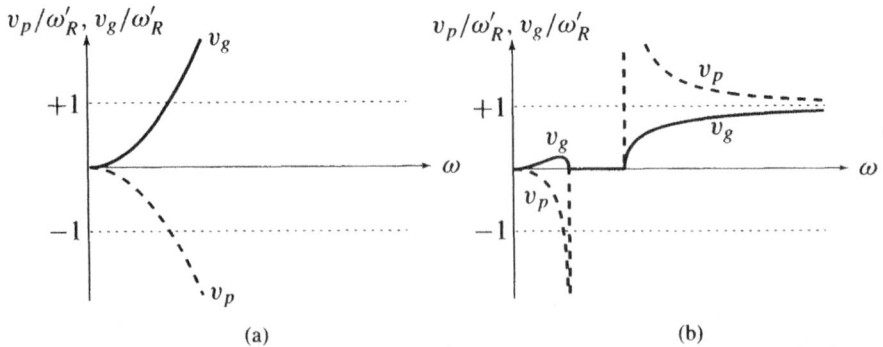

Fig. 3.5. Phase velocity [Eq. (3.19)] and group velocity [Eq. (3.20)]. (a) PLH TL ($L_R' = C_R' = 0$). (b) General (unbalanced) CRLH TL.

unbounded seems to violate Einstein's relativity according to which no modulated signal could propagate faster than the speed of light $c = 1/\sqrt{\varepsilon_0\mu_0}$.[4] This contradiction is automatically suppressed if we appropriately consider the parasitic RH contributions existing in a physical LH medium and recognize thereby that any LH medium is in effect a CRLH medium, where phase and group velocities are bounded at high frequencies to the velocity of the RH contributions only, $v_{p,g}^{PRH} = \omega_R'$ [Fig 3.5(b)], that would be equal to the velocity of light c if we had pure LH elements interconnected by air-filled waveguides.

3.1.2   Equivalent MTM Constitutive Parameters

The CRLH TL parameters (L_R, C_R, L_L, C_L) of the previous subsection can be related to the constitutive parameters (ε and μ) of a real material exhibiting the same propagation characteristics by mapping the telegraphist's equations [Eq. (3.3)] to Maxwell equations [Eq. (2.5)]. Such a mapping is rigorously possible only if the electromagnetic waves in the MTM are propagating in a *transverse electromagnetic mode* (TEM), where the longitudinal components of the electric and magnetic fields are both zero ($E_z = 0$ and $H_z = 0$, z direction of the TL) because it is only under this condition that Maxwell's equations exhibits the same form as the telegraphist's equations. To derive explicitly the MTM equivalent constitutive parameters, let us consider the example of the (TEM) parallel-plate waveguide structure shown in Fig 3.6[5] [7] filled with a hypothetical CRLH material. Maxwell equations [Eq. (2.5)] in this problem take the simple form

$$\frac{dE_y}{dz} = -Z'H_x = -j\omega\mu H_x, \tag{3.21a}$$

$$\frac{dH_x}{dz} = -Y'E_y = -j\omega\varepsilon E_y. \tag{3.21b}$$

These equations are identical to the telegraphist's equations [Eq. (3.3)] if the following mapping

$$E_y \to V, \tag{3.22a}$$

$$H_x \to I, \tag{3.22b}$$

---

[4]Note however that this statement needs to be considered with caution. Although v_g rigorously represents the velocity of propagation of energy in a *nondispersive medium* (where β is a linear function of ω and therefore $v_g = (d\beta/d\omega)^{-1}$ = is a constant), it is not the case in a *dispersive medium*, such as a PLH medium (Section 2.2). This is obvious if we consider that a pulse propagating through a nondispersive medium may be so distorted, due to frequency-dispersive group delay $t_g = -d\phi/d\omega = \mathrm{fct}(\omega)$, that it may be unrecognizable at the output of the medium, so that the group velocity could not even be defined. The appropriate energy velocity to use would then be the *front velocity*, described in [9]. In conclusion, v_g can be considered to represent the velocity of propagation of energy only in a weakly nondispersive medium.

[5]It will be shown in Section 4.5.1 that a CRLH TEM parallel-plate waveguide structure can indeed be realized in practice with a strip-line type mushroom architecture.

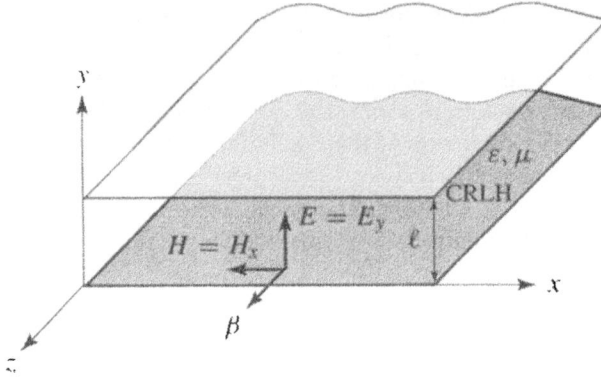

Fig. 3.6. Parallel-plate waveguide structure filled with a CRLH material excited in the fundamental TEM mode.

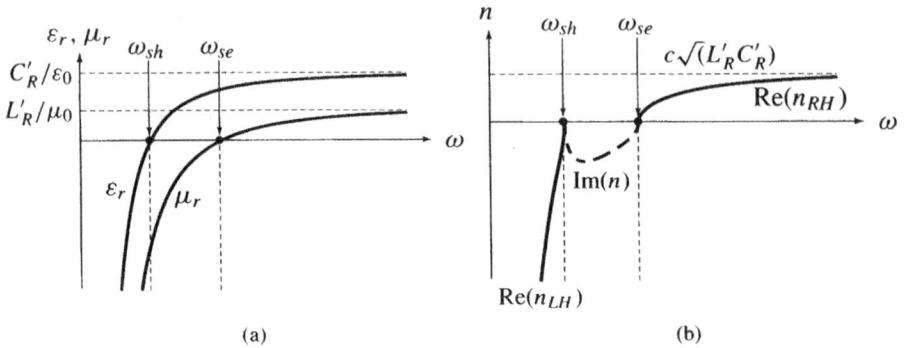

(a)   (b)

Fig. 3.7. Constitutive parameters of a CRLH material (case $\omega_{sh} < \omega_{se}$). (a) Permittivity and permeability computed by Eq. (3.23). (b) Refractive index computed by Eq. (3.25).

and

$$\mu = \mu(\omega) = L'_R - \frac{1}{\omega^2 C'_L}, \tag{3.23a}$$

$$\varepsilon = \varepsilon(\omega) = C'_R - \frac{1}{\omega^2 L'_L}, \tag{3.23b}$$

where $\mu = \mu_r \mu_0$ (H/m) and $\varepsilon = \varepsilon_r \varepsilon_0$ (F/m)[6]. Eqs. (3.23) represent the equivalent constitutive parameters of a CRLH TL MTM.

[6]These expression are valid in the loss-less case. In the presence of losses, associated with series resistance R' and shunt admittance G' in the CRLH model (Section 3.1.4), ε and μ naturally become complex [Eq. 2.3]. If losses are weak, the expressions in Eq. (3.23) are approximately equal to the real parts of ε and μ, ε', and μ'. The imaginary parts of ε and μ, ε'', and μ'' are generally positive for passive materials.

At highest frequencies, the equivalent parameters [Eq. (3.23)] tend to the PRH TL nondispersive equivalent parameters $\mu(\omega \to \infty) = L'_R$ and $\varepsilon(\omega \to \infty) = C'_R$; at lowest frequencies, they tend to the PLH TL dispersive equivalent parameters, $\mu(\omega \to 0) = -1/(\omega^2 C'_L)$ and $\varepsilon(\omega \to 0) = -1/(\omega^2 L'_L)$. It can be easily verified that the CRLH TL equivalent parameters [Eq. (3.23)] satisfy the entropy condition for dispersive media [Eq. (2.35)]

$$W = \frac{\partial (\omega\varepsilon)}{\partial \omega} E^2 + \frac{\partial (\omega\mu)}{\partial \omega} H^2$$

$$= \left(C'_R + \frac{1}{\omega^2 L'_L} \right) E^2 + \left(L'_R + \frac{1}{\omega^2 C'_L} \right) H^2 > 0,$$

(3.24)

since all the quantities in this expression are positive. We can finally derive the CRLH TL equivalent *refractive index* as

$$n = n(\omega) = \sqrt{\mu_r \varepsilon_r} = c\sqrt{\mu\varepsilon}$$

$$= c\frac{s(\omega)}{\omega} \sqrt{\left(\frac{\omega}{\omega'_R}\right)^2 + \left(\frac{\omega'_L}{\omega}\right)^2 - \kappa\omega'^2_L},$$

(3.25)

according to Eq. (3.10). This equivalent refractive index is related to the (TEM) propagation constant β by

$$\beta = nk_0,$$

(3.26)

where $k_0 = \omega/c$. Thus, the refractive index of a given material can be computed either by Eq. (3.25) after the CRLH TL LC parameters (L_R, C_R, L_L, C_L) have been extracted (Section 3.3.3) or from the dispersion relation $\beta(\omega)$, obtained analytically or by full-wave analysis using periodic boundary conditions, by (3.26), $n = \beta/k_0$.

The constitutive parameters [Eq. (3.23)] and refractive index [Eq. (3.25)] are plotted in Fig 3.7. Inspection of (3.23) and of Fig 3.7(a) reveals that $\mu(\omega)$ is negative when $\omega < \omega_{se}$ and that $\varepsilon(\omega)$ is negative when $\omega < \omega_{sh}$. Therefore, in the general case where $\omega_{se} \neq \omega_{sh}$, only one of the two quantities μ and ε is negative in the gap, from $\min(\omega_{se}, \omega_{sh})$ to $\max(\omega_{se}, \omega_{sh})$, which leads by Eq. (3.25) to the expected imaginary refractive index $\beta_{gap} = j\text{Im}(n)k_0 = -j\alpha$ associated with the attenuation constant $\alpha = -\text{Im}(n)k_0$. If $\omega_{sh} < \omega_{se}$ (case of Fig 3.7), only μ is negative in the gap, and the gap may therefore by referred to as a *magnetic gap*; if in contrast $\omega_{sh} < \omega_{se}$, only ε is negative in the gap, and the gap may therefore be referred to as an *electric gap*.

Another interesting observation is that the magnitude of the refractive index is *smaller than one* close to the frequencies ω_{se} and ω_{sh} (passbands) [Fig 3.7(b)]. Consequently, the phase velocity of the wave $v_p = c/n$ is larger than the velocity of light in these frequency regions, a phenomenon sometimes referred to as *superluminal propagation.*[7]

---

[7]This phenomenon of superluminal propagation is somewhat different from that occurring in waveguides [6]. In a waveguide, the phase velocity v_p is larger than the velocity of light and than the

3.1.3 Balanced and Unbalanced Resonances

The CRLH TL exhibits interesting properties in the particular case where the series and shunt resonant frequencies (3.9a) and (3.9b) are equal

$$\omega_{se} = \omega_{sh}, \tag{3.27a}$$

or equivalently

$$L'_R C'_L = L'_L C'_R, \tag{3.27b}$$

or equivalently

$$Z_L = Z_R, \tag{3.27c}$$

where Z_L and Z_R were defined in Eqs. (3.15) and (3.16), respectively. This case is called the *balanced case* and is opposed to the general *unbalanced case* (implicitly assumed in the previous section), where the series and shunt resonances are different

$$\omega_{se} \neq \omega_{sh}. \tag{3.28}$$

Let us first focus on the unbalanced case and inspect the corresponding characteristic impedance, represented in Fig 3.4. When the TL is fed by a signal with either frequency $\omega = \omega_{se}$ or $\omega = \omega_{sh}$, we have zero immittances from Eq. (3.2)

$$Z'(\omega = \omega_{se}) = 0, \tag{3.29a}$$

and

$$Y'(\omega = \omega_{sh}) = 0, \tag{3.29b}$$

introducing a zero and a pole, respectively, in the characteristic impedance (3.14)

$$Z_c(\omega = \omega_{se}) = 0, \tag{3.30a}$$

and

$$Z_c(\omega = \omega_{sh}) = \infty. \tag{3.30b}$$

group velocity v_g from the relation $v_p = c^2/v_g$ where causality requires $v_g < c$. This relation does *not* hold in the CRLH TL, as it may be seen in Eqs. (3.19) and (3.20) and in Fig 3.5. In addition, superluminal phase velocity in a waveguide is due to the fact that the waves are non-TEM and travel along the structure in a zigzagging manner as the superposition of two plane waves, each propagating in fact with the velocity c; $v_p > c$ because v_p refers to the *effective direction* of propagation along the axis of the guide and not to the real, zigzagging, direction of propagation of the physical waves. In contrast, propagation is TEM in a CRLH TL. Therefore, the effective and physical directions of propagation are the same, and superluminal propagation really refers to propagation of the phase of the physical wave.

These impedances conditions correspond to zero group velocities ($v_g = 0$) from Eq. (3.20) or zero slopes in the dispersion diagram, i.e., TL *resonances*, and result in the emergence of a stop band or gap, despite the assumed absence of loss. We have indeed from Eq. (3.14), assuming ports with constant and real impedance,

$$Z_c[0 < \omega < \min(\omega_{se}, \omega_{sh})] \in \mathbb{R} \quad \text{(LH pass band)}, \tag{3.31a}$$

$$Z_c[\min(\omega_{se}, \omega_{sh}) < \omega < \min(\omega_{se}\omega_{sh})] \in \mathbb{I}, \quad \text{(el. or mag. gap)}, \tag{3.31b}$$

$$Z_c[\omega > \max(\omega_{se}, \omega_{sh})] \in \mathbb{R}, \quad \text{(RH pass band)}. \tag{3.31c}$$

In contrast, in the balanced case ($\omega_{se} = \omega_{sh} = \omega_0$), we observe in Eq. (3.14) that the zero ω_{se} and the pole ω_{sh} in the characteristic impedance cancel each other,[8] suppressing resonance effect. As a consequence, the gap closes up and the characteristic impedance becomes a *frequency-independent* quantity

$$Z_c = Z_L = Z_R. \tag{3.32}$$

This means that the balanced condition allows *matching over an infinite band-width*.[9]

Let us examine what happens now for the propagation constant in the balanced case. The expression for κ in Eq. (3.8c) becomes

$$\kappa = \sqrt{L_R' C_R'} \left(C_L' \sqrt{\frac{L_R'}{C_R'}} + L_L' \sqrt{\frac{C_R'}{L_R'}} \right) = \frac{2}{\omega_R' \omega_L'}, \tag{3.33}$$

so that $\kappa \omega_L'^2 = \omega_L'/\omega_R'$, and the radicand in Eq. (3.10) forms a square, leading to the simplified propagation constant

$$\beta = \frac{\omega}{\omega_R'} - \frac{\omega_L'}{\omega}, \tag{3.34}$$

exhibiting the root

$$\omega_0 = \sqrt{\omega_R' \omega_L'}, \tag{3.35}$$

which is identical to Eq. (3.13). This indicates that the frequency of maximum gap attenuation of the unbalanced CRLH TL [Eq. (3.13)] becomes the frequency of gap-less transition between the LH and RH ranges if the line is balanced. This frequency ω_0 is called the *transition frequency*.

---

[8] After suppression of the 0/0 indetermination by L'Hospital rule.

[9] In real LC network implementations (Section 3.2), the balanced condition will correspond to *optimal matching condition over the passband of the resulting passband filter*.

Eq. (3.34) shows that, in a balanced CRLH TL, the propagation constant is simply expressed as the sum of the propagation constants of a (linear and positive) PRH TL

$$\beta^{PRH} = \frac{\omega}{\omega'_R} = \omega\sqrt{L'_R C'_R} \tag{3.36}$$

and of a (negative and hyperbolic) PLH TL

$$\beta^{PLH} = -\frac{\omega'_L}{\omega} = -\frac{1}{\omega\sqrt{L'_L C'_L}}. \tag{3.37}$$

Finally, the refractive index [Eq. (3.25)] takes the simpler form

$$n = c\left(\frac{1}{\omega'_R} - \frac{\omega'_L}{\omega^2}\right), \tag{3.38}$$

where negative sign for $\omega < \omega_0$ and positive sign for $\omega > \omega_0$ are immediately apparent. In addition, we note that the refractive index is *zero* at the transition frequency ω_0, which corresponds to infinite phase velocity and infinite guided wavelength.

The balanced transmission characteristics of Eqs. (3.34) and (3.32) correspond to the simplified circuit model shown in Fig 3.8.

The balanced CRLH TL has a number of advantages over the unbalanced CRLH TL, which may be summarized as follows:

- Its model is *simpler* than that of the general *unbalanced* model of Fig 3.2 because series combined LH and RH contributions are decoupled from each other.

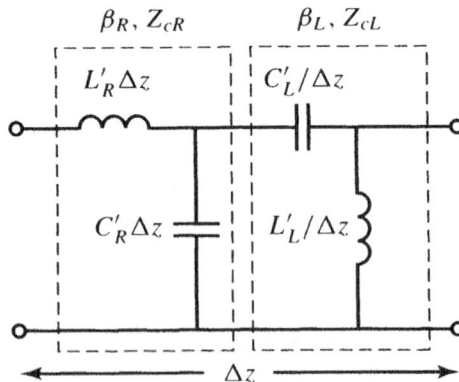

Fig. 3.8. Simplified equivalent circuit model for the unit cell of an ideal CRLH TL under the condition of balanced resonances (compared with the general case unbalanced circuit model shown in Fig 3.2).

- In contrast to the unbalanced TL, the balanced TL is *gapless* (no stop band). This can be seen by the fact that β in Eq. (3.34) is purely real and Z_c in Eq. (3.32) is purely real at all frequencies from $\omega = 0$ to $\omega = \infty$, whereas the unbalanced TL exhibits a usually undesired stop band in the frequency range extending from $\min(\omega_{se}, \omega_{sh})$ to $\max(\omega_{se}, \omega_{sh})$, where γ in Eq. (3.10) and Z_c in Eq. (3.14) are purely real and imaginary, respectively.

- The characteristic impedance of the balanced TL is a constant *frequency independent* quantity, $Z_c = \text{const}$ [Eq. (3.14)]. As a consequence, the balanced TL can be matched over a broad bandwidth, whereas the unbalanced TL can only be matched over a restricted bandwidth (strictly to one single frequency).

- In the unbalanced case, the gap, delimited by the series and shunt resonances ω_{se} and ω_{sh} at $\gamma = 0$ (i.e., $\beta = \alpha = 0$) (Fig 3.3), is associated with zero group velocity. In contrast, in the balanced case, there is no such gap and the group velocity [Eq. (3.20)], which reduces to

$$v_g = \frac{\omega^2 \omega'_R}{\omega^2 + \omega'_R \omega'_L},$$
(3.39)

is nonzero at ω_0

$$v_g(\omega = \omega_0) = \frac{\omega'_R}{2} = \frac{\omega_0^2}{2\omega'_L},$$
(3.40)

that is

$$v_g(\omega = \omega_0) = \frac{v_{g,R}(\omega = \omega_0)}{2} = \frac{v_{g,L}(\omega = \omega_0)}{2},$$
(3.41)

where $v_{g,R}$ and $v_{g,L}$ represent the group velocities of the PRH and PLH TLs, respectively. This unique property of nonzero group velocity at the transition frequency is a key feature of of CRLH structures, which is exploited in several applications presented in this book. Note also that the phase velocity [Eq. (3.19)] in the balanced CRLH TL takes the form

$$v_p = \frac{\omega^2 \omega'_R}{\omega^2 - \omega'_R \omega'_L},$$
(3.42)

and exhibits a pole at $\omega = \omega_0 = \sqrt{\omega'_R \omega'_L}$ where $v_g = \omega'_R/2$.

- Because the transition frequency ω_0 associated with the fundamental mode [Eq. (3.34)], supports wave propagation ($v_g \neq 0$) with $\beta = 0$, it is called the *phase origin* of the CRLH TL. At this frequency, the phase shift along a line of length ℓ is zero ($\phi = -\beta\ell = 0$). When frequency is decreased below ω_0 toward zero, the phase becomes positive, increases progressively, and tends to infinity as $\phi(\omega \to 0) = -\beta d \to +\omega'_L \ell/\omega$; when frequency is increased above ω_0 toward infinity, the phase becomes negative, increases progressively in magnitude, and tends to infinity as $\phi(\omega \to \infty) = -\beta d \to -\omega\ell/\omega'_R$.

- Translating the previous point in terms of *guided wavelength*, we have for the balanced CRLH TL[10]

$$\lambda_g = \frac{2\pi}{|\beta|} = \frac{2\pi}{|\omega/\omega_R' - \omega_L'/\omega|}.$$
(3.43)

This equation shows that guided wavelength in a CRLH TL exhibits the following behavior. At low frequencies, λ_g, which is essentially of LH nature, is small. As frequency is increased, λ_g *increases* roughly linearly (exactly linearly in a PLH TL) with frequency. This proportionality between guided wavelength and frequency, which is characteristic of a LH medium, is unusual, but perfectly physical, as be shown in Section 3.4. At the transition frequency ω_0, $\lambda_g = \infty$, which means that the electrical length of the line is zero, $\theta = \ell/\lambda_g = 0$. When frequency is further increased into the RH range, λ_g starts to decrease and decreases roughly hyperbolically (exactly hyperbolically in a PLH TL) with frequency, as in conventional TLs.

The propagation constant, characteristic impedance, constitutive parameters, refractive index and phase/group velocities shown in Figs. 3.3(a), 3.4, 3.7, and 3.5(b) for the unbalanced case are plotted in Fig 3.9 for the balanced case. Table 3.1 summarizes the main results for the balanced CRLH TL in comparison with the corresponding PRH and PLH TLs.

3.1.4 Lossy Case

The lossy CRLH TL model is shown in Fig 3.10. In this model, the per-unit-length immittances [Eq. (3.2)] can be written

$$Z' = R' + jX' \quad \text{with} \quad X' = \left(\omega L_R' - \frac{1}{\omega C_L'}\right),$$
(3.44a)

$$Y' = G' + jB' \quad \text{with} \quad B' = \left(\omega C_R' - \frac{1}{\omega L_L'}\right).$$
(3.44b)

Hence, the complex propagation constant and characteristic impedance read

$$\gamma = \alpha + j\beta = \sqrt{Z'Y'} = \sqrt{(R'G' - X'B') - j(R'B' + G'X')},$$
(3.45)

$$Z_c = R_0 + jX_0 = \sqrt{\frac{Z'}{Y'}} = \sqrt{\frac{R' + jX'}{G' + jB'}},$$
(3.46)

respectively, which reduce to their loss-less counterparts of Eqs. (3.10) and (3.14) when $R' = G' = 0$. The effects of losses in RH TL structures are well known. To obtain some insight into the effects of losses in the CRLH TL (particularly in the LH range), we will first consider, for simplicity, the PLH TL, which is a

---

[10]This point is also true for the unbalanced case in the passbands.

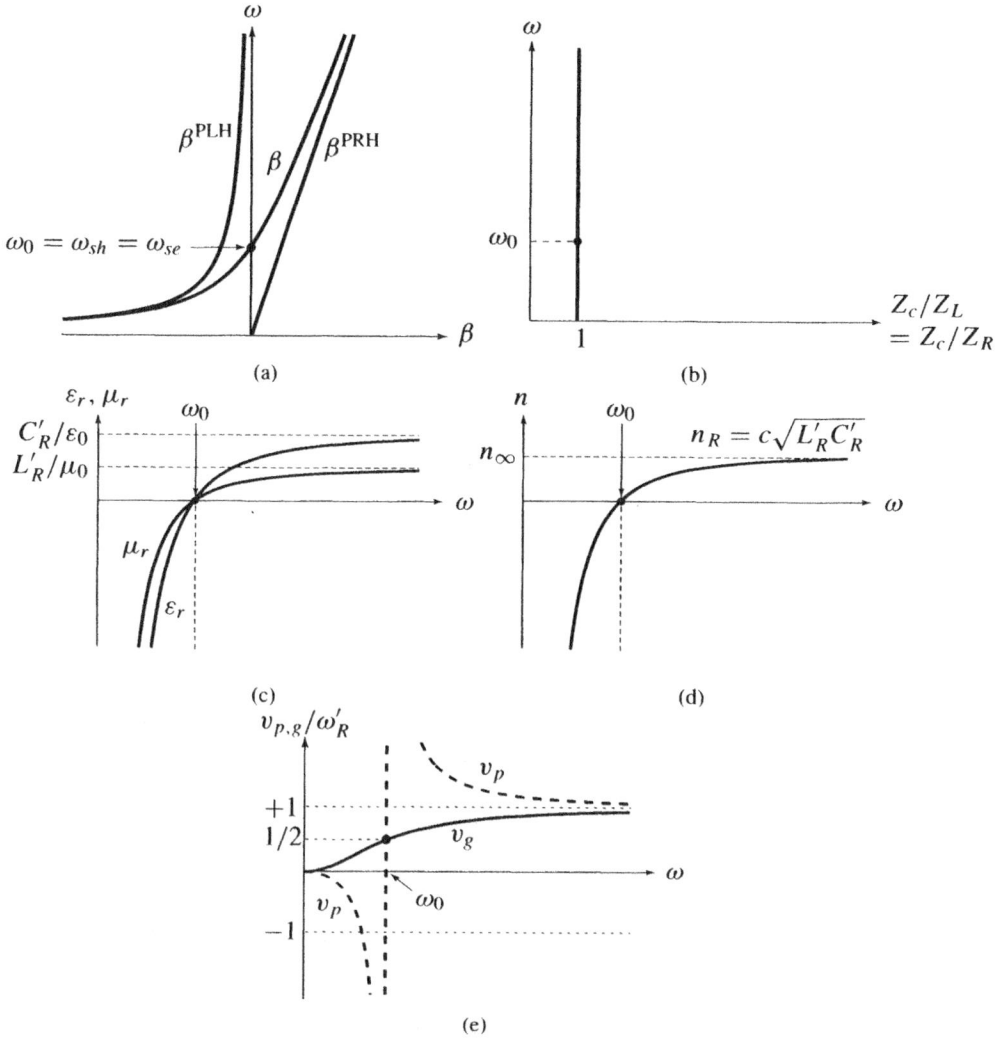

Fig. 3.9. Characteristics of a balanced CRLH material. (a) Propagation constant β [Eq. (3.34)], compared with those of PLH β^{PLH} [Eq. (3.37)] and of PRH β^{PRH} (3.36) materials, respectively. (b) Characteristic impedance Z_c [Eq. (3.32)]. (c) Relative permittivity ε_r and permeability μ_r [Eq. (3.23)]. (d) Refractive index [Eq. (3.38)]. (e) Phase and group velocities [Eqs. (3.39) and (3.42)].

good approximation of the CRLH TL at low frequencies. Using Eq. (3.45) with $L'_R = C'_R = 0$ in Eq. (3.44), we obtain the propagation constant γ_L for the PLH lossy TL

$$\gamma_L = -j \frac{\sqrt{1 - R'G'(\omega/\omega'_L)^2 + j\omega(C'_L R' + L'_L G')}}{\omega/\omega'_L} = \alpha_L + j\beta_L, \qquad (3.47)$$

TABLE 3.1. Summary of the Fundamental Characteristics of Ideal PLH, PRH, and Balanced CRLH TLs.

Quantity	PRH	PLH	CRLH (balanced)
Model	$L'_R \Delta z$, $C'_R \Delta z$, Δz	$C'_L/\Delta z$, $L'_L/\Delta z$, Δz	$L'_R \Delta z$, $C'_L/\Delta z$, $L'_L/\Delta z$, $C'_R \Delta z$, Δz
$\beta = \sqrt{Z'Y'}$	ω/ω'_R, with $\omega'_R = 1/\sqrt{L'_R C'_R}$	$-\omega'_L/\omega$, with $\omega'_L = 1/\sqrt{L'_L C'_L}$	$\omega/\omega'_R - \omega'_L/\omega$
$\omega - \beta$ diagram			$\omega_0 = \sqrt{\omega'_R \omega'_L}$
$\lambda_g = 2\pi/|\beta|$	$2\pi\omega'_R/\omega$	$2\pi\omega/\omega'_L$	$2\pi \big/ \left| \omega/\omega'_R - \omega'_L/|\omega| \right|$
$v_p = \dfrac{\omega}{\beta}$	ω'_R	$-\omega^2/\omega'_L$	$\left(\omega^2 \omega'_R\right) \big/ \left(\omega^2 - \omega'_R \omega'_L\right)$
$v_g = \left(\dfrac{d\beta}{d\omega}\right)^{-1}$	ω'_R	ω^2/ω'_L	$\left(\omega^2 \omega'_R\right) \big/ \left(\omega^2 + \omega'_R \omega'_L\right)$
$Z_c = \sqrt{\dfrac{Z'}{Y'}}$	$Z_{cR} = \sqrt{L'_R/C'_R}$	$Z_{cL} = \sqrt{L'_L/C'_L}$	$Z_{0C} = Z_{cR} = Z_{cL}$
ε	C'_R	$-1/\left((\omega^2)L'_L\right)$	$C'_R - 1/\left((\omega^2)L'_L\right)$
μ	L'_R	$-1/\left((\omega^2)C'_L\right)$	$L'_R - 1/\left((\omega^2)C'_L\right)$
$n = \pm c\sqrt{\varepsilon\mu}$	c/ω'_R	$-c\omega'_L/\omega^2$	$c\left(1/\omega'_R - \omega'_L/\omega^2\right)$

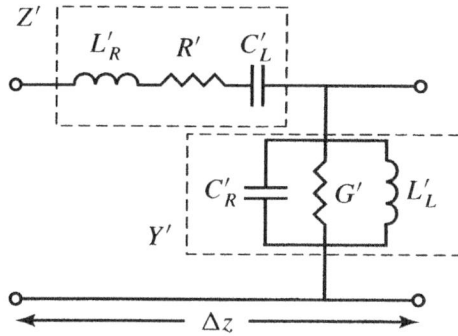

Fig. 3.10. Equivalent circuit model for the ideal lossy CRLH TL.

which has both a real and an imaginary part at all frequencies ($\alpha \neq 0$ and $\beta \neq 0$, $\forall \omega$). In practical low-frequency microwave structures, ohmic and dielectric losses are relatively small, and the parameters

$$A = R'G'(\omega/\omega'_L)^2 \tag{3.48}$$

$$B = \omega(C'_L R' + L'_L G') \tag{3.49}$$

are typically such that $A \ll B < 1$. Therefore, A can be neglected, and the remaining square root may be approximated by its first order Taylor expansion, so that

$$\gamma_L = -j\frac{1}{\omega/\omega'_L}\left[1 + \frac{j}{2}(C'_L R' + L'_L G')\right], \tag{3.50}$$

from which

$$\beta_L \approx -\frac{\omega'_L}{\omega}, \tag{3.51a}$$

$$\alpha_L \approx \frac{1}{2}\left[R'Y_0 + G'Z_c\right], \tag{3.51b}$$

where $Z_c = Z_L = \sqrt{L'_L/C'_L}$ and $Y_0 = Y_L = 1/Z_c$. We observe that a weakly lossy LH TL exhibits the same propagation constant as a loss-less LH TL [Eq. (3.34)] and the same attenuation factor as that of the conventional RH TL [6]. These two facts are not surprising since the loss mechanism is the same in a LH TL as in a RH TL. By the same order of approximation, we find that the characteristic impedance [Eq. (3.46)] becomes

$$Z_c = \sqrt{\frac{R' - j/(\omega C'_L)}{G' - j/(\omega L'_L)}} \overset{\beta \to 0}{\approx} \sqrt{\frac{L'_L}{C'_L}}, \tag{3.52}$$

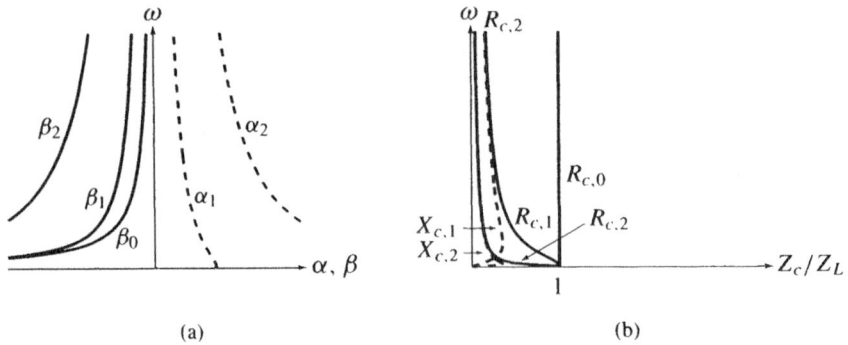

(a)   (b)

Fig. 3.11. Effects of losses in an ideal PLH lossy TL. (a) Dispersion and attenuation. (b) Characteristic impedance. The indexes "0," "1," "2" refer to a loss-less, weakly lossy, and strongly lossy TL, respectively.

which is identical to that of the loss-less LH TL [Eq. (3.32)]. Thus, a weakly lossy LH TL behaves essentially in the same manner as a loss-less LH TL. The same will naturally apply to a weakly lossy CRLH TL, including both LH and RH contributions.

The propagation constant and characteristic impedance of the PLH lossy TL are shown in Fig 3.11. It can be seen that the introduction of losses reduces group velocity over a bandwidth, which increases as the amount of losses is increased and nullifies the real part of the characteristic impedance at the transition frequency. As expected from Eq. (3.52), $Z_c(\omega \to 0) = Z_L$ and $Z_c(\omega \to \infty) = \sqrt{R'/G'}$.

The propagation constant [Eq. (3.45)] and characteristic impedance [Eq. (3.46)] of the CRLH lossy TL are shown in Figs. 3.12 and 3.13 for the balanced and unbalanced cases, respectively. We find from Eq. (3.44) that in the CRLH TL $X'(\omega_0) = B'(\omega_0) = 0$, so that from Eq. (3.46) $Z_c(\omega_0) = \sqrt{R'/G'}$. This value can

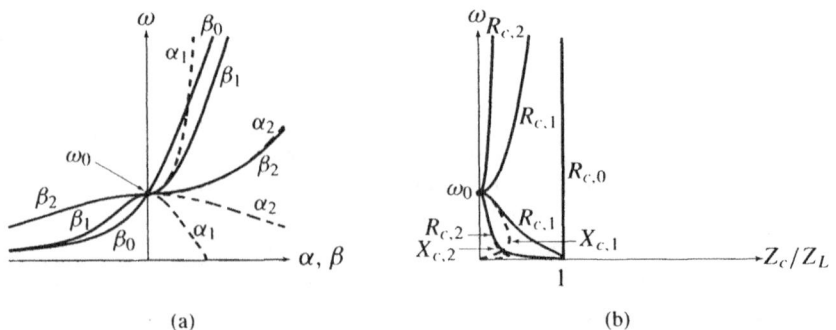

(a)   (b)

Fig. 3.12. Effects of losses in the ideal balanced CRLH TL. (a) Dispersion and attenuation. (b) Characteristic impedance. The indexes "0," "1," "2" refer to a loss-less, weakly lossy, and strongly lossy TL, respectively.

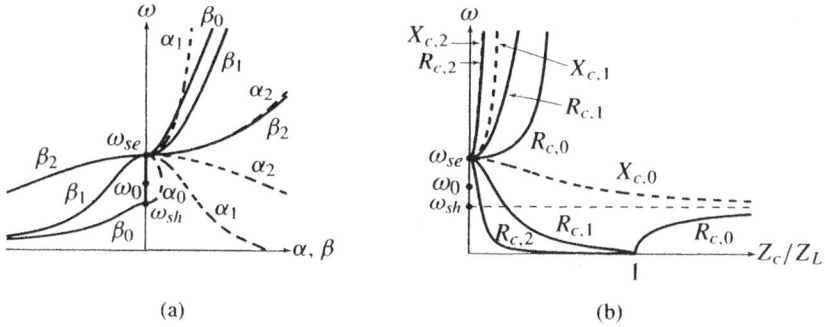

Fig. 3.13. Effects of losses in the ideal unbalanced CRLH TL for $\omega_{sh} < \omega_{se}$. (a) Dispersion and attenuation. (b) Characteristic impedance. The indexes "0," "1," "2" refer to a loss-less, weakly lossy, and strongly lossy TL, respectively.

be shown to represent a minimum of Z_c in the balanced case (Fig 3.12). We also observe that, in the unbalanced case, $X'(\omega_{se}) = 0$, leading to a minimum value of Z_c at $\omega = \omega_{sh}$.

3.2 LC NETWORK IMPLEMENTATION

The ideal CRLH TL described in Section 3.1 is not readily available from nature, but an *effectively homogeneous* CRLH TL operative *in a restricted frequency range*[11] can be realized the form of a ladder network circuit structure.

3.2.1 Principle

This network is obtained by cascading the LC unit cell shown in Fig 3.14 so as to obtain the ladder network shown in Fig 3.15.[12] The unit cell consists of an impedance Z (Ω) constituted by a RH inductance L_R (H) in series with a LH capacitance C_L (F) and of an admittance Y (S) constituted by a RH capacitance C_R (F) in parallel with a LH inductance L_L (H)

$$Z = j \left(\omega L_R - \frac{1}{\omega C_L} \right) = j \frac{(\omega/\omega_{se})^2 - 1}{\omega C_L}, \tag{3.53a}$$

$$Y = j \left(\omega C_R - \frac{1}{\omega L_L} \right) = j \frac{(\omega/\omega_{sh})^2 - 1}{\omega L_L}, \tag{3.53b}$$

[11]"Restricted" means here that there is limitation to the frequency range of the passband (in contrast to what we have in the ideal homogeneous TL case). However, this frequency range can be controlled by the LC parameter values (Section 3.2.5) and is incomparably larger to that of resonant-type structures (Section 1.3).

[12]Although periodicity is not mandatory in MTMs, we will generally consider periodic implementations for simplicity, which does not represent any restriction since only the average lattice size matters in terms of macroscopic parameters of MTMs (Fig 3.28).

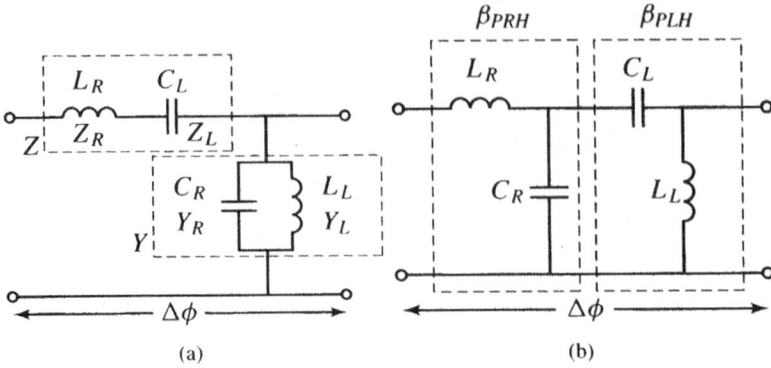

Fig. 3.14. Unit cell of an LC CRLH TL. (a) General (unbalanced). (b) Balanced $(L_R C_L = L_L C_R)$.

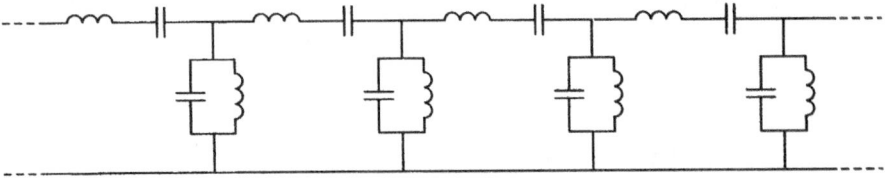

Fig. 3.15. Periodic ladder network implementation of a general (unbalanced) LC CRLH TL.

where the series and shunt resonance ω_{se} and ω_{sh} are defined in a manner similar to that for the ideal homogeneous case in Eq. (3.9)

$$\omega_{se} = \frac{1}{\sqrt{L_R C_L}}, \tag{3.54a}$$

$$\omega_{sh} = \frac{1}{\sqrt{L_L C_R}}. \tag{3.54b}$$

We note the phase shift induced by the unit cell is noted $\Delta\phi$. It should be understood that, whereas the infinitesimal models of Figs. 3.2 and 3.8 are associated with a *physical length* Δz (m), the circuit models of Figs. 3.14 and 3.15 are *dimensionless* and their size can be described only in terms of their *electrical length* $\theta = |\Delta\phi|$ (rad).[13] However, in a practical circuital implementation, the inductors and capacitors will occupy a physical length, varying with the technology used (Section 3.3).[14] If the footprint of the unit cell of Fig. 3.14 is p in

[13]The electrical length is a positive quantity by definition [7].
[14]For example, in microstrip technology, the capacitors C_L can be implemented as interdigital capacitors and the inductors L_L can be implemented as stub inductors (Section 3.3).

length, then we can write the immittances along Δz as

$$\frac{Z}{p} = j\left[\omega\left(\frac{L_R}{p}\right) - \frac{1}{\omega(C_L p)}\right], \tag{3.55a}$$

$$\frac{Y}{p} = j\left[\omega\left(\frac{C_R}{p}\right) - \frac{1}{\omega(L_L p)}\right]. \tag{3.55b}$$

Comparison of Eqs. (3.2) and (3.55) reveals that, if $p = \Delta z \to 0$, the immittances for the length p become $Z/p \to Z'$, $Y/p \to Y'$, which means that the LC implementation of Fig 3.14(a) is equivalent to the incremental model of Fig 3.2. The model of Fig 3.2 represents only a Δz-long piece of TL, and a real line of total length ℓ is naturally obtained by repeating this section an appropriate number of times N, such that $\ell = N\Delta z$.[15] Thus, cascading N LC unit cells as in the network of Fig 3.15 results in a TL equivalent to an ideal CRLH TL[16] of length ℓ under the condition $p \to 0$, as illustrated in Fig 3.16. This is called the *homogeneity condition* because it ensures that the artificial TL is equivalent to the ideal homogeneous TL. In practice, this condition can be translated into the rule-of-thumb *effective-homogeneity condition*, $p < \lambda_g/4$, introduced in Section 1.1. The homogeneity condition ensures the absence of Bragg-like

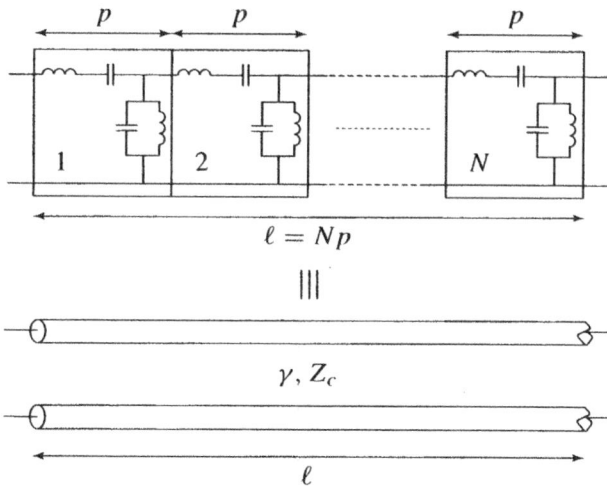

Fig. 3.16. Equivalence between a periodic ladder network and an ideal TL of length ℓ obtained in the limit $p = \Delta z \to 0$. If $p \neq 0$ but $p < \lambda_g/4$, the network still represents a good approximation of the ideal TL in a restricted frequency range.

[15]If $\Delta z \to 0$ and ℓ is finite, then $N \to \infty$. But in practice the homogeneity condition $\Delta z < \lambda_g/4$ is sufficient, so that an artificial TL can always be realized with a finite number of cells in a given frequency range.
[16]The reason why we need to synthesize such a line is of course the fact that it does not exist naturally.

TABLE 3.2. Summary of the Main Characteristics of LC Network PLH, PRH, and Balanced CRLH TLs With Very Small Unit Cell Electrical Length $|\Delta\phi| \ll \pi/2$ (approximation from ideal homogeneous case valid for $\Delta\phi \to 0$).

Quantity	PRH	PLH	CRLH (balanced)
Model	L_R — C_R — $\Delta\phi$	C_L — L_L — $\Delta\phi$	L_R C_L — C_R L_L — $\Delta\phi$
Filter type	Lowpass	Highpass	Band-Pass
ω_c	$\omega_{cR} = 2\omega_R$, with $\omega_R = 1/\sqrt{L_R C_R}$	$\omega_{cL} = \omega_L/2$, with $\omega_L = 1/\sqrt{L_L C_L}$	$\omega_0 = \sqrt{\omega_R \omega_L} = 1/\sqrt[4]{L_R C_R L_L C_L}$
Z_c	$Z_{cR} = \sqrt{L_R/C_R}$	$Z_{cL} = \sqrt{L_L/C_L}$	$Z_{0C} = Z_{cR} = Z_{cL}$
L', C' for N cells TL of length ℓ	$C'_R = C_R N/d,\ L'_R = L_R N/\ell$	$C'_L = C_L d/N,\ L'_L = L_L d/N$	Same values
$\Delta\phi$	$\Delta\phi_R = -\arctan\left[\dfrac{\omega\kappa_R}{2-(\omega/\omega_R)^2}\right]$ with $\kappa_R = L_R/Z_c + C_R Z_c$ $\Delta\phi_R < 0$: phase lag	$\Delta\phi_L = -\arctan\left[\dfrac{\omega\kappa_L}{1-2(\omega/\omega_L)^2}\right]$ with $\kappa_L = L_L/Z_c + C_L Z_c$ $\Delta\phi_L > 0$: phase advance	$\Delta\phi_c \approx \Delta\phi_R + \Delta\phi_L$
$t_g = -\dfrac{d\phi}{d\omega}$ (group delay)	$t_{gR} = -\dfrac{\kappa_R\left[2+(\omega/\omega_R)^2\right]}{\kappa_R^2\omega^2 + \left[2-(\omega/\omega_R)^2\right]^2}$	$t_{gL} = -\dfrac{\kappa_L\left[1+2(\omega/\omega_L)^2\right]}{\kappa_L^2\omega^2 + \left[1-2(\omega/\omega_L)^2\right]^2}$	$t_{gC} \approx t_{gR} + t_{gL}$

interferences (or resonances) along the discontinuities of the line.[17] In the limit $p \to 0$, all the results presented in Section 3.1 apply to the LC implementation, which is the object of this section.

In the present section, we consider only dimensionless LC ladder network constituted of ideal[18] inductors and capacitors, where the homogeneity condition becomes $|\Delta\phi| \to 0$ or $|\Delta\phi| < \pi/2$, since $\Delta\phi = \beta p = 2\pi p/\lambda_g$; real CRLH TL structures will be presented in Section 3.3.

Table 3.2 summarizes the main *approximate* characteristics of PRH, PLH, and CRLH LC network TLs obtained from the limit case of their ideal homogeneous counterparts, valid for very small unit cell electrical length, $|\Delta\phi| \ll \pi/2$. Exact relations for larger electrical length (but still smaller than $\pi/2$) will be derived further in this section.

3.2.2 Difference with Conventional Filters

Considered over all frequencies from $\omega = 0$ to $\omega \to \infty$, the CRLH networks of Figs. 3.2 and 3.8 are obviously *band-pass* filters. When $\omega \to 0$, $|Z| \to 1/(\omega C_L) \to \infty$ and $|Y| \to 1/(\omega L_L) \to \infty$, and therefore we have a stopband due to the *highpass* nature of the LH elements; when $\omega \to \infty$, $|Z| \to \omega L_R \to \infty$ and $|Y| \to \omega C_R \to \infty$, and therefore we have a stopband due to the *lowpass* nature of the RH elements. Between these two stop bands, a perfectly matched passband can exist under the balanced condition (Section 3.1.3), which we will extend in this section to the case of the LC network implementation. Thus, an LC network implementation of a CRLH TL is a band-pass filter, in the sense that it has LH highpass low-frequency and RH lowpass high-frequency stop bands, but it exhibits the very specific behavior of a CRLH TL, fully characterizable in terms of its propagation constant β and characteristic impedance Z_c, in its passband. These exact expressions for the cutoff frequencies delimiting the operational bandwidth will be derived in Section 3.2.5.

Because CRLH TLs may look at the first sight similar to conventional filters [10], let us now point out the essential distinctions existing between these two types of structures:

- A LH MTM structure exhibits a specific phase response, of the type Eq. (3.10) or Eq. (3.34),[19] leading to LH transmission at lower frequencies and RH transmission at higher frequencies. Conventional filters are generally designed to meet *magnitude* specifications and do not exhibit a LH range.[20]

---

[17] As a consequence, the LC components are electrically very small and are called *lumped*.

[18] The term "ideal" refers here to constant, nonfrequency-dependent elements.

[19] These expressions are for the ideal TL. The corresponding formulas for the ladder network implementation will be derived in this section.

[20] The proof of this statement is out of the scope of this book and is specific to the filter prototype considered. It may be done analytically. It may also be done heuristically in a circuit simulator by distributing voltage probes along the filter and exciting it with a harmonic signal switched on at a specific time; it can then be verified that (for standard filters) the observers closer to the

- A MTM structure is intended to be used as a transmission line or transmission structure. *Only the passband* is directly useful. The stopbands are usually parasitic effects limiting the operation bandwidth of the MTM. Thus, the filtering characteristic of the network are generally not used in MTM as in filters.

- A MTM structure is constituted of unit cells satisfying the *homogeneity condition* $|\Delta\phi| < \pi/2$. Conventional filters do not generally satisfy this condition; they may have node-to-node phase shifts larger than $\pi/2$. In fact, a MTM with $\Delta\phi \ll \pi/2$, or $\Delta z \ll \lambda_g/4$, would be an ideal (perfectly homogeneous) material, in the same manner as conventional dielectrics, made of molecules with dimensions many orders of magnitude smaller than wavelength. The only reason while Δz may be close to $\lambda_g/4$ at some frequencies in today's MTMs is the unavailability of more effective structures.

- A MTM structure *can be 2D or 3D and behave as bulk media*, whereas conventional filters are 1D and behave as electric circuits.

- A MTM structure *can be made of identical cells*, whereas in a conventional filter each "cell" has generally different LC values to match the specifications of a given prototype.

In connection with the last point, it is important to emphasize that MTMs *do not require periodicity*, as will be shown in Fig 3.28. Although they are generally implemented in periodic configurations, the reason for periodicity is not physics but convenience. From a fabrication point of view, it is easier to design and build a MTM by repeating periodically a unique cell than by using a collection of different cells. From an analysis point of view, making the MTM periodic allows one to use the well-established and powerful theory of periodic structures, whereas a nonuniform architecture would be much more difficult to analyze, without exhibiting different properties. The fact that periodicity is not a necessity in MTMs is obvious when considering the ideal TL (Section 3.1). This TL is nonperiodic in nature and supports a unique mode, shown in Figs. 3.3 and 3.9(a). In a periodic LC ladder network implementation, this mode will represent the *fundamental mode* of the periodic structure (Section 3.2.6). If periodicity is suppressed, the different space harmonics disappear and only this mode is left. Another way of understanding this fact is to consider that in the limit $p \to 0$, all space harmonics are shifted to very high frequencies, outside the frequency range of interest, leaving only the fundamental CRLH of the ideal TL. Thus a MTM, as defined in this book, is fundamentally different from photonic crystals or PBG structures, as already pointed out in Section 1.9.

---

output see, at all frequencies, the first maxima of the wave after at later times than the observers closer to the input, which indicates RH positive phase velocity. Note that these considerations are even not relevant in the case of the vast majority of conventional filters, where the phase shift of some "cells" of the structure are typically not inferior to the effective-homogeneity limit $\Delta\phi < \pi/2$ for a propagating wave, characterizable by a propagation constant β, to be defined (see third bullet).

3.2.3 Transmission Matrix Analysis

The transmission characteristics of an N-cell ladder network TL can be conveniently computed by using the $[ABCD]$ or transmission matrix formalism [7]. The $[ABCD]$ matrix for the two-port network represented in Fig 3.17 relates the input current and voltage to the output current and voltage in the following manner

$$\begin{bmatrix} V_{in} \\ I_{in} \end{bmatrix} = \begin{bmatrix} A & B \\ C & D \end{bmatrix} \begin{bmatrix} V_{out} \\ I_{out} \end{bmatrix}, \qquad (3.56)$$

or

$$\begin{bmatrix} V_{out} \\ I_{out} \end{bmatrix} = \frac{1}{AD - BC} \begin{bmatrix} D & -B \\ -C & A \end{bmatrix} \begin{bmatrix} V_{in} \\ I_{in} \end{bmatrix}, \qquad (3.57)$$

where it can be shown, after conversion from $ABCD$ parameters to scattering or S parameters, that $A = D$ if the two-port is symmetric ($S_{11} = S_{22}$), $AD - BC = 1$ if the two-port is reciprocal ($S_{21} = S_{12}$), and $A, D \in \mathbb{R}$ and $B, C \in \mathbb{I}$ if the two-port is loss-less.

It is well known and can be easily verified that the $[ABCD]$ matrix of the cascade connection of N two-port networks, $[A_N B_N C_N D_N]$, is equal to the product of the $[ABCD]$ matrices representing the individual two-ports, $[A_k B_k C_k D_k]$

$$\begin{bmatrix} A_N & B_N \\ C_N & D_N \end{bmatrix} = \prod_{k=1}^{N} \begin{bmatrix} A_k & B_k \\ C_k & D_k \end{bmatrix}, \qquad (3.58)$$

which reduces to

$$\begin{bmatrix} A_N & B_N \\ C_N & D_N \end{bmatrix} = \begin{bmatrix} A & B \\ C & D \end{bmatrix}^N \qquad (3.59)$$

if all the cells are identical, $[A_k B_k C_k D_k] = [ABCD], \forall k$. The $[ABCD]$ matrix for an *asymmetric* two-port network consisting of a series impedance Z and shunt admittance Y [7] is thus given by

$$\begin{bmatrix} A & B \\ C & D \end{bmatrix}_{asym} = \begin{bmatrix} 1 & Z \\ 0 & 1 \end{bmatrix} \begin{bmatrix} 1 & 0 \\ Y & 1 \end{bmatrix} = \begin{bmatrix} 1 + ZY & Z \\ Y & 1 \end{bmatrix}, \qquad (3.60)$$

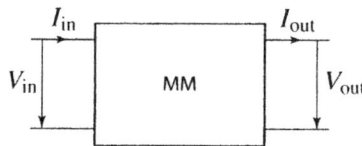

Fig. 3.17. Representation of a two-port network by the $[ABCD]$ matrix.

and takes for the CRLH unit cell of Fig 3.14(a) the specific form

$$
\begin{bmatrix} A & B \\ C & D \end{bmatrix}^{\text{CRLH}}_{\text{asym}} = \begin{bmatrix} 1 - \chi & j\dfrac{(\omega/\omega_{se})^2 - 1}{\omega C_L} \\ j\dfrac{(\omega/\omega_{sh})^2 - 1}{\omega L_L} & 1 \end{bmatrix}, \tag{3.61}
$$

where ω_{se} and ω_{sh} were defined in Eq. (3.54) and

$$
\chi = ZY = \left(\frac{\omega}{\omega_R}\right)^2 + \left(\frac{\omega_L}{\omega}\right)^2 - \kappa\omega_L^2, \quad \text{with} \tag{3.62a}
$$

$$
\omega_R = \frac{1}{\sqrt{L_R C_R}} \quad \text{rad/s}, \tag{3.62b}
$$

$$
\omega_L = \frac{1}{\sqrt{L_L C_L}} \quad \text{rad/s}, \tag{3.62c}
$$

$$
\kappa = L_R C_L + L_L C_R \quad \text{(s/rad)}^2. \tag{3.62d}
$$

The balanced condition for the LC network, as the counterpart of the ideal homogeneous TL expression (3.27), reads

$$
\omega_{se} = \omega_{sh} \quad \text{or} \quad L_R C_L = L_L C_R \quad \text{or} \quad Z_L = Z_R, \tag{3.63}
$$

where this time ω_{se} and ω_{sh} are given by Eqs. (3.54a) and (3.54b), respectively, and $Z_R = \sqrt{L_R/C_R}$, $Z_L = \sqrt{L_L/C_L}$. Following developments similar to those in Eqs. (3.33) and (3.34), we obtain thus $\kappa(\omega_{se} = \omega_{sh}) = 2/(\omega_R \omega_L)$ and

$$
\chi(\omega_{se} = \omega_{sh}) = \left(\frac{\omega}{\omega_R} - \frac{\omega_L}{\omega}\right)^2 \quad \text{(balanced resonances)}. \tag{3.64}
$$

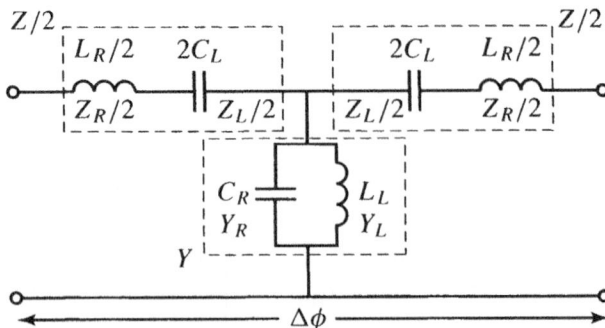

Fig. 3.18. Symmetric unit cell of an LC CRLH TL with identical infinitesimal immittances as the asymmetric unit cell of Fig 3.14(a).

Although the network TL of Fig 3.15, constituted by the repetition of the unit cell of Fig 3.14(a), is equivalent to the homogeneous TL represented in Fig 3.2 in the limit $\Delta\phi \to 0$, in practice $\Delta\phi \neq 0$ and the asymmetry of the unit cell of Fig 3.14(a) will introduce mismatch effects at the connections with external ports, as it will be explained shortly. For this reason, the *symmetric* unit cell shown in Fig 3.18,[21] including the same infinitesimal immittances as those of the symmetric unit cell of Fig 3.14(a), is generally preferable for MTM.

In the case of a symmetric T-network of the type shown in Fig 3.18, the $[ABCD]$ matrix reads

$$
\begin{bmatrix} A & B \\ C & D \end{bmatrix}_{\text{sym}} = \begin{bmatrix} 1 & Z/2 \\ 0 & 1 \end{bmatrix} \begin{bmatrix} 1 & 0 \\ Y & 1 \end{bmatrix} \begin{bmatrix} 1 & Z/2 \\ 0 & 1 \end{bmatrix}
$$

$$
= \begin{bmatrix} 1 + \dfrac{ZY}{2} & Z\left(1 + \dfrac{ZY}{4}\right) \\[2mm] Y & 1 + \dfrac{ZY}{2} \end{bmatrix}, \tag{3.65}
$$

leading to the symmetric CRLH unit cell matrix

$$
\begin{bmatrix} A & B \\ C & D \end{bmatrix}_{\text{sym}}^{\text{CRLH}} = \begin{bmatrix} 1 - \dfrac{\chi}{2} & j\dfrac{(\omega/\omega_{se})^2 - 1}{\omega C_L}\left[1 - \dfrac{\chi}{4}\right] \\[3mm] j\dfrac{(\omega/\omega_{sh})^2 - 1}{\omega L_L} & 1 - \dfrac{\chi}{2} \end{bmatrix}. \tag{3.66}
$$

Fig 3.19 compares asymmetric and symmetric CRLH TL networks. It is obvious that the asymmetric configuration [Fig 3.19(a)] has different input and output impedances ($Z_{\text{in}} \neq Z_{\text{out}}$), whereas these impedances are equal ($Z_{\text{in}} = Z_{\text{out}}$) in the symmetric configuration [Fig 3.19(b)]. As a consequence, the asymmetric network would require different port impedances for matching, which is impractical, and the symmetric network is recommended instead.

Once the matrix $[A_N B_N C_N D_N]$ for an N-cell network TL has been established, the corresponding *scattering parameters* $S_{ij,N}$ for terminations of impedance Z_c can be computed by using the well-known formula [7]

$$
\begin{bmatrix} S_{11,N} & S_{12,N} \\ S_{21,N} & S_{22,N} \end{bmatrix} = \frac{1}{A_N + B_N/Z_c + C_N Z_c + D_N} \cdot
$$

$$
\begin{bmatrix} A_N + B_N/Z_c - C_N Z_c - D_N & 2(A_N D_N - B_N C_N) \\[2mm] 2 & -A_N + B_N/Z_c - C_N Z_c + D_N \end{bmatrix}, \tag{3.67}
$$

---

[21]This is a T-network configuration. It is clear that the equivalent Π-network configuration could also be used instead.

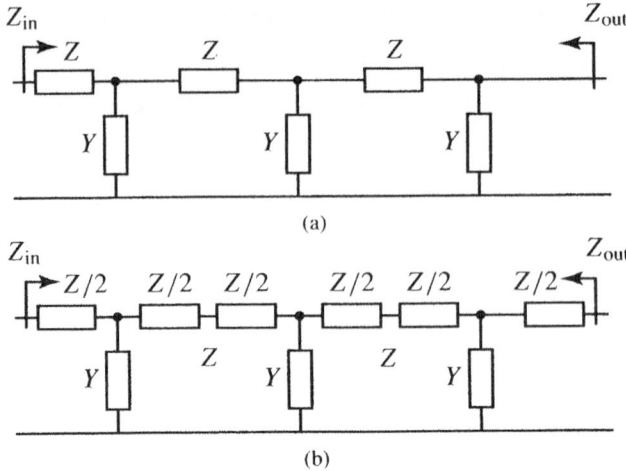

Fig. 3.19. Example of a three-cell network TL. (a) With asymmetric unit cell. (b) With symmetric unit cell.

where $S_{11.N} \neq S_{22.N}$ in the asymmetric configuration of Fig 3.19(a) and $S_{11.N} = S_{22.N}$ in the symmetric configuration of Fig 3.19(b). The propagation constant β and attenuation constant α of the network TL can be computed from the transmission parameter

$$S_{21.N} = |S_{21.N}|e^{j\varphi(S_{21.N})} = e^{-\alpha d}e^{-j\beta\ell}, \tag{3.68}$$

where ℓ represents the total physical length of the TL in a specific implementation, and read

$$\beta = -\varphi^{\text{unwrapped}}(S_{21.N})/\ell + \xi, \quad m \in \mathbb{Z}, \tag{3.69a}$$

$$\alpha = -\ln|S_{21.N}|/\ell. \tag{3.69b}$$

Although the computation of α by Eq. (3.69b) is straightforward, the computation of β comprises two difficulties. First, due to periodicity of the complex exponential function, the phase of $S_{21.N}$, $\varphi(S_{21.N})$, generally appears as a curve varying between $-\pi$ and $+\pi$, whereas the dispersion characteristic $\beta(\omega)$ is a continuous function of frequency. It is therefore necessary to *unwrap* the phase $\varphi(S_{21.N})$ in order to restore the continuous nature of β. Unwrapping means connecting the discontinuous $-\pi$ to $+\pi$ segments into a continuous curve, $\varphi^{\text{unwrapped}}(\omega)$, in frequency. Second, practical network TLs are in effect band-pass filters, as pointed out in Section 3.2.2. Consequently, the *phase origin*, defined as the frequency where the electrical length of the TL zero (or $\theta = |\beta|\ell = \ell/\lambda_g = 0$) has a Modulo 2π ambiguity. The exact shape of the dispersion curve is obtained immediately by phase unwrapping, but the phase offset from the unwrapped curve to the real physical dispersion, $2m\pi$ in Eq. (3) is not known a priori. To illustrate this problematic, we will first consider the conventional PRH case, then the PLH case, and finally we will examine the balanced and unbalanced CRLH cases.

The transmission characteristics of the PRH network TL, described in the first column of Table 3.2 and corresponding to the homogeneous TL described in the first column of Table 3.1, are shown in Fig 3.20. The PRH network TL is a *lowpass* filter with cutoff frequency

$$f_c^{PRH} = \frac{1}{\pi \sqrt{L_R C_R}} = \frac{\omega_R}{\pi} \quad \text{or} \quad \omega_c^{PRH} = 2\omega_R, \tag{3.70}$$

which will be derived in Section 3.2.5. Sufficiently *below* this cutoff, namely when the homogeneity condition $|\Delta\phi_R| < \pi/2$ (where $\Delta\phi_R$ is the electrical

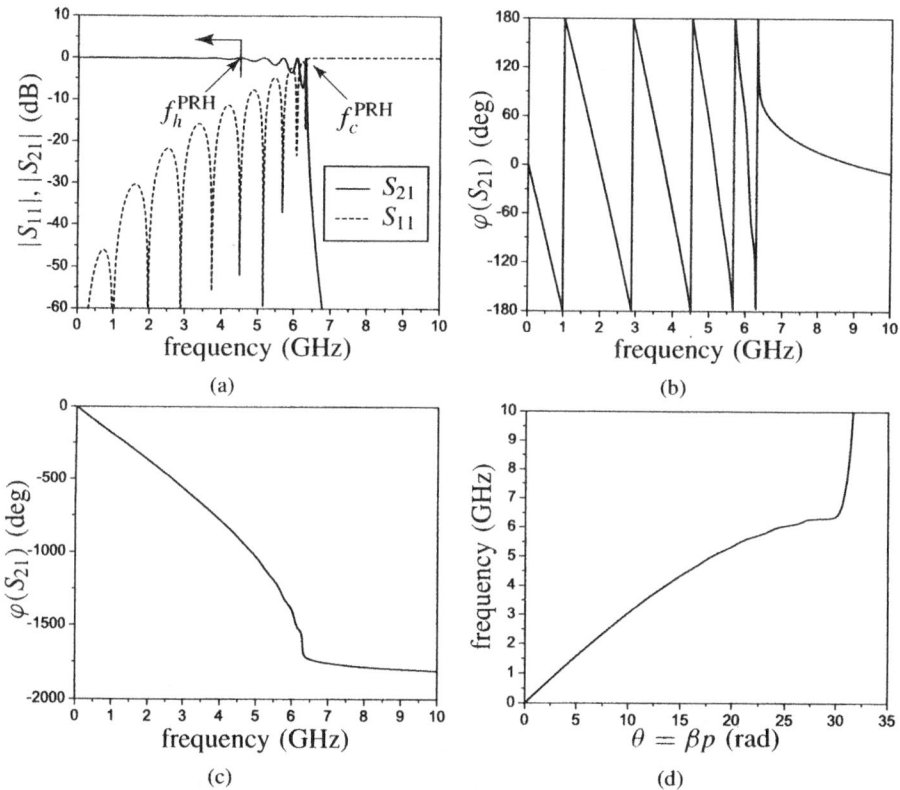

Fig. 3.20. PRH (symmetric unit cell) network TL (lowpass) characteristics computed by (3.67) and (3.66) with $L_L = C_L = 0$ for the parameters $N = 10$, $L_R = 2.5$ nH, $C_R = 1$ pF, $Z_c = 50\ \Omega$ ($Z_R = \sqrt{L_R/C_R} = Z_c$, matched condition). (a) Magnitude of $S_{21,N}$ and $S_{11,N}$. (b) Phase of $S_{21,N}$. (c) Unwrapped phase of $S_{21,N}$ (phase lag). (d) Dispersion relation computed by (3.69a). The -10 dB cutoff frequency is $f_{cR}^{-10dB} = 6.32$ GHz and the homogeneity limit is $f_{Rh} = 4.50$ GHz. The cutoff frequency computed by (3.97b) is $f_{cR} = 6.37$ GHz.

length of the unit cell) is satisfied, the network exhibits a broad passband where it is essentially equivalent to a homogeneous RH TL. The electrical length $\Delta\phi_R$ is obtained from Eq. (3.67) (with $N = 1$) with Eq. (3.66)[22]

$$\Delta\phi_R = \varphi(S_{21,R}) = -\arctan\left\{\frac{\frac{\omega}{2}\left\{\frac{L_R}{Z_{cR}}\left[1 - \frac{1}{4}\left(\frac{\omega}{\omega_R}\right)^2\right] + C_R Z_{cR}\right\}}{1 - \frac{1}{2}\left(\frac{\omega}{\omega_R}\right)^2}\right\}.$$

(3.71)

It can be easily verified that

$$\Delta\phi_R(\omega \ll \omega_R) \approx -\frac{\omega}{2}\left(\frac{L_R}{Z_{cR}} + Z_{cR}C_R\right) = -\omega\sqrt{L_R C_R},$$ (3.72)

where the result of Eq. (3.74) has been used in anticipation to establish the second equality, which is the counterpart of Eq. (3.36). Thus, $\Delta\phi_R(\omega = 0) = 0$, which means that $\omega = 0$ is the *phase origin* of the PRH network. In addition, $\Delta\phi_R(\omega \ll \omega_R) < 0$, which indicates that the PRH network exhibits *phase lag*. When frequency is increased from $\omega = 0$, $\Delta\phi_R$ remains *negative* until it reaches its pole

$$\omega = \omega_h^{PRH} = \sqrt{2}\omega_R.$$ (3.73)

above which it becomes positive. But this pole ω_h^{PRH} precisely corresponds to the limit of the homogeneity condition (hence the subscript h), $|\Delta\phi_R| = \pi/2$, which leads to the statement that the PRH network TL exhibits phase lag over the whole MTM frequency range of interest, which extends from $\omega = 0$ to $\omega = \omega_h^{PRH}$. In this range, the characteristic impedance is simply given by

$$Z_{cR} = \sqrt{\frac{L_R}{C_R}}.$$ (3.74)

Note that at cutoff $\Delta\phi_R(\omega = \omega_c^{PRH} = 2\omega_R) = \arctan(C_R\omega_R Z_c) = \arctan(1) = +\pi/4$, corresponding in fact to a cumulated phase of $-\pi/2 - \pi/4 = -3\pi/4$.

In the PRH network TL, phase unwrapping of S_{21} [Fig 3.20(c)] is straightforward, as we know that the phase origin is zero. The phase of S_{21} has not much significance above cutoff, since it corresponds there only to residual transmitted power, usually below the noise floor of measurement instruments. The *positive linear* dispersion curve $\beta(\omega)$ of the PRH network TL can be observed in Fig 3.20(d). It should also be noted in Fig 3.20(a) that the return loss S_{11} exhibits $N - 1$ reflection peaks, where N is the number of cells constituting the

---

[22]The electrical length $\Delta\phi_R$ appearing in Table 3.2 is for the asymmetric unit cell. It is equal to that of the symmetric cell in the limit $\omega/\omega_R \to 0$.

network. Labeling these peaks by the index k, where $k = 1, 2, \ldots, N - 1$, the kth peak starting from $\omega = 0$ corresponds to a weak resonance of the combination of $k + 1$ cells and is located at the frequency where the total phase shift along the line is $\phi_{Rk} = -90° - (k - 1) \cdot 180°$.[23] The magnitude of these reflection peaks decreases as frequency is decreased because the phase discontinuities are progressively reduced until the condition of perfect homogeneity is reached at $\omega = 0$. At low frequencies, the peaks are equidistant as a consequence of the fact that the unit cell electrical length is proportional to frequency ($\Delta\phi_R \propto \omega$) according to Eq. (3.72).

The transmission characteristics of the PLH network TL, described in the second column of Table 3.2 and corresponding to the homogeneous TL described in the second column of Table 3.1, are shown in Fig 3.21. The PLH network TL is a *highpass* filter with cutoff frequency

$$f_c^{PLH} = \frac{1}{4\pi\sqrt{L_L C_L}} = \frac{\omega_L}{4\pi} \quad \text{or} \quad \omega_c^{PLH} = \omega_L/2, \tag{3.75}$$

which will be derived in Section 3.2.5. Sufficiently *above* this cutoff, namely when the homogeneity condition $|\Delta\phi_L| < \pi/2$ (where $\Delta\phi_L$ is the electrical length of the unit cell) is satisfied, the network exhibits a broad passband where it is essentially equivalent to a homogeneous LH TL. The electrical length $\Delta\phi_L$ is obtained from Eq. (3.67) (with $N = 1$) with (3.66)[24]

$$\Delta\phi_L = \varphi(S_{21,L}) = +\arctan\left\{ \frac{\dfrac{1}{2\omega}\left\{ \dfrac{1}{Z_{cL}C_L}\left[1 - \dfrac{1}{4}\left(\dfrac{\omega_L}{\omega}\right)^2 \right] + \dfrac{Z_{cL}}{L_L} \right\}}{1 - \dfrac{1}{2}\left(\dfrac{\omega_L}{\omega}\right)^2} \right\}, \tag{3.76}$$

It can be easily verified that

$$\Delta\phi_L(\omega \gg \omega_L) \approx \frac{1}{2\omega}\left(\frac{1}{Z_c C_L} + \frac{Z_c}{L_L} \right) = \frac{1}{\omega\sqrt{L_L C_L}}, \tag{3.77}$$

where the result of Eq. (3.79) has been used in anticipation to establish the second equality, which is the counterpart of Eq. (3.37). Thus, $\Delta\phi_L(\omega \to \infty = 0)$, which means that $\omega \to \infty$ is the *phase origin* of the PLH network. In addition, $\Delta\phi_L(\omega \gg \omega_L) > 0$, which indicates that the PLH network exhibits *phase advance*. When frequency is decreased from $\omega \to \infty$, $\Delta\phi_L$ remains *positive* until it reaches its pole

$$\omega = \omega_h^{PLH} = \frac{\omega_L}{\sqrt{2}}, \tag{3.78}$$

---

[23] Alternately, the N zeros of $S_{11,N}$, which are associated with perfect transmission, occur at $\phi_{Rk} = -(k - 1) \cdot 180°$, $k = 1, 2, \ldots, N$.

[24] The unit cell electrical length $\Delta\phi_L$ appearing in Table 3.2 is for the asymmetric unit cell. It is equal to the of the symmetric cell in the limit $\omega/\omega_L \to \infty$.

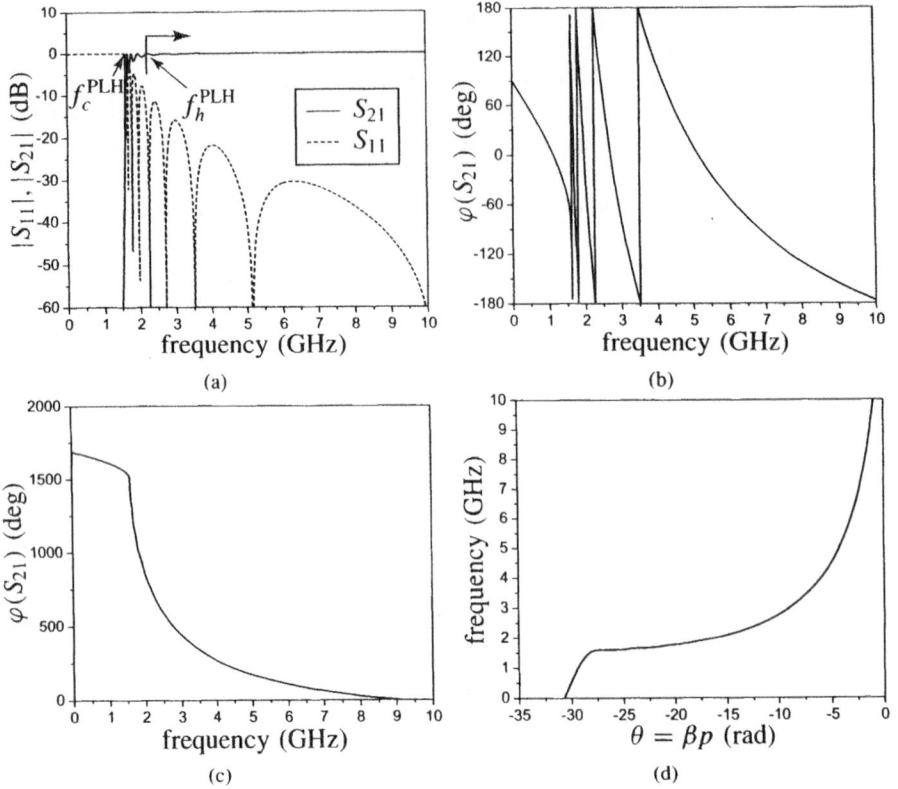

Fig. 3.21. PRH (symmetric unit cell) network TL (highpass) characteristics computed by (3.67) and (3.66) with $L_R = C_R = 0$ for the parameters $N = 10$, $L_L = 2.5$ nH, $C_L = 1$ pF, $Z_c = 50\ \Omega$ ($Z_L = \sqrt{L_L/C_L} = Z_c$, matched condition). (a) Magnitude of $S_{21,N}$ and $S_{11,N}$. (b) Phase of $S_{21,N}$. (c) Unwrapped phase of $S_{21,N}$ (phase advance). (d) Dispersion relation computed by (3.69a). The -10 dB cutoff frequency is $f_{cL}^{-10dB} = 1.60$ GHz, and the homogeneity limit is $f_{Lh} = 2.25$ GHz. The cutoff frequency computed by (3.97a) is $f_{cL} = 1.59$ GHz.

below which it becomes negative. But this pole ω_h^{PLH} precisely corresponds to the limit of the homogeneity condition (hence the subscript h), $|\Delta\phi_L| = \pi/2$, which leads to the statement that the PLH network TL exhibits phase advance over the whole MTM frequency range of interest, which extends from $\omega \to \infty$ to $\omega = \omega_h^{PLH}$. In this range, the characteristic impedance is simply given by

$$Z_c = \sqrt{\frac{L_L}{C_L}}. \tag{3.79}$$

Note that at cutoff $\Delta\phi_L(\omega = \omega_c^{PLH} = \omega_L/2) = -\arctan[Z_c/(L_L\omega_L)] = -\arctan(1) = -\pi/4$, corresponding in fact to a cumulated phase of $+\pi/2 + \pi/4 = +3\pi/4$.

In the PLH network TL, phase unwrapping of S_{21} [Fig 3.21(c)] has to start at $\omega = \infty$, as we know that the phase origin is infinity. The phase of S_{21} has not much signification below cutoff, since it corresponds only to residual transmitted power, usually below the noise floor of measurement instruments. The *negative hyperbolic* dispersion curve $\beta(\omega)$ of the PLH network TL can be observed in Fig 3.21(d). It should also be noted in Fig 3.21(a) that the return loss S_{11} exhibits $N - 1$ reflection peaks, where N is the number of cells constituting the network. Labeling these peaks by the index k, where $k = 1, 2, \ldots, N - 1$, the kth peak starting from $\omega \to \infty$ corresponds to a weak resonance of the combination of $k + 1$ cells and is located at the frequency where the total phase shift along the line is $\phi_{Rk} = +90° + (k - 1) \cdot 180°$.[25] The magnitude of these reflection peaks progressively decreases as frequency is increased because the phase discontinuities are progressively reduced until the condition of perfect homogeneity is reached at $\omega \to \infty$. The distance between the peaks is decreasing as frequency is decreased as a consequence of the fact that the electrical length is inversely proportional to frequency ($\Delta\phi_L \propto 1/\omega$) according to Eq. (3.77).[26] In other words, there is dramatic phase accumulation toward cutoff; this is a useful *slow-wave effect*, possibly associated with very low loss, which can be exploited in several practical applications, such as for instance phase shifters and diplexers.

Figs. 3.22 and 3.23 show the transmission characteristics of the CRLH TL network for the balanced and unbalanced cases, respectively. The CRLH TL network is qualitatively a combination of the PRH and PLH network TLs. It is thus a *band-pass* filter in condition that the LH highpass cutoff frequency f_{cL} is smaller than the RH lowpass cutoff frequency f_{cR}, $f_{cL} < f_{cR}$.[27] It can be easily anticipated from the study of the homogeneous CRLH TL that the CRLH TL network will also exhibit a low-frequency LH range and a high-frequency RH range, without or with interposed gaps between these two range, depending on whether the resonances are balanced or not (Section 3.1).

The exact LH highpass and RH lowpass cutoff frequencies of a general (unbalanced) CRLH TL network have relatively complex forms, which will be derived in Section 3.2.5 and will be given by Eq. (3.94). In the particularly useful balanced case, these cutoff frequencies will be shown to take the simpler forms

$$f_{cL}^{\text{bal}} = f_R \left| 1 - \sqrt{1 + \frac{f_L}{f_R}} \right|, \tag{3.80a}$$

$$f_{cR}^{\text{bal}} = f_R \left(1 + \sqrt{1 + \frac{f_L}{f_R}} \right), \tag{3.80b}$$

[25] Alternately, the N zeros of $S_{11,N}$, which are associated with perfect transmission, occur at $\phi_{Lk} = (k - 1) \cdot 180°$, $k = 1, 2, \ldots, N$.

[26] The peaks would appear equidistant at high frequency if the abscissa would be switched from ω to $1/\omega$.

[27] It will be shown in Section 3.2.5 that these cutoff frequencies are different from the cutoff frequencies f_c^{PRH} of the PRH and f_c^{PRH} of the PLH TLs, but relatively close to them. Another note is that no passband exists if $f_{cL} > f_{cR}$.

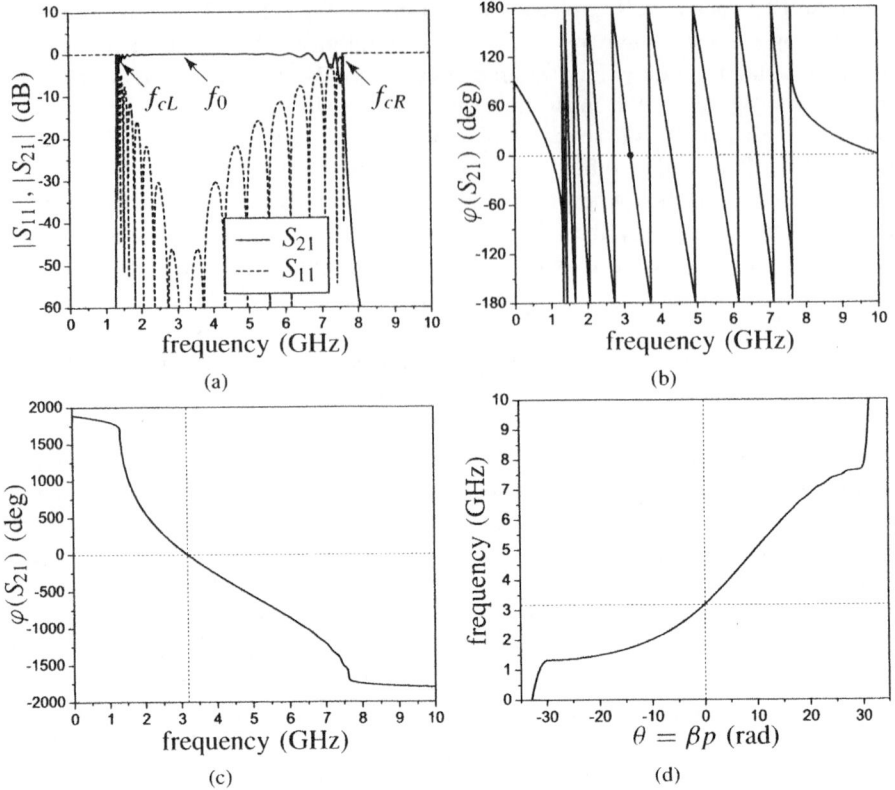

Fig. 3.22. Balanced CRLH (symmetric unit cell) network TL (band-pass) characteristics computed by Eqs. (3.67) and (3.66) for the parameters $N = 10$, $L_R = 2.5$ nH, $C_R = 1$ pF, $L_L = 2.5$ nH, $C_L = 1$ pF, $Z_c = 50$ Ω ($Z_L = \sqrt{L_L/C_L} = Z_R = \sqrt{L_R/C_R} = Z_c$, matched condition). (a) Magnitude of $S_{21.N}$ and $S_{11.N}$. (b) Phase of $S_{21.N}$. (c) Unwrapped phase of $S_{21.N}$. (d) Dispersion relation computed by Eq. (3.69a). The -10 dB cutoff frequencies are $f_{cL}^{-10\text{dB}} = 1.32$ GHz and $f_{cR}^{-10\text{dB}} = 7.64$ GHz, and the transition frequency is $f_0 = 1/(2\pi \sqrt[4]{L_R C_R L_L C_L}) = 3.18$ GHz. The cutoff frequencies computed by Eq. (3.96) are $f_{cL} = 1.32$ GHz and $f_{cR} = 7.68$ GHz.

where $f_R = 1/(2\pi\sqrt{L_R C_R})$ and $f_L = 1/(2\pi\sqrt{L_L C_L})$. The electrical length $\Delta\phi_C$ of the unit cell is obtained from Eq. (3.67) (with $N = 1$) with Eq. (3.66)

$$\Delta\phi_C = \varphi(S_{21.C})$$

$$= -\arctan\left\{ \frac{\dfrac{1}{\omega}\left\{ \dfrac{(\omega/\omega_{se})^2 - 1}{C_L Z_c}\left(1 - \dfrac{\chi}{4}\right) + \dfrac{Z_c}{L_L}\left[(\omega/\omega_{sh})^2 - 1\right]\right\}}{2 - \chi} \right\},$$

$$(3.81)$$

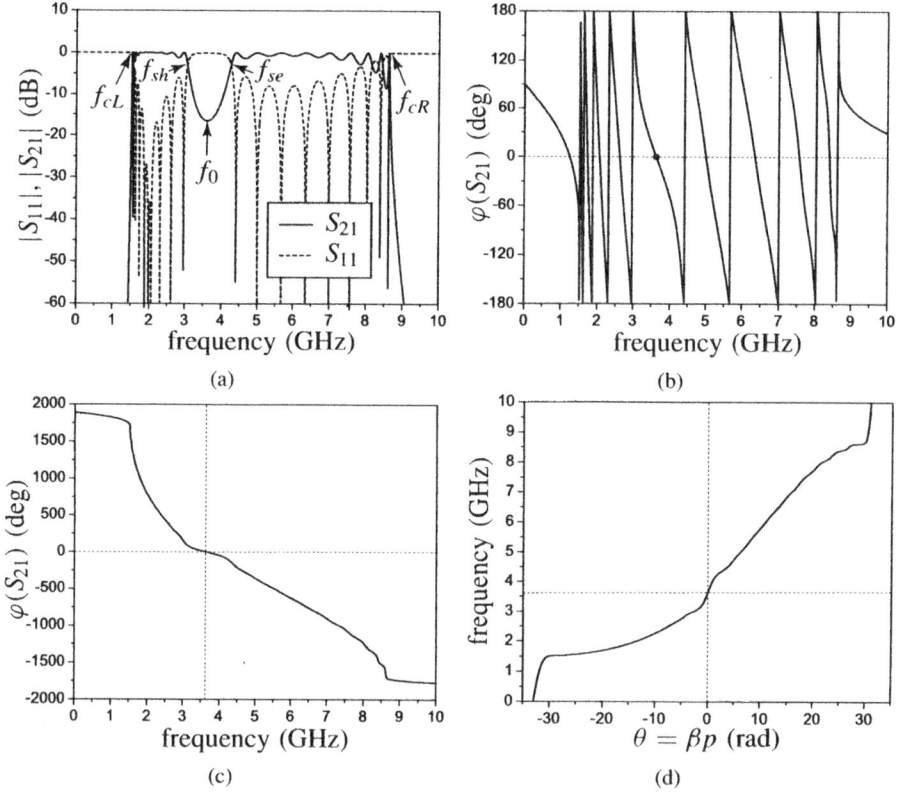

Fig. 3.23. Unbalanced CRLH TL (symmetric unit cell) network (band-pass with gap) characteristics computed by Eqs. (3.67) and (3.66) for the parameters $N = 10$, $L_R = 2$ nH, $C_R = 1$ pF, $L_L = 2.5$ nH, $C_L = 0.75$ pF, $Z_c = 50$ Ω, $Z_R = \sqrt{L_R/C_R} = 44.72,\Omega$ ($Z_L = \sqrt{L_L/C_L} = 57.54$ Ω). (a) Magnitude of $S_{21,N}$ and $S_{11,N}$. (b) Phase of $S_{21,N}$. (c) Unwrapped phase of $S_{21,N}$ ($Z_R \neq Z_c$ and $Z_L \neq Z_c$, mismatched case). (d) Dispersion relation computed by Eq. (3.69a). The frequencies of interest are $f_{cL}^{-10dB} = 1.51$ GHz, $f_{sh}^{-3dB} = 3.06$ GHz, $f_0 = 1/(2\pi \sqrt[4]{L_R C_R L_L C_L}) = 3.62$ GHz, $f_{se}^{-3dB} = 4.27$ GHz and $f_{cR}^{-10dB} = 8.64$ GHz. The cutoff frequencies computed by Eq. (3.94) are $f_{cL} = 1.51$ GHz and $f_{cR} = 8.69$ GHz.

where the different parameters were defined in (3.54) and (3.62). Let us now consider the frequency ω_0

$$\omega_0 = \sqrt{\omega_R \omega_L} = \frac{1}{\sqrt[4]{L_R C_R L_L C_L}} = \sqrt{\omega_{se}\omega_{sh}}, \qquad (3.82)$$

corresponding to the homogeneous TL formulas (3.13) (maximum gap attenuation for unbalanced case) and (3.35) (gap-less transition frequency between the LH and RH ranges for balanced case). In the case of balanced resonances, $\omega_{se} = \omega_{sh}$

(or $L_R C_L = L_L C_R = LC$), we have $\omega_0 = \omega_{se} = \omega_{sh}$. It follows that

$$\Delta\phi_C(\omega = \omega_0) = 0, \tag{3.83}$$

which indicates that $\omega = \omega_0$ is the *phase origin* of the CRLH network, corresponding to the phase origin of the homogeneous CRLH TL described in Section 3.1.3. It can be further established[28] that under the balanced condition

$$\Delta\phi_C(\omega \to \omega_0) \approx -\left[\omega\sqrt{L_R C_R} - \frac{1}{\omega\sqrt{L_L C_L}} \right], \tag{3.84}$$

leading for an N-cell TL to the total phase shift $\phi_C = N \cdot \Delta\phi_C$

$$\phi_C(\omega \to \omega_0) \approx -N\left[\omega\sqrt{L_R C_R} - \frac{1}{\omega\sqrt{L_L C_L}} \right]. \tag{3.85}$$

The matching condition, similar to Eq. (3.32),

$$Z_c = Z_L = Z_R, \quad \text{with} \quad Z_L = \sqrt{\frac{L_L}{C_L}} \text{ and } Z_R = \sqrt{\frac{L_R}{C_R}}, \tag{3.86}$$

where Z_c is the *characteristic impedance* of the CRLH TL network, was assumed in the derivation of this expression, which is the counterpart of the dispersion relation (3.34). As expected, the LC CRLH TL network is equivalent to the homogeneous CRLH TL in the frequency range centered at ω_0 [Eq. (3.84)].

Fig 3.22 illustrates the problematic of matching in a CRLH TL. If the line is balanced, it is intrinsically optimally matched over the whole bandwidth of the structure, as expected from Eq. (3.14) (with $\omega_{se} = \omega_{sh}$) and shown in Fig 3.24(a). If the line is unbalanced [Figs. 3.24(b), 3.24(c) and 3.24(d)], mismatching occurs. In the case of Fig 3.24(b), the ports impedance is equal to the geometric average of the LH and RH impedances, $Z_c = \sqrt{Z_L Z_R}$; consequently, mismatch is relatively comparable in both the LH and the RH bands. In the case of Fig 3.24(c), we have $Z_c = Z_L$ and therefore matching is favored in the LH band, while the RH band is penalized; the converse situation occurs in Fig 3.24(d), where $Z_c = Z_R$.

It can be easily verified that $\Delta\phi_C > 0$ (phase advance) for $\omega < \omega_0$ (LH range) and $\Delta\phi_C < 0$ (phase lag) for $\omega > \omega_0$ (RH range). The frequencies limiting the range of validity of the homogeneity condition $|\Delta\phi_C| < \pi/2$, ω_{hL} and ω_{hR} could be calculated analytically as the root of $2 - \chi$ in Eq. (3.81) but will be determined easier from the dispersion relations to be derived in the Section 3.2.6.

Since the phase origin of the CRLH TL is ω_0, it is obvious that phase unwrapping should be performed around this frequency, indicated by a dot in Fig 3.22(b), as shown in Fig 3.22(c). It is then an easy matter to compute the actual dispersion relation, plotted in Fig 3.22(d), which is identical to that of the homogeneous

---

[28]This requires recognizing that χ ($\omega_{se} = \omega_{sh}$) $<< 1$ in the case $\omega_L \ll \omega \ll \omega_R$, which prevails around $\omega = \omega_0$, according to Eq. (3.64).

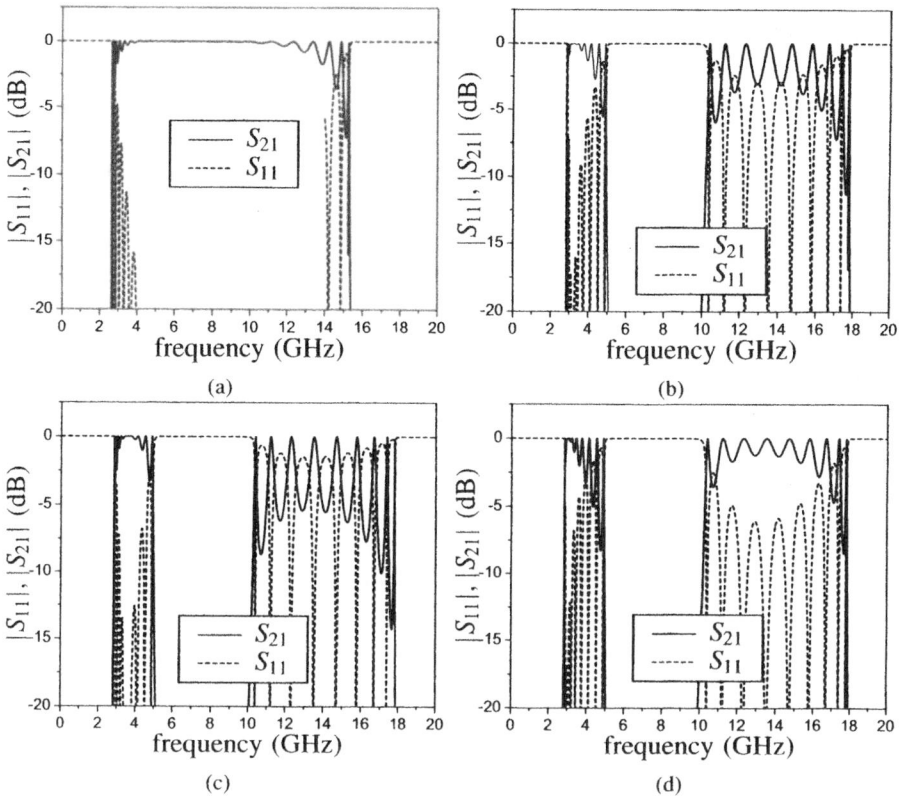

Fig. 3.24. Insertion and return losses in a 10-cell CRLH TL in different matched/mismatched condition. (a) Matched case, with $L_R = 1.25$ nH, $C_R = 0.5$ pF, $L_L = 1.25$ nH, $C_L = 0.5$ pF, i.e., $Z_L = \sqrt{L_L/C_L} = 50$ Ω, $Z_R = \sqrt{L_R/C_R} = 50$ Ω, and ports impedance $Z_c = 50$ Ω. (b) Mismatched case, with $L_R = 0.25$ nH, $C_R = 2.0$ pF, $L_L = 0.5$ nH, $C_L = 1.0$ pF, i.e., $Z_L = \sqrt{L_L/C_L} = 22.4$ Ω, $Z_R = \sqrt{L_R/C_R} = 11.2$ Ω, and ports impedance $Z_{av} = \sqrt{Z_L Z_R} = 15.8$ Ω. (c) Mismatched case with the same TL parameters as in (b) but port impedance $Z_c = Z_L = 22.4$ Ω. (d) Mismatched case with the same TL parameters as in (b) but port impedance $Z_c = Z_R = 11.2$ Ω.

CRLH TL [Fig 3.9(a)] in the vicinity of ω_0 and which will be derived rigorously for all frequencies in Section 3.2.6. The reflection peaks in Fig 3.22(a) follow the same rules as those for the PLH and PRH network TLs except that $\omega \to \infty$ and $\omega = 0$ phase origins are replaced by the phase origin $\omega = \omega_0$.

The characteristics of the unbalanced CRLH TL shown in Fig 3.23 are a priori of lesser interest than those of the balanced CRLH TL. But in practice, the balanced condition $\omega_{se} = \omega_{sh}$ may be sometimes difficult to satisfy exactly, and a small mismatch gap may therefore exist. The width and depth of this gap are proportional to the ratio $\max(\omega_{se})/\min(\omega_{sh})$ and may be consequently acceptable if ω_{se} and ω_{sh} are reasonably close to each other. Moreover, in some applications, only the LH range or only the RH range of the line may be needed.

It will then be appropriate to design the line such that $Z_L = Z_c$ or $Z_R = Z_c$, where Z_c represents the port impedance, to obtain good matching in the band of interest by sacrificing the other unused band. The properties of the homogeneous CRLH TL established in Section 3.1 also apply to the unbalanced CRLH TL network in the vicinity of ω_0 and will be rigorously derived in Section 3.2.6.

The phase origin and frequency evolution of the PLH, PRH, and CRLH TLs are shown in Fig 3.25. The phase origin translations from $\omega = 0$ in the RH TL to $\omega \to \infty$ in the LH TL and to $\omega = \omega_0$ in the CRLH TL are directly related to the conventional conversions from lowpass to highpass and to band-pass filter prototypes [10].

The single-tone time domain analyses of Fig 3.26 demonstrate the positive, negative, and infinite phase velocity v_g in the PRH, PRH, and CRLH (at the

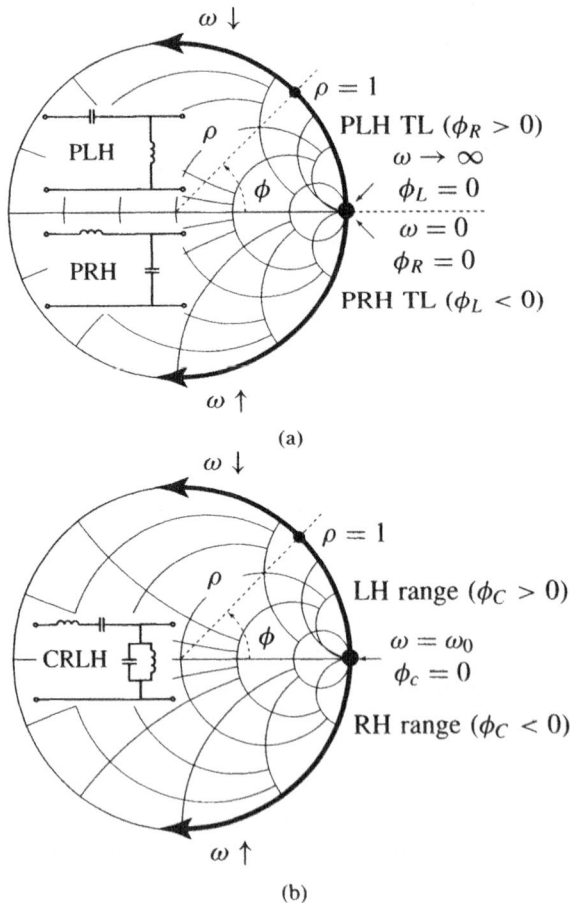

(a)

(b)

Fig. 3.25. Phase origin and phase evolution in frequency ($S_{21}(\omega) = \rho e^{j\phi(\omega)}$) for the different (matched) TLs in the Smith chart. (a) PRH and PLH TLs. (b) CRLH TL.

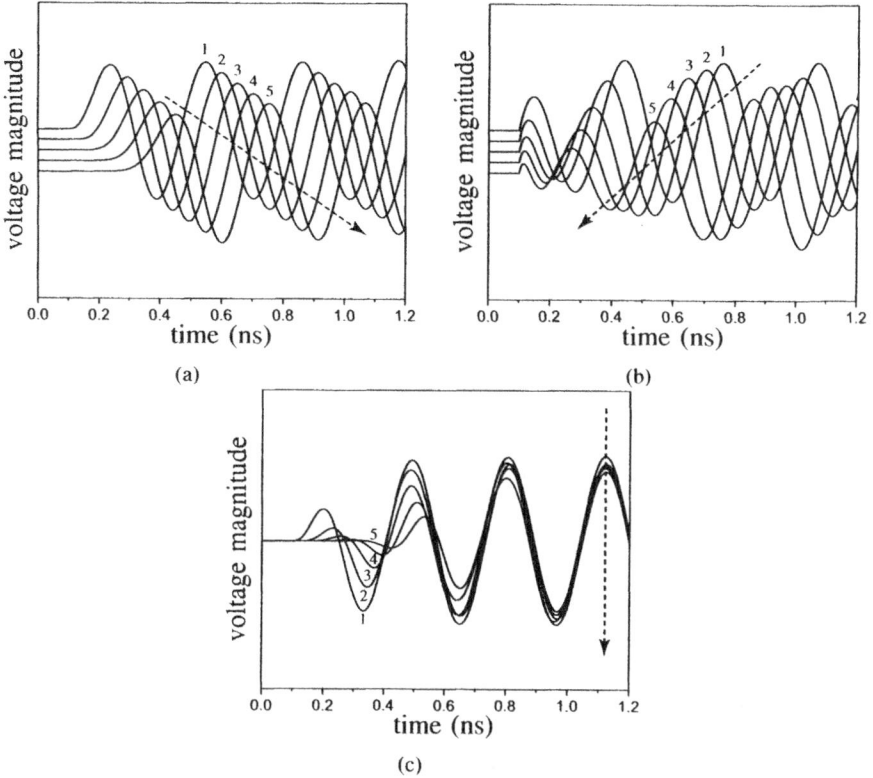

Fig. 3.26. Propagation of a single-tone (or harmonic) wave ($f = 3.81$ GHz) in the time domain. The labels $1, 2, \ldots, 5$ on the curves indicate the number of the node at which the voltage is calculated in increasing order from the input (5 first nodes for each 10-cell TL). (a) PRH TL of Fig 3.20; phase lag ($\Delta\phi_R < 0$), corresponding to positive phase velocity ($v_p^{PRH} > 0$). (b) PLH TL of Fig 3.21; phase advance ($\Delta\phi_L > 0$), corresponding to negative phase velocity ($v_p^{PLH} < 0$). (c) CRLH TL of Fig 3.22; here, because the $f = f_0$ (phase origin of the CRLH TL), phase is stationary ($\Delta\phi_C = 0$), which corresponds to infinite group velocity ($v_p \to \infty$) (but $v_g \neq 0$).

transition frequency) TLs, respectively. The results for the CRLH TL operated in the RH and LH ranges are qualitatively similar to those for the PRH [Fig 3.26(a)] and for the PLH [Fig 3.26(b)], respectively. The stationary phase behavior ($\beta = 0$) observed in the case of the CRLH TL at the transition frequency is not related to a standing wave regime since propagation occurs along the line (Fig 3.22, $v_g \neq 0$), but to the fact that this frequency corresponds to the phase origin ($\phi_C = 0$) of the line.

The two-tone time domain analyses of Fig 3.27 demonstrate the fact that the phase velocity v_p and group velocity v_g (envelope) are parallel and antiparallel in the PRH (or CRLH in the RH range) and in the PLH (or CRLH in the LH range), respectively.

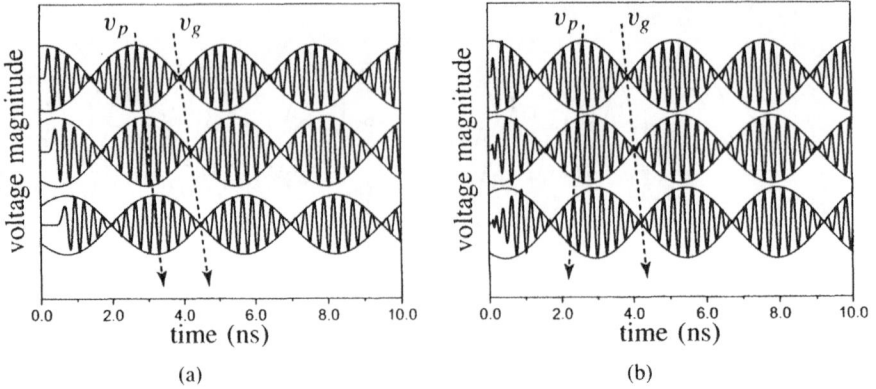

Fig. 3.27. Propagation of a two-tone (bi-harmonic) wave ($f_1 = f_0 - \Delta f$ and $f_2 = f_0 + \Delta f$, with $f_0 = 3.81$ GHz and $\Delta f = 200$ MHz) in the time domain. (a) PRH TL of Fig 3.20; positive group delay ($v_g^{PRH} > 0$) associated with positive phase velocity ($v_p^{PRH} > 0$). (b) PLH TL of Fig 3.21; positive group delay ($v_g^{PLH} > 0$) associated with negative phase velocity ($v_p^{PLH} < 0$).

TABLE 3.3. Random LC Values for the Nonuniform LC Network of Fig 3.28.

k	L_{Rk} (nH)	C_{Rk} (pF)	L_{Lk} (nH)	C_{Lk} (pF)
1	2.1	8.3	3.0	1.2
2	1.8	7.2	3.5	1.4
3	2.3	9.1	2.7	1.1
4	3.8	1.5	1.6	0.7
5	3.1	1.2	2.0	0.8
6	3.4	1.4	1.8	0.7
7	4.9	2.0	1.3	0.5
8	2.8	1.1	2.2	0.9
9	2.2	8.7	2.9	1.2
10	4.3	1.7	1.5	0.6

Finally, Fig 3.28 demonstrates that periodicity in a CRLH is not a necessity, by comparing a periodic or uniform (all cells identical) with a nonperiodic or nonuniform CRLH TL constituted of cells of random LC values (Table 3.3). Very similar transmission characteristics are observed in the nonperiodic TL due to the fact that each unit cell satisfies the homogeneity condition. The agreement is perfect at the phase origin, $f = f_0$, the two TLs are rigorously identical because perfectly homogeneous ($\Delta\phi = 0$).

3.2.4 Input Impedance

Consider the general infinite periodic ladder network transmission line shown Fig 3.29. Because the line is infinite and periodic, the input impedance Z_{in} is the

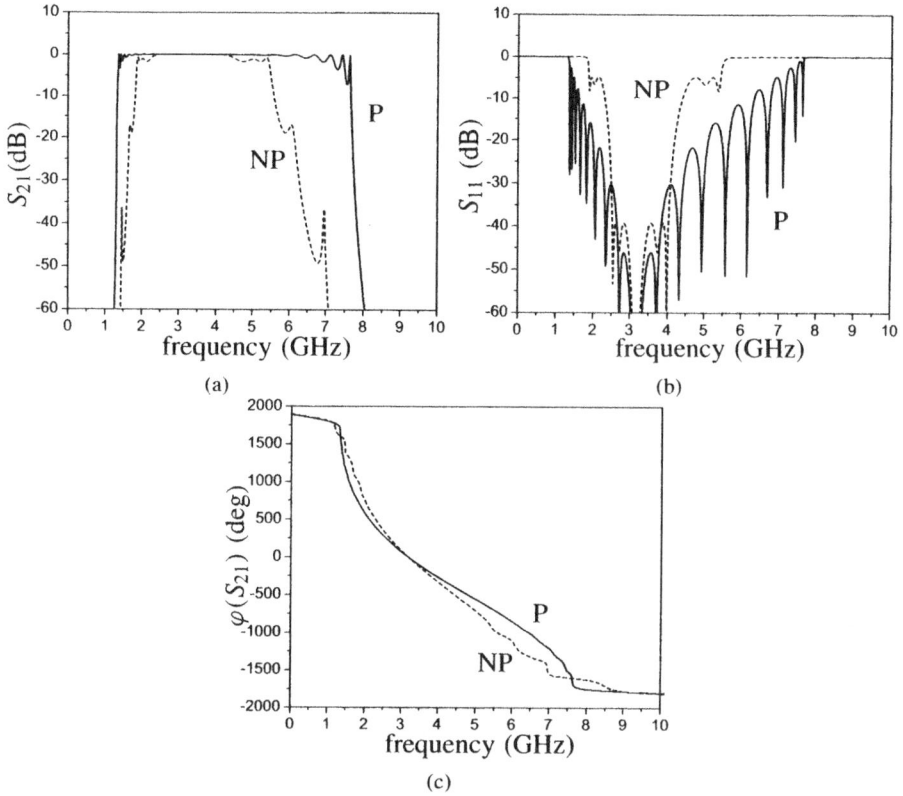

(a)

(b)

(c)

Fig. 3.28. Comparison between a periodic (P) (or uniform) and a nonperiodic (NP) (or nonuniform) CRLH TLs with $N = 10$ unit cells. The P TL has the parameters $L_R = 2.5$ nH, $C_R = 1$ pF, $L_L = 2.5$ nH, $C_L = 1$ pF, corresponding to the transition frequency $f_0 = 3.81$ GHz and characteristic impedance $Z_c = 50$ Ω (same parameters as in Fig 3.22). The NP TL has different LC parameters for each kth unit cell ($N = 1, 2, \ldots, 10$) obtained randomly with the three constraints $\Delta\phi_k(\omega_0) = 0$, $\sqrt{L_{Rk}/C_{Rk}} = Z_c$, and $\sqrt{L_{Lk}/C_{Lk}} = Z_c$; these parameters are given in Table 3.3. (a) Magnitude of S_{21}. (b) Magnitude of S_{11}. (c) Cumulated phase of S_{21}.

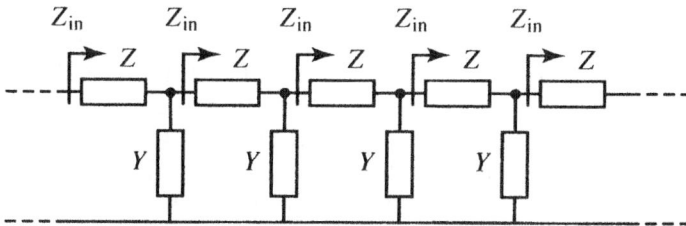

Fig. 3.29. Input impedance of an infinite periodic ladder network transmission line.

same at each node. Therefore,

$$Z_{in} = Z + \left[\left(\frac{1}{Y} \right) \| Z_{in} \right] = Z + \frac{Z_{in}/Y}{1/Y + Z_{in}}, \qquad (3.87)$$

were Z and Y represent arbitrary complex immittances. This yields a quadratic equation in Z_{in} with solution

$$Z_{in} = \frac{Z \pm \sqrt{Z^2 + 4(Z/Y)}}{2} = \frac{Z}{2} \left[1 \pm \sqrt{1 + \frac{4}{ZY}} \right] = R_{in} + j X_{in}. \qquad (3.88)$$

Because MTMs are constituted of electrically small unit cells, the immittances Z and Y of Eq. (3.53) are small and the input impedance (3.88) simplifies to

$$Z_{in} \approx \sqrt{\frac{Z}{Y}}. \qquad (3.89)$$

In the balanced CRLH TL, the RH and LH contributions split into RH and LH series-connected subcircuits, as shown in Fig 3.14(b), with total phase shift $\Delta\phi_C = \Delta\phi_R + \Delta\phi_L$. The unit cell is electrically small if $|\Delta\phi_C|$ is small, $|\Delta\phi_C| \to 0$. This homogeneity condition is automatically satisfied if both the phase lab $\Delta\phi_R = \omega\sqrt{L_R C_R}$ of the RH subcircuit and the phase advance $\Delta\phi_L = -1/\left(\omega\sqrt{L_L C_L}\right)$ of the LH subcircuit are small, $\Delta\phi_R \to 0$ and $\Delta\phi_L \to 0$. These two conditions require that $L_R, C_R \to 0$ and $L_L, C_L \to \infty$, which implies according to Eq. (3.53) that $Z \to 0$ and $Y \to 0$, as expected.[29] Because under the balanced condition we also have $\omega_{se} = 1/\sqrt{L_R C_L} = \omega_{sh} = 1/\sqrt{L_L C_R}$, we obtain from Eq. (3.89) the purely real frequency-independent input impedance

$$Z_{in} \approx \sqrt{\frac{L_R}{C_R}} = Z_R = \sqrt{\frac{L_L}{C_L}} = Z_L. \qquad (3.90)$$

The above rationale only holds in a limited frequency range. If $\omega \to 0$, $\Delta\phi_R \to 0$ and $\Delta\phi_L \to \infty$, and if $\omega \to \infty$, $\Delta\phi_R \to \infty$ and $\Delta\phi_L \to 0$. Therefore, the condition $\Delta\phi_C$ can be satisfied in neither of these two extreme cases. In reality, as frequency decreases or increases, resonances occur at some point, resulting in a passband behavior with low-frequency LH gap and high-frequency RH gap, immediately apparent in the circuit model of Figs. 3.14 and 3.15. The relation of Eq. (3.89) is clearly not valid anymore at these extreme frequencies,

---

[29]It is clear that $|\Delta\phi_C| \to 0$ is also achieved with large equal-magnitude values of $\Delta\phi_R$ and $\Delta\phi_L$, but the physical size of the resulting practical implementation is then of the order of or larger than the guided wavelength and is therefore not a MTM any more. So, this case is of no interest here.

and we must consider the general relation of Eq. (3.88), which becomes with Eq. (3.53) for the CRLH TL

$$Z_{in} = j \left[\frac{(\omega/\omega_{se})^2 - 1}{2\omega C_L} \right] \left\{ 1 \pm \sqrt{1 - \frac{4(\omega/\omega_L)^2}{[(\omega/\omega_{se})^2 - 1][(\omega/\omega_{sh})^2 - 1]}} \right\}, \quad (3.91)$$

simplifying in the balanced case to

$$Z_{in} = j \frac{1}{2\omega C_L} \left\{ [(\omega/\omega_0)^2 - 1] \pm \sqrt{[(\omega/\omega_0)^2 - 1]^2 - 4(\omega/\omega_L)^2} \right\}, \quad (3.92)$$

from which we retrieve the $\Delta\phi_c \to 0$ limit impedance (3.89) at the transition frequency $Z_{in} = Z_R = Z_L$.

3.2.5 Cutoff Frequencies

The cutoff frequencies of an infinite periodic TL can be determined from inspection of the input impedance [Eq. (3.88)]. To support propagation, a TL needs to have an input impedance $Z_{in} = R_{in} + jX_{in}$ with a nonzero real part, $R_{in} \neq 0$, assuming real (nonzero) impedance terminations, such as typically $Z_0 = 50\ \Omega$.[30] The line is then matched if $Z_{in} = Z_0$. In the loss-less case, Z is purely imaginary [e.g., $Z^{PRH} = j\omega L_R$ and $Z^{PLH} = 1/(j\omega C_L)$], and therefore the condition $R_{in} \neq 0$ is satisfied only at frequencies where the radicand of Z_{in} in Eq. (3.88), $R = 1 + 4/(ZY)$ is negative, yielding an imaginary root, which, combined with Z, produces the real part R_{in}. Thus, the range in which R is negative, is a passband and the range in which R is positive is a stopband. The zeros of $R = 0$, corresponding to $ZY = -4$, and the poles of R, corresponding to $ZY = 0$, represent the changes in sign of R, and therefore correspond to the cutoff frequencies. In a CRLH network TL, R in Eq. (3.91) exhibits two roots, which are obtained by solving the biquadratic equation

$$\omega^4 - \left[\omega_{se}^2 + \omega_{sh}^2 + 4 \left(\frac{\omega_0^2}{\omega_L} \right)^2 \right] \omega^2 + \omega_0^4 = 0, \quad (3.93)$$

where we have substituted $(\omega_{se}\omega_{sh})^2 = \omega_0^4$ from Eq. (3.82), yielding the solutions

$$\omega = \omega_{cL} = \omega_0 \sqrt{\frac{[\kappa + (2/\omega_L)^2]\omega_0^2 - \sqrt{[\kappa + (2/\omega_L)^2]^2 \omega_0^4 - 4}}{2}}, \quad (3.94a)$$

---

[30] If $R_{in} = 0$, then $Z_{in} = jX_{in}$ and the reflection coefficient at the port connection is $|\Gamma| = |(jX_{in} - Z_c)/(jX_{in} + Z_c)| = \sqrt{X_{in}^2 + Z_c^2}/\sqrt{X_{in}^2 + Z_c^2} = 1$.

$$\omega = \omega_{cR} = \omega_0 \sqrt{\frac{\left[\kappa + (2/\omega_L)^2\right]\omega_0^2 + \sqrt{\left[\kappa + (2/\omega_L)^2\right]^2 \omega_0^4 - 4}}{2}}, \qquad (3.94b)$$

where κ was defined in (3.62c) as $\kappa = L_R C_L + L_L C_R$. In the unbalanced case, we have in addition in Eq. (3.91) the two poles

$$\omega = \omega_{se}, \qquad (3.95a)$$

$$\omega = \omega_{sh}, \qquad (3.95b)$$

corresponding to the cutoff frequencies of the gap. In the balanced case ($\omega_{se} = \omega_{sh} = \omega_0$), the poles in Eq. (3.91) are canceled by zeros, corresponding to closure of the gap. Moreover, the remaining cutoff frequencies take the simpler form

$$\omega_{cL}^{bal} = \omega_R \left| 1 - \sqrt{1 + \frac{\omega_L}{\omega_R}} \right|, \qquad (3.96a)$$

$$\omega_{cR}^{bal} = \omega_R \left(1 + \sqrt{1 + \frac{\omega_L}{\omega_R}} \right). \qquad (3.96b)$$

In applications requiring only *phase lag/advance compensation* (e.g., dual-band components, Section 5.1), a real CRLH structure is not required. Instead, a RH and quasi-LH lines can be simply cascaded for a net CRLH phase effect, as shown in Fig 3.30. In this case, by superposition, the highpass cutoff is exclusively related to the PLH section and the lowpass cutoff is exclusively related to the PRH section, whereas in a real CRLH TL (Fig 3.15) phase interactions imply that the RH elements and LH elements slightly affect the highpass and lowpass cutoffs, respectively. The cutoff frequencies of the PRH and PLH TLs are easily obtained by repeating the above procedure with the immittances $Z = j\omega L_R, Y = j\omega C_R$ and $Z = -j/(\omega L_L), Y = -j/(\omega C_L)$, respectively. The results are

$$\omega_{cL}^{PLH} = \frac{\omega_L}{2}, \qquad (3.97a)$$

$$\omega_{cR}^{PRH} = 2\omega_R. \qquad (3.97b)$$

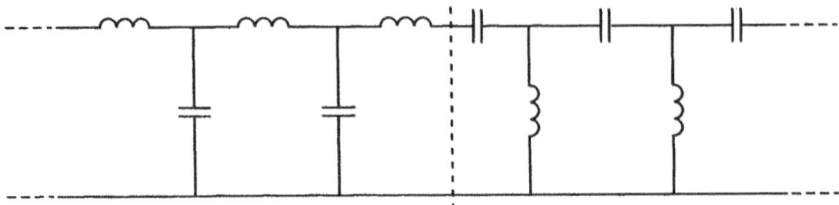

Fig. 3.30. Periodic ladder network implementation of a balanced LC CRLH TL.

In the typical situation where $\omega_L \ll \omega_R$ so that the available bandwidth is sufficient, the square root in the expressions for cutoff of the balanced CRLH TL may be approximated by its first order Taylor expansion $\sqrt{1 + \omega_L/\omega_R} \approx 1 + (\omega_L/\omega_R)/2$, leading to the approximations of (3.96)

$$\omega_{cL}^{\text{bal}}(\omega_L \ll \omega_R) \approx \frac{\omega_L}{2}, \tag{3.98a}$$

$$\omega_{cR}^{\text{bal}}(\omega_L \ll \omega_R) \approx 2\omega_R + \frac{\omega_L}{2} \approx 2\omega_R, \tag{3.98b}$$

which shows that the simple formulas (3.97) provide convenient rule-of-thumb estimation of the usable bandwidth to the engineer, Eq. (3.97a) being more accurate than (3.97b) to which the term $\omega_L/2$ should be added for the same order estimate as that of (3.97a).

In practice, no transmission line or structure is infinite. If the number of sections is small ($N < 2, 3$), cutoffs can even not be defined absolutely because of the weak-slope edges of the bands. However, the formulas derived in this section provide very accurate cutoff frequencies for a larger number of cells ($N > 3$ to 5), where the cutoffs are so sharp that they are unambiguously definable in absolute terms. The evolution of the cutoff as a function of the number of cells is illustrated in Fig 3.31. The excellent accuracy of the infinite-structure approximation for a sufficiently large number of cells can be verified in Figs. 3.20 to 3.23, where the infinite-approximation cutoff frequencies computed by Eqs. (3.94), (3.96), (3.97) are indicated in the captions in comparison with the -10 dB cutoffs obtained for the finite-size structure.

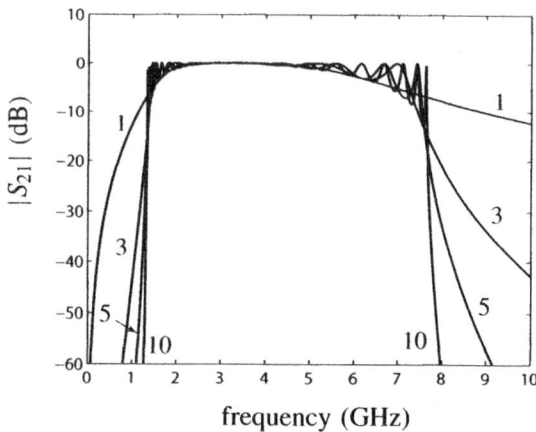

Fig. 3.31. Magnitude of S_{21} in the CRLH TL with the LC parameters of Fig 3.22 for different numbers N of unit cells, $N = 1, 3, 5, 10$.

3.2.6 Analytical Dispersion Relation

Consider an arbitrary *infinite periodic structure*, as depicted in Fig 3.32. The fields of a wave propagating along such a structure at any point z may differ from the fields one period p farther, at the point $z + p$, only by a complex constant $C = e^{-\gamma p}$, where $\gamma = \gamma(\omega) = \alpha(\omega) + j\beta(\omega)$ includes the phase shift $e^{-j\beta(\omega)p}$, related to the propagation constant $\beta(\omega)$, and the attenuation $e^{-\alpha(\omega)p}$, related to the attenuation coefficient $\alpha(\omega)$, which are experienced by the wave along the distance p. This fact can be formulated mathematically as

$$\frac{\psi(z + p)}{\psi(z)} = \frac{\psi(z + 2p)}{\psi(z + p)} = \frac{\psi(z + np)}{\psi[z + (n - 1)p]} = C = e^{-\gamma p}, \forall n. \tag{3.99}$$

Recursively applying this relation from $n = 0$, we obtain the sequence $\psi(z + p) = C\psi(z)$, $\psi(z + 2p) = C\psi(z + p) = C^2\psi(z)$, ..., from which we infer the generic term

$$\psi(z + np) = C^n\psi(z) = e^{-\gamma np}\psi(z) \quad (\forall n). \tag{3.100}$$

This relation represents the *periodic boundary condition* (PBCs) prevailing in a periodic structure. The wave function $\psi(z)$ may be written from Eq. (3.100)

$$\psi(z) = \psi(z + np)e^{+\gamma np}. \tag{3.101}$$

When multiplied by $e^{+\gamma z}$, this expression becomes $\psi(z)e^{+\gamma z} = \psi(z + np) e^{+\gamma(z+np)}$, which indicates that the function defined by

$$\xi_\gamma(z) = \psi(z)e^{+\gamma z} \tag{3.102}$$

is a *periodic function* of period p for any complex propagation constant γ. Consequently, this function can be expanded in Fourier series

$$\xi_\gamma(z) = \sum_{n=-\infty}^{+\infty} \xi_{\gamma n}e^{-j(2n\pi/p)z}, \tag{3.103}$$

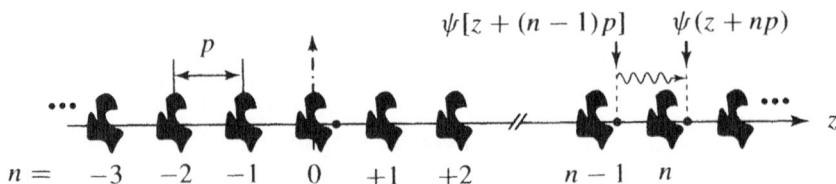

Fig. 3.32. Arbitrary infinite periodic structure.

with the Fourier expansion coefficients

$$\xi_{\gamma n} = \frac{1}{2\pi} \int_{-\pi}^{+\pi} \xi_\gamma(z) e^{+j(2n\pi/p)z} \, dz. \tag{3.104}$$

By inserting the expansion Eq. (3.103) into Eq. (3.102), we finally obtain the following expression for the waveform in a periodic structure with period p

$$\psi_\gamma(z) = e^{-\alpha(\omega)z} \sum_{n=-\infty}^{+\infty} \xi_{\gamma n} e^{-j[\beta(\omega)+2n\pi/p]z} = e^{-\alpha(\omega)z} \sum_{n=-\infty}^{+\infty} \xi_{\gamma n} e^{-j\beta_n z}, \tag{3.105}$$

where

$$\beta_n = \beta(\omega) + \frac{2n\pi}{p}, \tag{3.106}$$

and the subscript γ has been introduced to emphasize the fact that the wave form ψ is associated with the propagation constant $\beta(\omega)$ and the attenuation constant $\alpha(\omega)$, representing the dispersion and attenuation relation, respectively, of the periodic structure. The expression of Eq. (3.105) was obtained by considering only a positive-going wave $C = e^{-\gamma p}$. In general, if reflections are present, both positive-going $C^+ = e^{-\gamma p}$ and negative-going $C^- = e^{+\gamma p}$ waves exist along the line, which yields the more general expression

$$\psi_\gamma(z) = e^{-\alpha(\omega)z} \sum_{n=-\infty}^{+\infty} \xi_{\gamma n}^+ e^{-j\beta_n z} + e^{+\alpha(\omega)z} \sum_{n=-\infty}^{+\infty} \xi_{\gamma n}^- e^{+j\beta_n z}. \tag{3.107}$$

This expression is the mathematical representation of *Bloch-Floquet's theorem*,[31] which states that the wave in a periodic structure consists of the superposition of an infinite number of *plane waves*, called *space harmonics* or Bloch-Floquet's waves. The wave $\beta_0 = \beta(\omega)$ is called the *fundamental*, whereas the waves β_n with $n < 0$ and $n > 0$ are called the *negative* and *positive space harmonics*, respectively.

To illustrate the dispersion characteristics $\beta(\omega)$ of periodic structures, consider the classical problem of a rectangular waveguide periodically loaded with capacitive diaphragms, depicted in Fig 3.33(a). At frequencies below cutoff of the first higher order mode TE_{11} or TM_{11}, where only the dominant mode TE_{10} with propagation constant $\beta_{TE_{10}} = \sqrt{k_0^2 - (\pi/a)^2}$ propagates between the diaphragms[32], the dispersion relation for this structure is given by $\cos(\beta p) =$

---

[31] This theorem was first proved by Floquet in the 1D case and is therefore generally called "Floquet's theorem" in 1D electromagnetic structures. However, in solid-state physics [5], involving electronic waves propagating in 3D crystals, the name "Bloch's theorem" is used most often. To conciliate the two terminologies, we are referring to this theorem as the "Bloch-Floquet's" theorem.

[32] This implies that the spacing p between the diaphragms or period is large enough so that the evanescent modes excited at the discontinuity of each diaphragm have decayed to a negligible value at the positions of the adjacent diaphragms.

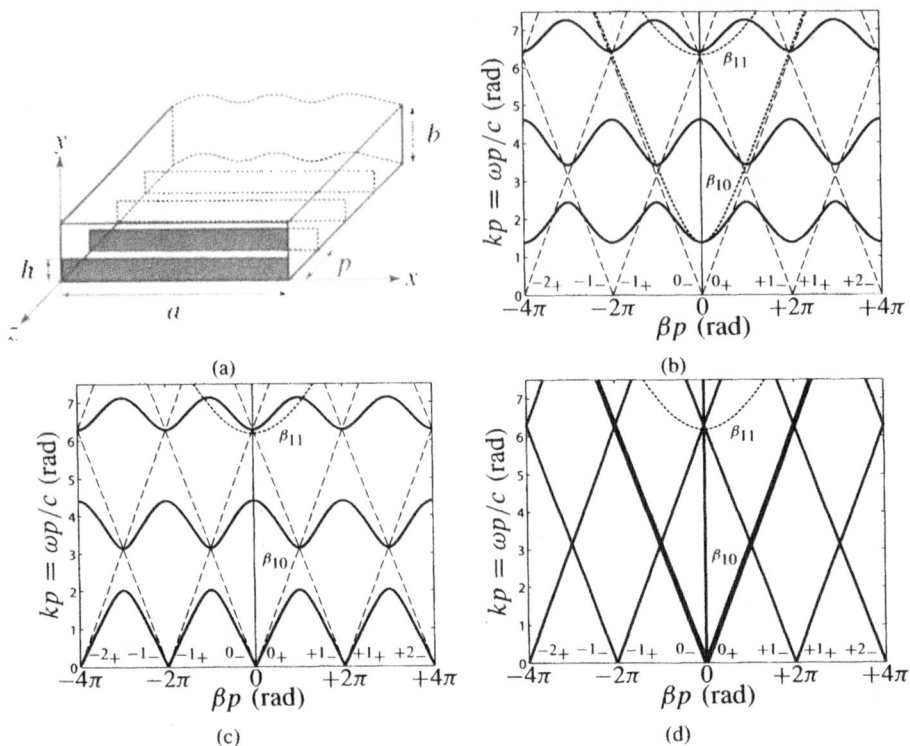

Fig. 3.33. Rectangular waveguide periodically loaded with capacitive diaphragms. (a) Structure. (b)-(c) Dispersion diagrams (solid lines) for different parameters. The short-dashed lines indicate the dominant TE_{10} mode and the TE_{11} or TM_{11} modes of the empty waveguide. The long-dashed lines are related to the space harmonics β_n (3.106), with the subscript $+$ for positive-going waves and the subscript $-$ for negative-going waves. (b) Parameters: $a = 22.86$ mm, $b = 5.08$ mm, $h = 3.81$ mmm, $p = 10$ mm. The cutoffs for TE_{10} and TE, M_{11} are 1.37 rad and 6.34 rad, respectively. (c) Very large waveguide width: $a = 1000$ mm, and other parameters as in (b). The cutoffs for TE_{10} and TE, M_{11} are 0.03 rad and 6.18 rad, respectively. (d) Negligible diaphragm height: $h = 0.001$ mm, and other parameters as in (c).

$\cos(\beta_{TE_{10}}p) - \frac{B}{2}\sin(\beta_{TE_{10}}p)$ [11], where B is the normalized shunt susceptance of the diaphragm given by the low-frequency formula $B = \frac{4b\beta_{TE_{10}}}{\pi}\ln\left[\sec\left(\frac{\pi h}{2b}\right)\right]$ [12]. Dispersion diagrams computed by this formula are plotted in Figs. 3.33(b)– 3.33(d). In conformity with (3.107), these diagrams are *periodic in* βp with period 2π (or periodic in β with period $2\pi/p$), so that the dispersion diagram can be restricted to the *Brillouin zone* (BZ) [5] $\beta p \in [0, 2\pi]$ without any loss of information.

Fig 3.33(b) represents a practical case of a periodically loaded rectangular waveguide, exhibiting a typical structure of alternating passbands and stopbands. The first (lowest) periodic mode shares the cutoff of the TE_{10} mode, $(kp)_c^{TE_{10}} =$

$\pi p/a$. Due to the assumption of mono/dominant-mode propagation, the computed curves are really meaningful only up to cutoff of the first higher-order mode TE_{11} or TM_{11}, $(kp)_c^{TE,M_{11}} = \sqrt{(\pi/a)^2 + (\pi/b)^2}\,p$. In Fig 3.33(c), the resonant dimension a of the waveguide has been increased so much that the structure of Fig 3.33(a) has become essentially equivalent to a periodically-loaded parallel-plate waveguide (PPWG) structure. Consequently, the cutoff $(kp)_c^{TE_{10}} = \pi p/a$ has decreased to zero and the TE_{10} mode has degenerated into the TEM mode $\beta = \pm\omega/c$ of the PPWG structure at low frequencies. In Fig 3.33(d), we have in addition decreased the height h of the diaphragms to such a small value that the structure of Fig 3.33(a) is now essentially equivalent to a simple PPWG structure (with negligible periodic perturbation). As a consequence, the dispersion curves cannot be distinguished any more from the space harmonics curves. In fact, if h is exactly zero (no diaphragms), Bloch-Floquet analysis is artificial and unappropriate, because it yields a *discrete spectrum* of modes, whereas the PPWG structure is perfectly *continuous*. Therefore, the dispersion curves of Fig 3.33(d) reduce exclusively to fundamental the mode $n = 0_{\pm}$ in Eq. (3.106) (thick curves), corresponding to the PPWG TEM mode, $k = \pm\beta = \pm\omega/c$.

The phase and group velocities of the space harmonics [Eq. (3.107)] are straightforwardly obtained from their propagation constants β_n given in Eq. (3.106)

$$v_{pn} = \frac{\omega}{\beta_n} = \frac{\omega}{\beta + 2n\pi/p}, \quad v_{gn} = \frac{1}{d\beta_n/d\omega} = \frac{1}{d\beta/d\omega} = v_{g0}. \quad (3.108)$$

These expressions show that the different space harmonics have different phase velocities v_{pn} but share the same group velocity $v_{gn} = v_{g0}$, $\forall n$ (slope in the $\omega - \beta$ diagram). The phase velocities can be either positive or negative, corresponding to forward and backward waves, respectively, and their magnitudes can range from zero to infinity. The group velocity v_{g0}, corresponding to the energy velocity in a moderately dispersive medium, is always smaller than the velocity of light c.

In 2D or 3D structures, it may be advantageous to use the solid-state physics techniques [5] based on the *direct lattice* and *reciprocal lattice* concepts, the former representing the (infinite) set of vectors defining the spatial domain lattice sites and the latter representing the (infinite) set of vectors defining the spectral domain lattice sites. In this formalism, Eq. (3.107) takes the compact and general (any dimension and any lattice type) form

$$\psi_{\vec{\gamma}}(\vec{r}) = e^{-\vec{\alpha}\cdot\vec{r}} \sum_{\vec{G}} \psi_{\vec{\beta}+\vec{G}}\, e^{j(\vec{\beta}+\vec{G})\cdot\vec{r}}, \quad (3.109)$$

where $\vec{G}$ is the generic vector for the reciprocal lattice and the complex propagation constant is a vector $\vec{\gamma} = \vec{\alpha} + j\vec{\beta}$ [5].

To determine the dispersion relation or dispersion diagram of a periodic LC network TL (such as PRH, PRH or CRLH TLs), we apply PBCs of Eq. (3.100) to the unit cell of the line represented by its $[ABCD]$ matrix, as shown in Fig 3.34. As a consequence, the output current and voltage variables, related to

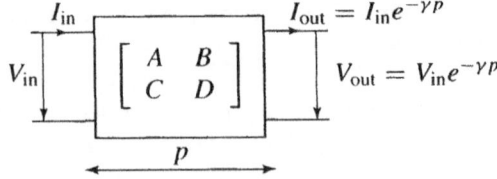

Fig. 3.34. Application of PBCs to a two-port network represented by its unit cell [$ABCD$] matrix to describe an infinite periodic network structure. The physical length of the unit cell or period is p.

the input current and voltage by the propagation term $e^{-\gamma p}$, where $\gamma = \alpha + j\beta$ is the complex propagation constant of the wave traveling along the line, can be eliminated

$$\begin{bmatrix} A & B \\ C & D \end{bmatrix} \begin{bmatrix} V_{\text{in}} \\ I_{\text{in}} \end{bmatrix} = \psi \begin{bmatrix} V_{\text{in}} \\ I_{\text{in}} \end{bmatrix}, \quad \text{with} \quad \psi = e^{+\gamma p}. \qquad (3.110)$$

This is an eigensystem with eigenvalues $\psi_n = e^{+\gamma_n p}$ ($n = 1, 2$), yielding $\gamma_n = \ln[\psi_n(\omega)]/p$. The propagation constant $\beta(\omega)$ and attenuation constant $\alpha(\omega)$ at *each frequency* ω can be computed numerically from these eigenvalues by

$$\alpha_n(\omega) = \text{Re}[\gamma_n(\omega)], \qquad (3.111a)$$

$$\beta_n(\omega) = \pm\text{Im}[\gamma_n(\omega)]. \qquad (3.111b)$$

The dispersion/attenuation diagrams are then be obtained by computing the discrete solutions for β and α over the Brillouin zone [5]. To derive an analytical dispersion relation, we rewrite the eigensystem of Eq. (3.110) in the form of the following homogeneous linear system

$$\begin{bmatrix} A - e^{\gamma p} & B \\ C & D - e^{\gamma p} \end{bmatrix} \begin{bmatrix} V_{\text{in}} \\ I_{\text{in}} \end{bmatrix} = \begin{bmatrix} 0 \\ 0 \end{bmatrix}, \qquad (3.112)$$

which must have a zero determinant to provide a nontrivial solution

$$AD - (A + D)e^{\gamma p} + e^{2\gamma p} - BC = 0. \qquad (3.113)$$

This leads, using either Eq. (3.60) or Eq. (3.65) for the T network unit cell, to the general determinantal equation

$$4\sinh^2\left(\frac{\gamma p}{2}\right) - ZY = 0. \qquad (3.114)$$

The particular relation for the CRLH TL can be obtained by replacing the immittances Z and Y by their CRLH expressions as in Eq. (3.66), noting that $ZY = \chi$

and using the identity $\sinh(x/2) = [\cosh(x) - 1]/2$, which yields

$$\gamma = \frac{1}{p}\cosh^{-1}\left(1 - \frac{\chi}{2}\right), \tag{3.115}$$

where χ was defined in Eq. (3.62a). Eq. (3.115) can be split into the following two equations

$$\alpha = \frac{1}{p}\cosh^{-1}\left(1 - \frac{\chi}{2}\right) \quad \text{if} \quad \chi < 0 \quad \text{(stopband)}, \tag{3.116a}$$

$$\beta = \frac{1}{p}\cos^{-1}\left(1 - \frac{\chi}{2}\right) \quad \text{if} \quad \chi > 0 \quad \text{(passband)}. \tag{3.116b}$$

Eq. (3.115) constitutes the dispersion relation of a CRLH TL, which is solved for the different values of β in the Brillouin zone to yield the eigenfrequencies $\omega_n(\beta)$ ($n = 1, 2$) building up the dispersion diagram. Note that, if the electrical length of the unit cell is small, $|\Delta\phi| = |\beta p| \ll 1$, which is always assumed in MTMs, we obtain $\cos(\beta p) \approx 1 - (\beta p)^2/2$ using Taylor expansion approximation to the second order, and Eq. (3.116b) becomes

$$1 - \frac{(\beta p)^2}{2} \approx 1 - \frac{\chi}{2}, \tag{3.117}$$

or

$$\beta = \frac{s(\omega)}{p}\sqrt{\left(\frac{\omega}{\omega_R}\right)^2 + \left(\frac{\omega_L}{\omega}\right)^2 - \kappa\omega_L^2} = s(\omega)\sqrt{\left(\frac{\omega}{\omega_R'}\right)^2 + \left(\frac{\omega_L'}{\omega}\right)^2 - \kappa'\omega_L'^2}, \tag{3.118}$$

which, if the balanced condition $\omega_{se} = \omega_{sh}$ is satisfied, simplifies to

$$\beta = \frac{1}{p}\left(\frac{\omega}{\omega_R} - \frac{\omega_L}{\omega}\right) = \left(\frac{\omega}{\omega_R'} - \frac{\omega_L'}{\omega}\right). \tag{3.119}$$

In Eq. (3.118), the sign function $s(\omega)$, to be determined by physics considerations, is introduced because of the loss of sign information due to the double-valued square root function. The second equalities in Eqs. (3.118) and (3.119) are obtained by noticing that for $p = \Delta_z \to 0$

$$\omega_R p = \frac{p}{\sqrt{L_R C_R}} = \frac{1}{\sqrt{(L_R/p)(C_R/p)}} = \frac{1}{\sqrt{L_R' C_R'}} = \omega_R', \tag{3.120a}$$

$$\frac{\omega_L}{p} = \frac{1}{p\sqrt{L_R C_R}} = \frac{1}{\sqrt{(L_L p)(C_L p)}} = \frac{1}{\sqrt{L_L' C_L'}} = \omega_L', \tag{3.120b}$$

$$\kappa\left(\frac{\omega_L}{p}\right)^2 = (L_R C_L + L_L C_R)\omega_L'^2 = \left[\frac{L_R}{p} \cdot C_L p + L_L p \cdot \frac{C_R}{p}\right]\omega_L'^2$$

$$= (L_R' C_L' + L_L' C_R')\omega_L'^2 = \kappa'\omega_L'^2. \tag{3.120c}$$

The expressions of Eqs. (3.118) and (3.119) are exactly identical to Eqs. (3.10) and (3.34), respectively, which confirms the fact, already proven in Section 3.2.3, that the CRLH LC network TL is equivalent to the homogeneous CRLH TL for small electrical lengths. The dispersion relation of Eq. (3.115), obtained by solving the eigenvalue problem of Eq. (3.110) for an infinite periodic structure with PBCs, provides results equivalent to those of Eq. (3.69a), obtained from the phase of the transmission parameter S_{21} of a network of finite length when the number of cells N is large enough. This relation is ideal when the LC parameters of the structure under consideration are known, for instance after parameters extraction (Section 3.3.3), because it immediately provides the information on the sign $s(\omega)$ of β and on the transition frequency in the balanced CRLH TL or frequency of maximum gap attenuation in the unbalanced CRLH TL, ω_0. However, because this relation pertains to an infinite structure, it is not appropriate for some practical LC network implementations to include too small of a number of cells. In this case, the relation of Eq. (3.69a) can be used instead, but with the difficulties of ω_0 determination, for phase unwrapping, and of β sign determination, discussed in Section 3.2.3.

The dispersion relation of Eq. (3.115) is plotted in Fig 3.35 and can be compared with the results obtained by phase unwrapping of S_{21} shown in Figs 3.23 and 3.22. The curves for the homogeneous TL are also shown in Fig 3.35 for comparison. The (loss-less) homogeneous CRLH TL exhibits a gap only between $\min(\omega_{se}, \omega_{sh})$ and $\max(\omega_{se}, \omega_{sh})$ in the unbalanced case (Section 3.1), whereas the CRLH TL network always exhibits a low-frequency LH stopband and a high-frequency RH stopband (Section 3.2).

The phase and group velocities of the infinite periodic CRLH network TL derived from Eq. (3.116b) read

$$v_p = \frac{\omega}{\beta} = \frac{\omega p}{\cos^{-1}\left\{1 - \frac{1}{2}\left[(\omega/\omega_R)^2 + (\omega_L/\omega)^2 - \kappa\omega_L^2\right]\right\}} \qquad (3.121)$$

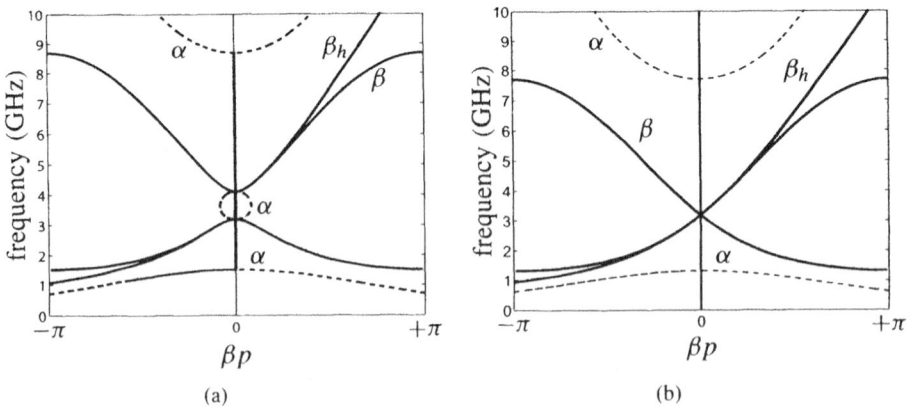

Fig. 3.35. Dispersion/attenuation diagram for the CRLH TL computed by Eq. (3.115). The results for the homogeneous case, β_h, and α_h [see Eqs. (3.10)/(3.34) and Figs. 3.3/3.9(a))] are also shown. (a) Balanced case with parameters of Fig 3.22. (b) Unbalanced case with parameters of Fig 3.23.

and

$$v_g = \left(\frac{d\beta}{d\omega}\right)^{-1} = \frac{p\sin(\beta p)}{(\omega/\omega_R^2) - (\omega_L^2/\omega^3)},$$ (3.122)

and exhibit shapes essentially similar to those shown for the homogeneous case in Figs. 3.5 and 3.9(e) in the vicinity of ω_0. In the balanced case ($\omega_{se} = \omega_{sh}$) at the transition frequency $\omega_0 = \sqrt{\omega_R \omega_L}$, the expression of Eq. (3.122) gives an indetermination of the type 0/0, since $\beta = 0$ and $(\omega_0/\omega_R^2) - (\omega_L^2/\omega_0^3) = 0$. This indetermination can be removed by applying L'Hospital rule

$$v_g(\omega = \omega_0)\Big]_{\omega_{se}=\omega_{sh}} = \frac{p^2\cos(\beta p)(d\omega/d\beta)^{-1}}{(1/\omega_R^2) + 3(\omega_L/\omega_0^{-2})^2}\Bigg]_{\omega_{se}=\omega_{sh}}$$

$$= \frac{\omega_R p}{2} = \frac{v_g^{PRH}(\omega_0)}{2}$$ (3.123)

$$= \frac{\omega_0^2 p}{2\omega_L} = \frac{v_p^{PLH}(\omega_0)}{2},$$

in agreement with the result of Eq. (3.41) obtained for the homogeneous CRLH TL.

Finally, the characteristic impedance can be written, from Eq. (3.14) with $L_R \to L_R' p$, $C_R \to C_R' p$, $L_L \to L_L'/p$, $C_L \to C_L'/p$,

$$Z_c = Z_L \sqrt{\frac{(\omega/\omega_{se})^2 - 1}{(\omega/\omega_{sh})^2 - 1}},$$ (3.124)

with ω_{se}/ω_{sh} defined in Eq. (3.54) and Z_R/Z_L defined in Eqs. (3.16)/(3.15).

3.2.7 Bloch Impedance

A last concept of importance in periodic TL structures is the one of Bloch impedance [7, 8]. We have defined in Eqs. (3.7) and (3.14) the characteristic impedance of an ideal CRLH TL as the square root of the ratio of the immittances of an infinitesimal section Δz of line, $Z_c = \sqrt{Z'/Y'}$. In an LC network implementation of a TL, such a quantity cannot be defined uniquely stricto senso because of the discontinuities of the network. This is obvious by considering that the ratio of voltages and currents in the network of Fig 3.15 vary along the unit cell of Fig 3.14. However, if the network is periodic, this ratio is constant at the terminals of the unit cell at any kth point of the TL. This constant is called the *Bloch impedance* Z_B and is obtained from Eq. (3.112)

$$Z_B = \frac{V_k}{I_k} = \frac{V_{in}}{I_{in}} = -\frac{B}{A - e^{\gamma p}} = -\frac{D - e^{\gamma p}}{C} \quad (\Omega).$$ (3.125)

By solving Eq. (3.113) for $e^{\gamma p}$, with $AD - BC = 1$ assuming a reciprocal unit cell, we get $e^{\gamma p} = \left[(A + D) \pm \sqrt{(A + D)^2 - 4} \right]/2$ and we obtain the explicit form of the Bloch impedance as a function of the $ABCD$ parameters

$$Z_{B\pm} = \frac{-2B}{A - D \mp \sqrt{(A + D)^2 - 4}} = \frac{(A - D) \pm \sqrt{(A + D)^2 - 4}}{2C}, \qquad (3.126)$$

which reduces to

$$Z_{B\pm} = \frac{\pm B}{\sqrt{A^2 - 1}} = \frac{\pm \sqrt{D^2 - 1}}{C}, \qquad (3.127)$$

if the unit cell is symmetric ($A = D$). In these expressions, the $\pm$ solutions correspond to positively and negatively traveling waves, respectively. In practical *truncated periodic network* TLs terminated in a load impedance Z_{load}, matching is naturally achieved by avoiding reflection at the end of the periodic structure, that is by the conjugate-matching condition $Z_B = Z_L^*$, and the reflection coefficient at the load is given by

$$\gamma_L = \frac{Z_{\text{load}} - Z_B}{Z_{\text{load}} + Z_B}. \qquad (3.128)$$

The relation between the Bloch impedance and the characteristic impedance in MTMs can be best understood by considering the periodic $Z/2 - Y - Z/2$ T-network TL with the general $[ABCD]$ matrix of Eq. (3.65) for the unit cell. The Bloch impedance [Eq. (3.127)] then takes the form

$$Z_B = \frac{Z(1 + ZY/4)}{\sqrt{(ZY/2)^2 + ZY}} = \frac{\sqrt{(ZY/2)^2 + ZY}}{Y}. \qquad (3.129)$$

Since, in a MTM, the physical length p of the unit cell is very small, the corresponding admittances Z and Y are very small as well. At the limit, $p = \Delta z \to 0$, we have thus $Z, Y \to 0$ and the Bloch impedance becomes

$$\lim_{p \to 0} Z_B = \lim_{p \to 0} \sqrt{\frac{Z}{Y}} = \lim_{p \to 0} \sqrt{\frac{Z/p}{Y/p}} = \sqrt{\frac{Z'}{Y'}} = Z_c, \qquad (3.130)$$

where Z_c is the characteristic impedance of the ideal TL defined in Eq. (3.7). This indicates that *the Bloch impedance for a periodic network reduces to the characteristic impedance of the homogeneous TL in the homogeneity limit.* In practice, Z_B may slightly differ from Z_c and should be the quantity used for matching, according to Eq. (3.128).

The Bloch impedance [Eq. (3.126)] for the particular case of the symmetric unit cell CRLH network TL of Fig 3.18 becomes with Eq. (3.66)

$$Z_{B_\pm} = \pm \frac{j[(\omega/\omega_{se})^2 - 1](1 - \chi/4)/(\omega C_L)}{\sqrt{\chi(\chi/4 - 1)}}, \tag{3.131}$$

which can also be written with Eq. (4.11) as

$$Z_B = Z_L \sqrt{\frac{(\omega/\omega_{se})^2 - 1}{(\omega/\omega_{sh})^2 - 1} - \left\{ \frac{\omega_L}{2\omega} \left[\left(\frac{\omega}{\omega_{se}} \right)^2 - 1 \right] \right\}^2}, \tag{3.132}$$

where, as in the characteristic impedance, ω_{se} and ω_{sh} [see Eq. (3.17)] correspond to a zero and a pole, respectively. It can be verified that

$$Z_B(\omega = \omega_0) = Z_L \sqrt{\frac{(\omega_0/\omega_{se})^2 - 1}{(\omega_0/\omega_{sh})^2 - 1}}, \quad \text{and} \quad Z_B = Z_c \quad \text{if} \quad \omega_{se} = \omega_{sh}, \tag{3.133}$$

since the electrical length is zero at the transition frequency, and that $Z_B^{\text{PRH}}(\omega = 0) = \sqrt{L_R/C_R} = Z_c$ and $Z_B^{\text{PLH}}(\omega \to \infty) = \sqrt{L_L/C_L} = Z_c$, as expected from the fact that the phase origins of the PRH and PLH TLs are at the origin and at infinity, respectively.

3.2.8 Effect of Finite Size in the Presence of Imperfect Matching

In Sections 3.2.4 to 3.2.7, we treated the CRLH TL as an *infinitely periodic* structure, claiming that PBCs yielded in general accurate results also for finite-size TLs. This was indeed verified for the cutoff frequencies in Figs. 3.20 to 3.23 and for the dispersion diagrams in Fig 3.35. However, if the TL becomes electrically very small, its dispersion characteristics may significantly depart from those of the corresponding infinitely periodic structure if the TL is imperfectly matched to its ports *imperfect matching*.

If the structure is perfectly matched to its terminations, the wave propagating along it does not "see" the terminations and can therefore not distinguish a finite-size structure from an infinite structure. Thus, a finite-size perfectly matched periodic structure exhibits exactly the same propagation characteristics as its infinitely periodic counterpart. In contrast, if there is some amount of mismatch between the TL and the terminations, the resulting standing wave regime affects the dispersion characteristics of the line.

To illustrate this phenomenon, let us consider a periodic TL constituted by the repetition of N identical TL sections perfectly matched to each other, as illustrated in Fig 3.36. Fig 3.37 shows the transmission characteristics of this line in the case of perfect matching [Figs. 3.37(a) and 3.37(b)] and in the case

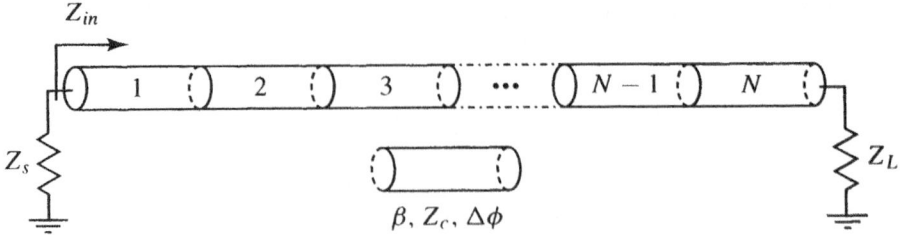

Fig. 3.36. Finite-size TL constituted by the repetition of N TL sections and terminated by a source of impedance Z_s and a load of impedance Z_L. Each section is characterized by a propagation constant β, a characteristic impedance Z_c, and an electrical length $\Delta\phi$.

of strong mismatch [Figs. 3.37(c) and 3.37(d)] for different numbers of sections. For simplicity, we will consider a TL made of PRH sections.[33]

The case of perfect matching is trivial. Because there is no reflection at the terminations, transmission is total ($|S_{21} = 0|$ dB) for any number of sections and the *effective phase* per section, obtained by unwrapping the phase of S_{21} and dividing it by the number of sections N,

$$\Delta\phi_{eff} = \varphi^{\text{unwrapped}}(S_{21})/N, \tag{3.134}$$

is the same for all Ns. In other words, the effective phase per section is equal to the electrical length of the isolated section. Moreover, the phase curve is a straight line for PRH sections since PRH TLs are dispersionless.

It should be noted that relation of Eq. (3.134) is equivalent to Eq. (3.69a) after the LC network has been implemented into a specific technology so that a physical length of the line ℓ can be defined (Section 3.3). Thus, the effective phase $\Delta\phi_{eff}$ is directly related to the effective propagation constant β_{eff} along the line in a real structure.

The case of mismatch is more subtle. In general, the voltage or current along a TL can be expressed as the sum of an incident and a reflected wave [7],

$$\psi(z) = \psi^+ e^{-j\beta z} + \psi^- e^{+j\beta z} = \psi^+ e^{-j\beta z}\left[1 + \Gamma_L^{+2j\beta z}\right] \tag{3.135}$$

where Γ_L is the reflection coefficient at the load

$$\Gamma_L = \frac{\psi^-}{\psi^+}. \tag{3.136}$$

---

[33] In this case, the phase origin is at DC and unwrapping from DC is straightforward. In the case of CRLH sections, phase unwrapping has to be performed around the nonzero phase origin frequency f_0 given by Eq. (3.82), as shown in Eq. (3.69a). Another difference between PRH and CRLH sections is that the dispersion curve is a straight line in the former case while it is a curve the type shown in Fig 3.22(d) in the latter case.

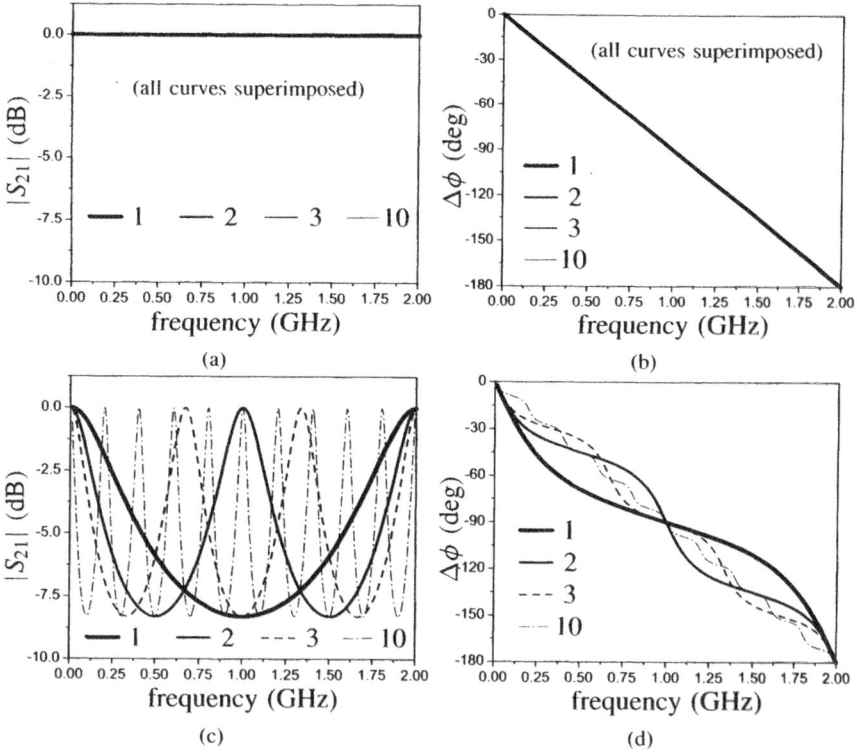

Fig. 3.37. Transmission characteristics of the finite size periodic TL represented in Fig 3.36 for different number of cells $N = 1, 2, 3, 10$, each of which is an ideal PRH TL with electrical length $\Delta\phi = 90°$ (intensionally a large value, at the limit of homogeneity, to emphasize the effect of mismatch) at the frequency $f = 1$ GHz and with characteristic impedance $Z_c = 50$ Ω. (a) Perfectly matched case ($Z_s = Z_L = Z_c$) transmission coefficient $|S_{21}|$. (b) Perfectly matched case effective phase per unit length $\Delta\phi_{eff}$. (c) Strongly mismatched case S_{21} with $Z_s = Z_L = 10$ Ω and $Z_c = 50$ Ω ($\Gamma^{Z_c \to Z_s, Z_l} = -0.67$). (d) Strongly mismatched case (also with $Z_t = 10$ Ω) effective phase per unit length $\Delta\phi_{eff}$.

The first expression in Eq. (3.135) may also be rearranged in terms of the superposition of a traveling wave and a standing wave as

$$\psi(z) = (\psi^+ - \psi^-)e^{-j\beta z} + 2\psi^- \cos(\beta z). \qquad (3.137)$$

At frequencies where the electrical length of the *total* TL is $\Delta\phi_{tot} = m\pi$ ($m \in \mathbb{N}$), the input impedance is equal to the load impedance $Z_{in} = Z_L$, since m complete turns have been accomplished in the Smith chart from the load. Consequently, if the source impedance is equal to the load impedance $Z_s = Z_L$, the TL is perfectly matched; only the traveling wave exists from $\psi^- = 0$ in Eq. (3.137), so that $S_{11} = 0$ and $|S_{21}| = 1$. At frequencies where $\Delta\phi_{tot} = \pi/2 + m\pi$ ($m \in \mathbb{N}$), the impedance is inverted with respect to the load (quarter wavelength transformation), $Z_{in} = Z_c^2/Z_L$, which results in maximum mismatch, given by $S_{11} = (Z_{in} - Z_s)/(Z_{in} + Z_s) = (Z_c^2/Z_L - Z_s)/(Z_c^2/Z_L + Z_s)$, which is equal to S_{22} when $Z_s = Z_L$. Therefore, ψ^- is nonzero and maximum, and the physical wave along the TL is a combination of a traveling wave and a standing wave according to Eq. (3.137). As a consequence, *in the presence of mismatch*, the physical or effective propagation constant along the line, related to Eq. (3.134) by $\beta_{eff} = -\Delta\phi_{eff}/\ell$ for a physical length ℓ, will be different from the propagation constant of the perfectly matched TL. This is clearly apparent in the comparison of Figs. 3.37(b) and 3.37(d). Whereas the effective phase per section of the perfectly matched TL is a unique straight curve, the effective phase per section of the mismatched TL exhibits ripples, with a period following the period of the magnitude oscillations and varying with the number of sections constituting the TL.[34] However, the *average slope* is identical to that of the perfectly matched TL because the mismatched TL phase is necessarily identical to that of the matched TL at the frequencies of perfect transmission. As N is increased, the electrical length of the TL in a given frequency range of interest (e.g., DC to 2.0 GHz in Fig 3.37) is increased, which results in increase in the period and decrease in the magnitude of the phase oscillations. Therefore, the effective propagation constant (and effective phase) of a mismatched TL tends to the same straight phase curve as that of the matched TL the number of sections tends to infinity, $\lim_{N\to\infty} \beta_{eff}(N) = \beta, \forall N$.

The exact relation between β_{eff} and β of the isolated sections can be derived by writing the wave form as

$$\psi(z) = \psi^+ e^{-\gamma_{eff} z}, \quad \text{with} \quad \gamma_{eff} = \alpha_{eff} + j\beta_{eff}, \tag{3.138}$$

and equating this expression to the second expression in Eq. (3.135). This yields, by taking the magnitude and the phase of the resulting equation, the effective (distorted) attenuation and propagation coefficients for a TL of length ℓ as

$$\alpha_{eff} = \frac{\ln\left|1 + \Gamma_L e^{+2j\beta d}\right|}{d}, \tag{3.139a}$$

$$\beta_{eff} = \beta + \frac{\angle\left(1 + \Gamma_L e^{+2j\beta d}\right)}{d}. \tag{3.139b}$$

which are represented in Fig 3.38. The oscillations in the phase response can be related to the space harmonics [Eq. (3.106)] of the infinitely periodic structure. As $N \to \infty$, all the space harmonics (infinite number) contribute in the Bloch-Floquet expansion [Eq. (3.105)], and this results in a periodic-structure continuous (nonwiggly) dispersion curve, as typically obtained by application of PBCs.

---

[34]These ripples correspond physically to periodic increase and decrease of the *effective phase velocity* in frequency. When animated in the time domain along the line axis, this fact results in a kind of "hiccup" effect in the wave motion.

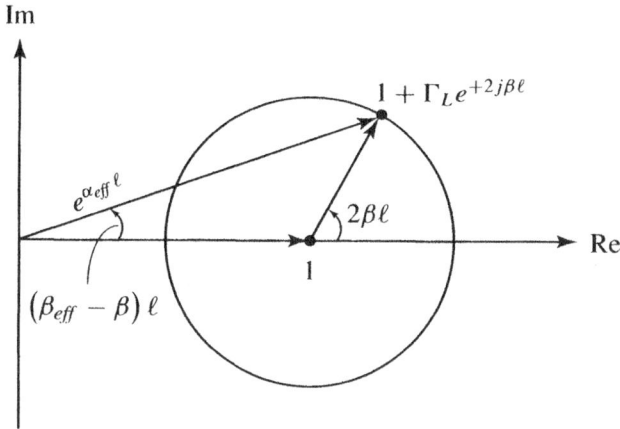

Fig. 3.38. Representation of the effective attenuation constant α_{eff} and propagation constant β_{eff} given by Eq. (3.139) as a function of the isolated line sections propagation constant β.

It should be noted that the situation shown in Figs. 3.37(c) and 3.37(d) corresponds to strongly exaggerated mismatch to emphasize the effect. In practice, excellent matching can be obtained in balanced CRLH structures by satisfying the matching condition of Eq. (3.86), and the ripples observed in Fig 3.37(c) are so negligibly small that they cannot even be observed for typical number of unit cells ($N > 10$) (e.g., Fig 3.44). An example of a practical situation where the effects of finite size MTMs have to be understood and taken into account is the design of low-directivity leaky-wave antennas (Chapter 6). If a relatively fat beam is required, the aperture of the radiator needs to be small and therefore the length of the leaky-wave structure is small. In this case, the radiation angle predicted from the propagation constant by application of PBCs may be significantly different from the real radiation angle, especially if mismatch is significant, because the effective propagation constant is different from the infinitely periodic structure propagation constant due to the contribution of standing waves. It is then preferable to use Eq. (3.134), after full-wave determination of the scattering parameters of the unit cell, rather than resorting to PBCs.

3.3 REAL DISTRIBUTED 1D CRLH STRUCTURES

Let us first recall that, in most applications, the balanced ($\omega_{se} = \omega_{sh}$) CRLH design is preferable to the unbalanced ($\omega_{se} \neq \omega_{sh}$) CRLH design because there is no gap at the transition between the LH and RH ranges. Therefore, we will first consider the balanced case here. However, imperfections in the actual design may introduce some amount of unbalance, in which case the formulas derived for the unbalanced CRLH TL can be used.

3.3.1 General Design Guidelines

Consider the periodic CRLH LC network TL of Fig 3.15 with the unit cell of Fig 3.14.[35] Typical guidelines for the design of a real CRLH TL structure may be the following:

- First select an appropriate transition frequency, as given by Eq. (3.82)

$$\omega_0 = \frac{1}{\sqrt[4]{L_R C_R L_L C_L}} = \sqrt{\omega_R \omega_L} = \sqrt{\omega_{se} \omega_{sh}}, \qquad (3.140)$$

where we defined

$$\omega_R = \frac{1}{\sqrt{L_R C_R}}, \; \omega_L = \frac{1}{\sqrt{L_L C_L}}, \; \omega_{se} = \frac{1}{\sqrt{L_R C_L}}, \; \omega_{sh} = \frac{1}{\sqrt{L_L C_R}}. \qquad (3.141)$$

The transition frequency ω_0 will play various roles depending on the application, but it generally represents the center of the operational bandwidth.

- Apply the matching condition of Eq. (3.86) to ports of impedance Z_0

$$Z_R = \sqrt{\frac{L_R}{C_R}} = Z_0, \qquad (3.142a)$$

$$Z_L = \sqrt{\frac{L_L}{C_L}} = Z_0. \qquad (3.142b)$$

- Eqs. (3.140) and (3.142) represent three equations in the four unknowns L_R, C_R, L_L, and C_L. Thus, there is one remaining degree of freedom for one of these four variables, which may be exploited to meet some technological constraint or application requirement, such as bandwidth. The bandwidth was shown to extend from the highpass LH cutoff frequency ω_{cL} to the lowpass RH cutoff frequency ω_{cR} given by Eq. (3.96)

$$\omega_{cL} = \omega_R \left| 1 - \sqrt{1 + \frac{\omega_L}{\omega_R}} \right|, \qquad (3.143a)$$

$$\omega_{cR} = \omega_R \left(1 + \sqrt{1 + \frac{\omega_L}{\omega_R}} \right), \qquad (3.143b)$$

in the balanced case and by the more complicated formulas of Eq. (3.94) in the unbalanced case. Note that these cutoffs do not depend on the number of

---

[35]This design procedure to be presented here will be extendable to structures with more complicated unit cells, such tunable structures, including varactors with biasing networks (Section 6.4.3).

cells but only on the value of the LC parameters. The *fractional bandwidth* may thus be written as

$$FBW = 2\frac{(\omega_{cR} - \omega_{cL})}{(\omega_{cL} + \omega_{cR})}, \tag{3.144}$$

which may constitute the fourth equation to determine L_R, C_R, L_L, and C_L.

- Choose an appropriate number of cells N for the specific application. For instance, in leaky-wave (LW) antenna, directivity depends on the radiation aperture and hence on the total length of the line $\ell = N \cdot p$, where p is the length of the unit cell, so that N will have to be properly adjusted for the desired directivity (Chapter 6).

- At this step, we have determined the set of parameters $\{L_R, C_R, L_L, C_L\}$ and the number of cells N. The network model of Figs. 3.15 and 3.14, yielding transmission characteristics of the type shown in Figs. 3.22 or 3.23, are therefore readily available to evaluate in a fast and efficient manner the potential performances of the future device, as will be largely illustrated in Chapters. 5 and 6.

- The network can now be implemented practically by generating the inductances L_R, L_L and capacitances C_R, C_L with real inductors and capacitors. Two main options are available for these reactive elements: *surface mount technology (SMT)* chip components or *distributed*[36] components. The SMT chip have important limitations for MTMs: 1) they are limited to low frequencies (typically $3 - 6$ GHz, depending on their values) due to their self-resonance; 2) they are available only in discrete values provided by manufacturers; 3) they cannot be implemented in MIC/MMIC-compatible technologies; 4) their EM characteristics are difficult to control; 5) they cannot be conveniently used for radiation applications; 6) they lead to circuit-type structures while we eventually want to create materials, which should be manufactured by semiconductor or nanotechnology processes in the future. For these reasons, distributed components are generally preferable. A *technology* has to be selected for the implementation of the line. This technology can be microstrip, coplanar waveguide, stripline, or any other type.

- Once a technology has been chosen, we have to decide how to fabricate the *inductors and capacitors*. For instance in microstrip technology, inductors can be implemented in the form of spiral inductor strips or simply stub strips, while capacitors may be implemented in interdigital or metal-insulator-metal (MIM) structures. Unfortunately, only very inaccurate synthesis formulas are available for distributed inductors and capacitors, and we will therefore

---

[36]Here "distributed" refers to components that are metal package-less "patterned" (e.g., interdigital capacitors and spiral inductors), as opposed to discrete (commercial) commercial chip (SMT) components. However, we must be very careful to understand that these distributed components are "electromagnetically discrete" in the sense that their size is smaller than $\lambda_g/4$.

not describe them in detail. A comprehensive text on these components is available in [13]. Eventually, an accurate design requires full-wave simulations. LC parameters are then extracted, following the procedure that will be presented in Section 3.3.3, and inserted into the LC network model. The design procedure may have to be reiterated a few times to produce and optimal design. Although synthesis only by iterative-analysis is available, a good extraction model provides the tools for accurate and efficient design.

- The characteristic parameters of the effectively homogeneous CRLH network TL are finally determined by the relations

$$ L'_R = \frac{L_R}{p}, \quad C'_R = \frac{C_R}{p}, \quad L'_L = L_L \cdot p, \quad C'_L = C_L \cdot p, \quad (3.145) $$

where p still represents the physical length of the unit cell. For instance, the C'_L of a MIM capacitor will be typically *smaller* than that of an interdigital capacitor, because a given capacitance (e.g., 1 pF) can be obtained with a smaller length in an MIM structure (with small interspacing and high permittivity) than in an interdigital structure where only weak edge coupling is present. Once the characteristic parameters have been found, the approximate homogeneous TL analysis of Section 3.1 can be used if necessary.

3.3.2 Microstrip Implementation

Fig 3.39 depicts a typical *microstrip CRLH TL* constituted by *interdigital capacitors* and *stub inductors* shorted to the ground plane by a *via*.[37] This structure is the first distributed TL MTM structure. It was introduced by Caloz et al. in 2002 [14, 15] and subsequently used in various applications [16, 17]. The unit cell, centered in the plane defined by the axis of the stub, represents a T network constituted by two impedance branches with capacitance $2C_L$ and inductance $L_R/2$[38] and by an admittance branch with inductance L_L and capacitance C_R. This T-network is modeled by the equivalent circuit of Fig 3.18. The contributions C_L and L_L are provided by the interdigital capacitors and stub inductors whereas the contributions C_R and L_R come from their parasitic reactances (see Section 3.3.3) and have an increasing effect with increasing frequency. The parasitic inductance L_R is due to the magnetic flux generated by the currents flowing along the digits of the capacitor and the parasitic capacitance C_R is due to the parallel-plate voltage gradients existing between the trace and the ground plane.

Many other lumped (using SMT chip components) or distributed implementations (stripline, CPW, etc.) are possible for CRLH TLs. For instance, in

---

[37] The interdigital capacitor could be replaced by an MIM capacitor and the stub inductor could be replaced by a spiral inductor for more compact unit cell footprint.

[38] In fact, there is simply a capacitance of value C_L and an inductance of value L_R between each stub, corresponding in effect to the series connection of two capacitors of value $2C_L$ and two inductors of value $L_R/2$, except at each end of the line (last cell), where actual capacitors of value $2C_L$ with inductance $L_R/2$ are used.

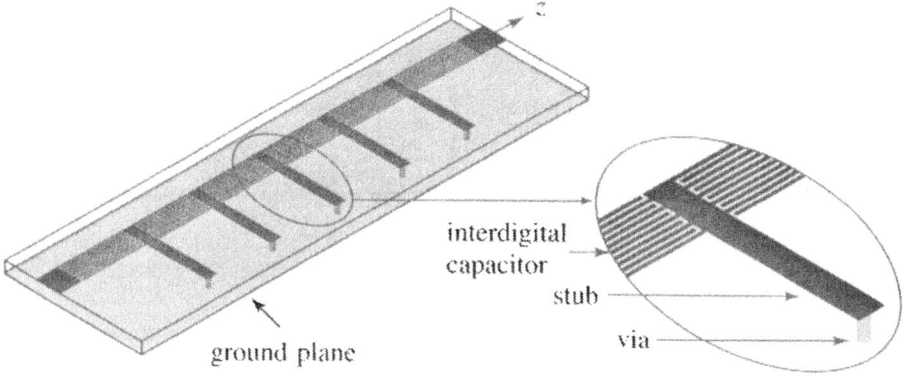

Fig. 3.39. Microstrip CRLH TL using interdigital capacitors and shorted stub inductors [14, 15].

Sections 5.1 and 5.2, applications using lumped-element implementations are presented, whereas Section 5.3 describes a multilayer structure with propagation perpendicular to the plane of the substrate.

It should be noted that the structure of Fig 3.39 is *open* to air. Consequently, because the CRLH dispersion curves (e.g., Fig 3.35) penetrate into the *fast-wave region* of the dispersion diagram, defined by $v_p = \omega/\beta > \omega/k_0 = c$, the line is potentially radiative, a characteristic that will exploited in leaky-wave (LW) antenna applications, to be presented in Chapter 6. In the case of guided-wave applications, radiation should be minimized by mismatching the line to the air impedance ($Z_c^{air} = \sqrt{\mu_0/\varepsilon_0} = 376.7\ \Omega$) as much as possible.

First approximation formulas for the inductance L_L and capacitance C_L may be obtained as follows. The shorted stub corresponds to a shorted TL, with input impedance $Z_{in}^{si} = j Z_c^{si} \tan(\beta^{si}\ell^{si})$, where Z_c^{si}, β^{si}, and ℓ^{si} represent the characteristic impedance, the propagation constant, and the length of stub, respectively. Equating this expression with the ideal impedance of an inductor, $j\omega L_L$, we obtain the following low-frequency approximation for the inductance of the stub

$$L_L \approx \frac{Z_c^{si}}{\omega} \tan(\beta^{si}\ell^{si}), \qquad (3.146)$$

which is seen to be frequency dependent.[39] One (among other) rough approximation for the capacitance of the interdigital capacitor is the empirical formula

$$C_L \approx (\varepsilon_r + 1)\ell^{ic}[(N - 3)A_1 + A_2] \quad \text{(pF)}, \qquad (3.147)$$

---

[39]It is obvious that if the electrical length of the shorted stub is longer than $\pi/2$, $|\Delta\phi^{si}| = \beta^{si}\ell > \pi/2$, the reactance of the stub becomes negative and therefore capacitive. In that case, an open stub, with impedance $Z_{in}^{os} = -j Z_c^{os}\cot(\beta^{os}\ell)$, must be considered instead, yielding $L_L = -Z_c^{os}\cot(\beta^{os}\ell)/\omega$, where $\cot(\beta^{os}\ell) < 0$. Using an open stub represents the disadvantage of requiring a longer strip (but this strip may be meandered to minimize length) and a virtual large-value ground capacitor (Section 6.6). The advantage of the open stub is that it does not require a via conductor.

with

$$A_1 = 4.409 \tanh \left[0.55 \left(\frac{h}{w^{\text{ic}}} \right)^{0.45} \right] \cdot 10^{-6} \quad \text{(pF/}\mu\text{m)}, \qquad (3.148a)$$

$$A_2 = 9.92 \tanh \left[0.52 \left(\frac{h}{w^{\text{ic}}} \right)^{0.5} \right] \cdot 10^{-6} \quad \text{(pF/}\mu\text{m)}, \qquad (3.148b)$$

where ℓ^{ic}, w^{ic}, and h represent the length of the capacitor, the overall width of its finger, and the height of the substrate, respectively.

The approximate formulas of Eqs. (3.146) and (3.147) may be used only as a starting guess for design.[40] In the next section, we will present a more accurate approach based on parameters extraction from full-wave simulation or measurement.

3.3.3 Parameters Extraction

For the purpose of extraction of the parameters L_R, C_R, L_L, and C_L in the CRLH implementation of Fig 3.39, we consider the unit cell shown in Fig 3.40.

The equivalent circuit of this unit cell, constituted by the series connection of the interdigital capacitor and the shorted stub inductor, is shown in Fig 3.41(a), whereas Fig 3.41(b) shows an auxiliary $T - \Pi$ network that will be used for extraction.

First, the scattering parameters of the interdigital capacitor and of the stub inductor, taken separately, are determined by either full-wave simulation or measurement. For this purpose, a short section of microstrip TL has to be added at each end of the component to ensure extinction of the higher-order modes generated by the coaxial connector-to-microstrip transition discontinuity. Because the most important properties of MTMs are related to *phase*, it is essentially to

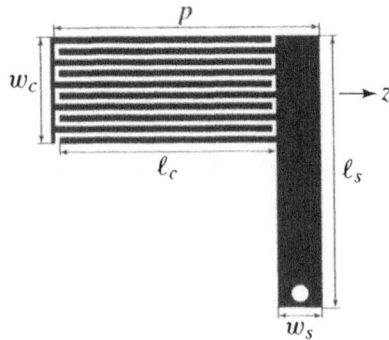

Fig. 3.40. Unit cell of the microstrip CRLH TL of Fig 3.39 for parameters extraction.

[40]For further information on detailed modeling of quasi-lumped components, the reader is referred to textbooks such as [13].

interdigital capacitor (ic)   stub inductor (si)

(a)

interdigital capacitor (ic)   stub inductor (si)

(b)

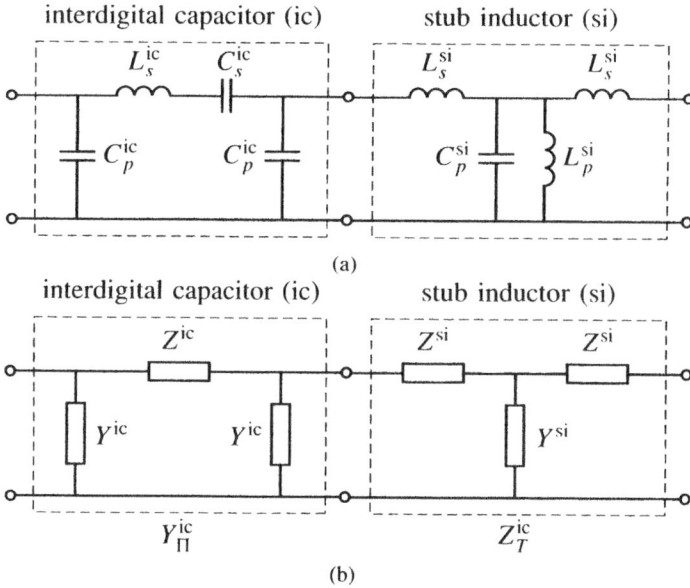

Fig. 3.41. Circuit models for the parameters extraction of the unit cell of Fig 3.40. (a) Equivalent circuit. (b) Auxiliary equivalent Π and T networks.

de-embed the component, that is, to subtract the phase shifts due to the additional microstrip lines by appropriate reference plane positioning or calibration (e.g., TRL calibration). Practically, if the lengths of these lines are ℓ_1 and ℓ_2, the de-embedded transmission parameters is

$$S_{21}^{\text{de-embedded}} = S_{21}^{\text{sim/meas}} e^{-j\Delta\phi^{\mu\text{strip}}}, \tag{3.149}$$

with

$$\Delta\phi^{\mu\text{strip}} = -k_0\sqrt{\varepsilon_{eff}}(\ell_1 + \ell_2), \tag{3.150}$$

where ε_{eff} is the effective permittivity of the microstrip lines. Failing to de-emb the components will result in an erroneous unit cell, with extra lines at the ends, inducing a wrong phase lag in addition to the phase shift [Eq. (3.81)] of the actual CRLH structure.

The scattering or S parameters of the interdigital capacitor and sub inductors are then converted into admittance or Y parameters and impedance or Z parameters, respectively, using standard conversion formulas [7]. The corresponding T and Π matrixes, that we note $\left[Y_\Pi^{ic}\right]$ and $\left[Z_T^{si}\right]$, are related to the elements of the circuits in Fig 3.41 as follows

$$\left[Y_\Pi^{ic}\right] = \begin{bmatrix} Y_{11}^{ic} & Y_{12}^{ic} \\ Y_{12}^{ic} & Y_{22}^{ic} \end{bmatrix} = \begin{bmatrix} \dfrac{1}{Z^{ic}} + Y^{ic} & -\dfrac{1}{Z^{ic}} \\ -\dfrac{1}{Z^{ic}} & \dfrac{1}{Z^{ic}} + Y^{ic}, \end{bmatrix} \tag{3.151}$$

with

$$Z^{\text{ic}} = j \left[\omega L_s^{\text{ic}} - \frac{1}{\omega C_s^{\text{ic}}} \right], \tag{3.152a}$$

$$Y^{\text{ic}} = j\omega C_p^{\text{ic}}, \tag{3.152b}$$

and

$$[Z_T^{\text{si}}] = \begin{bmatrix} Z_{11}^{\text{si}} & Z_{12}^{\text{si}} \\ Z_{12}^{\text{si}} & Z_{22}^{\text{si}} \end{bmatrix} = \begin{bmatrix} \dfrac{1}{Y^{\text{si}} + Z^{\text{si}}} & -\dfrac{1}{Y^{\text{si}}} \\ -\dfrac{1}{Y^{\text{si}}} & \dfrac{1}{Y^{\text{si}} + Z^{\text{si}}} \end{bmatrix}, \tag{3.153}$$

where

$$Y^{\text{si}} = j \left[\omega C_p^{\text{si}} - \frac{1}{\omega L_p^{\text{si}}} \right], \tag{3.154a}$$

$$Z^{\text{si}} = j\omega L_s^{\text{si}}. \tag{3.154b}$$

Next, we have to determine the LC parameters in Fig 3.41(a). These parameters are obtained by comparison of Figs. 3.41(a) and 3.41(b). Determination of the expressions of the isolated reactances C_p^{ic} and L_s^{si} is immediate, whereas calculations of the reactances L_s^{ic}- C_s^{ic} and C_p^{si}-L_p^{si}, appearing resonant/antiresonant tanks, require derivation with respect to ω to provide one additional equation for determination of all unknowns. The results are

$$C_p^{\text{ic}} = \frac{\left(Y_{11}^{\text{ic}}\right)^{-1} + \left(Y_{21}^{\text{ic}}\right)^{-1}}{j\omega}, \tag{3.155a}$$

$$L_s^{\text{ic}} = \frac{1}{2j\omega} \left[\omega \frac{\partial(1/Y_{21}^{\text{ic}})}{\partial\omega} - \frac{1}{Y_{21}^{\text{ic}}} \right], \tag{3.155b}$$

$$C_s^{\text{ic}} = \frac{2}{j\omega} \left[\omega \frac{\partial(1/Y_{21}^{\text{ic}})}{\partial\omega} + \frac{1}{Y_{21}^{\text{ic}}} \right], \tag{3.155c}$$

and

$$L_s^{\text{si}} = \frac{\left(Z_{11}^{\text{si}}\right)^{-1} + \left(Z_{21}^{\text{si}}\right)^{-1}}{j\omega}, \tag{3.156a}$$

$$C_p^{\text{si}} = \frac{1}{2j\omega} \left[\omega \frac{\partial(1/Z_{21}^{\text{si}})}{\partial\omega} + \frac{1}{Z_{21}^{\text{si}}} \right], \tag{3.156b}$$

$$L_p^{\text{si}} = \frac{2}{j\omega} \left[\omega \frac{\partial(1/Z_{21}^{\text{si}})}{\partial\omega} - \frac{1}{Z_{21}^{\text{si}}} \right]. \tag{3.156c}$$

Finally, neglecting the extremely small inductance L_s^{si}, we obtain the four CRLH parameters

$$L_R = L_s^{\text{ic}}, \tag{3.157a}$$

$$C_R = 2C_p^{\text{ic}} + C_p^{\text{si}}, \tag{3.157b}$$

$$L_L = L_p^{\text{si}}, \tag{3.157c}$$

$$C_L = C_s^{\text{ic}}. \tag{3.157d}$$

It should be noted that this parameter extraction procedure considers complex S, Y, and Z parameters. Therefore, it is a rigorous technique describing *both the magnitude and the phase* behavior of the waves traveling along the structure.[41]

A second important point to note is that the expressions for the extracted parameters depend on frequency, since Eqs. (3.155) and (3.156) are explicit functions of ω, as well as the Y and Z parameters. Consequently, a frequency at which extraction is performed, or *extraction frequency*, ω_e, must be chosen. If the whole bandwidth (LH and RH) of the CRLH structure is used, a judicious choice for the extraction frequency is the transition frequency ω_0. Performing extraction at other frequencies will yield slightly different responses. Fortunately, these variations are moderate, and it will be shown in the next section that using *constant values* for L_R, C_R, L_L and C_L provides remarkably accurate results, which is consistent with our implicit assumption of *nondispersive reactances*[42] in the previous sections.

3.4 EXPERIMENTAL TRANSMISSION CHARACTERISTICS

Fig 3.42 shows a typical prototype of a CRLH microstrip structure, and Fig 3.43 shows the frequency characteristics of such a structure both for a balanced [Fig 3.43(a)] and an unbalanced [Fig 3.43(b)] designs. Although it does not include higher order modes between adjacent cells, the LC network model exhibits excellent agreement with experimental and full-wave simulation results over a very broad bandwidth. This confirms that the three-step TL approach developed

---

[41] In this sense, this procedure is more elaborated than wide-spread magnitude curve fitting techniques, which may provide inaccurate results in terms of phase.

[42] There is no contradiction in stating that a structure is dispersive and constituted of nondispersive elements. The CRLH *is intrinsically dispersive* since, as seen in Eqs. (3.10) or Eq. (3.115), the CRLH propagation constant is a nonlinear function of frequency, and therefore group delay is frequency dependent. This dispersion is a physical requirement, expressed by the entropy condition of [Eq. (2.35)]. However, there is no requirement for the LC parameters involved in Eqs. (3.10) and (3.115) to be frequency-dependent. So, a physical CRLH structure is always dispersive but may constituted by nondispersive reactances in the frequency range of interest. Dispersion in CRLH structures (in particular in the LH range) comes from the unusual way the L and C components are arranged (series C / shunt L instead of series L / shunt C) in the unit cell.

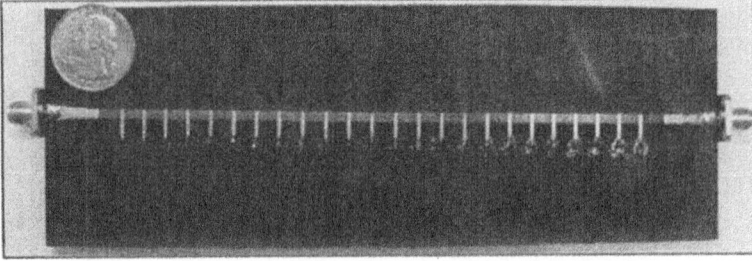

Fig. 3.42. Top view of a microstrip CRLH prototype of the type of Figs. 3.39 and 3.40. This prototypes includes $N = 24$ cells.

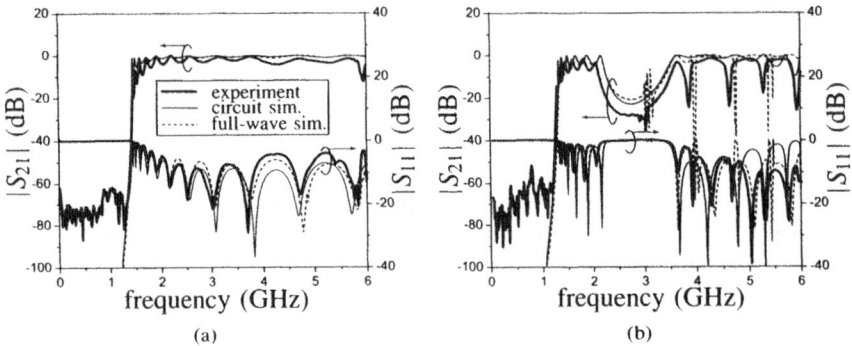

(a)   (b)

Fig. 3.43. Frequency characteristics of the microstrip CRLH structure of Figs. 3.39 and 3.42 obtained by measurement, by full wave (MoM) simulation, and by circuit simulation using Eqs. (3.151) to (3.157). (a) $N = 9$-cell *balanced* design with parameters (Fig 3.40): $p = 6.1$ mm, $\ell_c = 5.0$ mm, $w_c = 2.4$ mm, $\ell_s = 8.0$ mm, $w_s = 1.0$ mm, width of digits 0.15 mm, and all spacings 0.1 mm. Extracted parameters: $L_R = 2.45$ nH, $C_R = 0.50$ pF, $L_L = 3.38$ nH, $C_L = 0.68$ pF; $Z_L/Z_R = 1.01$, $f_0 = 3.9$ GHz. (b) $N = 7$-cell *unbalanced* design with parameters (Fig 3.40): $p = 12.2$ mm, $\ell_c = 10.0$ mm, $w_c = 4.8$ mm, $\ell_s = 6.0$ mm, $w_s = 2.0$ mm, width of digits 0.3 mm, and all spacings 0.2 mm. Extracted parameters: $L_R = 2.22$ nH, $C_R = 1.55$ pF, $L_L = 1.39$ nH, $C_L = 1.57$ pF; $Z_L/Z_R = 0.46$, $f_{se} = 2.3$ GHz, $f_0 = 3.0$ GHz, $f_{sh} = 3.4$ GHz. In both (a) and (b) designs, the number of pairs of interdigited fingers is 5 and the substrate used is Rogers RT/duroid 5880 with dielectric constant $\varepsilon_r = 2.2$ and thickness $h = 62$ mil (loss tangent = 0.0009).

in this chapter, ideal TL (Section 3.1), LC network (Section 3.2), and real distributed structure (Section 3.3), is appropriate to describe TL MTMs.

The experimental dispersion relation for the balanced CRLH microstrip structure of Fig 3.43(a), obtained by Eq. (3.69a) with the phase unwrapping technique exposed in Section 3.2.3, is shown in Fig 3.44. In addition, the analytical homogeneous TL relation of Eq. (3.34) (Section 3.1) and LC network relation of Eq. (3.116b) (Section 3.2.6) are also shown for comparison. Results obtained

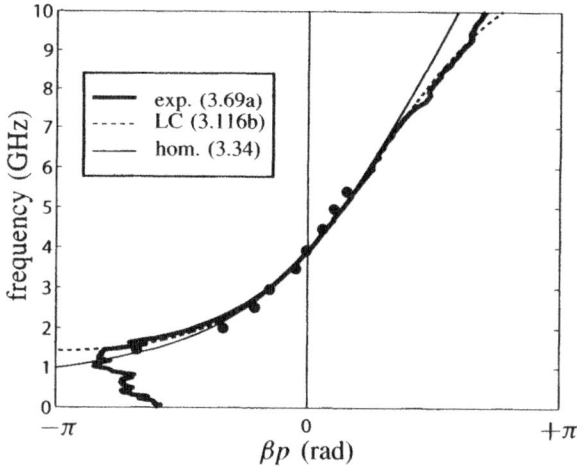

Fig. 3.44. Experimental dispersion diagram obtained by Eq. (3.69a) with phase unwrapping, compared with the analytical homogeneous TL relation (3.34) and LC network relation (3.116b) for balanced CRLH the microstrip structure of Fig 3.43(a). The thick dots show results obtained from near-field.

from near-field measurement on top of the structure are also shown.[43] Remarkable agreement may be observed between experiments and the theoretical predictions.

The nature of propagation in the CRLH microstrip structure is *quasi-TEM*,[44] which means that the components of the electric and magnetic field in the direction of propagation (longitudinal components) are much smaller than the components perpendicular to this direction (transverse components). The field distribution is essentially the same as in the conventional microstrip structure [7, 8], with electric field dominantly perpendicular to the strip and magnetic field circulating around it, except in the cross section through the stub inductors, where the magnetic field has a circulating longitudinal component around the

---

[43]The following method is used [18]. A component of the field ψ is near-field measured at K different points z_k along the structure, which provides the data $\psi_{meas}(z_k)$ $(k = 1, \ldots, K)$. In addition, the general wave form of a propagating wave is considered, at the same points, $\psi_{th}(z_k) = \psi^+(\omega)e^{-j\beta(\omega)z_k} + \psi^+(\omega)e^{-j\beta(\omega)z_k}$ $(k = 1, \ldots, K)$; at each frequency ω_l this expression has the two unknowns $\psi^+(\omega_l)$ and $\psi^-(\omega_l)$, in addition to the sought dispersion point $\beta(\omega_l)$. Next, the difference between the measured and theoretical fields is defined as the error function $\epsilon(z_k)(\omega) = \psi_{meas}(z_k) - \psi_{th}(z_k)$ $(k = 1, \ldots, K)$. This represents an overdetermined system (2 unknowns/$K > 2$ equations), which can be solved by standard singular value decomposition (SVD) in order to determine $\psi^+(\omega_l)$ and $\psi^-(\omega_l)$ for a given guess value $\beta(\omega_l)$. By searching (in the range of interest within the Brillouin zone) the value of $\beta(\omega_l)$ minimizing $|\epsilon(z_k)(\omega)|$ with an appropriate root-searching procedure such as Newton-Raphson, each value $\beta(\omega_l)$ can be obtained easily, which leads, by scanning frequency, to the dispersion relation $\beta(\omega)$.

[44]Otherwise, the LC model presented in this chapter, which is based on the assumption of TEM propagation, would provide so accurate results and would thus have to be further elaborated.

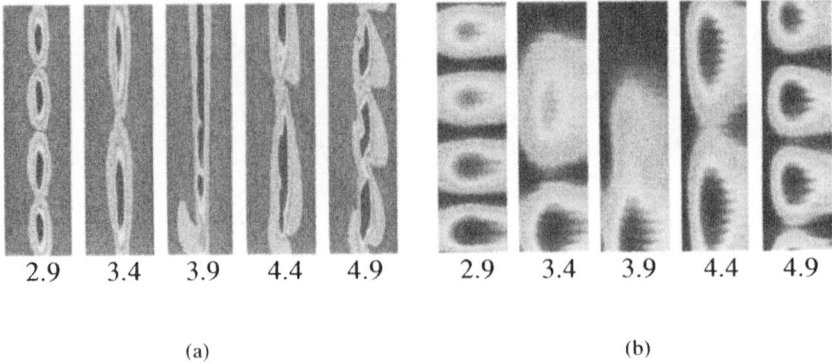

Fig. 3.45. Distribution of the magnitude of the electric field in the transverse plane of the line of Figs. 3.43(a) and 3.44 just above the structure. (a) Simulation (MoM). (b) Near-field measurement with a vertical monopole. The authors thank Drs. Laurin and Ouardirhi, École Polytechnique de Montréal, for performing these near-field measurements.

part of the stub exceeding the width of the capacitor (section $\ell_s - w_s$ referring to Fig 3.40).

The variations of he dispersion curves [Figs. 3.3(a), 3.35 and 3.44] induced by variations of the microstrip design parameters are the following.[45] If the substrate thickness is increased, C_R is decreased and L_R is increased, in a factor roughly proportional to the factor of thickness increase. The value of L_L is increased as L_R, whereas C_L is weakly affected because most of the quasi-static electric field generating the capacitance is concentrated between the fingers of the interdigital capacitors. Consequently, the RH lowpass cutoff f_{cR} tends to be relatively unchanged, whereas the LH highpass cutoff f_{cL} tends to be lowered, which results into wider CRLH bandwidth. In addition, ω_{sh} is weakly changed while ω_{se} is decreased. If the permittivity of the substrate is increased, C_R is increased proportionally, whereas L_R and L_L are unchanged, and C_L may be slightly decreased due to higher concentration of the electric field inside the substrate to the ground plane. The consequences are a significant decrease in f_{cR} and minor decrease in f_{cL}, resulting in reduction of the CRLH bandwidth, significant decrease in ω_{sh}, and minor decrease in ω_{se}.

Finally, Fig 3.45 demonstrates the guided wavelength behavior described in Section 3.1.3. Both full-wave and experimental results show how guided wavelength increases with frequency in the LH range, becomes very large near the transition frequency (theoretically infinite at exactly $f = f_0$), and then decreases as frequency increases in the RH range.

---

[45]The consideration of per-unit-length capacitance and inductance in a conventional wide microstrip line may help to understand these trends: $C = \varepsilon w/h$ and $L = \mu h/w$, where ε and μ are the permittivity and permeability of the substrate, respectively, h is the thickness of the substrate, and w is the width of the line.

TABLE 3.4. Conversion between the Material Constitutive Parameters ε_r, μ_r and the CRLH (balanced) TL Parameters β, Z_c of a Material.

	ε_r, μ_r	β, Z_c	L_R, C_R, L_L, C_L
ε_r, μ_r	—	$\varepsilon_r = \sqrt{\dfrac{n}{\eta_r}} = \dfrac{c\beta}{\omega Z_c}$ $\mu_r = n\eta_r = \dfrac{c\beta Z_c}{\omega}$	via β and Z_c $\downarrow$
β, Z_c	$\beta = n\dfrac{\omega}{c} = \sqrt{\varepsilon_r \mu_r}\dfrac{\omega}{c}$ $Z_c = \eta_r \eta_0 = \sqrt{\mu_r/\varepsilon_r}\,\eta_0$	—	$\beta(L_R, C_R, L_L, C_L)$ $Z_c =$ $\sqrt{\dfrac{L_R}{C_R}} = \sqrt{\dfrac{L_L}{C_L}}$
$L_R, C_R,$ L_L, C_L	$\dfrac{L_R}{C_R} = \dfrac{L_L}{C_L} \overset{(*)}{=} \left(\dfrac{\mu_r}{\varepsilon_r}\right)^2$ $\beta^{-1}(C_R, C_L)$ with $(*)$ + 1 degree of freedom	$\dfrac{L_R}{C_R} = \dfrac{L_L}{C_L} \overset{(**)}{=} Z_c^2$ $\beta^{-1}(C_R, C_L)$ with $(**)$ + 1 degree of freedom	—

The relation $\beta = \beta(L_R, C_R, L_L, C_L)$ is available from the dispersion diagram computed by Eq. (3.116b). The inverse of this function, $\beta^{-1}(L_R, C_R, L_L, C_L)$ is found numerically after applying the two conditions on η_r or Z_c and using the available degree of freedom.

3.5 CONVERSION FROM TRANSMISSION LINE TO CONSTITUTIVE PARAMETERS

With the fundamental constitutive parameters and TL relations

$$n = \sqrt{\varepsilon_r \mu_r} \tag{3.158a}$$

$$\eta = \eta_r \eta_0 = \sqrt{\frac{\mu_r}{\varepsilon_r}}\,\eta_0, \quad \text{with} \quad \eta_0 = \sqrt{\frac{\mu_0}{\varepsilon_0}}, \tag{3.158b}$$

$$\beta = nk_0 = \frac{\omega}{c}, \quad \text{(TEM assumption)}, \tag{3.158c}$$

$$Z_c = nk_0 = \sqrt{\frac{L}{C}}, \tag{3.158d}$$

the conversion formulas between the constitutive material parameters (ε_r, μ_r), the TL MTM parameters (β, Z_c), and the LC CRLH network implementation of the TL MTM (L_R, C_R, L_L, C_L) are easily found as given in Table 3.4.

REFERENCES

1. C. Caloz and T. Itoh. "Novel microwave devices and structures based on the transmission line approach of meta-materials," *IEEE-MTT Int'l Symp.*, vol. 1, pp. 195–198, Philadelphia, PA, June 2003.

2. A. Sanada, C. Caloz, and T. Itoh. "Characteristics of the composite right/left-handed transmission lines," *IEEE Microwave Wireless Compon. Lett.*, vol. 14, no. 2, pp. 68–70, February 2004.

3. C. Caloz, A. Sanada, and T. Itoh. "A novel composite right/left-handed coupled-line directional coupler with arbitrary coupling level and broad bandwidth" *IEEE Trans. Microwave Theory Tech.*, vol. 52, no. 3, pp. 980–992, March 2004.

4. L. B. Felsen and N. Marcuvitz. *Radiation and Scattering of Waves*, Reissue edition, Ieee/Oup Series on Electronic Wave Theory, 1996.

5. C. Kittel. *Introduction to Solid State Physics*, Eighth Edition, Wiley, 2005.

6. S. Ramo, J. R. Whinnery, and T. Van Duzer. *Fields and Waves in Communication Electronics*, Third Edition, John Wiley & Sons, 1994.

7. M. D. Pozar. *Microwave Engineering*, Third Edition, John Wiley & Sons, 2004.

8. R. E. Collin. *Foundations for Microwave Engineering*, Second Edition, McGraw-Hill, 1992.

9. L. Brillouin. *Wave Propagation and Group Velocity*, Academic Press, 1960.

10. G. L. Matthaei, L. Young, and E. M. T. Jones. *Microwave Filters, Impedance-Matching Networks, and Coupling Structures*, Artech House, Dedham, 1964.

11. R. E. Collin. *Field Theory of Guided Waves*, Second Edition, Wiley-Interscience, 1991.

12. N. Marcuvitz. *Waveguide Handbook*, McGraw-Hill, London, 1951.

13. I. Bahl. *Lumped Elements for RF and Microwave Circuits*, Artech House, Boston, 2003.

14. C. Caloz, H. Okabe, T. Iwai, and T. Itoh. "Transmission line approach of left-handed (LH) materials," *USNC/URSI National Radio Science Meeting*, San Antonio, TX, vol. 1, p. 39, June 2002.

15. C. Caloz and T. Itoh. "Transmission line approach of left-handed (LH) structures and microstrip realization of a low-loss broadband LH filter," *IEEE Trans. Antennas Propagat.*, vol. 52, no. 5, May 2004.

16. C. Caloz, A. Sanada, and T. Itoh. "Microwave applications of transmission-line based negative refractive index structures," *in Proc. of Asia-Pacific Microwave Conf.*, Seoul, Korea, vol. 3, pp. 1708–1713, Nov. 2003.

17. A. Lai, C. Caloz and T. Itoh. "Transmission line based metamaterials and their microwave applications," *Microwave Mag.*, vol. 5, no. 3, pp. 34–50, Sept. 2004.

18. Z. Ouardirhi and J.-J. Laurin. "A new near-field based noncontacting measurement technique for S-parameters," *Int. Conf. Electromag. Near Field Caract.(ICONIC)*, Rouen, France, pp. 147–152, June 2003.

4

TWO-DIMENSIONAL MTMs

The TL theory and fundamental properties of CRLH MTMs structures have been established in Chapter 3 for the 1D case. Chapter 4 extends these 1D concepts to 2D structures, or metasurfaces, where additional refractive effects involving angles, such as those described in Chapter 2, occur.[1]

Due to *effective-homogeneity*, the theory of ideal homogeneous 2D MTMs is absolutely identical to that developed in Section 3.1 for the 1D case, where the direction of propagation in the 1D TL can represent any direction in the 2D structure and is not replicated here. In contrast, the analysis of 2D LC, or more generally TL, network structures,[2] where a discrete TL unit cell is defined, somewhat differs from that of 1D TL network structures, given in Section 3.2 and is therefore presented.

Because the fundamental properties of CRLH MTMs have already been established in Chapter 3, this chapter only deals with numerical methods for 2D MTMs and with illustration of refractive effects computed by these methods.

Section 4.1 solves the 2D TL network *eigenvalue problem*, where no source is present and periodic boundary conditions (PBCs) are applied along two

---

[1]This book does not include 3D MTMs structures, as they have been little explored to date. But the TL analysis for future 3D MTMs is straightforward from that of 2D MTM, requiring only introduction of an additional TL axis and extension of the Brillouin zone domain to 3D spectral domain [1].

[2]Some principles of modeling of 2D and 3D TL networks are presented in [2], but this book provides neither the explicit dispersion relations for unbounded networks nor the propagation characteristics for finite-size networks and is therefore found of little use for MTMs.

Electromagnetic Metamaterials: Transmission Line Theory and Microwave Applications,
By Christophe Caloz and Tatsuo Itoh
Copyright © 2006 John Wiley & Sons, Inc.

orthogonal directions at the edges of the unit cell. Section 4.2 solves the *driven-solution* problem, where the frequency characteristics of a finite-size $M \times N$-cell structure are determined in terms of scattering parameters, by a simple 2D transmission matrix method (TMM) algorithm. A more sophisticated and powerful transmission line method (TLM) approach, particularly well suited for TL MTMs, is presented in Section 4.3. Some refractive effects, such as cylindrical wave propagation, negative refraction and negative focusing, RH/LH surface plasmons (SPs) and LH prism reflection and LH perfect absorbing reflection, are illustrated in Section 4.4. Finally, Section 4.5 is dedicated to real distributed meta-structures.

Since, as pointed out in Chapter 3, it is both a computational and fabrication advantage to treat MTMs as periodic structures (even if they are not, see Fig 3.28), the configuration considered in this chapter are periodic. However, extension to aperiodic and anisotropic structures is straightforward.

4.1 EIGENVALUE PROBLEM

The resolution of the 2D TL network CRLH eigenvalue problem, leading to an analytical formula for the dispersion relation, is very similar to that presented in Section 3.2.6 for the 1D case [3, 4]. Fig 4.1 shows the unit cell of a 2D TL, where the four impedance branches and the admittance branch are represented by their [$ABCD$] matrix. This unit cell is the 2D counterpart of the 1D unit cell of Fig 3.34. An essential difference between the 2D and 1D cells however is that the complex propagation constant γ, and corresponding propagation constant β and attenuation constant α are *vectors* in the 2D case

$$\overline{\gamma} = \overline{\alpha} + j\overline{\beta} = k_x \hat{x} + k_y \hat{y}. \tag{4.1}$$

4.1.1 General Matrix System

We will consider a general structure where all the immittances are arbitrary,[3] and particularize it later to the case of the CRLH 2D network, which exhibits the unit cell shown in Fig 4.2 and is shown in Fig 4.3.

Using the [$ABCD$] matrix [Eq. (3.56)] and its inverse [Eq. (3.57)], we can relate the input voltages and currents to the output voltages and currents to obtain five different expressions for the center node voltage V_0

$$V_x^{i0} = V_0 = \frac{D_x^i V_x - B_x^i I_x}{A_x^i D_x^i - B_x^i C_x^i}, \tag{4.2a}$$

$$V_y^{i0} = V_0 = \frac{D_y^i V_y - B_y^i I_y}{A_y^i D_y^i - B_y^i C_y^i}, \tag{4.2b}$$

$$V_x^{o0} = V_0 = \left(A_x^o V_x + B_x^o I_x \right) e^{-jk_x p_x}, \tag{4.2c}$$

---

[3]This can also be useful for other structures than MTMs, such as general periodic structures, operated in the Bragg regime.

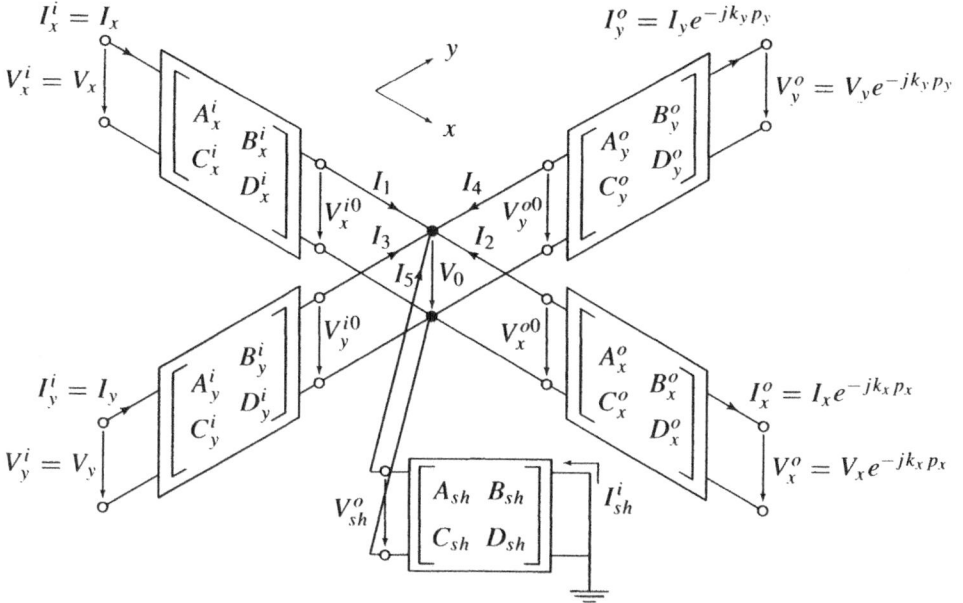

Fig. 4.1. Unit cell of a general 2D periodic network TL. The output currents and voltages are related to the input currents and voltages by PBCs [Eq. (3.100)]. The periods along x and y are p_x and p_y, respectively. If the network is constituted of ideal and therefore dimensionless, LC components, the phase shifts $\phi_x = -k_x p_x$ and $\phi_y = -k_y p_y$ have to be considered instead of the wavenumbers k_x and k_y, which are then undefined.

$$V_y^{o0} = V_0 = \left(A_y^o V_y + B_y^o I_y \right) e^{-jk_y p_y}, \tag{4.2d}$$

$$V_{sh}^o = V_0 = \frac{-B_{sh} I_{sh}^i}{A_{sh} D_{sh} - B_{sh} C_{sh}}, \tag{4.2e}$$

where PBCs were used, with periods p_x and p_y along x and y, respectively. In addition, Kirchoff's current law may be applied at the center node, yielding

$$\sum_{k=1}^{5} I_k = \frac{-C_x^i V_x + A_x^i I_x}{A_x^i D_x^i - B_x^i C_x^i} + \frac{-C_y^i V_y + A_y^i I_y}{A_y^i D_y^i - B_y^i C_y^i}$$
$$- \left(C_x^o V_x + D_x^o I_x \right) e^{-jk_x p_x} - \left(C_y^o V_y + D_y^o I_y \right) e^{-jk_y p_y} \tag{4.3}$$
$$+ \frac{A_{sh} I_{sh}^i}{A_{sh} D_{sh} - B_{sh} C_{sh}} = 0.$$

Eq. (4.2) together with Eq. (4.3) represents a linear system of six equations in the six unknowns V_x, I_x, V_y, I_y, I_{sh}^i, and V_0. By eliminating I_{sh}^i from Eq. (4.3) and equating each of the expressions from Eq. (4.2a) to Eq. (4.2d) with the resulting expression for Eq. (4.2e) to also eliminate V_0, we obtain the following linear

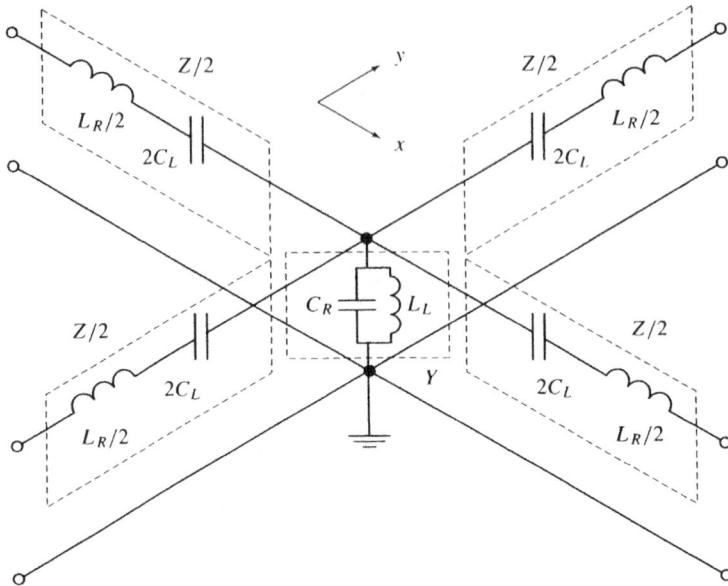

Fig. 4.2. Unit cell of a CRLH 2D periodic network TL. This is the 2D counterpart of the CRLH 1D unit cell shown in Fig. 3.18.

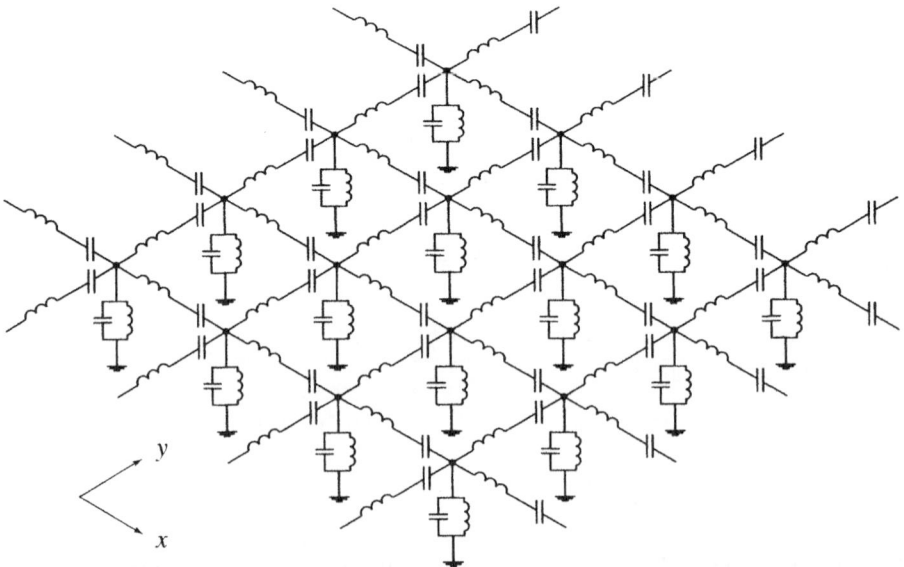

Fig. 4.3. 2D CRLH periodic network consisting by the repetition of the unit cell of Fig 4.2 along both the x and y directions.

matrix system

$$[M] \begin{bmatrix} V_x \\ I_x \\ V_y \\ I_y \end{bmatrix} = \begin{bmatrix} 0 \\ 0 \\ 0 \\ 0 \end{bmatrix}, \tag{4.4a}$$

where

$$[M] = \begin{bmatrix} \alpha - \dfrac{D_x^i A_{sh}}{B_{sh}\Delta_x^i} & \beta + \dfrac{B_x^i A_{sh}}{B_{sh}\Delta_x^i} & \gamma & \delta \\[2ex] \alpha & \beta & \gamma - \dfrac{D_y^i A_{sh}}{B_{sh}\Delta_y^i} & \delta + \dfrac{B_y^i A_{sh}}{B_{sh}\Delta_y^i} \\[2ex] \alpha - \dfrac{A_x^o A_{sh} e_x}{B_{sh}} & \beta - \dfrac{B_x^o A_{sh} e_x}{B_{sh}} & \gamma & \delta \\[2ex] \alpha & \beta & \gamma - \dfrac{A_y^o A_{sh} e_y}{B_{sh}} & \delta - \dfrac{B_y^o A_{sh} e_y}{B_{sh}} \end{bmatrix},$$

$$\tag{4.4b}$$

with

$$\alpha = -C_x^o e_x - \frac{C_x^i}{\Delta_x^i}, \ \beta = -D_x^o e_x - \frac{A_x^i}{\Delta_x^i}, \ \gamma = -C_y^o e_y - \frac{C_y^i}{\Delta_y^i} \delta = -D_y^o e_y - \frac{A_y^i}{\Delta_y^i},$$

$$\tag{4.4c}$$

$$\Delta_u^i = A_u^i D_u^i - B_u^i C_u^i \ (u = x, y), \tag{4.4d}$$

$$e_x = e^{-jk_x p_x}, \quad \text{and} \quad e_y = e^{-jk_y p_y}. \tag{4.4e}$$

This system admits a nontrivial solution if

$$\det\{[M]\} = \det\{[M(k_x, k_y; \omega)]\} = 0, \tag{4.5}$$

which is in general a transcendental equation yielding the eigenfrequencies ω_n (k_x, k_y) or the *dispersion relation* when the spectral parameters (k_x, k_y), defining the propagation constant $\vec{\beta} = k_x\hat{x} + k_y\hat{y}$,[4] sweep the edge of the irreducible BZ [1].

Eq. (4.4) is a completely general relation that can be solved, numerically or analytically with some further developments, for any kind of periodic 2D network, either perfectly *isotropic* (with four identical series branches), *orthogonally anisotropic* (different periods p_x and p_y [4], or different impedances along x and y, or *arbitrary* (four different series branches).

---

[4]Although this expression is different from Eq. (4.1), it can also represent the general lossy case if $\vec{\beta}$ is allowed to be complex. Then, we can write $\vec{\beta} = \vec{\beta}' - j\vec{\alpha}'$, which results, by setting $\vec{\alpha} = 0$ in Eq. (4.1), into $\vec{\gamma}' = \vec{\alpha}' + j\vec{\beta}'$, where the primes can be left out.

4.1.2 CRLH Particularization

Let us now consider the case of an isotropic network with impedances $Z/2$ branches and admittance Y, such as the CRLH network of Fig 4.2. In this case, Eq. (4.5) is dramatically simplified after substitution of the appropriate $ABCD$ parameters [5]

$$
\begin{aligned}
A_x^i &= A_y^i = A_x^o = A_y^o = 1, \\
B_x^i &= B_y^i = B_x^o = B_y^o = Z/2, \\
C_x^i &= C_y^i = C_x^o = C_y^o = 0, \\
D_x^i &= D_y^i = D_x^o = D_y^o = 1, \\
A_{sh} &= 1, \quad B_{sh} = 0, \quad C_{sh} = Y, \quad D_{sh} = 1, \\
\Delta_u^i &= 1, \ (u = x, y) \quad \text{(reciprocal network branches)},
\end{aligned}
\tag{4.6}
$$

from which we obtain, after some calculations,

$$
\frac{\left(1 - e^{-jk_x p_x}\right)^2}{e^{-jk_x p_x}} + \frac{\left(1 - e^{-jk_y p_y}\right)^2}{e^{-jy_y p_y}} - ZY = 0.
\tag{4.7}
$$

This equation with the substitutions $k_u \to \beta_u - j\alpha_u$ $(u = x, y)$ clearly reduces to the 1D dispersion relation [Eq. (3.114)] if $k_x = 0$ or $k_y = 0$. In the particular case of the CRLH network TL of Fig 4.2, it becomes, with the immittances of Eq. (3.53),

$$
\frac{\left(1 - e^{-jk_x p_x}\right)^2}{e^{-jk_x p_x}} + \frac{\left(1 - e^{-jk_y p_y}\right)^2}{e^{-jy_y p_y}} + \chi = 0,
\tag{4.8}
$$

which can also be written

$$
\sin^2\left(\frac{k_x p_x}{2}\right) + \sin^2\left(\frac{k_y p_y}{2}\right) - \frac{\chi}{4} = 0,
\tag{4.9}
$$

or

$$
2 - \cos\left(k_x p_x\right) - \cos\left(k_y p_y\right) - \frac{\chi}{2} = 0,
\tag{4.10}
$$

where we recall χ defined in Eq. (3.62a)

$$
\chi = \left(\frac{\omega}{\omega_R}\right)^2 + \left(\frac{\omega_L}{\omega}\right)^2 - \kappa\omega_L^2, \quad \text{with}
\tag{4.11a}
$$

$$
\omega_R = \frac{1}{\sqrt{L_R C_R}} \quad \text{rad/s},
\tag{4.11b}
$$

$$
\omega_L = \frac{1}{\sqrt{L_L C_L}} \quad \text{rad/s},
\tag{4.11c}
$$

$$
\kappa = L_R C_L + L_L C_R \quad \text{(s/rad)}^2.
\tag{4.11d}
$$

4.1.3 Lattice Choice, Symmetry Points, Brillouin Zone, and 2D Dispersion Representations

In Figs. 4.1 to 4.3, a *square or rectangular lattice* network interconnection was implicitly assumed. In 2D, a triangular (hexagonal) lattice network is also possible [6].[5] However, the triangular lattice yields results identical to those of the square lattice in effectively homogeneous media such as MTMs.[6] Consequently, we naturally choose the simpler, square lattice network here.[7]

In a square lattice, the *direct (or spatial) lattice*, shown in Fig 4.4(a), is represented by the generic vector $\vec{R}_{uv}$

$$\vec{R}_{uv} = u\vec{a}_1 + v\vec{a}_2, \quad (u, v \in \mathbb{Z}), \tag{4.12a}$$

where

$$\vec{a}_1 = p_x\hat{x} \quad \text{and} \quad \vec{a}_2 = p_y\hat{y}, \tag{4.12b}$$

while the *reciprocal (or spectral) lattice*, shown in Fig 4.4(b), is represented by the generic vector $\vec{G}_{hk}$

$$\vec{G}_{hk} = h\vec{b}_1 + k\vec{b}_2, \quad (h, k \in \mathbb{Z}). \tag{4.13a}$$

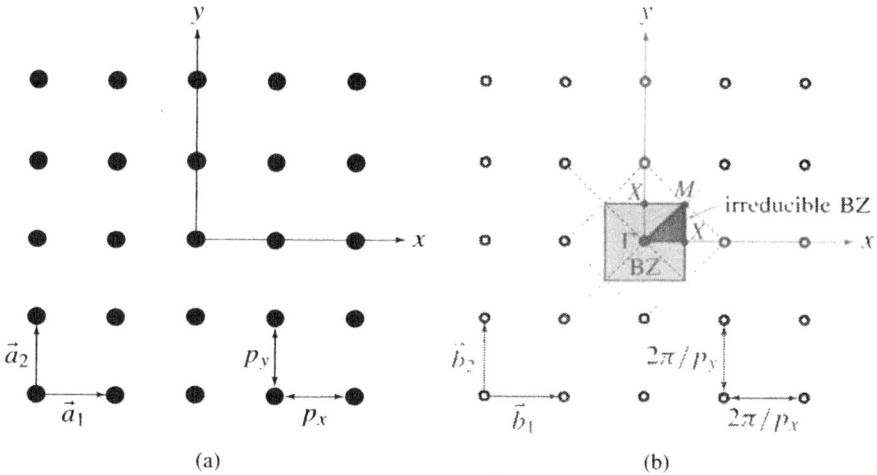

(a)   (b)

Fig. 4.4. Square lattices for a 2D structure. (a) Direct (spatial) lattice. (b) Reciprocal (spectral) lattice. The extension to a non square rectangular lattice ($p_x \neq p_y$) is straightforward.

---

[5]In 3D, even more diverse and complex lattices, such as simple cubic, face-centered cubic, and body-centered cubic lattices, well-known in solid-state physics, are possible [1].

[6]This statement does not naturally apply to photonic crystals and PBG structures, where different lattices exhibit very different properties.

[7]Similarly, in 3D structures, the simplest, simple-cubic lattice would be the most indicated.

where

$$\vec{b}_1 = k_x \hat{x} = \left(\frac{2\pi}{p_x}\right)\hat{x} \quad \text{and} \quad \vec{b}_2 = k_y \hat{y} = \left(\frac{2\pi}{p_y}\right)\hat{y}. \qquad (4.13b)$$

Fig 4.4(b) also shows the *Brillouin zone* (BZ) of the periodic lattice, which corresponds to one (spectral) period centered at the origin of the Floquet function $\xi(x, y) = \psi(x, y)e^{k_x x + k_y y}$, given in Eq. (3.102) for the 1D case. This Brillouin zone is the square surface $[-\pi/p_x < k_x < +\pi/p_x, -\pi/p_y < k_y < +\pi/p_y]$, and the dispersion relation is a 2D function $\omega_n(k_x, k_y)$, where n labels the number of the mode. However, due to symmetry, this square surface can be reduced, without loss of information, to the triangular surface shown in the figure, which is called the *irreducible Brillouin zone*. For computation of the dispersion diagram, this irreducible zone is generally further reduced to the segment $\Gamma - X - M - \Gamma$ because this segment includes the most important symmetry points Γ, X (and Y in case of anisotropy) and M, where maximum diffraction occurs in the Bragg regime [1]. These symmetry points are given by

$$\Gamma : \beta = 0, \qquad (4.14a)$$

$$X : \beta = \frac{\pi}{p_x}\hat{x} \quad \left(Y : \beta = \frac{\pi}{p_y}\hat{y}\right) \qquad (4.14b)$$

$$M : \beta = \pi\left(\frac{1}{p_x}\hat{x} + \frac{1}{p_y}\hat{y}\right). \qquad (4.14c)$$

The wavenumbers k_x and k_y along the BZ can be derived analytically from Eq. (4.10)

$$\Gamma - X \ (0 < k_x p_x < \pi, k_y = 0) : \ k_x = \frac{1}{p_x}\cos^{-1}\left(1 - \frac{\chi}{2}\right), \qquad (4.15a)$$

$$X - M \ (k_x p_x = \pi, 0 < k_y p_y < \pi) : \ k_y = \frac{1}{p_y}\cos^{-1}\left(3 - \frac{\chi}{2}\right), \qquad (4.15b)$$

$$M - \Gamma \ (0 < k_u p_u < \pi, u = x, y) : \ k_u = \frac{1}{p_u}\cos^{-1}\left(1 - \frac{\chi}{4}\right), \qquad (4.15c)$$

and the eigenfrequencies at the symmetry points can also be determined from Eq. (4.8)

$$\omega_{\Gamma 1} = \min(\omega_{se}, \omega_{sh}), \quad \omega_{\Gamma 2} = \max(\omega_{se}, \omega_{sh}), \qquad (4.16a)$$

$$\left.\begin{array}{c}\omega_{X1} \\ \omega_{X2}\end{array}\right\} = \sqrt{\frac{\omega_{se}^2 + \omega_{sh}^2}{2} + 2\omega_R^2 \mp \sqrt{\left(\frac{\omega_{se}^2 + \omega_{sh}^2}{2} + 2\omega_R^2\right)^2 - (\omega_{se}\omega_{sh})^2}},$$

$$(4.16b)$$

$$\left.\begin{array}{c}\omega_{M1}\\\omega_{M2}\end{array}\right\} = \sqrt{\frac{\omega_{se}^2 + \omega_{sh}^2}{2} + 4\omega_R^2 \mp \sqrt{\left(\frac{\omega_{se}^2 + \omega_{sh}^2}{2} + 4\omega_R^2\right)^2 - (\omega_{se}\omega_{sh})^2}}.$$

(4.16c)

Figs. 4.5 and 4.6 show the balanced and unbalanced dispersion relation of the CRLH network TL of Figs. 4.2 and 4.3 in different representations, the conventional $\Gamma - X - M - \Gamma$ diagram, the 2D curve $f_n(k_x, k_y)$ and the isofrequency maps over the BZ. These figures can be compared with the 1D-case dispersion diagrams of Fig 3.35 and are verified to be exactly identical to them on the $\Gamma - X$ segment.

The isofrequency maps of Figs. 4.5 and 4.6 clearly show the difference between the *long-wavelength regime*, of interest in MTMs, close to the transition

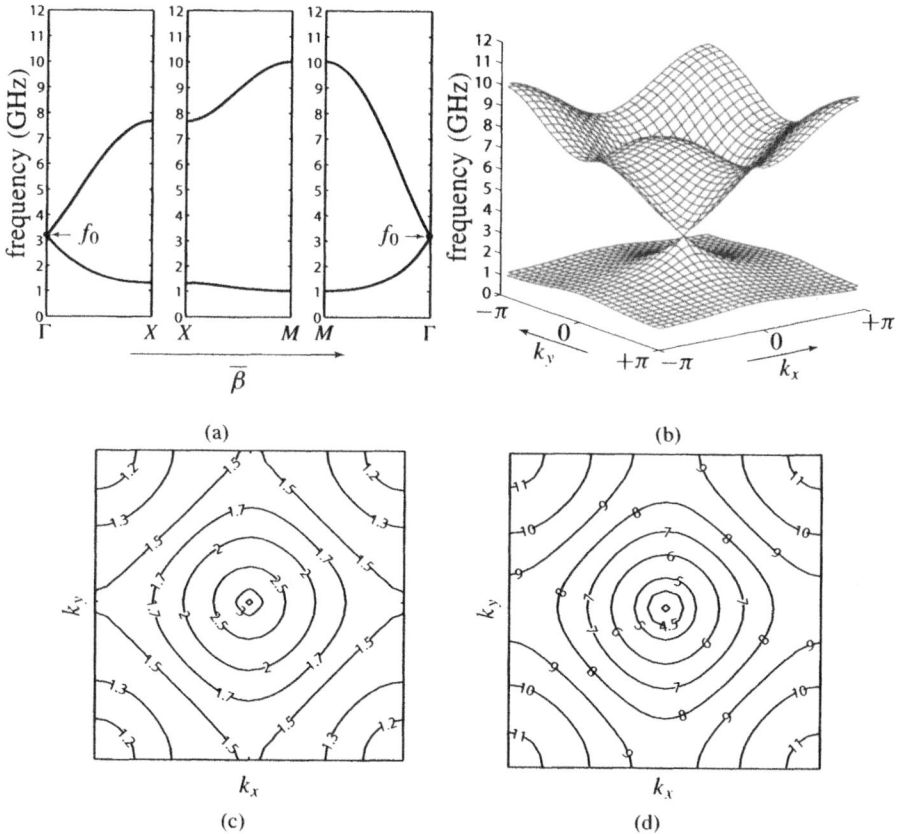

Fig. 4.5. Dispersion diagram for a 2D balanced CRLH network TL computed by Eq. (4.8) with the parameters of Figs. 3.22 and 3.35(b). (a) Conventional representation along the irreducible BZ. (b) 3D representation of the function $\omega_n(k_x, k_y)$ $(n = 1, 2)$ over the BZ. (c) Isofrequency curves for $f < f_0$. (d) Isofrequency curves for $f > f_0$.

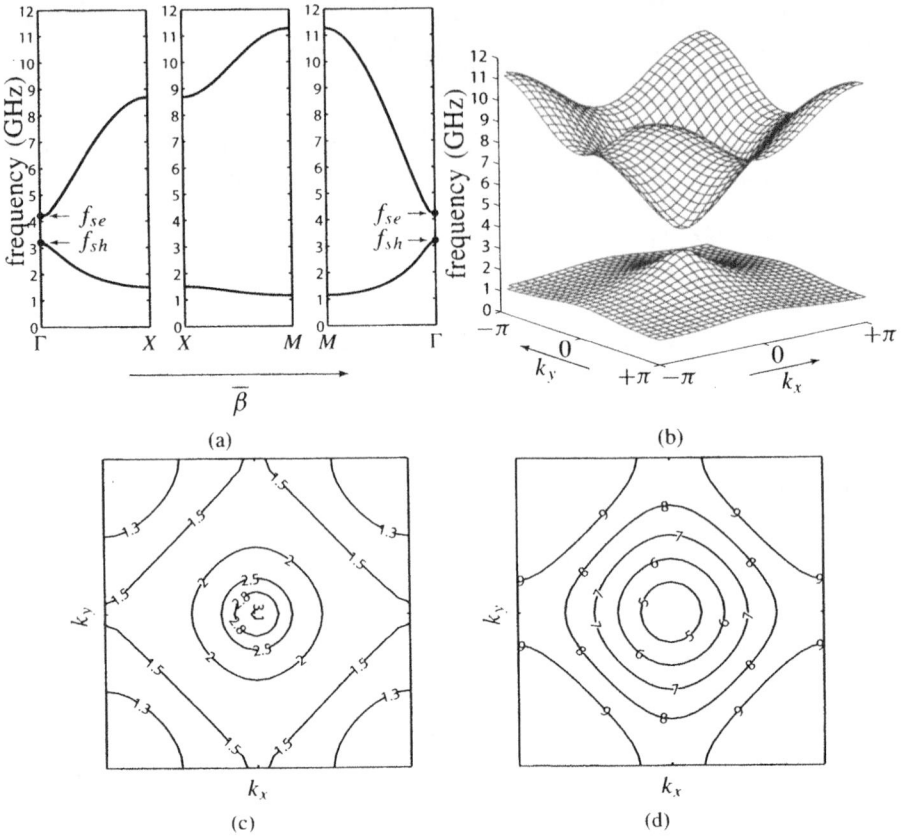

Fig. 4.6. Dispersion diagram for a 2D unbalanced CRLH network TL computed by Eq. (4.8) with the parameters of Figs. 3.23 and 3.35(a). (a) Conventional representation along the irreducible BZ. (b) 3D representation of the function $\omega_n(k_x, k_y)$ ($n = 1, 2$) over the BZ. (c) Isofrequency curves for $f < f_{sh}$. (d) Isofrequency curves for $f > f_{se}$.

frequency at the Γ point, and the *Bragg regime*. In the former case, the isofrequency curves are regular circles, indicating azimuthal isotropy and homogeneity, whereas in the latter case, either in the low LH band or in the high RH band, the circles are degenerated into non azimuthally symmetric contours (with $90°$ rotational symmetry), indicating anisotropy.[8]

The refractive index can be immediately obtained from the dispersion relation of [Eq. (4.10)]

$$n) = \frac{c_0 \beta}{\omega} = \begin{cases} \dfrac{c_0}{\omega p} \cos^{-1}\left(1 - \dfrac{\chi}{2}\right) & (\Gamma - X) \\[3mm] \dfrac{c_0}{\omega p} \cos^{-1}\left(1 - \dfrac{\chi}{4}\right) & (M - \Gamma). \end{cases} \qquad (4.17)$$

---

[8]Consider for instance the frequency 3 GHz in Fig 4.5. At this frequency, propagation is allowed along the diagonal axes of the structure but prohibited along its main axes.

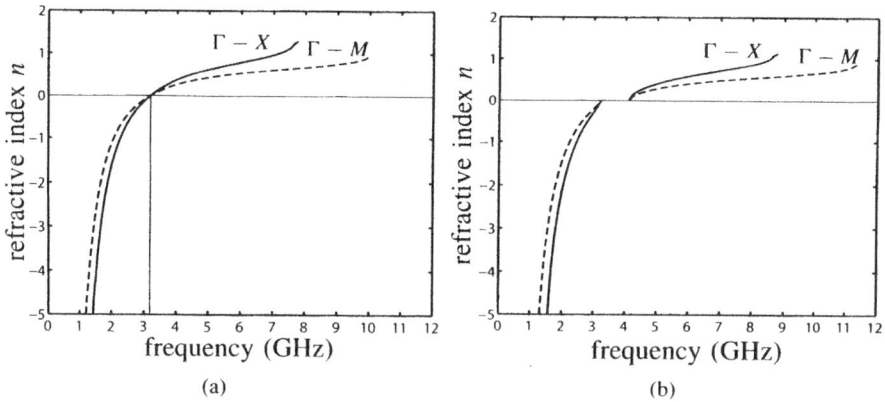

Fig. 4.7. Refractive index (real part) computed by Eq. (4.17). (a) Balanced design of Fig 4.5. (b) Unbalanced design of Fig 4.6.

The refractive index was not given for the $X - M$ spectral segment because this segment corresponds to Bragg regime ($p_x = \lambda_g/2$ and $p_y = 0, \ldots, \lambda_g/2$), where the medium is inhomogeneous and the refractive index can therefore not be defined. The refractive indexes of the balanced and unbalanced designs of Figs. 4.5 and 4.6 are plotted in Figs. 4.7(a) and 4.7(b), respectively. It can be seen that the refractive index along the $\Gamma - X$ and $\Gamma - M$ are exactly equal at the transition frequency (balanced case), where perfect effective-homogeneity is achieved, and tend to become progressively more and more different as we move away from the transition frequency, which indicates increasing anisotropy, as observed in the isofrequency charts of Figs. 4.5 and 4.6. As already pointed out for 1D CRLH structures in Chapter 3, the magnitude of the refractive index is smaller then 1 in the vicinity of the transition frequency (superluminal propagation) and is zero at the transition frequency in the balanced case, which corresponds to infinite phase velocity and infinite guided wavelength.

4.2 DRIVEN PROBLEM BY THE TRANSMISSION MATRIX METHOD (TMM)

Because the electrical length of the unit cell in a MTM is very small compared with guided wavelength, a piece of MTM with overall size of a few guided wavelengths includes a very large number of cells.[9] Determining the propagation characteristics of such a MTM structure [$\beta(x, y)$ and $n(x, y)$] is a conceptually simple and computationally fast problem (sourceless, eigenvalue problem) consisting in computing the dispersion relation for the corresponding infinitely periodic structure by PBCs analysis of *one unit cell only*, as presented in Section 4.1. In contrast, determining the response of a practical finite-size MTM

---

[9]For instance, a 2D structure of overall size $10\lambda_g \times 10\lambda_g$ with unit cell of period $p_x = p_y = p = \lambda_g/10$ includes 10,000 cells; if the more homogeneous solution $p_x = p_y = p = \lambda_g/100$ could be attained, the total number if cells would increase to 1,000,000!

to a given source is a problem (driven problem) for which analysis and design by conventional numerical methods may be extremely difficult, if possible at all!, due to the huge number of cells that need to be considered. New numerical tools are therefore needed to solve practical problems and develop useful applications of 2D (and possibly also 3D) MTMs.

In this section, we present a *transmission matrix method* (TMM) capable of dealing with finite-size 2D networks TLs. This method is essentially an extension of [7].

4.2.1 Principle of the TMM

Consider the arbitrary 2D (or planar) metastructure, depicted in Fig 4.8(a), constituted by a square lattice of periodic or nonperiodic unit cells.[10] The TMM consists of the division of the structures into *columns* of unit cells, each of which being represented by a general circuit model, as shown in Fig 4.8(b),[11] similar to that of Figs. 4.1 and 4.2. First, the *generalized transmission matrix*, relating

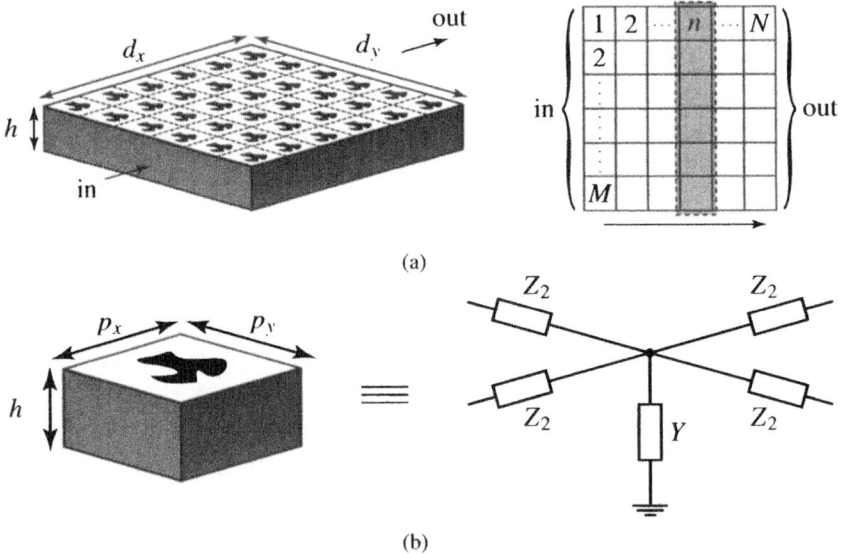

(a)

(b)

Fig. 4.8. Principle of the transmission matrix method (TMM). (a) Arbitrary 2D metastructure and decomposition in *column* cells. (b) Unit cell and its general T network equivalent circuit. The unit cell for a CRLH structure is the one shown in Fig 4.2.

[10]The TMMs can also be used for conventional continuous RH structures, such as for instance microstrip or parallel-plate waveguide structures [7]. In this case, the structure is "sampled" or divided into an arbitrary lattice with small unit cell electrical length satisfying the homogeneity condition.

[11]Although we will consider here only a T-network unit cell circuit, for consistency with the models of Figs. 4.1 and 4.2, a Π network model is naturally also possible [7].

the input voltages and currents to the output voltages and currents of this column cell, is established. The transmission matrix of the complete structure is then obtained by simple multiplication of the column cells constituting the network, in the same manner as for simple two-port networks [5]. Next, this transmission matrix is converted into the scattering matrix of the structure, which provides the scattering coefficient for any combination of ports at the edges of the structure. Finally, for a source exciting the structure, at an arbitrary point (edge or internal), a "folded" equivalent network is established and the overall current/voltage distributions are determined by letting the observer points scan the structure.

4.2.2　Scattering Parameters

The equivalent network for a column cell with termination impedances Z_{tv} is shown in Fig 4.9. A column cell constituted of M unit cells can be represented by a $2M \times 2M$ transmission matrix $[T]$ associated with M input ports and M output ports

$$\left[\begin{array}{c} [V_{\text{in}}] \\ [I_{\text{in}}] \end{array} \right] = [T] \cdot \left[\begin{array}{c} [V_{\text{out}}] \\ [I_{\text{out}}] \end{array} \right],$$

(4.18a)

where

$$[V_{\text{in}}] = \left[\begin{array}{c} V_1 \\ V_2 \\ \vdots \\ V_M \end{array} \right], \quad [I_{\text{in}}] = \left[\begin{array}{c} I_1 \\ I_2 \\ \vdots \\ I_M \end{array} \right],$$

(4.18b)

$$[V_{\text{out}}] = \left[\begin{array}{c} V_{M+1} \\ V_{M+2} \\ \vdots \\ V_{2M} \end{array} \right], \quad [I_{\text{out}}] = \left[\begin{array}{c} I_{M+1} \\ I_{M+2} \\ \vdots \\ I_{2M} \end{array} \right],$$

(4.18c)

and

$$[T] = \left[\begin{array}{cc} [A] & [B] \\ [C] & [D] \end{array} \right].$$

(4.18d)

By decomposing the column cell into three distinct subcolumn cells, as shown in Fig 4.9, we can easily obtain the transmission matrix $[T]$ as the product of the three matrixes corresponding to the three subcell columns

$$[T] = \left[\begin{array}{cc} [A] & [B] \\ [C] & [D] \end{array} \right] = [T_{\text{h}}] \cdot [T_{\text{v}}] \cdot [T_{\text{h}}],$$

(4.19a)

where

$$[T_{\text{h}}] = \left[\begin{array}{cc} [I] & (Z/2)[I] \\ [O] & [I] \end{array} \right]$$

(4.19b)

$$[T_{\text{v}}] = \left[\begin{array}{cc} [I] & [O] \\ [X] & [I] \end{array} \right]$$

(4.19c)

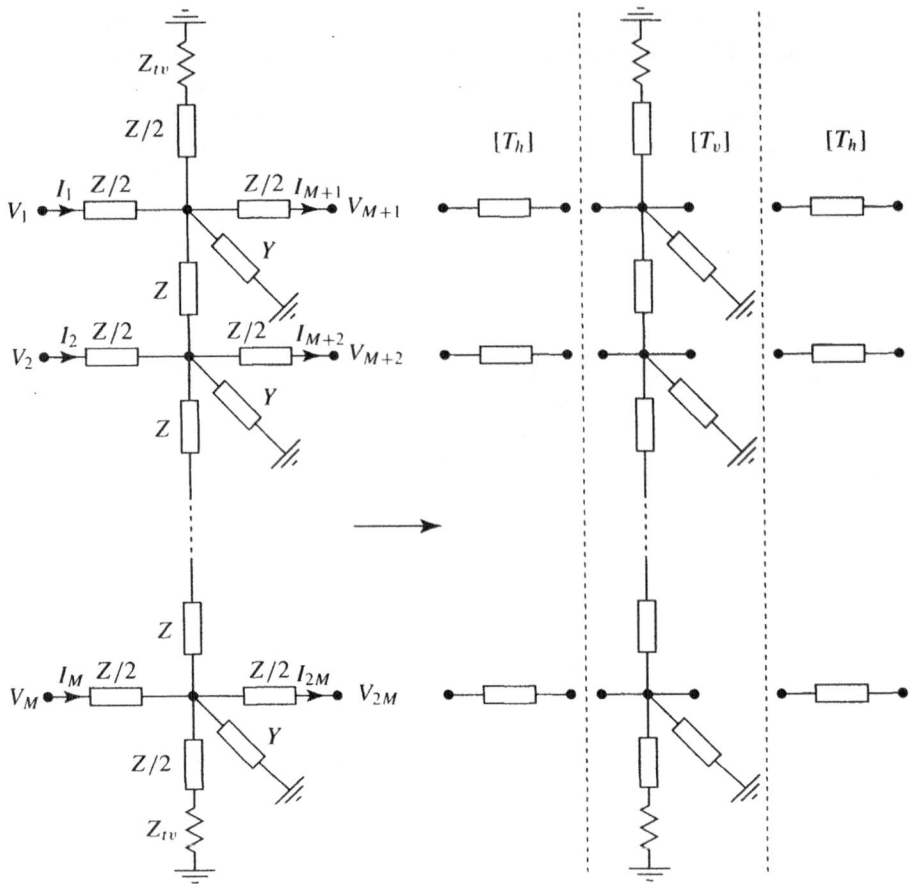

Fig. 4.9. Equivalent circuit T network for a column ($M \times 1$) cell and corresponding decomposition into three subcolumn cells.

with

$$[X] = \begin{bmatrix} Y_a & Y_b & 0 & \cdots & 0 & 0 & 0 \\ Y_b & Y_c & Y_b & \cdots & 0 & 0 & 0 \\ 0 & Y_b & Y_c & \cdots & 0 & 0 & 0 \\ \vdots & \vdots & \vdots & \ddots & \vdots & \vdots & \vdots \\ 0 & 0 & 0 & \cdots & Y_c & Y_b & 0 \\ 0 & 0 & 0 & \cdots & Y_b & Y_c & Y_b \\ 0 & 0 & 0 & \cdots & 0 & Y_b & Y_a \end{bmatrix} \qquad (4.19d)$$

and

$$Y_a = Y + \frac{1}{Z} + \frac{1}{Z/2 + Z_{tv}}, \quad Y_b = -\frac{1}{Z}, \quad Y_c = Y + \frac{2}{Z}, \qquad (4.19e)$$

and where $[I]$ and $[O]$ represent the $M \times M$ unit and zero matrixes, respectively. Note that the information on the vertical connections and terminations Z_{tv} is included in the submatrix $[X]$ via the admittance Y_a only. If the structure is open-ended ($Z_{tv} = \infty$), the only modification is to remove the term $1/(Z/2 + Z_{tv})$ in Y_a, since no current is flowing in the termination branch. The transmission matrix of the total structure constituted of N columns is then simply obtained by taking the Nth power of transmission matrix of the column cell

$$[T_{tot}] = [T]^N = \begin{bmatrix} [A_{tot}] & [B_{tot}] \\ [C_{tot}] & [D_{tot}] \end{bmatrix}, \qquad (4.20)$$

where the matrixes $[A_{tot}]$, $[B_{tot}]$, $[C_{tot}]$, and $[D_{tot}]$ are all of size $M \times M$. It appears that the TMM is not an "isotropically 2D method," in the sense that the product of Eq. (4.20) of the individual transmission matrixes is performed only along one direction from one input edge to the opposite output edge of the structure. However, the network is really 2D and arbitrary terminations Z_{tv} can be used at the other two edges. Therefore, in effect, fully 2D problems can be handled by the TMM.

In a simple two-port network, the conversion formulas between the transmission matrix and the impedance matrix are $Z_{11} = A/C$, $Z_{12} = (AD - BC)/C$, $Z_{21} = 1/C$ and $Z_{22} = D/C$, where $AD - BC = 1$ in the case of a reciprocal network [5]. The extension of these formulas to the *impedance matrix* of the 2D network structure yields

$$[Z_{tot}] = \begin{bmatrix} [Z_{11}] & [Z_{12}] \\ [Z_{21}] & [Z_{22}] \end{bmatrix} = \begin{bmatrix} [A_{tot}] \cdot [C_{tot}]^{-1} & [C_{tot}]^{-1} \\ [C_{tot}]^{-1} & [C_{tot}]^{-1} \cdot [D_{tot}] \end{bmatrix}. \qquad (4.21)$$

The scattering matrix of the complete structure is then obtained by the conventional conversion formula [5]

$$[S_{tot}] = \big[[Z_{tot}]/Z_c + [I] \big] \cdot \big[[Z_{tot}]/Z_c - [I] \big]^{-1}, \qquad (4.22)$$

where Z_c represents the characteristic impedance of the ports.

Fig 4.10 shows an $M \times N = 12 \times 12$-cell network structure, and Fig 4.11 compares a few scattering parameters obtained by Eq. (4.22) and by a standard circuit simulator for this network in the CRLH case. Eq. (4.22) is exact, and therefore the results obtained are perfectly superimposed with those obtained by a standard circuit simulator.

4.2.3 Voltage and Current Distributions

The technique just described provides the scattering parameters for ports located *at the edges* of the structure. In MTM applications, we also need to probe the

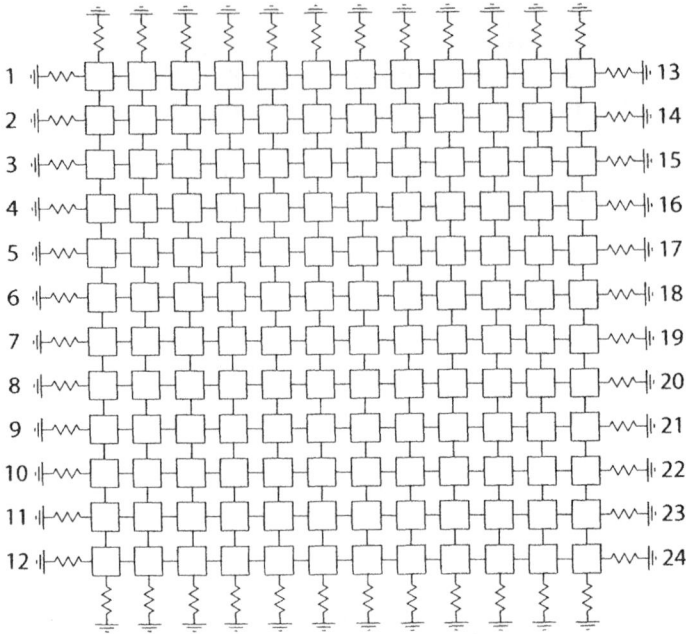

Fig. 4.10. 2D $M \times N = 12 \times 12$-cell network structure with ports designation for TMM analysis. Each 4-port block corresponds to the unit cell shown in Fig 4.2 for a CRLH MTM. The horizontal termination impedances at the left and at the right of the structure are Z_{th}, while the vertical termination impedances at the bottom and at the top of the structure are Z_{tv}.

inside of the structure to determine the refractive characteristics of the propagating waves. This can be accomplished with the help of the transmission matrix representations of Fig 4.12.[12]

We consider the case of an $M \times N$ network with a voltage (or current) source located at any point of it, at the kth row between ith and $(i + 1)$th column, and we wish to determine the resulting voltages and currents at any point of the network. This operation will be performed column by column, as suggested in Fig 4.12(a), where the observer (load R_L) is located between the $(i + m)$th and $(i + m + 1)$th columns of the network. The equivalent "folded" network representation of Fig 4.12(b), which represents the source and the load at each end of the network and the parts of the network on each side of the source-to-load network by shunt networks, should ease the forthcoming developments.

Refer to Fig 4.12(b). First, the input and output matrixes $[T_{in}]$ and $[T_{out}]$ are modified into the matrixes $[T'_{in}]$ and $[T'_{out}]$ including the horizontal terminations

---

[12]The formulation that will be presented is not unique. For instance, the currents and voltages are determined *between* the nodes of the network, but they could also be determined *at* the nodes by a different decomposition than that of Fig 4.9. However, the results will not differ between the different formulations if the homogeneity condition is satisfied.

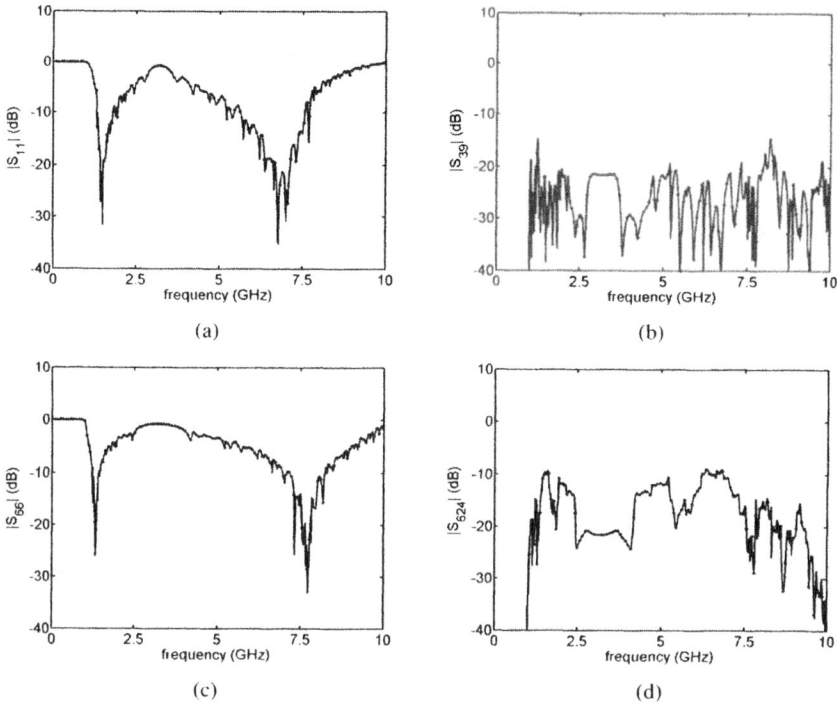

Fig. 4.11. Scattering parameters (solid lines) for a 2D $M \times N = 12 \times 12$-cell CRLH MTM with parameters $L_R = L_L = 2.5$ nH, $C_L = C_R = 1$ pF, $Z_{th} = 50$, $Z_{h'} = \infty$. Dots (perfectly superimposed with the TMM curves) are the results obtained with a standard circuit simulator. (a) $S_{1,1}$. (b) $S_{3,9}$. (c) $S_{6,6}$. (d) $S_{6,24}$.

Z_{th}[13]

$$\left[T'_{in}\right] = [T_{in}] \cdot [T_t],\tag{4.23a}$$

$$\left[T'_{out}\right] = [T_{out}] \cdot [T_t],\tag{4.23b}$$

where

$$[T_t] = \begin{bmatrix} [I] & [O] \\ Z_{th}[I] & [I] \end{bmatrix}.\tag{4.24}$$

Then, by matricially applying Kirchoff's currents law at the input and output nodes, we obtain the currents at the reference planes A and A' located at each side of the center section $[T_c]$

$$[I_1] = [I_{in}] - [I_2],\tag{4.25a}$$

where

$$[I_2] = [Y_{in}] \cdot [V_{in}].\tag{4.25b}$$

---

[13]Note that the terminations Z_{th} appear in shunt.

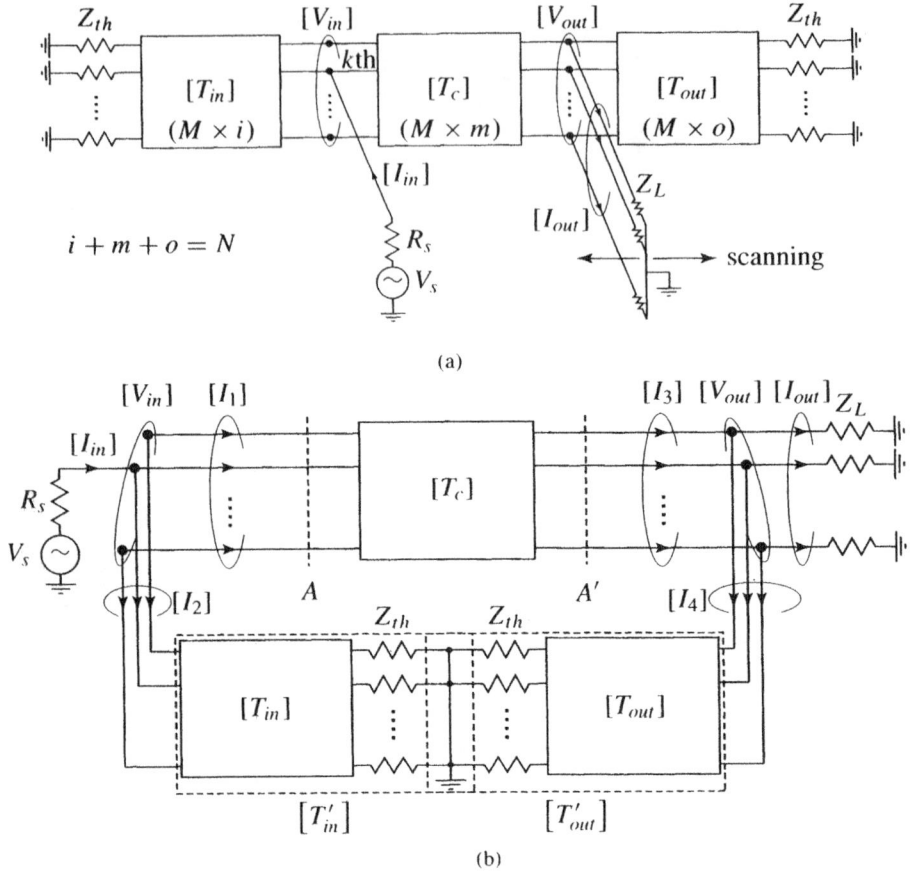

Fig. 4.12. Transmission matrix representations used to determine the currents and voltages at any (inside) node of the structure with source arbitrary location of the source (kth row and between *i*th and $(i + 1)$st column of the network). (a) Original network, including *M* inputs and *M* outputs, divided into an input network T_{in} (left-hand side of the source), a center network T_c (between the source and the observer column) and an output network T_{out} (right-hand of the observer column). The locations of the source and the observer can be interchanged. (b) Equivalent "folded" representation facilitating the TMM analysis.

and

$$[I_3] = [I_{out}] + [I_4], \qquad (4.26a)$$

where

$$[I_4] = [Y_{out}] \cdot [V_{out}]. \qquad (4.26b)$$

Note that in the case of a single excitation (one point in the network), illustrated in Fig 4.12, the vector $[I_{in}]$ reduces to a vector with all elements set to zero except for the kth row where the excitation is applied. But multiple column excitations, corresponding to the existence of multiple nonzero elements in $[I_{in}]$, can also be straightforwardly considered if necessary. The case of sources distributed between different columns can be handled by superposition of the corresponding single-column excitation cases.

The admittance matrixes $[Y_{in}]$ and $[Y_{out}]$ in Eqs. (4.25) and (4.26) can be derived from the definition of the *impedance* parameter of a 2-port network (with input/output ports labeled 1/2), $1/Z_{11} = I_1/V_1\big|_{I_2=0}$, which translates here into $[Z_{22}]^{-1} = [I_2]/[V_2]\}_{[I_{ie}=0]} = [Y_{in}]$ and $[Z_{44}]^{-1} = [I_4]/[V_4]\}_{[I_{oe}=0]} = [Y_{out}]$, where $[I_{ie}]$ and $[I_{oe}]$ represent the currents, effectively zero, at the outputs of the matrixes $[T'_{in}]$ and $[T'_{out}]$, respectively. Using conventional conversion formulas from Z to $ABCD$ parameters [5] then yields

$$[Y_{in}] = [T'_{iA}]^{-1} \cdot [T'_{iC}], \tag{4.27a}$$

$$[Y_{out}] = [T'_{oC}] \cdot [T'_{oA}]^{-1}, \tag{4.27b}$$

where the $M \times M$ A and C blocks of the transmission matrixes

$$[T'_{in}] = \begin{bmatrix} [T'_{iA}] & [T'_{iB}] \\ [T'_{iC}] & [T'_{iD}] \end{bmatrix} \tag{4.28a}$$

and

$$[T'_{out}] = \begin{bmatrix} [T'_{oA}] & [T'_{oB}] \\ [T'_{oC}] & [T'_{oD}] \end{bmatrix}. \tag{4.28b}$$

have been used. By inserting Eq. (4.27) into Eqs. (4.25) and (4.26) and noting that $[V_1] = [V_{in}]$, $[V_3] = [V_{out}]$, we obtain the following relation between the input and output current voltages at the planes A and A'

$$\begin{bmatrix} [I] & [O] \\ -[Y_{in}] & [I] \end{bmatrix} \cdot \begin{bmatrix} [V_{in}] \\ [I_{in}] \end{bmatrix} = [T_c] \begin{bmatrix} [I] & [O] \\ [Y_{out}] & [I] \end{bmatrix} \cdot \begin{bmatrix} [V_{out}] \\ [I_{out}] \end{bmatrix}, \tag{4.29}$$

or

$$\begin{bmatrix} [V_{in}] \\ [I_{in}] \end{bmatrix} = [T_{io}] \cdot \begin{bmatrix} [V_{out}] \\ [I_{out}] \end{bmatrix}, \tag{4.30a}$$

where

$$[T_{io}] = [T_{Yi}] \cdot [T_c] \cdot [T_{Yo}], \tag{4.30b}$$

with

$$[T_{Yi}] = \begin{bmatrix} [I] & [O] \\ [Y_i] & [I] \end{bmatrix}, \tag{4.31a}$$

$$[T_{Yo}] = \begin{bmatrix} [I] & [O] \\ [Y_o] & [I] \end{bmatrix}. \tag{4.31b}$$

Eq. (4.30) constitutes the key transmission relation between the source (input) and the observer (output). In a practical MTM problem, we will typically consider a fixed source and scan the network by moving the observer from the input to the output column of the full network represented in Fig 4.12(a), as suggested by the arrows in this figure.

In general, Eq. (4.30) represents a linear system of $2M$ equations with $4M$ unknowns, $V_{in,1}, \ldots, V_{in,M}, I_{in,1}, \ldots, I_{in,M}, V_{out,1}, \ldots, V_{out,M}, I_{out,1}, \ldots, I_{out,M}$. The number of unknowns can be reduced by a factor 2 with the boundary condition equations

$$I_{in,i} \begin{cases} = 0 & \text{if } i \neq k \\ \neq 0 & \text{if } i = k \end{cases}, \tag{4.32a}$$

$$V_{in,k} = V_s - R_s I_{in,k}, \tag{4.32b}$$

$$V_{out,i} = Z_L I_{out,i}, \quad \forall i = 1, \ldots, M, \tag{4.32c}$$

where k still corresponds to the index of the row where the excitation is applied. By combining Eqs. (4.30) and (4.32), we obtain after some algebraic developments, denoting the elements of $[T_{io}]$ in Eq. (4.30) t_{ij} for convenience,

$$[v_{sol}] = \left[[R] \quad \begin{matrix} -Z_L[I] \\ -[I] \end{matrix} \right]^{-1} \cdot [v_{exc}], \tag{4.33}$$

where

$$[R] = \begin{bmatrix} t_{11} & t_{12} & \cdots & t_{1,M} & t_{1,M+k} - R_s t_{1,k} \\ t_{21} & t_{22} & \cdots & t_{2,M} & t_{2,M+k} - R_s t_{2,k} \\ \vdots & \vdots & \cdots & \ddots & \vdots \\ t_{2M,1} & t_{2M,2} & \cdots & t_{2M,M} & t_{2M,M+k} - R_s t_{2M,k} \end{bmatrix}, \tag{4.34}$$

$t_{i,k}^{\text{th}}$ column skipped $\downarrow$

where the column $\left[t_{i,k}^{\text{th}} \right]$ has been skipped in $[R]$ ($2M \times M$ matrix), with the solution and excitation vectors

$$[v_{sol}] = \begin{bmatrix} V_{in,1} \\ V_{in,2} \\ \vdots \\ V_{in,M} \\ I_{in,k} \\ \hline I_{out,1} \\ I_{out,2} \\ \vdots \\ I_{out,M-1} \\ I_{out,M} \end{bmatrix} \begin{matrix} V_{in,k} \\ \text{skipped} \end{matrix} \quad \text{and} \quad [v_{exc}] = -V_s \begin{bmatrix} t_{1k} \\ t_{2k} \\ \vdots \\ t_{M-1,k} \\ t_{M,k} \\ \hline t_{M+1,k} \\ t_{M+2,k} \\ \vdots \\ t_{2M-1,k} \\ t_{2M,k} \end{bmatrix}, \tag{4.35}$$

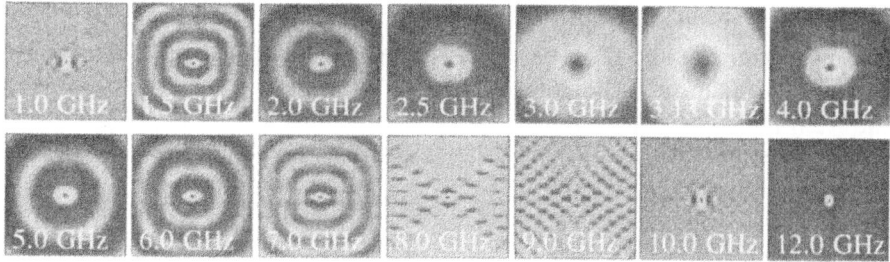

Fig. 4.13. TMM current distribution along an $M \times N = 31 \times 31$-cell 2D CRLH MTM (source placed in the center of the structure) with the same parameters as in Fig 4.5, where $f_{cL} = 1.32$ GHz, $f_0 = 3.18$ GHz and $f_{cR} = 7.68$ GHz.

where the row $V_{in,k}$, known from Eq. (4.32b), has been skipped in the first half of the vector $[v_{sol}]$ ($2M \times 1$ vector). Eq. (4.33) provides the solution to the problem of determining the currents and voltages at any point of a 2D MTM network.

Fig 4.13 shows current distributions obtained by the TMM from Eq. (4.33). These distributions verify the guided wavelength behavior described in Section 3.1.3. Below the highpass LH cutoff frequency f_{cL}, there is a stopband and consequently no propagation. Just above f_{cL}, propagation starts to occur, but it is very anisotropic due to the fact the isofrequency dispersion curves are not circular (Fig 4.5(c)). When frequency increases toward the transition frequency f_0 (LH range), guided wavelength increases progressively and, correspondingly, propagation becomes more isotropic. Close to f_0, guided wavelength is extremely large (in principle ∞ exactly at f_0). When frequency is further increased above f_0 (RH range), guided wavelength decreases again. Far from f_0, anisotropy is observed again (Fig 4.5(d)). Finally, above the lowpass RH cutoff frequency f_{cL}, we have another stopband and there is therefore no more propagation.

Finally, Fig 4.14 shows an illustration of the negative focusing effect (Section 2.6) at the interface between a RH medium and a LH medium obtained by TMM analysis.

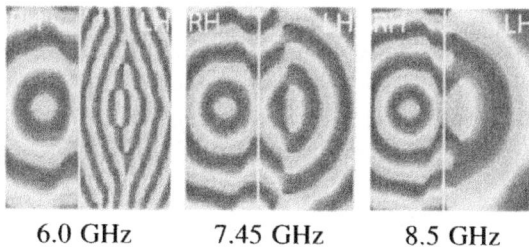

6.0 GHz   7.45 GHz   8.5 GHz

Fig. 4.14. Negative focusing at the interface between a RH and CRLH medium (in its LH range) observed by TMM-computed phase at three different frequencies. The parameters are $n_{RH} = 0.007$, and $n_{LH}(6.0\,\text{GHz}) = -0.015$, $n_{LH}(7.45\,\text{GHz}) = -0.007$, $n_{LH}(8.5\,\text{GHz}) = -0.003$.

4.2.4   Interest and Limitations of the TMM

The main interest of the TMM is the fact that it provides a simple and straightforward 2D extension ($M \times N$-ports) of the transmission matrix principles well-known in 1D networks (two-port). It is a fast technique, which can be applied easily to CRLH MTMs, once the parameters L_R, C_R, L_L, C_L, have been extracted (Section 4.5.3).

However, as the size of the structure becomes very large, the TMM in the formulation presented becomes unstable, due to the emergence of ill-conditioned matrixes, and alternative formulations of the TMM have to be considered to handle the problem.[14]

In addition, the TMM is a *frequency domain* technique. As such, it does not provide direct information on the "real-life" wave propagation in time, while time-domain visualization of field behavior is often very useful in MTM phenomena.

A more elaborate method, which is both intrinsically stable and time-domain based, the transmission line matrix (TLM) method, is presented in the next section.

4.3   TRANSMISSION LINE MATRIX (TLM) MODELING METHOD[15]

As pointed out in the previous section, the modeling of large 2D CRLH arrays, such as the network shown in Fig 4.3, can be a daunting task. A practical problem could easily contain thousands or even millions of cells, the boundaries could be of complex shape, and the properties of the medium could be inhomogeneous. To handle such large problems with so many unknowns, one must express the wave properties of the individual cells in the simplest possible way. Such a simplification is indeed possible, provided that the guided wavelength in the network is much larger than the cell dimensions, which is the homogeneity condition always valid in MTMs by definition (Section 1.1).

Consider the unit cell of a 2D CRLH network shown in Fig 4.2. If we remove, for the time being, the series capacitors $2C_L$ in the four branches, and the shunt inductor L_L across the node, we are left with the PRH (Sections 1.7 and 3.1.1) LC host network shown in Fig 4.15(a), which can be approximated by two orthogonal shunt-connected TL sections, as shown in Fig 4.15(b).

At wavelengths much larger than the cell size $\Delta\ell$, the inductance L' and capacitance C' per unit length of the TL sections, henceforth called "link lines", are related to the lumped elements of the LC network as

$$L' \approx \frac{L_R}{\Delta\ell} \quad \text{and} \quad C' \approx \frac{C_R}{\Delta\ell}. \tag{4.36}$$

---

[14]Such methods may typically consist in partitioning the too large matrices into smaller matrixes and applying iterative techniques for the matrix inversion process. These methods can be implemented relatively easily, but they are considered out of the scope of this book and are therefore not presented here.

[15]This section was written by Wolfgang Hoefer, University of Victoria, British Columbia, Canada.

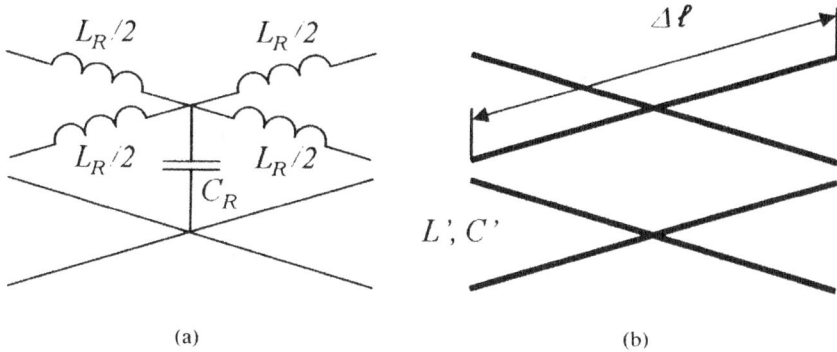

Fig. 4.15. Unit cells used as basis of the TLM. (a) LC network obtained by removing the series capacitors and the shunt inductor from the unit cell of the CRLH shown in Fig 4.2. (b) Approximation of that LC network by two orthogonal shunt-connected TL sections.

The phase velocity and characteristic impedance of the link lines are

$$v_p = \frac{1}{\sqrt{L'C'}} \approx \frac{\Delta\ell}{\sqrt{L_R C_R}} \quad \text{and} \quad Z_c = \sqrt{\frac{L'}{C'}} \approx \sqrt{\frac{L_R}{C_R}}. \tag{4.37}$$

We can reconstitute the CRLH properties by embedding the series capacitors and the shunt inductor into the TL model in Fig 4.15(b). We will show that this TL model allows us to describe wave propagation in such a structure with high computational efficiency and that this forms the basis for an efficient, versatile, and robust numerical technique known as the Transmission Line Matrix (TLM) modeling method [8].

4.3.1 TLM Modeling of the Unloaded TL Host Network

The key advantage in replacing the lumped element host cell by a pair of shunt-connected nondispersive TLs resides in their ability to transmit short impulses without distortion and to scatter the impulses at the shunt nodes according to a simple local scattering algorithm; this approach was first proposed by Johns and Beurle in 1971 [9]. Inspired by earlier work on TL networks to experimentally characterize waveguide discontinuities [10], Johns and Beurle developed a time domain algorithm to track the propagation and scattering of very short impulses in such a network. They showed that the impulses correctly sampled electromagnetic field propagation in 2D space and time. Frequency domain information can be extracted from the impulse response of the network via discrete or fast Fourier transform. Johns and Beurle referred to this technique as transmission line matrix (TLM) modeling, where the term "matrix" refers to the physical array of TLs and not to a mathematical object.

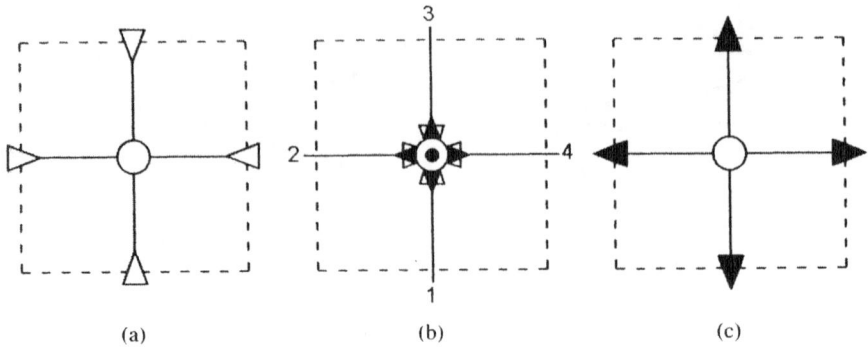

Fig. 4.16. Elementary process of impulse propagation and scattering in a 2D TLM cell. (a) Incident impulses enter the cell at $t = k\Delta t$. (b) Scattering occurs at $t = (k + 1/2)\Delta t$. (c) Reflected impulses leave (and new impulses enter) the cell at $t = (k + 1)\Delta t$. The cell size is $\Delta\ell \times \Delta\ell$. The numbering of the ports is indicated in (b).

Fig 4.16 shows the elementary process of impulse propagation and scattering in a 2D TLM cell.

The time interval Δt is the time required for the impulses to travel across a distance $\Delta\ell$ along the TLs (link lines). The scattering event is best described by a matrix equation relating the reflected voltage impulses at time $(k + 1)\Delta t$ to the incident voltage impulses at the previous time step $k\Delta t$

$$_{k+1}[v]^r = [S] \cdot {}_k[v]^i, \tag{4.38}$$

where $[S]$ is the impulse scattering matrix of the shunt node given by

$$[S] = \frac{1}{2} \begin{bmatrix} -1 & 1 & 1 & 1 \\ 1 & -1 & 1 & 1 \\ 1 & 1 & -1 & 1 \\ 1 & 1 & 1 & -1 \end{bmatrix}, \tag{4.39}$$

and $_{k+1}[v]^r$ and $_k[v]^i$ are the vectors of reflected and incident voltage impulses, respectively,

$$_{k+1}[v]^r = \begin{bmatrix} _{k+1}v_1^r \\ _{k+1}v_2^r \\ _{k+1}v_3^r \\ _{k+1}v_4^r \end{bmatrix} \quad \text{and} \quad {}_k[v]^r = \begin{bmatrix} _kv_1^r \\ _kv_2^r \\ _kv_3^r \\ _kv_4^r \end{bmatrix}. \tag{4.40}$$

The subscripts 1 to 4 designate the port number as indicated in Fig 4.16(b). When the reflected impulses leave the cell, four impulses $_{k+1}[v]^i$ are incident simultaneously upon each node from its four neighbors. This transfer process can also be written in matrix form as

$$_{k+1}[v]^i = [C] \cdot {}_{k+1}[v]^{r*}, \tag{4.41}$$

where $[C]$ is the connection matrix describing the topology of the network. For the 2D TLM shunt network, we have

$$[C] = \frac{1}{2} \begin{bmatrix} 0 & 0 & 1 & 0 \\ 0 & 0 & 0 & 1 \\ 1 & 0 & 0 & 0 \\ 0 & 1 & 0 & 0 \end{bmatrix}, \tag{4.42}$$

and $_{k+1}[v]^i$ and $_{k+1}[v]^r$ are the vectors of the voltage impulses incident on a node situated at point (x, y) and of the voltage impulses reflected from the neighboring nodes toward it,

$$_{k+1}[v]^i = \begin{bmatrix} _{k+1}v_1^i(x, y) \\ _{k+1}v_2^i(x, y) \\ _{k+1}v_3^i(x, y) \\ _{k+1}v_4^i(x, y) \end{bmatrix} \quad \text{and} \quad _{k+1}[v]^{r*} = \begin{bmatrix} _{k+1}v_1^r(x, y + 1) \\ _{k+1}v_2^r(x + 1, y) \\ _{k+1}v_3^r(x, y - 1) \\ _{k+1}v_4^r(x - 1, y) \end{bmatrix}. \tag{4.43}$$

When Eqs. (4.39) and (4.41) are executed alternately for all nodes in an unbounded network consisting of such cells, an explicit update process results. Fig 4.17 shows the result of this algorithm for three consecutive time steps. At the initial step, a short impulse of 1 V is injected at the center node. The impulses spread out at successive time steps. In the general case, multiple excitations in space and time can be applied and tracked using this algorithm.

In a physical network, the propagation of impulses between nodes is a continuous process, but in the computational model the state of the network is considered

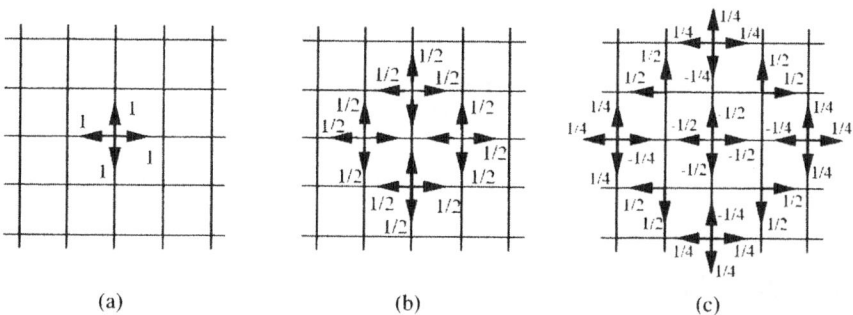

(a)   (b)   (c)

Fig. 4.17. Three consecutive scatterings in a two-dimensional TLM network excited by a unit voltage impulse. (a) Initial impulse excitation. (b) First update. (c) Second update.

only at discrete points in space (nodes) and time (steps). In other words, TLM is a discretized procedure, yielding impulsive "samples" in space and time of the continuous wave function it models.

The node voltage at time $k\,\Delta t$ is the sum of the incident and reflected impulses

$$_kV_{node}\left(1-\frac{1}{2}\right)\left(_kv_1^i+_kv_2^i+_kv_3^i+_kv_4^i\right)=\frac{1}{2}\sum_{m=1}^{4}{_kv_m^i}. \tag{4.44}$$

The time and space envelope of the node voltages emulates a continuous (analog) wave function. It is quite obvious that the rather simple and robust TLM update algorithm is much faster than a solution in which a large system of analog network equations is solved, particularly when the number of cells is very large.

4.3.2 TLM Modeling of the Loaded TL Host Network (CRLH)

To model the unit cell of the CLRH material in Fig 4.2, we must extend the TLM algorithm such that it can account for the series capacitances C_L and shunt inductance L_L embedded in the host cell. The best way to accomplish this is to represent the capacitance in each branch by an open-circuited TL stub, and the inductance by a short-circuited stub, as shown in Fig 4.18. All stubs have the same length $\Delta\ell/2$ so that the impulses scattered into the stubs return to the node in synchronism with the impulses incident from neighboring nodes.

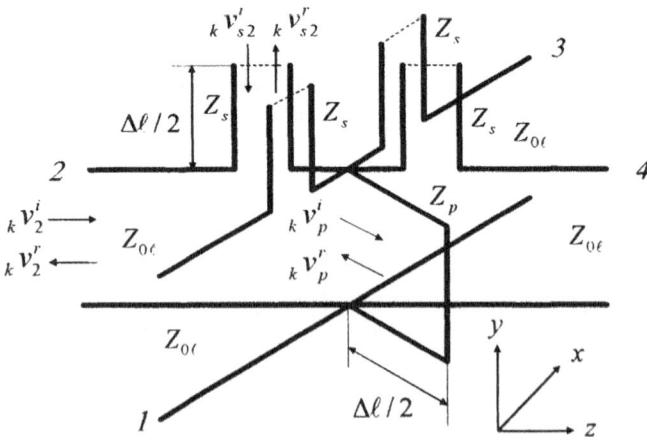

Fig. 4.18. Modified 2D shunt TLM cell of size $\Delta\ell\times\Delta\ell$ equipped with four open-circuited series stubs and one short-circuited shunt stub that model the reactive elements $2C_L$ and L_L of the CLRH cell. Scattered impulses and line characteristic impedances are also defined in this figure.

The characteristic impedances of these stubs are related to C_L and L_L by the following expressions

$$Z_s = \frac{\Delta\ell}{4C_L v_p} \quad \text{and} \quad Z_p = \frac{2L_L v_p}{\Delta\ell}, \tag{4.45}$$

where v_p is the phase velocity on the stub and link lines. These expressions are approximately valid for electrically short TLs ($\Delta\ell/\lambda_m \ll 1$).

The voltage impulse scattering matrix of this 9-port junction is of the form

$$
\begin{bmatrix} v_1 \\ v_3 \\ v_4 \\ v_{s1} \\ v_{s2} \\ v_{s3} \\ v_{s4} \\ v_p \end{bmatrix}^r_k
=
\begin{bmatrix}
p & d & d & d & e & -d & -d & -d & p \\
d & a & d & d & -d & e & -d & -d & p \\
d & d & a & d & -d & -d & e & -d & p \\
d & d & a & -d & -d & -d & -d & e & p \\
f & -g & -g & -g & b & g & g & g & -q \\
-g & f & -g & -g & g & b & g & g & -q \\
-g & -g & f & -g & g & g & b & g & -q \\
-g & -g & -g & f & g & g & g & b & -q \\
h & h & h & h & -h & -h & -h & -h & c
\end{bmatrix}
\cdot
\begin{bmatrix} v_1 \\ v_3 \\ v_4 \\ v_{s1} \\ v_{s2} \\ v_{s3} \\ v_{s4} \\ v_p \end{bmatrix}^i_k
$$

$$\tag{4.46}$$

The elements of this scattering matrix have been derived in [11] and will thus not be repeated here. The connection matrix $[C]$ given in Eq. (4.42) is not changed by the stubs. Hence, by replacing the impulse scattering matrix of Eq. (4.39) by the expanded scattering matrix of Eq. (4.46), we obtain a numerical CLRH model with the same wave properties and dispersion characteristics as the lumped periodic network of Fig 4.3, provided that the cell size is small compared with the wavelength on the network.

4.3.3 Relationship between Material Properties and the TLM Model Parameters

It is convenient to relate the parameters of the TLM model to the constitutive parameters of the MTM by considering the generic network model of 2D wave propagation in an isotropic homogeneous medium shown in Fig 4.19.

The series and shunt elements Z' and Y', specified in units of complex impedance (Ω/m) and complex shunt admittance per unit length (S/m), are related to the constitutive parameters μ and ε of the medium by the following equivalences

$$j\omega\mu = Z' \quad \text{and} \quad j\omega\varepsilon = Y'. \tag{4.47}$$

For infinitesimal cell size, the network equations and the 2D Maxwell equations for the TM-to-y case are isomorphic when the following identities are introduced

$$E_y \equiv v_y, \quad H_x \equiv -i_z, \quad H_z \equiv i_x. \tag{4.48}$$

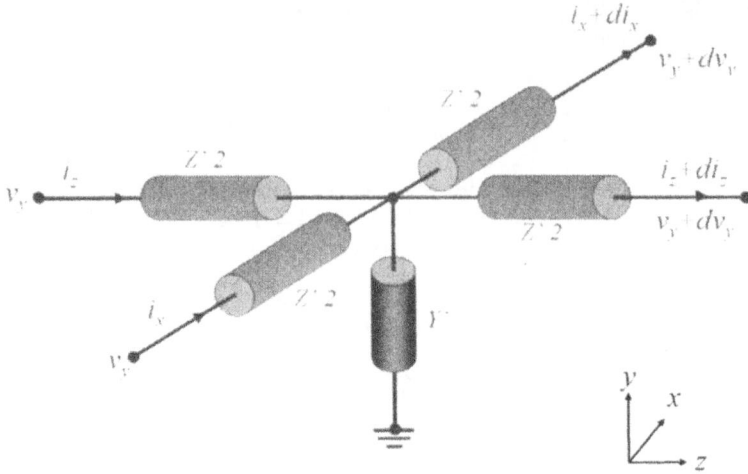

Fig. 4.19. Unit cell of a generic 2D distributed network that models a continuous medium at low frequencies. The elements are labeled in units of impedance and susceptance per unit length, respectively.

A regular material with positive constitutive parameters μ_n and ε_n is represented by Fig 4.15 if

$$j\omega\mu_n = Z' = j\omega L_R/\Delta\ell \quad \text{and} \quad j\omega\varepsilon_n = Y' = j\omega C_R/\Delta\ell, \tag{4.49}$$

resulting in the following relationships

$$L_R = \mu_n\Delta\ell \quad \text{and} \quad C_R = \varepsilon_n\Delta\ell. \tag{4.50}$$

However, in the case of a CRLH MTM with constitutive parameters μ_m and ε_m that is modeled by the equivalent network in Fig 4.2, we obtain

$$j\omega\mu_m = Z' = \frac{j\omega}{\Delta\ell}\left(L_R - \frac{1}{\omega^2 C_L}\right), \quad j\omega\varepsilon_m = Y' = \frac{j\omega}{\Delta\ell}\left(L_R - \frac{1}{\omega^2 L_L}\right), \tag{4.51}$$

and we can now determine the reactive loading elements from the parameters of the MTM and the host network

$$C_L = -\frac{1}{\omega^2\left(\Delta\ell\mu_m - L_R\right)} = -\frac{1}{\omega^2\Delta\ell\left(\mu_m - \mu_n\right)}, \tag{4.52a}$$

$$L_L = -\frac{1}{\omega^2\left(\Delta\ell\varepsilon_m - C_R\right)} = -\frac{1}{\omega^2\Delta\ell\left(\varepsilon_m - \varepsilon_n\right)}, \tag{4.52b}$$

where μ_n and ε_n are the intrinsic effective constitutive parameters of the host network without the embedded lumped elements. Note that the presence of the ω^2

term in the denominator of Eqs. (4.52a) and (4.52b) does not imply that C_L and L_L are dispersive. ω is a design parameter that is chosen a priori, but, once the design frequency has been selected, the resulting μ_m and ε_m are indeed frequency dispersive.

Alternatively, the MTM properties can be defined in terms of the refractive index n_m and intrinsic impedance Z_{cm} which yield μ_m and ε_m by means of the following expressions

$$\mu_m = \frac{n_m Z_{cm}}{c} \quad \text{and} \quad \varepsilon_m = \frac{n_m}{Z_{cm} c}. \tag{4.53}$$

Eqs. (4.52) and (4.53) complement the TL-material relations summarized in Table 3.4.

Finally, we must select the cell size $\Delta\ell$ of the discrete network; it should be much smaller than the guided wavelength at the highest frequency of interest. The choice should satisfy the condition

$$\frac{2\pi\Delta\ell}{\lambda_m} = \beta\Delta\ell = \Delta\ell\omega\sqrt{\mu_m\varepsilon_m} \ll 1. \tag{4.54}$$

With Eqs. (4.52a) and (4.52b), we can now determine the lumped elements that must be embedded into the TLM cell with the chosen size $\Delta\ell$, and Eqs. (4.37) and (4.43) yield the parameters of the link and stub lines of the TLM implementation shown in Fig 4.18.

4.3.4 Suitability of the TLM Approach for MTMs

The TLM algorithm of numerical analysis is based on a network model of Maxwell's field equations. It is thus a natural candidate for modeling the field properties of MTMs that are realized as loaded TL networks. Indeed, the modified TLM algorithm presented here is a numerical incarnation of the 2D CRLH arrays shown in Fig 4.3, and it therefore exhibits all the salient wave properties of these arrays.

Faustus Scientific Corporation [12] has incorporated the above algorithm as a standard material property feature in the commercial electromagnetic simulator MEFiSTo. The computational efficiency, accuracy, resolution, and flexibility of the TLM model and its implementation in MEFiSTo provide powerful CAD capabilities. The MEFiSTo tool is indeed a virtual test bed for researching, visualizing, and demonstrating the properties of media with negative refractive index and for developing innovative components based on CLRH materials. The combination of time domain modeling and interactive computer graphics allows us to observe processes that, previously, could only be imagined or described in abstract mathematical language; it enriches the perception to an extent rarely achieved by other tools in science or engineering and enhances our creative potential by leveraging the synergy of intellect, intuition, and experience.

The next section presents a number of examples and field plots for MTMs refractive effects computed by TLM with MEFiSTo. These examples demonstrate the wave properties of the loaded TL model of 2D CLRH materials and confirm Veselago's [13] and Pendry's [14] theoretical predictions. In particular, the phenomena of negative phase velocity and negative refraction in CLRH MTMs will be demonstrated.

4.4 NEGATIVE REFRACTIVE INDEX (NRI) EFFECTS[16]

It should be noted that, while the figures that will be shown in this section provide informative snapshots of field distributions at a given time, only a transient dynamic computer simulation can fully convey the complex field evolution in space and time, particularly the slow buildup of the plasmon-like surface wave resonances that can form at the interface between regular media and MTMs. Such transient simulations clearly illustrate the physics of wave propagation in MTMs and clarify some of the controversies that gave rise to heated exchanges on the subject of the "perfect lens" by Pendry [14]. Readers who wish to download movies of these numerical experiments or wish to explore the features of the simulator MEFiSTo may visit the web site of Faustus Scientific Corporation [12]. MEFiSTo also has a full 3D MTM modeling capability; the theoretical foundations of that algorithm can be found in [15].

4.4.1 Negative Phase Velocity

We first demonstrate the phenomenon of negative phase velocity (Sections 1.1, 2.1 and 3.1) by transmitting a uniform monochromatic plane wave through a slab of MTM, as shown in Fig. 4.20. At 10 GHz, the refractive index of the MTM is designed to be $n_m = -2$, ($\mu_r = -2$, $\varepsilon_r = -2$), and its characteristic impedance is 377 Ω. The slab is embedded in air, and the 10-GHz plane wave source is located on the left. The wave impedance of the slab is thus matched to the air, but its phase velocity is negative and half that in air. The electric field is polarized in the z-direction.

Fig 4.21 shows two snapshots of the electric field propagating across the MTM slab shown in Fig 4.20. The power flows from left to right in all subsections; the phase velocity, however, is positive in the air sections and negative in the MTM. Note that it takes many periods of oscillation before the steady-state shown in Fig 4.21 is reached. Quantitative analyses of the above waveforms and their displacement clearly show that

1. The wavelength in the MTM is half that in air.
2. Its phase velocity is negative and half that in air.
3. All subsections are matched since no scattering occurs at the material interfaces.

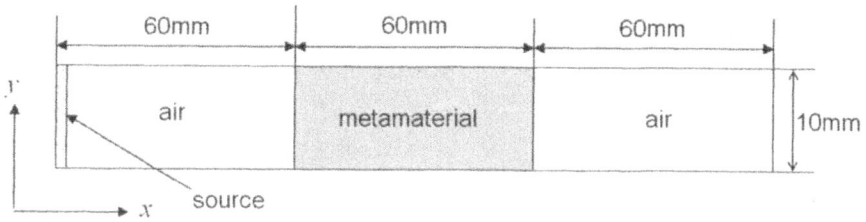

Fig. 4.20. Metamaterial slab (refractive index $n_m = -2$ and wave impedance $Z_{cn} = 377\ \Omega$ at 10-GHz) embedded in air. A z-polarized uniform 10 GHz TEM wave is incident normally from the left.

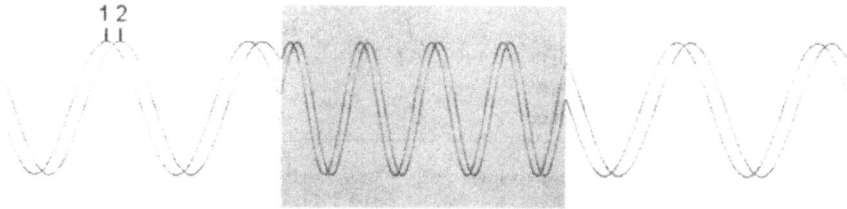

Fig. 4.21. Two snapshots of the electric field that propagates across the air and MTM sections at 10 GHz. Trace 1 changes into trace 2 after 10 ps. Clearly, the wave pattern in the air regions has shifted toward the right, while the pattern in the MTM has moved toward the left by half that distance. No scattering occurs at the interfaces by virtue of the impedance match. (*Waveforms generated with the MEFiSTo-3D Pro Electromagnetic Simulator by Faustus Scientific Corporation.*)

This simple experiment validates the accurate modeling of the characteristic impedance and the negative phase velocity by the modified TLM algorithm. The material and modeling parameters for this example are given in Table 4.1.

4.4.2 Negative Refraction

The phenomenon of negative refraction (Section 2.6) at an air-MTM interface is demonstrated in Fig 4.22. A monochromatic Gaussian beam is incident from the top upon a MTM slab at an angle of $14°$.[17] The MTM has the same properties as in the previous example ($n = -2$, $Z = 377\ \Omega$). The beam is refracted at an angle of $-7°$, as predicted by Snell's law. The impedance match between the media and the compression of the wavelength in the MTM slab are also clearly visible.

---

[16]This section was written by Wolfgang Hoefer, University of Victoria, British Columbia, Canada.
[17]Note that the problem of incidence of a Gaussian beam on LH interfaces has also been studied by Ziolkowski with the FDTD [16].

TABLE 4.1. Material and Modeling Parameters for the 1D Example Shown in Figs. 4.21 and 4.20.

	Parameter	Symbol	Value	Dim.
	Permeability	$\mu_n = \mu_0$	$1.2566E-06$	H/m
	Permittivity	$\varepsilon_n = \varepsilon_0$	$8.8542E-12$	F/m
Air	Refractive index	n_n	1	
	Phase velocity	c	$2.9979E+08$	m/s
	Wave impedance	η_0	$3.7673E+02$	Ω
	Permeability	$\mu_n = -2\mu_0$	$-2.5133E-06$	H/m
	Permittivity	$\varepsilon_n = -2\varepsilon_0$	$-1.7708E-11$	F/m
MTM	Refractive index	n_n	-2	
	Phase velocity	$-c/2$	$-1.4990E+08$	m/s
	Wave impedance	η_0	$3.7673E+02$	Ω
	Frequency	f	10	GHz
TL model	Cell size (Mesh)	$\Delta\ell$	0.5	mm
	Shunt inductance	L_L	$1.9072E-08$	H
	Series capacitance	C_L	$1.3438E-13$	F

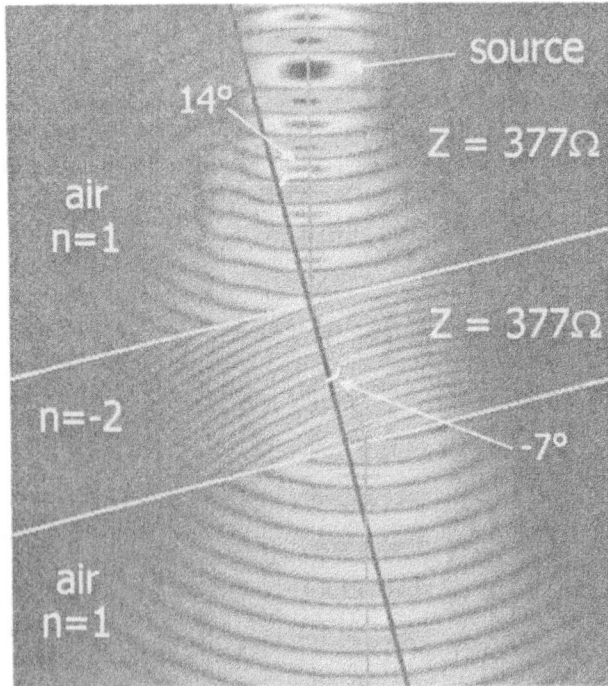

Fig. 4.22. Negative refraction of a monochromatic Gaussian beam incident on a MTM interface at an angle of 14°. The refracted angle is 14° predicted by Snell's law. (*Field pattern generated with the MEFiSTo-3D Pro Electromagnetic Simulator by Faustus Scientific Corporation.*)

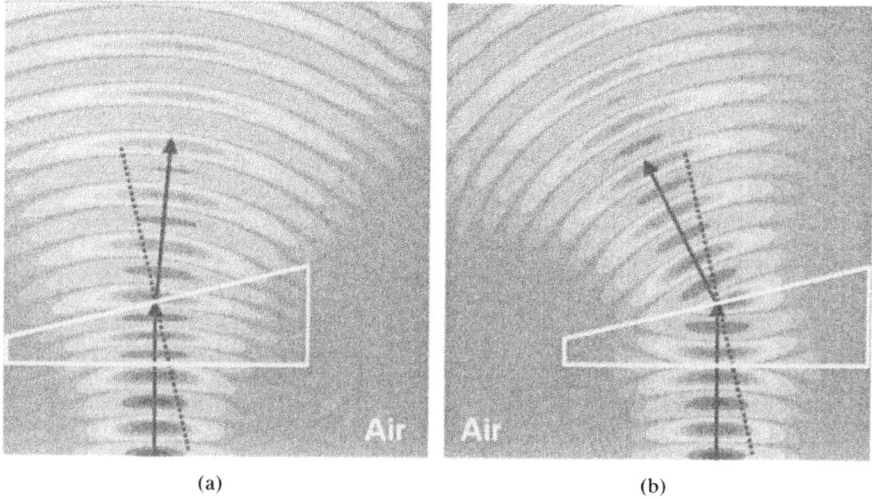

(a)   (b)

Fig. 4.23. Refraction of a monochromatic Gaussian beam. (a) Regular prism with $\varepsilon_r = \mu_r = n = 1.4142$. (b) MTM prism with $\varepsilon_r = \mu_r = n = -1$. (*Field pattern generated with the MEFiSTo-3D Pro Electromagnetic Simulator by Faustus Scientific Corporation.*)

A similar numerical experiment is depicted in Fig 4.23. Two prisms are shown side by side. The prism on the left is made of regular dielectric ($\varepsilon_r = \mu_r = n = 1.4142$), and the prism on the right is made of MTM ($\varepsilon_r = \mu_r = n = -1.00$). Both prisms are embedded in air. Identical monochromatic Gaussian beams, incident from the bottom, are refracted at the upper slanted face of the prisms. The incident angle at the upper surface is $13.5°$ in both cases. The refracted angle is $+19.3°$ for the regular prism (a) and $-13.5°$ for the MTM prism (b), as predicted by Snell's law. Both Figs. 4.22 and 4.23 not only confirm Veselago's prediction of negative refraction, but they also quantitatively validate the CRLH 2D network model against Snell's law.

4.4.3 Negative Focusing

The previous examples clearly demonstrate that the negative refractive index of a MTM causes a monochromatic beam to change direction in such a way that the refracted angle is negative. Closer inspection of Figs. 4.22 and 4.23(b) also reveals a slight focusing of the Gaussian beams. To show this focusing effect (Sections 2.6 and 2.7) more clearly, we consider the refraction of a cylindrical wave at a plane air-MTM interface, when the source is only a short distance away from that interface, perhaps of the order of one wavelength or less. This situation is shown in Fig 4.24 where two rays emanating from the source are incident on the interface between two semi-infinite media. Refracted rays are shown for the two cases of positive and negative refractive index of $+2$ and -2, respectively, as predicted by Snell's law. The situation is actually more complex than the ray diagram suggests since the interface is in the near-field of the source. Fig 4.25

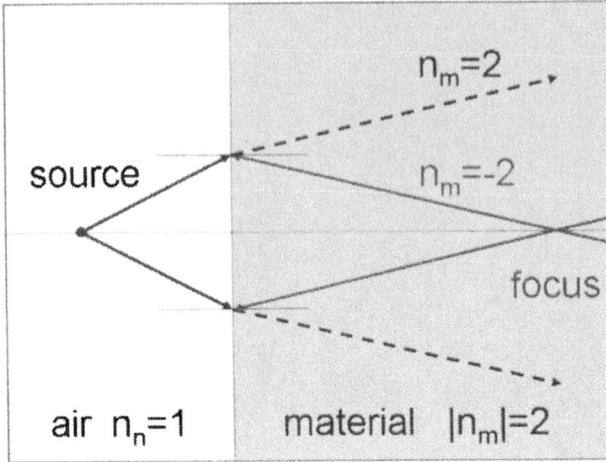

Fig. 4.24. Refraction of two rays of a cylindrical wave incident on an air-material interface for a positive (dashed line) and a negative refractive index of $+2$ and -2, respectively, as predicted by Snells law. In the second case, a focusing effect is predicted [see also Fig 4.25(b)].

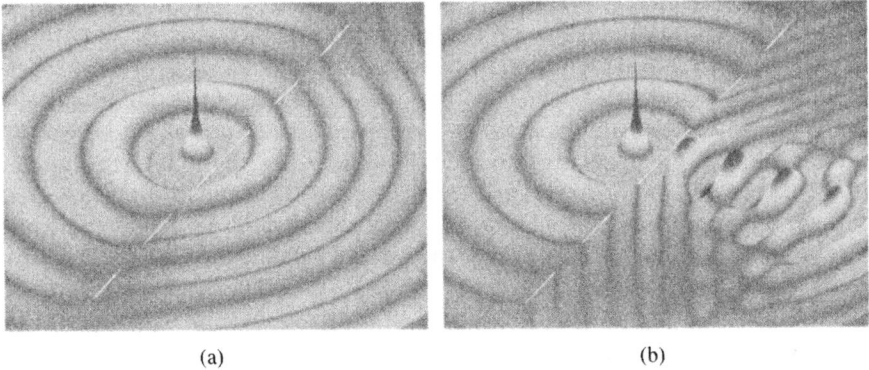

(a)   (b)

Fig. 4.25. Refraction of a 10 GHz cylindrical wave emanating from a point source in air at 9.5 mm from the air-material interface (white line). The TLM grid size is 0.5 mm. (a) Regular (RH) material with positive refractive index $n_m = +2$ ($\mu_r = \varepsilon_r = +2$). (b) MTM (LH) with negative refractive index $n_m = -2$ ($\mu_r = \varepsilon_r = -2$). (*Field pattern generated with the MEFiSTo-3D Pro Electromagnetic Simulator by Faustus Scientific Corporation.*)

shows the full-wave 2D solution for both cases. The instantaneous electric field normal to the plane of incidence is shown, and the focusing effect in the LH material is clearly visible. The guided wavelength is identical in both materials since the magnitude of the refractive index is the same but the phase velocities are of opposite signs, causing the wave pattern in the MTM to move toward the interface rather than away from it. This is also indicated in Fig 4.24 by the direction of the arrows that represent the wave vectors in the material. Note that the intrinsic impedances of both materials are equal to that of the air so that there is no reflection due to impedance mismatch.

4.4.4   RH-LH Interface Surface Plasmons

Fig 4.26 shows a thin slab of LH MTM of refractive index -1 embedded in an infinite RH material of refractive index $+1$. Their intrinsic impedances are identical. This case has been discussed and treated analytically by Pendry [14] who called it a "perfect lens." A simplified ray analysis predicts that the foci of this lens are situated in planes that are parallel to the faces of the slab and at distances equal to half its thickness. However, unlike traditional lenses with curved surfaces, Pendry's lens has no optical axis, and a point source can thus be placed anywhere in a focal plane, giving rise to two images, one in the center of the slab and one on the other side at equal distance. The resolution of the image produced by these rays is limited to about half a wavelength, just as in a conventional lens. However, if the slab is in the near field of the source, resonant surface waves are excited along the interfaces that resemble the surface plasmons (Section 2.11) observed on the surface of metals such as silver [17].[18] The TLM full-wave solution for Pendry's lens is presented in Fig 4.27. It takes many periods of oscillation before these resonances build up to a steady state,

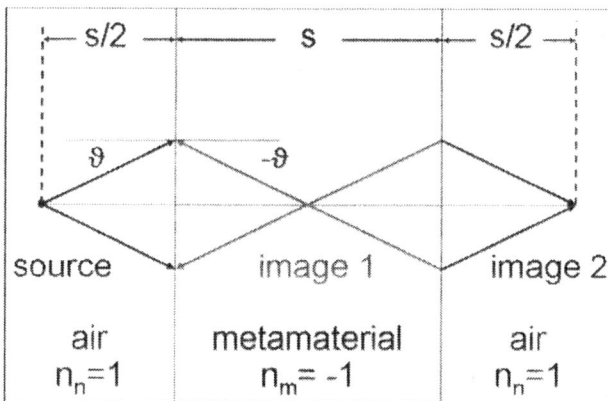

Fig. 4.26. Double refraction of two rays of a cylindrical wave incident on a MTM slab with negative refractive index of -1. The slab acts like a planar lens (Pendrys perfect lens), creating one inner and one outer image of the source.

[18]Surface plasmons at a RH/LH interface were first discussed by Ruppin [18]. Full-wave demonstration of these plasmons (pointed out to be similar to so-called Zenneck waves in the microwave regime) and the idea of exploiting them in compact microwave components was suggested by Caloz et al. in [19].

Fig. 4.27. Full-wave solution of the Pendry lens problem showing the source (white dot) on the left, two images, and the plasmon-like surface mode resonances. The TLM grid size is 1 mm. The MTM (LH) slab is 21 mm thick, with $n - \mu_r = \varepsilon_r = -1$. The electric field tangential to the slab is shown at 16.31 GHz in the steady-state. (*Field pattern generated with the MEFiSTo-3D Pro Electromagnetic Simulator by Faustus Scientific Corporation.*)

but, when it is reached, the near-field of the source is reconstituted at image 2 by the global interaction of the resonating "plasmons."

This phenomenon is often referred to in the literature as evanescent wave amplification, a term that is somewhat misleading since no power is generated in the MTM. In fact, resonant energy is deposited in the MTM over time by the source until the strong resonant fields reconstruct the image detail contained in the near field of the source. The high-Q resonant nature of this super-focusing effect implies that it is essentially a narrowband phenomenon and that it will be very sensitive to damping by losses in the slab. Pendry solved this problem analytically in the frequency domain, and Grbic and Eleftheriades [20] confirmed the super-resolution capability of the lens by experiment.

A second numerical experiment that shows the formation of highly localized resonances and the focusing effect in a MTM is presented in Fig 4.28. Here, we compare the inner reflection of a Gaussian beam by a regular RH prism and a MTM (LH) prism placed in air. Although the regular prism exhibits considerable radiation leakage and disperses the beam, the MTM prism not only reverses the beam with almost no leakage, but it also refocuses it as it exits downward. When the reflected beam passes the source plane, its width has shrunk virtually to its original girth at the source.

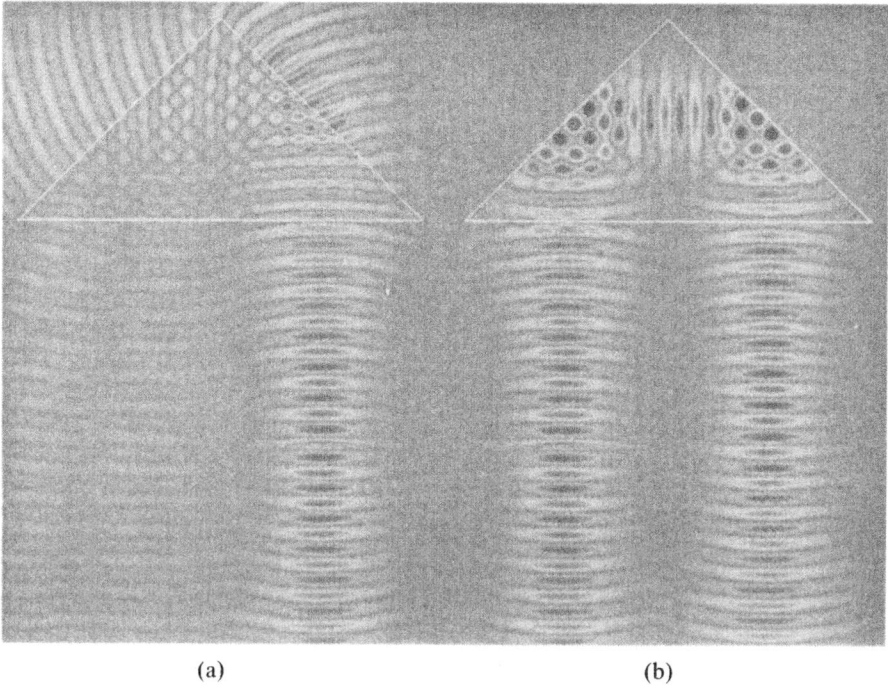

(a)　　　　　　　　　　　　　　　　　　(b)

Fig. 4.28. Inner reflection of monochromatic Gaussian beams by a regular and a MTM prism. The sources are located in air below the horizontal faces of the prisms. Note the standing wave resonances in the MTM and the clear focusing effect in (b). (a) Regular (RH) prism: $n = \mu_r = \varepsilon_r = 1.4142$. (b) MTM (LH) prism: $n = \mu_r = \varepsilon_r = -1$. The TLM grid size is 2 mm, and the frequency is 20 GHz. (*Field pattern generated with the MEFiSTo-3D Pro Electromagnetic Simulator by Faustus Scientific Corporation.*)

4.4.5 Reflectors with Unusual Properties

By extending the example of the planar lens in Figs. 4.26 and 4.27, we can create novel components with unconventional properties, such as the perfectly absorbing reflector, the perfectly retro-focusing reflector, and radiation-free directional interconnects. Such components have been proposed recently by So and Hoefer [21], and we will demonstrate here only one of them, which exhibits particularly strange and counter-intuitive properties, namely, the perfectly absorbing reflector. We observe in Fig 4.29 that Pendry's perfect lens effectively transposes the source into a symmetrical point on the other side of the slab so that the far-field radiated into the right half-space appears to emanate from that image. In fact, the lens simply shifts the field distribution along the left focal plane exactly into the right focal plane and vice versa. Hence, if we place a second source of equal amplitude and frequency on the opposite side of the lens but with a 180° phase shift, we should annihilate the far-field of the sources on both sides of the slab. The same effect can be produced with a single source by placing a perfect electric wall in the vertical symmetry plane of the lens, resulting in the "perfectly

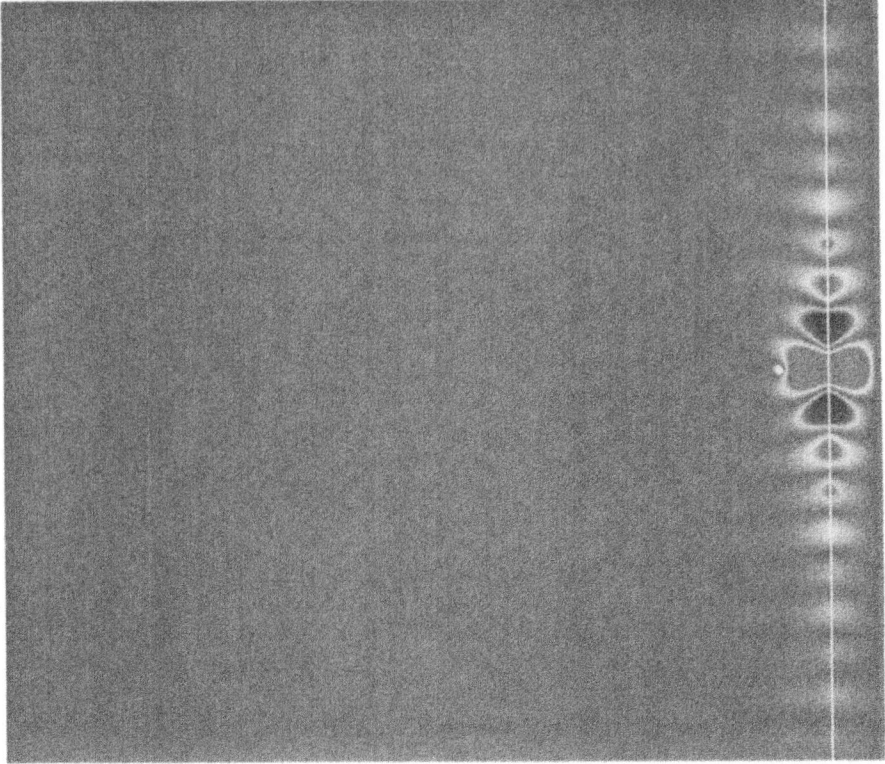

Fig. 4.29. Full-wave solution of the "perfectly absorbing reflector" created by placing a perfect conducting sheet across the middle of the Pendry lens in Fig 4.27. The negative image of the source and the "plasmons" extinguish the far-field, while the energy propagates as a surface wave along the slab. Modeling parameters as in Fig 4.27. (*Field pattern generated with the MEFiSTo-3D Pro Electromagnetic Simulator by Faustus Scientific Corporation.*)

absorbing reflector" shown in Fig 4.29. It shows half of the MTM slab, its right face clad with a perfectly conducting sheet. The resulting negative mirror image of the source and of the excited surface mode resonances cancel its far-field and the total power emitted by the source propagates as a surface waves along the slab. This is a very strange effect because the source is made invisible by placing a MTM-covered reflector behind it. Such a component can also be interpreted as a radiation-free interconnect between an omnidirectional source and a surface waveguide consisting of a metal-backed MTM slab guide. Further examples of such interconnects can be found in [11].

4.5 DISTRIBUTED 2D STRUCTURES

After establishing analytical tools and demonstrating fundamental phenomena for general 2D CRLH MTM networks, we will consider in this section specific

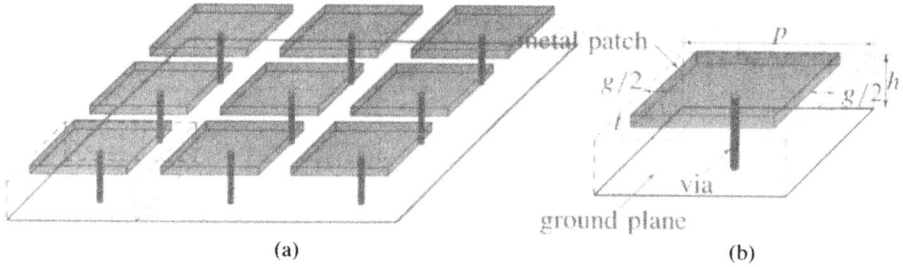

Fig. 4.30. Mushroom structure of [22]. (a) Overall structure. (b) Unit cell.

distributed 2D structures, compute their dispersion diagrams, present a method for extraction of their LC parameters, and finally demonstrate a fully distributed implementation of Pendry's lens.

4.5.1 Description of Possible Structures

A real 2D CRLH structure should correspond to the CRLH model depicted in Figs. 4.2 and 4.3. Various types of structures may be envisioned. In microstrip technology, a 2D CRLH structure would have a configuration of the type shown in Fig 4.30.

The configuration of Fig 4.30 is referred to as the *mushroom structure*. This structure was first introduced by Sievenpiper et al. [22] in the very different context (than that of NRI materials) of high-impedance surfaces.[19] An immediate difficulty with this mushroom structure for CRLH effects is the fact that the series capacitance C_L is very low because it is provided by edge coupling between the patches, which represents very weak capacitive coupling in general due to the extremely small electrical thickness of metal $t/\lambda_g \ll 1$.[20] As a consequence, the LH contributions may be so much weaker than the RH contributions that the dominant-mode LH bandwidth may be extremely narrow or even nonexistent and replaced by a RH band.

Fig 4.31 shows an alternative interdigital CRLH mushroom structure, which may be considered as a 2D extension of the 1D structure shown in Fig 3.39. This structure is *qualitatively* equivalent to the original mushroom structure of Fig 4.30, in the sense that is has the same equivalent circuit (Fig 4.30), but it tends to be *quantitatively* different because of the strong enhancement of C_L provided by the interdigital patterns. As a consequence, the interdigital structure may exhibit a dominant mode LH frequency band with parameters (period, substrate

---

[19]When this structure was introduced, it was essentially considered for its stopband characteristics (high-impedance surface) at *Bragg frequencies*, with applications such as surface waves suppression in planar antennas. It was *not* considered in its passbands and *not* recognized as a possible negative refractive index (NRI) structure in the *long wavelength regime* (although the dispersion diagram shown in [22] does show a very slightly negative slope dispersion curve in the dominant mode).

[20]This applies to microwave frequencies, but at millimeter-wave frequencies the ratio t/λ_g may become sufficiently significant for good CRLH operation.

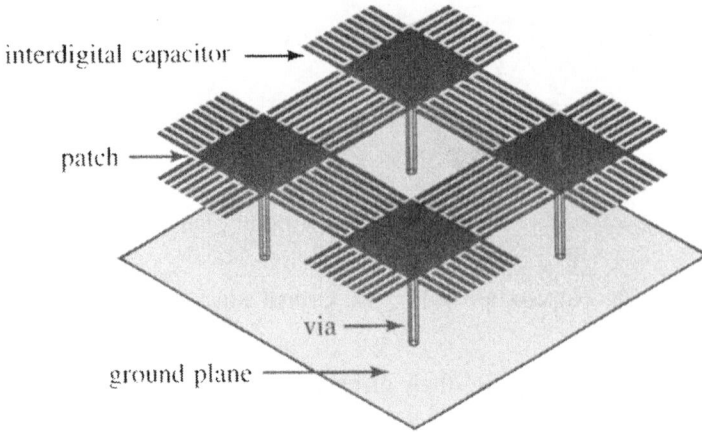

Fig. 4.31. 2D interdigital CRLH structure.

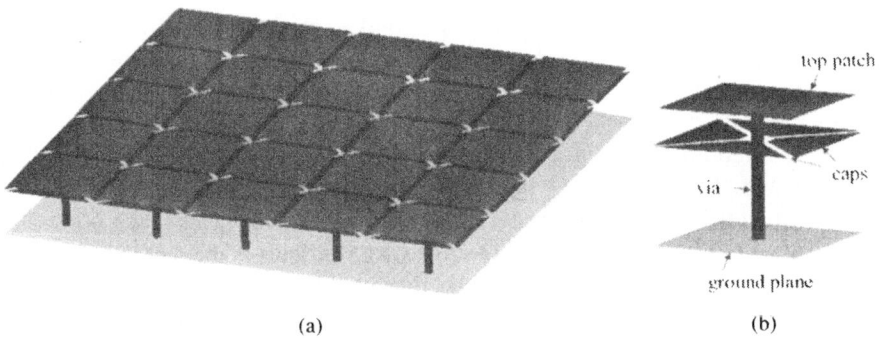

(a)   (b)

Fig. 4.32. Open 2D MIM CRLH structure. (a) Overall structure. (b) Unit cell. The caps are floating (not connected to the vias) patches located at a short distance of the connected patch for high MIM capacitance value. From [23], © IEEE; reprinted with permission.

permittivity and height) for which the mushroom structure fails to exhibit left-handedness.

Another *capacitively-enhanced* (C_L) 2D CRLH structure is the metal-insulator-metal (MIM) structure, shown in Figs. 4.32 and 4.33, which was proposed by Sanada et al. in [23]. This structure typically exhibits even greater left-handedness than that of Fig 4.31 because high series capacitance C_L can be easily achieved with small plates interspacing and/or high permittivity spacer.

Note that the structure of Fig 4.32 is *open* to the air, as the structures of Figs. 4.30 and 4.31. Fig 4.33 shows in contrast a *closed* structure, sandwiched between two metal plates, constituting a combination of two mirrored open structures. The open structure may be considered as a *microstrip-type* CRLH configuration, whereas the closed structure would be the CRLH counterpart of a *stripline-type* configuration.

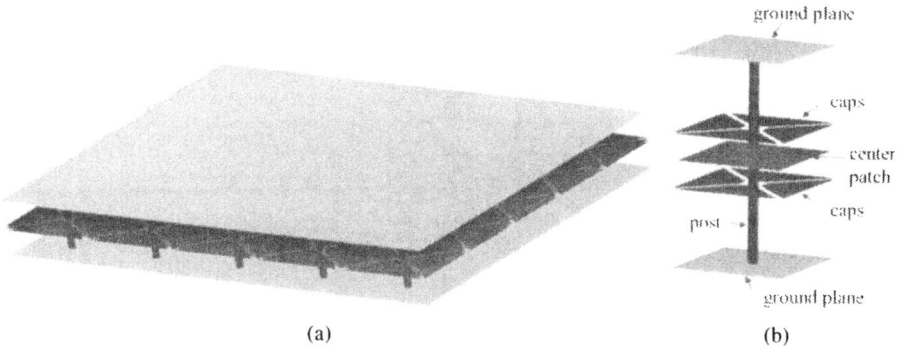

Fig. 4.33. Closed 2D MIM CRLH structure. (a) Overall structure. (b) Unit cell.

The open and closed mushroom structures present a fundamental difference in terms of dispersion characteristics. This difference will be explained in the next sections, but it can already be anticipated by inspection of their architectures. The dominant mode of the open structures will always couple to the (RH) TM air mode $\omega = \beta c$ of a metal plane [24], because as frequency is decreased toward zero the phase difference between adjacent cells becomes negligible so that EM waves do not "see" any difference between this structure and a simple flat metal plate. This will result in a *mixed dominant RH-LH mode* [see Fig 4.35(a)]. It is clear that the equivalent circuit of Fig 4.2, which is based on a TEM or quasi-TEM assumption, will then be meaningful only in the LH spectral region of this mixed mode. In contrast, the dominant mode of closed structures naturally does not suffer such parasitic coupling because, as frequency is decreased toward zero, the electrical plate spacing also decreases to zero and the structure becomes equivalent to a parallel-plate waveguide structure shorted by the inductive vias separated from each other by the capacitive patches. The corresponding equivalent circuit of the closed structure is thus *exactly* the one of Fig 4.2 at any frequency. Therefore, the closed mushroom structure exhibits exactly the CRLH behavior described in Sections 4.1 and 4.2, with the expected (nonmixed) LH characteristic at low frequency.

4.5.2 Dispersion and Propagation Characteristics

In this section, we will consider the open and closed MIM structures of Figs. 4.32 and 4.33 with the parameters indicated in Fig 4.34.

The dispersion diagrams are computed by full-wave (FEM) analysis with PBCs applied at the edges of the unit cell along both the x and y directions. The simulated dispersion diagrams for the open and closed structures are shown in Figs. 4.35(a) and 4.35(b), respectively, and the fields distributions of each mode are shown in Fig 4.36.

The fundamental difference announced in the previous section between the open and closed structures can be observed in the dispersion diagrams. In the

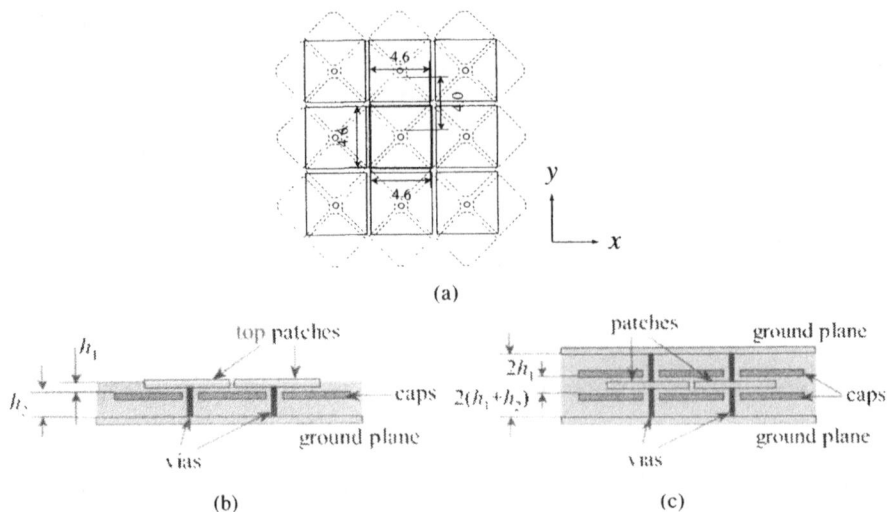

Fig. 4.34. Parameters of the MIM structures to be analyzed. The period of the unit cell is 5.0 mm × 5.0 mm (square lattice $p_x = p_y = p$). The two substrates used have a relative permittivity of 2.2 and thicknesses of $h_1 = 0.127$ mm and $h_2 = 1.57$ mm, respectively. The diameter of the vias is 0.2 mm (their length is $h_1 + h_2 = 1.697$ mm). (a) Dimensions of the patches and caps. (b) Side view of the open structure. (c) Side view of the closed structure. From [23], © IEEE; reprinted with permission.

case of the open structure, the first mode exhibits the expected negative slope, corresponding to left-handedness, from around one fourth of the $\Gamma - X$ and $\Gamma - M$ segments to the points X and M, respectively, but couples the TM air mode in the vicinity of the Γ point. Consequently, a pure LH operation is not possible; the dominant mode is *mixed RH-LH*. If the structure is excited in the LH range (e.g., $f \approx 3.7$ GHz), the positive-slope RH part of the dispersion curve is also intercepted,[21] and can therefore potentially induce RH effects if it is strongly excited. However the metastructure can be *spectrally matched* to the adjacent RH structures at a LH isofrequency curve of the dispersion diagram,[22] as will be shown in Fig 4.35 In contrast, in the case of the closed structure, the first mode is perfectly LH (purely negative slope from Γ to X and from Γ to M). This presents the twofold advantage over the open structure that no parasitic RH contributions can affect the LH behavior of the structure and that operation very close the Γ point, where the structure is highly effectively homogeneous, is possible.

If the capacitance enhancement caps are removed as in the high-impedance surface of [22], the LH band is shifted to higher frequencies by a factor of approximately 2 [Fig 4.35(b)], which suggests that the capacitance C_L is enhanced by a factor of four in the proposed structure.

---

[21] Tow isofrequency curves (see Figs. 4.5 and 4.6) exist in this case.

[22] Therefore sufficient mismatch of the RH part may be obtained to reduce the RH contributions to negligible effects in comparison with those of the LH contributions.

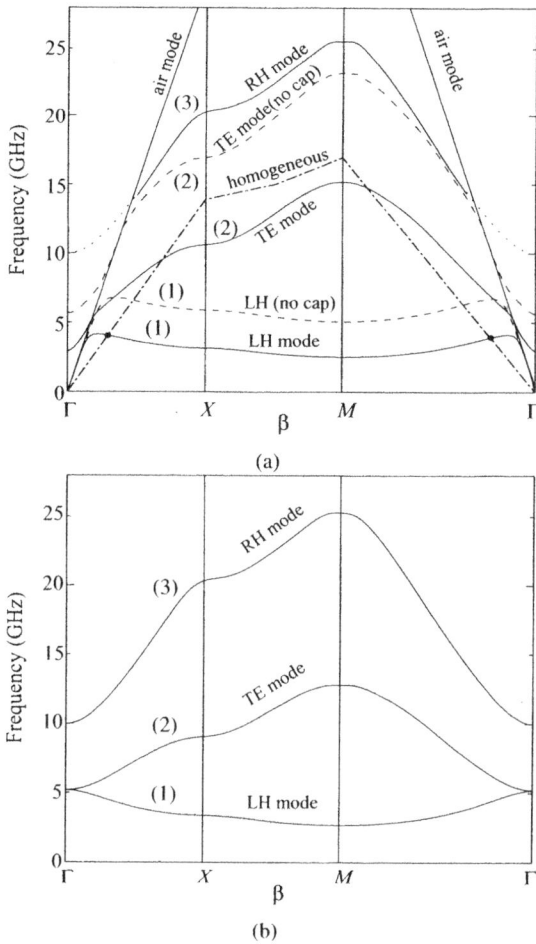

Fig. 4.35. Dispersion diagrams (first 2 or 3 modes) of the MIM mushroom structures of Fig 4.34 obtained (FEM) by full-wave simulation with PBCs. (a) Open structure [Fig 4.34(b)]. Dashed lines are for the open structure without the caps. The dashed-dotted line represents the dispersion curve of an homogeneous medium with $\varepsilon_r > 1$, such as for instance a PPWG structure, which would be interfaced with the MIM mushroom structure; this curve intercepts the mixed RH-LH dispersion curve at a LH spectral contour (Figs. 4.5 and 4.6) (spectral matching); consequently, only LH effects will be excited in the structure, which will therefore be well-matched to other RH (interfaced) structures at the corresponding frequency. (b) Closed structure (Fig 4.34(c)). From [23], © IEEE; reprinted with permission.

The observation of the field distribution in Fig 4.36, for the closed structure, provides precious information on the nature of the different modes. The field distributions in the open structure are qualitatively identical and are thus not shown here. The first (LH) mode is *quasi-TEM* in nature: although the magnetic field *locally* circulates around the vias, it is globally perpendicular to the direction of

(a)

(b)

(c)

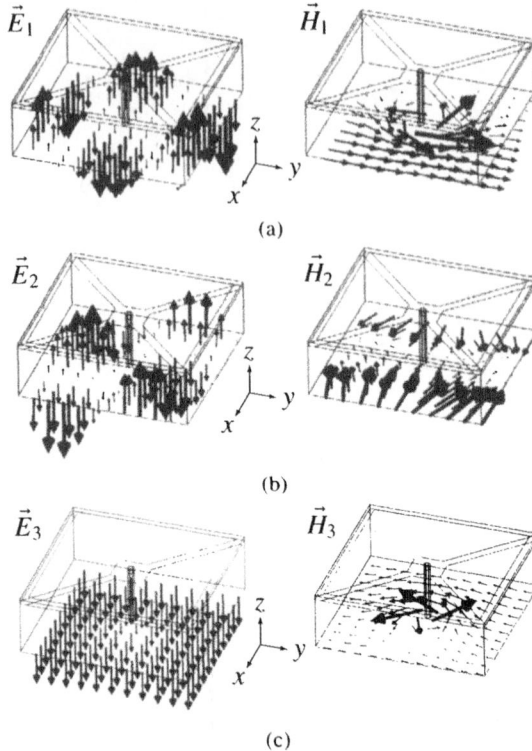

Fig. 4.36. Field distribution on the ground plane of the closed structure of Fig 4.33 at the spectral point $k_x p_x = \pi/20$, $k_y = 0$, corresponding to $p_x = \lambda_g/40$ (1/20th of the $\Gamma - X$ segment in Fig 4.35(b) (propagation along x). Only the lower half of the symmetric is shown for clarity. (a) First (dominant) quasi-TEM LH mode ($f = 5.206$ GHz). (b) Second TE RH mode, degenerated with the dominant quasi-TEM LH mode at the Γ point ($f = 5.286$ GHz). (c) Third quasi-TEM RH mode ($f = 10.012$ GHz). From [23], © IEEE; reprinted with permission.

propagation ($\perp \hat{x}$ and $\| \hat{y}$), as is the electric field ($\perp \hat{x}$ and $\| \hat{z}$). This mode corresponds to the dominant LH mode of the CRLH structure, as in the circuit theory (Fig 4.6). The second (RH) mode is completely different: the electric field is still perpendicular to the direction of propagation, but the magnetic field has a significant component along this direction[23]; this mode is therefore *TE* in nature; it is a *degenerate* mode which does *not* correspond to the CRLH RH continuation of the LH mode (Fig 4.6), in contrast to what Fig 4.35(b) could possibly suggest.[24] Thus, the design of Fig 4.35(b) is unbalanced. The CRLH RH continuation of

[23]In addition, there is practically no magnetic field circulating around the vias. Therefore, there is no equivalent shunt inductance (L_L), and the circuit model of Fig 4.2 clearly fails to describe this mode.

[24]This could be expected from the fact that the slopes of both the 1st and 2nd mode dispersion curves are zero at the Γ point, which corresponds to zero group velocity and therefore to stop bands (above for the 1st mode and below for the 2nd mode).

the LH 1st mode corresponding to CRLH circuit theory (Fig 4.6) is in fact the 3rd mode, as this mode is seen to also exhibit *quasi-TEM* fields. In summary, the 1st mode is the dominant quasi-TEM LH mode, the 3rd mode is the CRLH quasi-TEM RH counterpart of the first mode, with a gap due to unbalanced resonances from around 5 to 10 GHz, and the 2nd mode is a degenerate, TE, mode, which would require a different feeding mechanism. The fields distributions in Fig 4.36 are computed at a spectral point very close to the Γ axis, where effective-homogeneity is ensured.[25] The variations of the dispersion curves for the 1st and 2nd modes of these 2D structures induced by variations of the design parameters are identical to those of the 1D CRLH structures described in Section 3.4.

Fig 4.37 shows the refractive indexes of the open and closed MIM structures of Fig 4.34 computed from the dispersion characteristics of Fig 4.35. An alternative way to compute these refractive indexes is to use the analytical formulas of Eq. (4.17), as in Fig 4.7, after the extracted LC parameters following the procedure that will be exposed in Section 4.5.3. Both structures clearly exhibit a NRI. The closed structure has a refractive index with magnitude smaller than one (superluminal propagation) for $f > 4.95$ GHz. In contrast, the magnitude of the refractive index in the open structure is always larger than one due to coupling

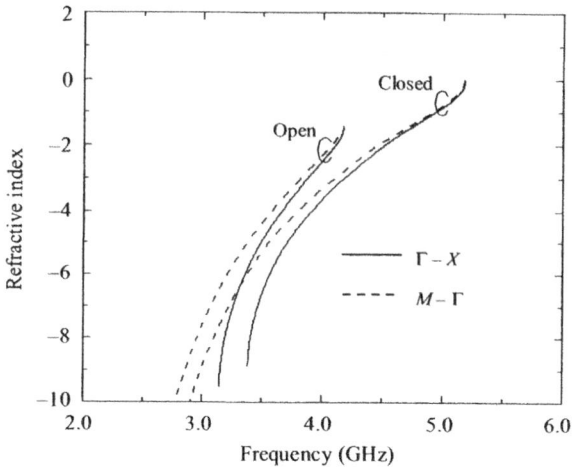

Fig. 4.37. Refractive indexes of the open and closed MIM structures of Fig 4.34 computed from the dispersion characteristics of Fig 4.35 with the formula $n = c\beta/\omega$. In the case of the open structure, only the negative part of the refractive index, corresponding to the negative slopes in Fig 4.35(a), is shown. The refractive index can also have a positive contribution (from $f = 0$ to $f \approx 4.2$), but this contribution will be minimized by mismatch (Fig 4.35). From [23], © IEEE; reprinted with permission.

---

[25]As we move away from the Γ point to the X and M points, the quasi-TEM nature of the 1st and 3nd modes is progressively lost, and expected (Bragg regime) resonant distributions are observed at the X and M points.

with the TM surface mode. The difference between the refractive indexes along the $\Gamma - X$ and $\Gamma - M$ directions is smaller than 7.5% for $\beta p < \pi/2$, which shows that the structure is essentially homogeneous and isotropic.

4.5.3 Parameter Extraction

The parameter extraction procedure for a 2D structure is significantly different from that of a 1D structure, presented in Section 3.3.3 for the case of the microstrip CRLH structure of Fig 3.39.

A useful quantity for the parameters extraction is the *Bloch impedance* (Section 3.2.7), which is defined in Eq. (3.125) for the 1D case. In the 2D case, the Bloch impedance in one direction (x or y) of the network may be defined as the 1D Bloch impedance [Eq. (3.125)] along that direction with open-circuited ports in the other direction,[26] which results in the same expression as that of Eq. (3.132)

$$Z_B = Z_L \sqrt{\frac{(\omega/\omega_{se})^2 - 1}{(\omega/\omega_{sh})^2 - 1} - \left\{ \frac{\omega_L}{2\omega} \left[\left(\frac{\omega}{\omega_{se}} \right)^2 - 1 \right] \right\}^2}, \qquad (4.55)$$

where

$$Z_L = \sqrt{\frac{L_L}{C_L}}, \qquad (4.56)$$

and where we recall that $Z_B(\omega_{se}) = 0$ and $Z_B(\omega_{sh}) = \infty$ [Eq. (3.132)]. Referring to Fig 4.38, the Bloch impedance can be computed from full-wave analysis by any of the following three formulas, combining the voltage V, the current I, and the power P at the unit cell port,

$$Z_{B.PI} = \frac{2P}{|I|^2}, \qquad (4.57a)$$

$$Z_{B.VI} = \frac{|V|}{I}, \qquad (4.57b)$$

$$Z_{B.PV} = \frac{|V|^2}{2P}, \qquad (4.57c)$$

where the values of P, I, and V may be computed from the $\overline{E}$ and $\overline{H}$ fields in the structure by

$$P = \frac{1}{2} \iint_S \left(\overline{E} \times \overline{H}^* \right) \cdot \hat{n}_p \, dS, \qquad (4.58a)$$

---

[26]In MTMs, due to effective-homogeneity, the Bloch impedance can be defined in any azimuthal direction. If in addition the material is isotropic, then Z_B is the same in all directions.

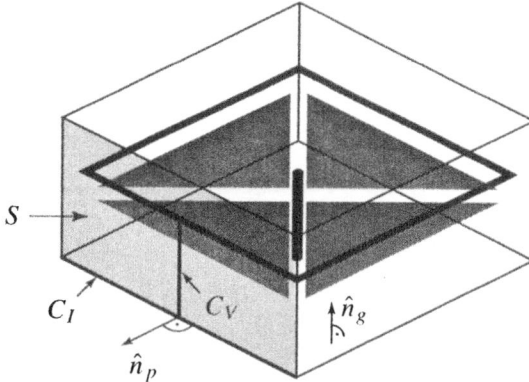

Fig. 4.38. Integration surface S and paths C_I and C_V for power, current, and voltage computation in the unit cell for the formulas of Eq. (4.58). The structure shown is that of the MIM mushroom structure of Fig 4.32, but these definitions hold for any 2D structure.

$$P = \frac{1}{2} \iint_S \left(\overline{E} \times \overline{H}^* \right) \cdot \hat{n}_p \, dS, \qquad (4.58b)$$

$$I = \int_{C_I} \left(\hat{n}_g \times \overline{H}^* \right) \cdot \hat{n} \, dl, \qquad (4.58c)$$

where S is the cross section surface of the port, C_I and C_V are the integration paths for the current and voltage, respectively, and $\hat{n}_p$ and $\hat{n}_g$ are the unit vectors normal to the port and the ground plane, respectively, as shown in Fig 4.38. I corresponds to the total current on the ground plane across the input port, and V corresponds to the voltage at the center of the patch from the ground in the reference plane of the port.

Eqs. (4.8) and (4.55) provide the dispersion relation and the Bloch impedance as a function of the circuit parameters L_R, C_R, L_L, and C_L, respectively. By inverting these relations and identifying them with the dispersion diagram (similar to that shown in Fig 4.6) and Bloch impedance computed by full-wave simulation, we can determine the values of L_R, C_R, L_L and C_L.

For the sake of clarity, let us recall the relations between the parameters L_R, C_R, L_L, and C_L and the frequencies ω_R, ω_L, ω_{se}, and ω_{sh}

$$\omega_R = \frac{1}{\sqrt{L_R C_R}}, \qquad (4.59a)$$

$$\omega_L = \frac{1}{\sqrt{L_L C_L}}, \qquad (4.59b)$$

$$\omega_{se} = \frac{1}{\sqrt{L_R C_L}}, \qquad (4.59c)$$

$$\omega_{sh} = \frac{1}{\sqrt{L_L C_R}}, \qquad (4.59d)$$

where note that only three of these four equations are independent because of the relation

$$\omega_R \omega_L = \omega_{se} \omega_{sh}, \qquad (4.60)$$

and the relation

$$Z_L = \sqrt{\frac{L_L}{C_L}}. \qquad (4.61)$$

The frequencies ω_R, ω_L, ω_{se} can be calculated from the frequencies $\omega_{\Gamma 1}$, ω_{X1}, and ω_{M1} (where the subscript 1 refers to the first mode) obtained by solving the eigenvalue problem for the structure using full-wave analysis. From the dispersion relations of Eqs. (4.15a) and (4.15b), $\omega_{\Gamma 1}$, ω_{X1}, and ω_{M1} satisfy

$$\chi(\omega_{\Gamma 1}) = 0, \qquad (4.62a)$$

$$\chi(\omega_{X1}) = 4, \qquad (4.62b)$$

$$\chi(\omega_{M1}) = 8, \qquad (4.62c)$$

since $k_x(\omega_{\Gamma l}) = 0$, $k_x p_x(\omega_{Xl}) = \pi$, and $k_u p_u(\omega_{Ml}) = \pi)$ $(u = x, y)$, $\forall l$.

Eq. (4.62) with Eq. (4.11a) represents a linear system of three equations with the three unknowns ω_R, ω_L, ω_{se}, after eliminating ω_{se} using Eq. (4.60), expressed in terms of the known (from the dispersion diagram) variables $\omega_{\Gamma 1}$, ω_{X1}, and ω_M. The solution to this system is

$$\omega_{se} = \omega_{\Gamma 1}, \qquad (4.63a)$$

$$\omega_L^2 = 4 \frac{\left(\omega_{M1}^2 - \omega_{\Gamma 1}^2\right) - 2\left(\omega_{X1}^2 - \omega_{\Gamma 1}^2\right)}{\left(\omega_{M1}^2 - \omega_{\Gamma 1}^2\right)\left(\frac{1}{\omega_{X1}^2} - \frac{1}{\omega_{\Gamma 1}^2}\right) - \left(\omega_{X1}^2 - \omega_{\Gamma 1}^2\right)\left(\frac{1}{\omega_{M1}^2} - \frac{1}{\omega_{\Gamma 1}^2}\right)}, \qquad (4.63b)$$

$$\omega_R^2 = \frac{\omega_{X1}^2 - \omega_{\Gamma 1}^2}{4 - \omega_L^2\left(\frac{1}{\omega_{X1}^2} - \frac{1}{\omega_{\Gamma 1}^2}\right)}. \qquad (4.63c)$$

Here, we have assumed $\omega_{\Gamma 1} = \omega_{se}$ (and $\omega_{\Gamma 2} = \omega_{sh}$), by anticipating the case of the experimental structure which will be presented in the next section. However, $\omega_{\Gamma 1} = \omega_{sh}$ is also possible a priori (depending on the LC values). A simple test to determine which of ω_{se} or ω_{sh} corresponds to the first eigenfrequency, $\omega_{\Gamma 1}$, is to look at the value obtained for Z_B at the Γ point. According to Eq. (4.55), if $Z_B = 0$, then $\omega_{\Gamma 1} = \omega_{se}$; if $Z_B = \infty$, then $\omega_{\Gamma 1} = \omega_{sh}$.

In addition, Z_L is expressed in terms of the Bloch impedance [computed by Eq. (4.57)] at any point of the dispersion diagram, for instance ω_{X1}, as

$$Z_L^2 = \frac{Z_B^2 (\omega_{X1})}{\dfrac{(\omega_{X1}/\omega_{se})^2 - 1}{(\omega_{X1}/\omega_{sh})^2 - 1} - \left\{ \dfrac{\omega_L}{2\omega_{X1}} \left[\left(\dfrac{\omega_{X1}}{\omega_{se}} \right)^2 - 1 \right] \right\}^2}. \tag{4.64}$$

In summary, the procedure for extraction of the parameters L_R, C_R, L_L, and C_L is as follows:

1. Full-wave compute the first eigenfrequency at the symmetry points Γ, X, and M of the dispersion diagram, $\omega_{\Gamma 1}$, ω_{X1}, and ω_{M1}, using the PBCs $k_x p_x = k_y p_y = 0$, $k_x p_x = \pi$ and $k_y p_y = 0$, and $k_x p_x = k_y p_y = \pi$, respectively.

2. Compute ω_{se} by Eq. (4.63a), ω_L by Eq. (4.63b), and ω_R by Eq. (4.63c). By Eq. (4.60), ω_{sh} is then also determined.

3. Introduce the fields obtained by full-wave analysis at the X point into the two relations of Eq. (4.58) required to compute $Z_B (\omega_{X1})$ by a selected formula in Eq. (4.57).

4. Insert the obtained value for $Z_B (\omega_{X1})$ into Eq. (4.64) to determine Z_L.

5. Then all the parameters are obtained by Eqs. (4.61) and (4.59) by using successively

$$L_L = \frac{Z_L}{\omega_L}, \tag{4.65a}$$

$$C_L = \frac{1}{L_L \omega_L^2}, \tag{4.65b}$$

$$L_R = \frac{1}{C_L \omega_{se}^2}, \tag{4.65c}$$

$$C_R = \frac{1}{L_R \omega_R^2}. \tag{4.65d}$$

With the availability of the four CRLH parameters L_R, C_R, L_L, and C_L, all of the efficient circuit-based eigenvalue and driven-solution techniques described in Sections 4.1, 4.2, and 4.3 can be straightforwardly used.

The Bloch impedance and the Z_L impedance of the *closed* structure of Fig 4.32 as a function of the spectral variable β along the $\Gamma - X$ segment ($\beta = k_x$) are shown in Fig 4.39. The three impedances $Z_{B.PI}$, $Z_{B.VI}$, and $Z_{B.PV}$ agree well with each other. The impedance Z_L is almost constant, as expected,[27] with

---

[27]It was already pointed out in Chapter 3 that, although a CRLH TL is naturally dispersive, it may be constituted by essentially nondispersive (independent on frequency) LC components over the frequency range of interest; this assumption is made successfully in all the applications presented in the next chapters.

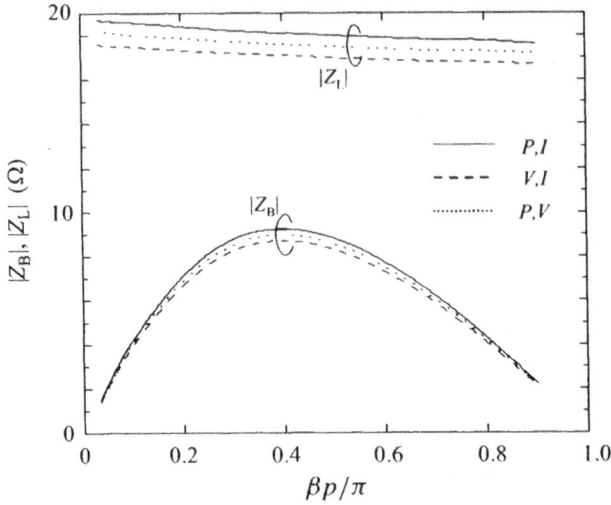

Fig. 4.39. Bloch impedance and Z_L of the closed structure of Fig 4.32 as a function of the spectral variable β along the $\Gamma - X$ segment ($\beta = k_x$). The Bloch impedances are computed from (FEM) full-wave simulation using the formulas of Eq. (4.57) with Eq. (4.58) and Z_L is computed from the Bloch impedances using Eq. (4.64). Solid line: using Eq. (4.58a); dashed line: using Eq. (4.58b); dotted line: using Eq. (4.58c). From [23], © IEEE; reprinted with permission.

an average value of $Z_L = 17.30$ Ω. The fact that the Bloch impedance tends to zero at $\omega_{\Gamma1}$ indicates, according to Eq. (4.64), that $\omega_{\Gamma1} = \omega_{se}$, and consequently $\omega_{\Gamma2} = \omega_{sh}$. The equivalent circuit parameters L_R, C_R, L_L, and C_L are computed following the procedure described above. Fig 4.35(b) provides the parameters $\omega_{\Gamma1}/(2\pi) = 5.199$ GHz, $\omega_{\Gamma X}/(2\pi) = 3.318$ GHz, and $\omega_{\Gamma M}/(2\pi) = 2.695$ GHz. From these values we obtain from Eqs. (4.63b) and (4.63c) $\omega_L/(2\pi) = 9.453$ GHz and $\omega_R/(2\pi) = 4.442$. Therefore, all the circuit parameters can be obtained with the formulas of Eq. (4.65). The results are $L_R = 0.963$ nH, $C_R = 1.333$ pF, $L_L = 0.291$ nH, and $C_L = 0.973$ pF.

The Bloch impedance, the Z_L impedance, and the LC equivalent parameters of the *open* structure of Fig 4.33 are computed in the same manner. The Bloch and Z_L impedances are shown in Fig 4.40. Here, the Γ point frequency cannot be directly obtained from the dispersion curve because of the couplings of the LH/RH modes with the TM surface wave mode. However, one can determine this frequency by extrapolating the dispersion curve of the $\Gamma - X$ segment with the condition that Z_L calculated from the Bloch impedance has to be constant. The Bloch impedances corresponding to different Γ points are shown in Fig 4.40, and the Γ frequency yielding a constant Z_L is $\omega_{\Gamma1}/(2\pi) = 4.771$ GHz. The other X and M points are readily available at $\omega_{X1}/(2\pi) = 3.046$ GHz and $\omega_{\Gamma1}/(2\pi) = 2.445$ GHz. From these results, we find $\omega_L/(2\pi) = 8.307$ GHz and $\omega_R/(2\pi) = 5.748$, and the resulting circuit parameters are $L_R = 1.851$ nH, $C_R = 0.414$ pF, $L_L = 0.611$ nH, and $C_L = 0.601$ pF.

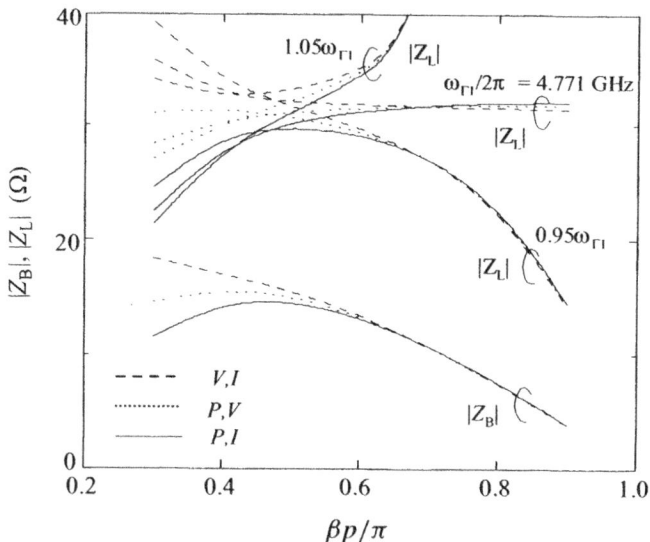

Fig. 4.40. Bloch impedance and Z_L of the open structure of Fig 4.33 as a function of the spectral variable β along the $\Gamma - X$ segment ($\beta = k_x$). The Bloch and Z_L impedances are computed in the same manner as for the closed structure in Fig 4.39. From [23], © IEEE; reprinted with permission.

   The dispersion diagrams computed from the extracted LC parameters are compared with the full-wave analysis results in Fig 4.41. Excellent agreement is observed.[28]

4.5.4   Distributed Implementation of the NRI Slab

We will now present a distributed implementation of the NRI focusing slab or "Pendry's lens," described in Section 2.7 and illustrated by TLM fields distributions in Section 4.4.3, using the planar MIM mushroom structure characterized in Section 4.5.2 [23].

   The LH slab is constituted by a MIM mushroom rectangular section sandwiched between two parallel-plate waveguides (PPWG) and excited by a coaxial probe, as shown in Fig 4.42. The design parameters and dispersion diagram of the considered structure are those of Fig 4.34(b) and Fig 4.35, respectively, corresponding to the extracted CRLH parameters given in Fig 4.41. The MIM mushroom structure is operated in its mixed LH-RH range (Sections 4.5.1 and 4.5.2) and matched to the PPWG structure, as shown in Fig 4.35, with an operation frequency of $f = 3.737$ GHz. At this frequency, the Bloch impedance

---

[28]The RH mode (3rd quasi-TEM mode in Fig 4.35) is not shown (higher frequencies) but can also be accurately predicted by the circuit theory. In contrast the (2nd TE mode in Fig 4.35), not corresponding to the CRLH model of Fig 4.2, will not be obtained.

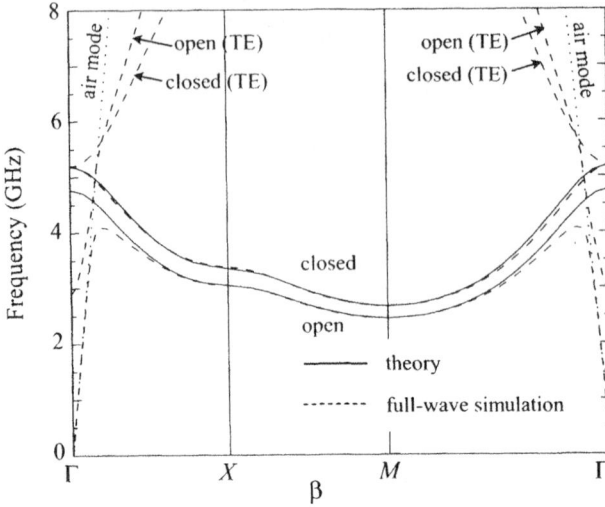

Fig. 4.41. Comparison between the theoretical and full-wave dispersion diagrams (dominant mode) of the open and closed structure (full-wave results reported from Fig 4.35). Solid line: theoretical curves computed by Eq. (4.15) with the LC values obtained by parameter extraction using formulas Eq. (4.65). The circuit parameters are $L_R = 0.963$ nH, $C_R = 1.333$ pF, $L_L = 0.291$ nH, $C_L = 0.973$ pF for the closed structure and $L_R = 1.851$ nH, $C_R = 0.414$ pF, $L_L = 0.611$ nH, $C_L = 0.601$ pF for the open structure. From [23], © IEEE; reprinted with permission.

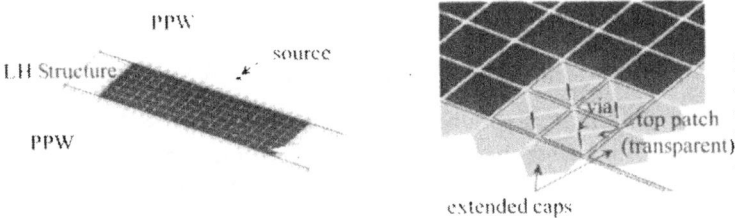

Fig. 4.42. Planar-distributed MIM mushroom NRI slab. The permittivity of the PPWG structure is $\varepsilon_r = 10.2$. The mushroom structure includes 20×6 unit cells. From [23], © IEEE; reprinted with permission.

[Eq. (4.55)] and the refractive index [Eq. (4.17)] calculated from the circuit parameters are $Z_B = 16.38$ Ω and $n_L = -3.40$, respectively.

Before resorting to heavy and time-consuming full-wave simulation of the overall structure, it is a good practice to perform circuit simulations using the extracted LC parameters of the unit cell using the TMM (Section 4.2) or the TLM (Section 4.3). The corresponding network architecture is shown in Fig 4.43.

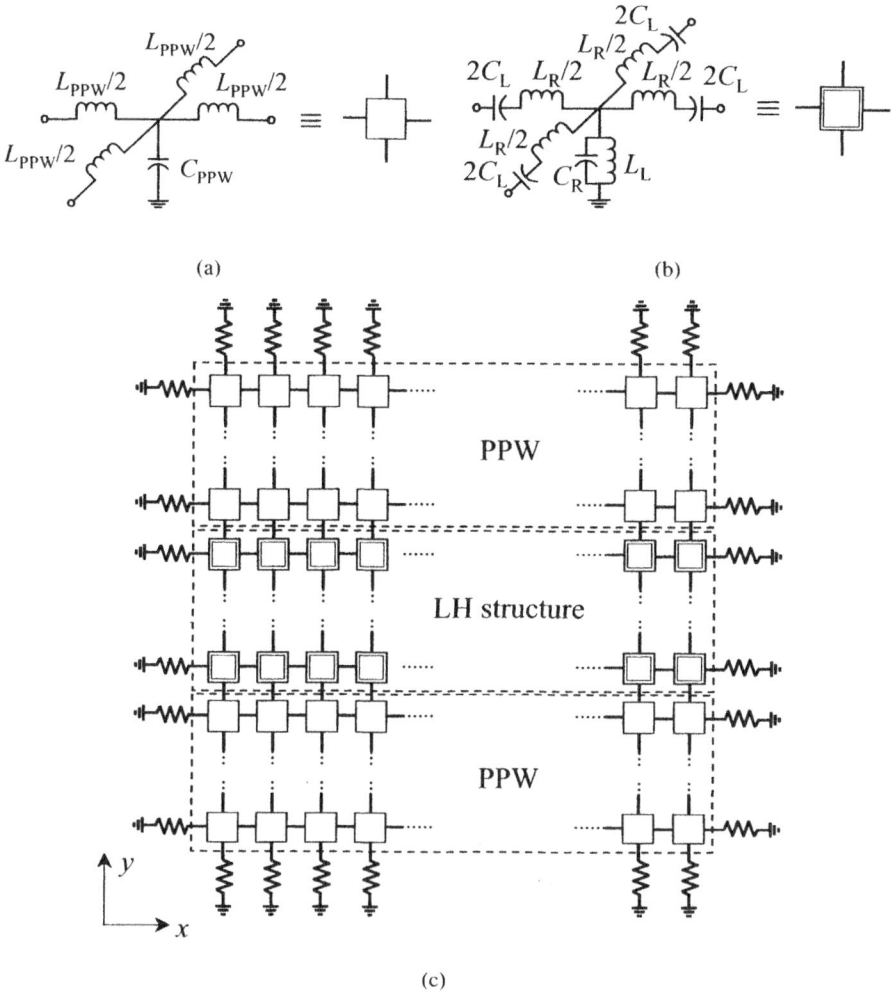

(a)　　　　　　　　　　　(b)

(c)

Fig. 4.43. LC network circuits for the TMM or TLM analysis of the NRI slab. (a) PPWG unit cell. (b) CRLH unit cell. (c) Complete NRI network corresponding to the structure shown in Fig 4.42. A voltage source is connected to a node in the middle of one of the RH circuits. All the unit cells at the edges are terminated by Z_B ($= 16.38$ Ω). From [23], © IEEE; reprinted with permission.

Referring to Fig 3.6, the (TEM) PPWG characteristic impedance Z_{PPWG}, per-unit-length inductance L'_{PPWG}, and per-unit-length capacitance $C'_{0,PPWG}$ are [5]

$$Z_{c,PPWG} = \frac{\eta d}{w} \left(= \sqrt{\frac{L'_{PPWG}}{C'_{PPWG}}} \right),$$

(4.66a)

$$L'_{\text{PPWG}} = \frac{\mu d}{w} = n_{\text{PPWG}} \frac{X_{c,\text{PPWG}}}{c},$$

(4.66b)

$$C'_{\text{PPWG}} = \frac{\varepsilon w}{d} = \frac{n_{\text{PPWG}}}{c X_{c,\text{PPWG}}},$$

(4.66c)

where η is the intrinsic impedance of the dielectric, d is the spacing between the plates, and w is the width of the PPWG structure in the direction perpendicular to propagation. In the last two relations, the refractive index n_{PPWG} is consistently obtained from to the RH terms of Eq. (3.23) as

$$n_{\text{PPWG}} = \sqrt{\varepsilon_r \mu_r} = \frac{\sqrt{\varepsilon \mu}}{c} = \frac{\sqrt{L'_{\text{PPWG}} C'_{\text{PPWG}}}}{c}.$$

(4.67)

Matching between the PPWG structure and the MIM mushroom structure requires the fulfillment of two conditions. First, the characteristic impedance of the PPWG must be equal to the Bloch impedance of the MIM mushroom structure.[29] This is the *Bloch / characteristic impedance matching condition*, which is related to the overall finite size (here w) of the structure and depends on the *ratio* of the TL parameters

$$Z_{c,\text{PPWG}} = \sqrt{\frac{L'_{\text{PPWG}}}{C'_{\text{PPWG}}}} = Z_{B,\text{(mushroom)}}.$$

(4.68)

Second, the refractive indexes of the two structures must be equal in magnitude. This is the *refractive index matching condition* or intrinsic impedance matching condition, which is related only to the internal parameters of the structure (not to PPWG structure width w here) considered of infinite extent and depend on the *product* of the TL parameters

$$n_{\text{PPWG}} = \frac{\sqrt{L'_{\text{PPWG}} C'_{\text{PPWG}}}}{c} = |n_{\text{(mushroom)}}| = |n_L|.$$

(4.69)

Applying these two conditions to Eqs. (4.66b) and (4.66c) provides the unit cell inductance and capacitance of the PPWG

$$L_{\text{PPWG}} = L'_{\text{PPWG}} \cdot p = |n_L| \frac{Z_B}{c} p,$$

(4.70a)

$$C_{\text{PPWG}} = C'_{\text{PPWG}} \cdot p = |n_L| \frac{1}{c Z_B} p.$$

(4.70b)

[29]The reason why matching to the MIM mushroom structure is rigorously related to the Bloch impedance and not to the characteristic impedance is that matching is a local phenomenon at the interface, involving the local impedance which is, by definition, the Bloch impedance. However, it was pointed out in Eq. (3.130) that in a metamaterial Z_B is close to Z_c because p is electrically small.

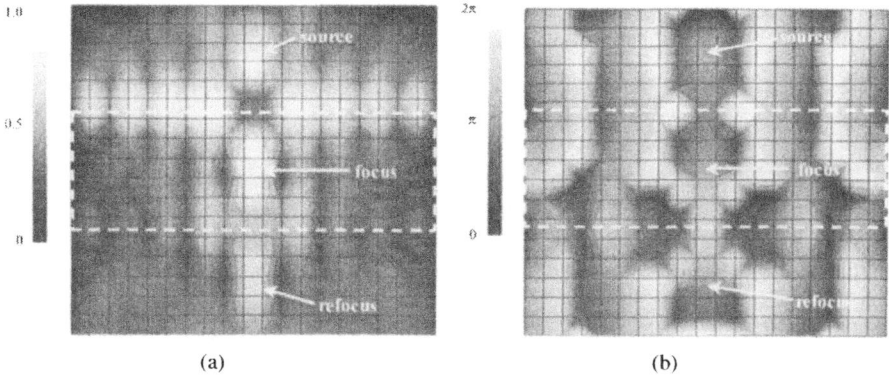

(a)   (b)

Fig. 4.44. Voltage distribution in the NRI slab network circuit of Fig 4.43 in the case of perfect matching obtained by Eq. (4.70) with $L_{PPWG} = 0.928$ nH and $C_{PPWG} = 3.459$ pF. (a) Normalized magnitude. (b) Phase. From [23], © IEEE; reprinted with permission.

Fig 4.44 shows the magnitude and phase distributions of the voltages, respectively, at the nodes of the network circuit. High-voltage magnitude concentrations are clearly observed at the expected locations of focus and refocus predicted by Eqs. (2.56) and (2.58), and relatively concentric voltage phase distributions are observed around the focus and refocus points, which verifies the NRI characteristics of the LH TL slab. Some degree of aberration (Section 2.7 and Fig 2.7), due to the slight anisotropy of the structure (Fig 4.37), is apparent. In addition to focus and refocus concentration, field concentrations, corresponding to *surface plasmons* (Section 4.4.4), are also visible along the interfaces.

For comparison, Fig 4.45 shows the voltage distribution along the network circuit with the slight mismatch that is present in the experimental structure due to fabrication constraints [see Fig 4.47(a)]. For continuity with the upper substrate of the mushroom structure and optimal matching, the experimental PPWG is constituted by the combination of an upper substrate of thickness $h_1 = 0.127$ mm and permittivity $\varepsilon_{r1} = 2.2$ and of a lower substrate of thickness $h_2 = 1.57$ mm and permittivity $\varepsilon_{r2} = 10.2$, resulting in an effective permittivity of $\varepsilon_{eff,PPWG} = \varepsilon_1\varepsilon_1(h_1 + h_2)/(\varepsilon_1 h_2 + \varepsilon_2 h_1) = 8.02$, which is not perfectly matched to the PPWG in terms of refractive index ($n_{PPWG} = \sqrt{\varepsilon_{eff,PPWG}} = 2.8 \neq |n_L| = 3.4$). The corresponding characteristic impedance of the PPWG is given by Eq. (4.66a) with $\eta = \sqrt{\mu_0/(\varepsilon_0\varepsilon_{eff,PPWG})}$, $d = h_1 + h_2$, and $w = N \cdot p$ (with $N = 19$ cells along the interface), as $Z_{c,PPWG} = 2.38$ Ω, which corresponds to an impedance per port of $Z_{c,PPWG}^{port} = N Z_{c,PPWG} = 45.19$ Ω. This represents considerable Bloch/characteristic impedance mismatch $|\Gamma| = |(Z_{c,PPWG}^{port} - Z_B)/(Z_{c,PPWG}^{port} + Z_B)| = \approx -6$ dB. The PPWG unit cell inductance and capacitance are then obtained as $L_{PPWG} = n_{PPWG} Z_{c,PPWG}^{port} p/c = 2.13$ nH and $C_{PPWG} = n_{PPWG} p/(c Z_{c,PPWG}^{port}) = 1.05$ pF. Despite the mismatch imposed by design constraints, focusing and refocusing effects can also be observed in Fig 4.45.

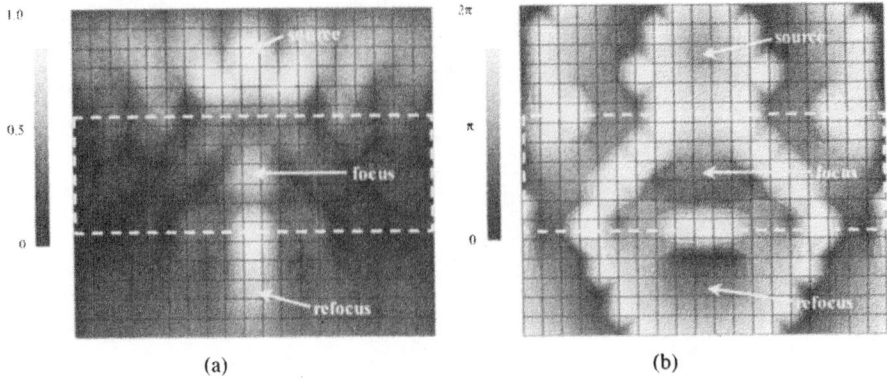

(a)　　　　　　　　　　　　　　　　(b)

Fig. 4.45. Voltage distribution in the NRI slab network circuit of Fig 4.43 in the case of the slight mismatch present in the experimental structure, corresponding to the PPWG parameters $L_{PPWG} = 2.13$ nH and $C_{PPWG} = 1.05$ pF. (a) Normalized magnitude. (b) Phase. From [23], © IEEE; reprinted with permission.

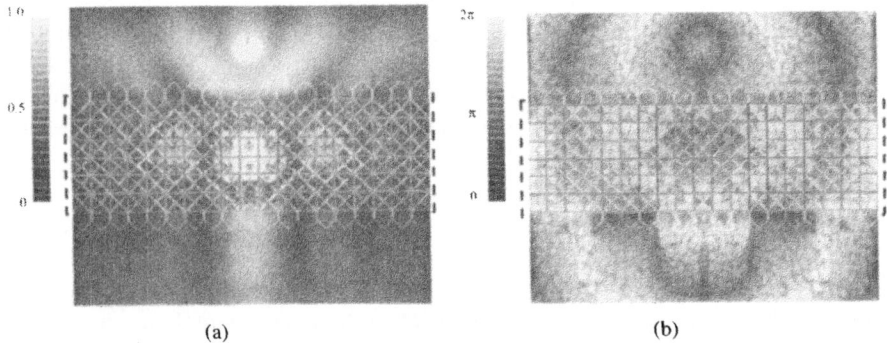

(a)　　　　　　　　　　　　　　　　(b)

Fig. 4.46. Full-wave (FEM) simulated electric field distributions in the planar distributed NRI slab of Fig 4.47 ($f = 3.737$ GHz) on the ground plane. (a) Normalized magnitude. (b) Phase. From [23], © IEEE; reprinted with permission.

Next, the field distributions along the real NRI slab of Fig 4.42 computed by FEM full-wave analysis are shown in Fig 4.46. A perfect magnetic conductor (PMC) boundary condition is introduced at the symmetry plane of the structure to reduce computation time. Radiation boundary conditions are applied along the x and y directions, and the open top of the structure is simulated by a perfectly matched layer (PML) boundary condition. A vertical current line source is set at 15 mm away from the interface to excite the PPWG mode. The observed magnitude and phase field distributions qualitatively agree with the LC network predictions of Fig 4.45.

Finally, the experimental prototype and results are shown in Figs. 4.47 and 4.48, respectively. The MIM caps are etched on the bottom side of the upper

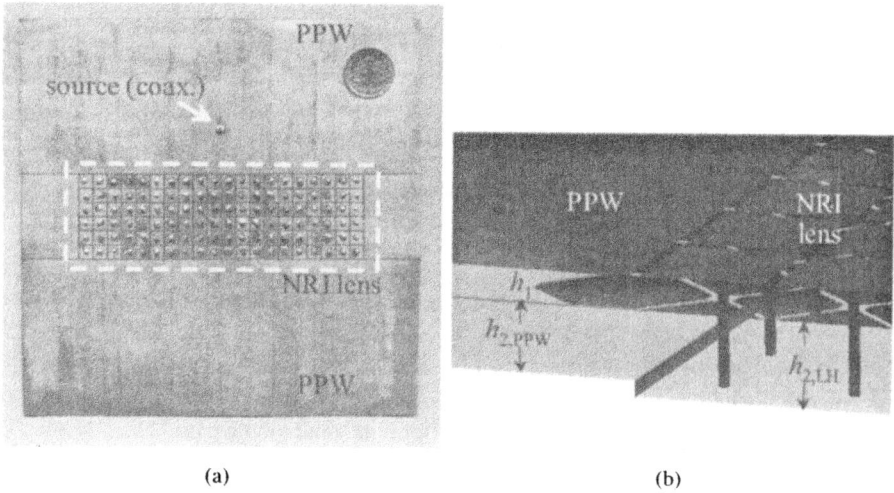

Fig. 4.47. Prototype of a 20×3-cell NRI slab ($f = 3.737$ GHz). (a) Top view. (b) Cross section. From [23], © IEEE; reprinted with permission.

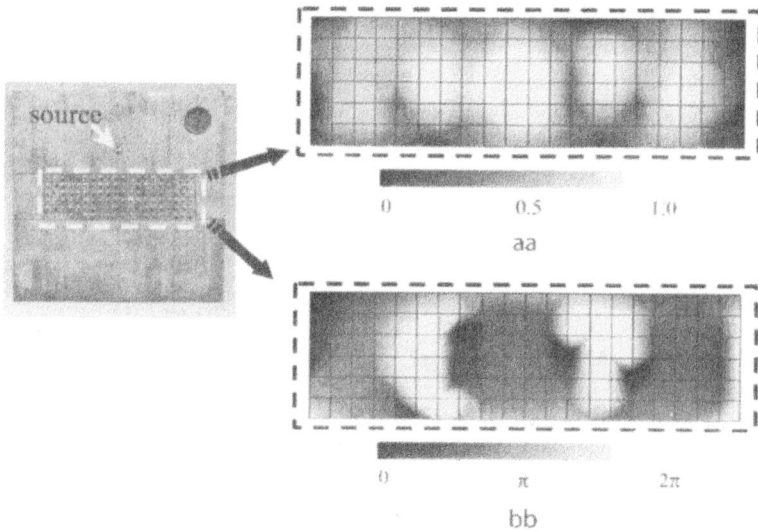

Fig. 4.48. Measured electric field distribution on the LH structure ($f = 3.737$ GHz). A vertical electric probe is used. (a) Magnitude. (b) Phase. From [23], © IEEE; reprinted with permission.

substrate. The lower substrates are laminated separately in the LH and PPWG regions. The PPWG is excited by a coaxial cable through a hole in the ground plane located at 15 mm of the interface. The size of the overall structure is 140×140 mm^2. The refractive index and per-port characteristic impedances of the PPWG are $n_{\text{PPWG}} = \sqrt{\varepsilon_{\textit{eff},\text{PPWG}}} = 2.77$ and $Z_{c,\text{PPWG}}^{\text{port}} = (w/p)Z_{c,\text{PPWG}} =$

27.17 Ω. The vertical electric field distribution is measured at 1 mm above the NRI slab using a vertical coaxial probe fixed to an automated moving stage to picking up the capacitors voltages. The observed distributions of the electric field exhibits the expected NRI magnitude concentration and phase concentring fronts in agreement with LC theory and full-wave simulation results.

REFERENCES

1. C. Kittel. *Introduction to Solid State Physics,* Seventh Edition, Wiley Text Books, 1964.

2. C. Christopoulos. *The Transmission-Line Modeling Method TLM,* IEEE Press, 1995.

3. A. Grbic and G. V. Eleftheriades. "Periodic analysis of a 2-D negative refractive index transmission lines," *IEEE Trans. Antennas Propagat.,* vol. 51, no. 10, pp. 2604–2611, Oct. 2003.

4. C. Caloz and T. Itoh. "Positive / negative refractive index anisotropic 2D meta-materials," *IEEE Microwave Wireless Compon. Lett.,* vol. 13, no. 12, pp. 547–549, Dec. 2003.

5. M. D. Pozar. *Microwave Engineering,* Third Edition, John Wiley & Sons, 2004.

6. J. D. Joannopoulos, R. D. Meade, and J. .N. Win. *Photonic Crystals, Molding the Flow of Light,* Princeton University Press, 1995.

7. J. -H. Kim and M. Swaminathan. "Modeling of irregular shaped power distribution planes using transmission matrix method," *IEEE Trans. Advanced Packag.,* vol. 24, no. 3, pp. 334–346, August 2001.

8. W. J. R. Hoefer. "The transmission line matrix (TLM) method," in *Numerical Techniques for Microwave and Millimeter-Wave Passive Structures,* edited by T. Itoh, Wiley-Interscience, 1989.

9. P. B. Johns and R. L. Beurle. "Numerical solution of 2-dimensional scattering problems using a transmission line matrix," *Proc. IEE,* vol. 118, no. 9, pp. 1203–1208, Sept. 1971.

10. J. R. Whinnery and S. Ramo. "A new approach to the solution of high frequency field problems," *Proc. I.R.E.,* vol. 32, pp. 284–288, May 1944.

11. P. P. M. So and W. J. R. Hoefer. "Time domain TLM modeling of metamaterials with negative refractive index," *IEEE-MTT Int'l Symp.,* Fort Worth, TX, pp. 1779–1782, June 2004.

12. Multipurpose Electromagnetic Field Simulation Tool, *Faustus Scientific Corporation,* http://www.faustcorp.com.

13. V. Veselago. "The electrodynamics of substances with simultaneously negative values of ε and μ," *Soviet Physics Uspekhi,* vol. 10, no. 4, pp. 509–514, Jan., Feb. 1968.

14. J. B. Pendry. "Negative refraction makes a perfect lens," *Phys. Rev. Lett.,* vol. 85, no. 18, pp. 3966–3969, Oct. 2000.

15. P. P. M. So, H. Du, and W. J. R. Hoefer. "Modeling of metamaterials with negative refractive index using 2D-shunt and 3D-SCN TLM Networks," *IEEE Trans. Microwave Theory Tech.,* vol. 53, no. 4, pp. 1496–1505, April 2005.

16. R. W. Ziolkowski. "Pulsed and CW Gaussian beam interactions with double negative metamaterial slabs," *Optics Express.,* vol. 11, no. 7, April 2003.

17. H. Raether. *Surface Plasmons,* Springer-Verlag, 1988.

18. R. Ruppin. "Surface polaritons of a left-handed material slab," *J. Phys. Condens. Matter,* vol. 13, pp. 1811–1818, 2001.

19. C. Caloz, C.-J. Lee, D. R. Smith, J. B. Pendry, and T. Itoh. "Existence and properties of microwave surface plasmons at the interface between a right-handed and a left-handed media," *IEEE AP-S USNC/URSI National Radio Science Meeting,* Monterey, CA, June 2004.

20. A. Grbic and G. V. Eleftheriades. "Subwavelength focusing using a negative-refractive-index transmission line lens," *IEEE Antennas Wireless Propagat. Lett.,* vol. 2, pp. 186–189, 2003.

21. P. P. M. So and W. J. R. Hoefer. "Microwave CAD environment for metamaterials with negative refractive index," *in Proc. European Microwave Conf.,* Amsterdam, The Netherlands, Oct. 2004.

22. D. Sievenpiper, L. Zhang, R. F. J. Broas, N. G. Alexópolous, and E. Yablonovitch. "High-impedance surface electromagnetic surfaces with a forbidden frequency band," *IEEE Trans. Microwave Theory Tech.,* vol. 47, no. 11, pp. 2059–2074, Nov. 1999.

23. A. Sanada, C. Caloz, and T. Itoh. "Planar distributed structures with negative refractive index" *IEEE Trans. Microwave Theory Tech.,* vol. 52, no. 4, pp. 1252–1263, April 2004.

24. R. E. Collin. *Field Theory of Guided Waves,* Second Edition, Wiley-Interscience, 1991.

5

GUIDED-WAVE APPLICATIONS

Many novel *guided-wave* applications can be devised on the foundation of the CRLH MTM concept established in Chapters 3 and 4. The guided-wave applications presented in this chapter essentially refer to 1D CRLH (Chapter 3) components, where electromagnetic energy remains confined in the metal and dielectric media constituting the components. Although 1D CRLH structures can be considered as TLs embedded in a NRI medium when operated in their LH range [Eq. (3.25)], they do not utilize negative refraction in the sense of Snell's law, since they do not involve angles. Rather, they exploit the unusual frequency dependence of the guided wavelength and the low-frequency antiparallelism/high-frequency parallelism existing between the phase and the group velocities in CRLH structures.

Chapter 5 describes specific guided-wave properties of CRLH structures and presents subsequent examples of practical applications. The properties include dual-band operation, bandwidth enhancement, multilayer-architecture super-compactness, arbitrary coupling level, and negative/zeroth-order resonance. The corresponding applications are dual-band components (TLs, stubs, hybrid couplers, Wilkinson power divider, subharmonically-pumped mixer) (Section 5.1), enhanced-bandwidth components (with rat-race coupler example) (Section 5.2), super-compact multilayer TLs (with diplexer application example) (Section 5.3), phase and impedance couplers with arbitrary coupling level (asymmetric CRLH/RH and symmetric CRLH/CRLH) (Section 5.4), and negative/zeroth order

Electromagnetic Metamaterials: Transmission Line Theory and Microwave Applications,
By Christophe Caloz and Tatsuo Itoh
Copyright © 2006 John Wiley & Sons, Inc.

resonator (Section 5.5). Other guided-wave applications based on the CRLH concept are possible.

5.1 DUAL-BAND COMPONENTS

A dual-band (DB) component is a component accomplishing the *same function at two different arbitrary frequencies* ω_1 and ω_2. Such a component is therefore constituted of TL sections inducing equivalent[1] phase shifts $\phi_1 = -\beta_1\ell$ and $\phi_2 = -\beta_2\ell$, where ℓ is the length of the TL sections, at these two frequencies. In other words, a DB component should exhibit a dispersion relation $\beta(\omega)$ satisfying the double condition

$$\beta(\omega_1) = \beta_1, \tag{5.1a}$$

$$\beta(\omega_2) = \beta_2, \tag{5.1b}$$

where the two frequencies (ω_1, ω_2) and the two propagation constants (β_1, β_2) are arbitrary. A CRLH TL possesses this DB property and can therefore be used to transform virtually any microwave TL-based component into a DB component. This DB property of CRLH TL is extremely beneficial to modern wireless communication systems covering two bands because it reduces the number of required circuit components for a given functionality.

5.1.1 Dual-Band Property of CRLH TLs

The DB property of CRLH TLs will be explained with the help of Fig 5.1, where the phase curves of both a PRH TL and a CRLH TL are shown.

Let us first consider the case of the PRH TL, which is the building brick of most conventional microwave components. A PRH TL is characterized by the propagation constant and characteristic impedance given by Eqs. (3.36) and (3.32), respectively,

$$\beta^{\mathrm{PRH}} = \omega\sqrt{L_R' C_R'}, \tag{5.2a}$$

$$Z_c^{\mathrm{PRH}} = \sqrt{\frac{L_R'}{C_R'}}. \tag{5.2b}$$

For this line to be matched to ports of impedance Z_t and to provide the appropriate phase shift at the first frequency ω_1, it must satisfy the conditions

$$Z_c^{\mathrm{PRH}} = Z_t, \tag{5.3a}$$

$$\beta^{\mathrm{PRH}}(\omega = \omega_1) = \beta_1, \tag{5.3b}$$

---

[1]As will be seen in the forthcoming examples, "equivalent" does not mean exactly equal and does not necessarily mean equal to the modulo 2π. For instance, in the case of a quadrature hybrid, the equivalent phase shifts will be $\phi_1 = -90°$ and $\phi_2 = -270°$ (Section 5.1.3.1).

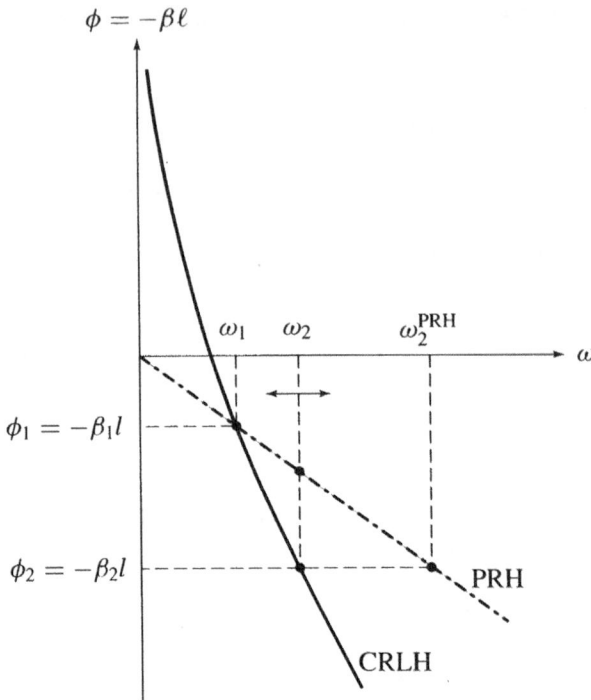

Fig. 5.1. Dual-band property of a CRLH TL.

respectively. Inserting Eqs. (5.3) into Eq. (5.2) yields a linear system of two equations with the two unknowns L'_R and C'_R, which are easily found to be

$$L'_R = \frac{Z_t \beta_1}{\omega_1}, \tag{5.4a}$$

$$C'_R = \frac{\beta_1}{\omega_1 Z_t}. \tag{5.4b}$$

Consequently, the PRH TL is fully determined by the matching condition and by the specification of a single operation frequency. The frequency at which the required second band propagation constant of β_2 is achieved is fixed to the frequency

$$\omega_2^{PRH} = \frac{\beta_2}{\beta_1} \omega_1, \tag{5.5}$$

which is obtained by Eqs. (5.2a) with Eq. (5.4). In practical applications, it is highly unlikely that ω_2^{PRH} would coincide with the required second band frequency ω_2 (Fig 5.1). Therefore, a PRH TL is essentially monoband, and components made of PRH TL are thus monoband.

Let us now consider the case of the balanced CRLH TL, for which the propagation constant and characteristic impedance are given by (Eqs. 3.34) and (3.32), respectively,

$$\beta^{\text{CRLH}} = \omega\sqrt{L'_R C'_R} - \frac{1}{\omega\sqrt{L'_L C'_L}}, \tag{5.6a}$$

$$Z_c^{\text{CRLH}} = \sqrt{\frac{L'_R}{C'_R}} = \sqrt{\frac{L'_L}{C'_L}}. \tag{5.6b}$$

Matching to terminations of impedance Z_t and phasing to β_1 for a required operation at frequency ω_1 require

$$Z_c^{\text{CRLH}} = Z_t, \tag{5.7a}$$

$$\beta^{\text{CRLH}}(\omega = \omega_1) = \beta_1, \tag{5.7b}$$

respectively. These equations correspond in fact to three equations because the matching condition splits into two equations from Eq. (5.6b). Thus, insertion of Eq. (5.7) into Eq. (5.6) yields a linear system of *three equations* with the *four unknowns* L'_R, C'_R, L'_L, C'_L, and there is therefore *one available degree of freedom that can be exploited to satisfy the second band condition* [Eq. (5.1b)]. From the resulting system of four equations, we obtain, after some straightforward calculations, the DB CRLH parameters

$$L'_R = \frac{Z_t \left[\beta_2 - \beta_1 (\omega_1/\omega_2)\right]}{\omega_2 \left[1 - (\omega_1/\omega_2)^2\right]}, \tag{5.8a}$$

$$C'_R = \frac{\beta_2 - \beta_1 (\omega_1/\omega_2)}{\omega_2 Z_t \left[1 - (\omega_1/\omega_2)^2\right]}, \tag{5.8b}$$

$$L'_L = \frac{Z_t \left[1 - (\omega_1/\omega_2)^2\right]}{\omega_1 \left[(\omega_1/\omega_2) \beta_2 - \beta_1\right]}, \tag{5.8c}$$

$$C'_L = \frac{1 - (\omega_1/\omega_2)^2}{\omega_1 Z_t \left[(\omega_1/\omega_2) \beta_2 - \beta_1\right]}, \tag{5.8d}$$

depending only on $(\omega_1, \omega_2) - (\beta_1, \beta_2)$ and corresponding to the dispersion curve shown in Fig 5.1.

The argument presented above is based on the ideal homogeneous CRLH TL, described in Section 3.1. As pointed out in Chapter 3, such a TL does not exist naturally and needs therefore to be synthesized in the form of an LC ladder network with identical responses in the frequency range of interest,[2] as it was shown in Section 3.2[3] and verified in Section 3.4 for a real distributed structure. The exact phase shift produced by an LC network unit cell takes in general the form [Eq. (3.81)]

$$
\Delta\phi = -\arctan\left\{\frac{\dfrac{1}{\omega}\left\{\dfrac{(\omega/\omega_{se})^2 - 1}{C_L Z_c}\left(1 - \dfrac{\chi}{4}\right) + \dfrac{Z_c}{\omega L_L}\left[(\omega/\omega_{sh})^2 - 1\right]\right\}}{2 - \chi}\right\},
$$

(5.9a)

$$
\text{with}\quad \chi = \omega^2 L_R C_R + \frac{1}{\omega^2 L_L C_L} - \frac{L_R C_L + L_L C_R}{L_L C_L}.
$$

(5.9b)

As shown in Section 3.2.3, in the frequency range where $|\Delta\phi| = \beta p \ll 1$, this complicated LC network expression reduces to the much simpler expression [Eq. (3.84)]

$$
\Delta\phi(\omega \to \omega_0) = -\left[\omega\sqrt{L_R C_R} - \frac{1}{\omega\sqrt{L_L C_L}}\right],
$$

(5.10)

which can be shown to be identical to the ideal homogeneous TL expression Eq. [(5.6a)] in the following manner. An artificial CRLH TL of physical length ℓ is constituted by the repetition of N unit cells of length p, so that

$$
\ell = N \cdot p.
$$

(5.11)

Because each cell induces a phase shift of $\Delta\phi$, the total phase shift along the line is then

$$
\phi = N \cdot \Delta\phi.
$$

(5.12)

With this relation, the relation $\beta = -\phi/\ell$ and the fundamental substitutions

$$
L_R = L'_R \cdot p, C_R = C_R \cdot p, L_L = L_L/p, C_L = C_L/p,
$$

(5.13)

the expression of Eq. (5.10) is reduced exactly to Eq. (5.6a).

It is of practical interest to relate the real inductance and capacitance values of the LC components constituting the CRLH with the required phase shifts for dual

---

[2]In this range, the homogeneity condition $|\Delta\phi| = \beta p \ll 1$, where p is the physical length of the unit cell, should be satisfied.

[3]The equivalence is rigorously shown analytically in Eqs. 3.115 to 3.118 of Section 3.2.6.

band operation. This is accomplished by substituting $\beta_i = -\phi_i/(Np)$ $(i = 1, 2)$ and Eq. (5.13) into Eq. (5.8) and yields the DB CRLH parameters

$$L_R = \frac{Z_t \left[\phi_1 \left(\omega_1/\omega_2\right) - \phi_2\right]}{N\omega_2 \left[1 - (\omega_1/\omega_2)^2\right]}, \tag{5.14a}$$

$$C_R = \frac{\phi_1 \left(\omega_1/\omega_2\right) - \phi_2}{N\omega_2 Z_t \left[1 - (\omega_1/\omega_2)^2\right]}, \tag{5.14b}$$

$$L_L = \frac{N Z_t \left[1 - (\omega_1/\omega_2)^2\right]}{\omega_1 \left[\phi_1 - (\omega_1/\omega_2)\,\phi_2\right]}, \tag{5.14c}$$

$$C_L = \frac{N \left[1 - (\omega_1/\omega_2)^2\right]}{\omega_1 Z_t \left[\phi_1 - (\omega_1/\omega_2)\,\phi_2\right]}, \tag{5.14d}$$

depending only on $(\omega_1, \omega_2) - (\phi_1, \phi_2)$ and also corresponding to the dispersion curve shown in Fig 5.1.

5.1.2 Quarter-Wavelength TL and Stubs

The simplest DB application is the DB quarter-wavelength[4] TL, and its corresponding open- and short-circuited stubs [1].

A conventional *quarter-wavelength* TL is a PRH TL of length $\ell = \lambda/4$. Such a TL induces a phase shift of $\phi = -\beta\ell = -\beta(\lambda/4) = -(\beta\pi c)/(2\omega) = -\pi/2$ at the operating frequency ω. If this frequency is called ω_1, in reference with Fig 5.1, it is thus associated with the first band relation [Eq. (5.1a)]

$$\beta_1 = \beta(\omega_1) = \frac{\omega_1}{c} \quad \text{or} \quad \phi_1 = \phi(\omega_1) = -\frac{\pi}{2}. \tag{5.15}$$

The next available frequency at which the PRH TL provides an equivalent response is obtained after one additional full turn in the Smith chart, which corresponds to the phase shift $\phi_2 = -\beta_2\ell = -\beta_2(3\lambda/4) = -(3\beta_2\pi c)/(2\omega) = -3\pi/2$. At this point, $\beta_2 = 3\beta_1$ and the frequency is $\omega_2^{PRH} = 3\omega_1$ according to Eq. (5.5).[5] As explained in the previous section, ω_2^{PRH} cannot be tuned without affecting the first band, and it is unlikely to coincide with the second frequency ω_2 of interest in a specific application.

---

[4]In fact, the term "'quarter-wavelength' here is abusive in the sense that it strictly applies only to the first band. But it expresses the fact that the electrical lengths $\ell = \lambda/4$ and $\ell = 3\lambda/4$, which are indistinguishable from the point of view of impedance, can be achieved at any arbitrary pair of frequencies ω_1 and ω_2 (arbitrary ratio ω_2/ω_1).

[5]A response equivalent to that of $\phi = -\pi/2$ is obtained at all the *odd harmonics* of the design frequency ω_1, since the equivalent phase shifts $\phi_{n+1} = -(\pi/2 + n\pi)$ $(n \in \mathbb{N})$ occur at the frequencies $\omega_{n+1} = (2n + 1)\omega_1$.

In contrast, the CRLH TL can be designed to meet both bands conditions Eq. (5.1). In the case of the quarter-wavelength transformer, the explicit expression for Eq. (5.1a) was already given in Eq. (5.15) and the explicit expression for Eq. (5.1b) reads

$$\beta_2 = \beta(\omega_2) = \frac{3\omega_1}{c} \quad \text{or} \quad \phi_2 = \phi(\omega_2) = -\frac{3\pi}{2}. \quad (5.16)$$

The LC values of the CRLH TL for DB operation at the arbitrary pair of frequencies (ω_1, ω_2) are then directly available from Eq. (5.14) with the phase substitutions of Eqs. (5.15) and (5.16)

$$L_R = \frac{Z_t \pi \left[1 - 3\left(\omega_1/\omega_2\right)\right]}{2N\omega_2 \left[1 - \left(\omega_1/\omega_2\right)^2\right]}, \quad (5.17a)$$

$$C_R = \frac{\pi \left[1 - 3\left(\omega_1/\omega_2\right)\right]}{2N\omega_2 Z_t \left[1 - \left(\omega_1/\omega_2\right)^2\right]}, \quad (5.17b)$$

$$L_L = \frac{2N Z_t \left[1 - \left(\omega_1/\omega_2\right)^2\right]}{\pi \omega_1 \left[3\left(\omega_1/\omega_2\right) - 1\right]}, \quad (5.17c)$$

$$C_L = \frac{2N \left[1 - \left(\omega_1/\omega_2\right)^2\right]}{\pi \omega_1 Z_t \left[3\left(\omega_1/\omega_2\right) - 1\right]}. \quad (5.17d)$$

We have considered here the pair of phases $(\phi_1, \phi_2) = (-\pi/2, -3\pi/2)$, but any pair of phases with π difference between ω_1 and ω_2 can be used in principle. For instance, the pairs $(\phi_1, \phi_2) = (+3\pi/2, +\pi/2)$ (f_1 and f_2 both in LH range) and $(\phi_1, \phi_2) = (+\pi/2, -\pi/2)$ (f_1 in LH range and f_2 in RH range) would yield similar results. The choice of a pair of phases mostly depends on design constraints, such as location of the CRLH cutoff frequencies with respect to the two bands of interest and characteristics of the available chip components. Bandwidth considerations may also be important.[6]

In DB components, as in some other 1D applications of MTMs, the key property of the transmission medium does not lie in its *distributed* characteristics of the transmission medium, but in the *net phase shift between its input and its output*. This cognizance allows the following practical simplification. We remember from Section 3.2.5 that the real distributed CRLH TL of Fig 3.15 is indistinguishable from the series RH-LH TL combination of Fig 3.30 in terms of port-to-port transmission characteristics if the structure is balanced, because under this condition the equivalent circuit model of the CRLH TL decouples into separated RH and LH subcircuits. This means that a TL configuration of the type shown in Fig 3.30 produces exactly the same port-to-port response as a real CRLH TL. Therefore, the RH section of the CRLH TL can be replaced by a naturally existing TL, such as a simple microstrip, for simplicity. In this case, the DB total

---

[6]As RH TLs, CRLH TLs induce smaller bandwidths when they are larger.

phase shift ϕ is split into the LH TL phase shift ϕ^{LH} and the RH TL phase shift ϕ^{RH}: $\phi = \phi^{LH} + \phi^{RH}$. The LH section is still directly designed by using the last two expressions of Eq. (5.17), which provide the values of L_L and C_L leading to the phase advance ϕ^{LH}. The natural homogeneous RH section is designed by first computing L_R and C_R using the first two expressions of Eq. (5.17), which determines the characteristic impedance, and then applying Eq. (5.12) with the RH part only of Eq. (5.10) [or Eq. (3.72)] to determine the required RH phase delay, which yields[7]

$$Z_{cR} = \sqrt{\frac{L_R}{C_R}}, \tag{5.18a}$$

$$\phi_1^{RH} = -N\omega_1\sqrt{L_R C_R}. \tag{5.18b}$$

From these two quantities, the design parameters of the homogeneous RH TL (e.g., line width and length in the microstrip case) can be straightforwardly obtained [2].

Fig 5.2 shows the photograph and equivalent circuit of a CRLH TL implemented in the form of the series connection of an SMT-chips LH section and of a conventional (RH) microstrip line. The design parameters of this DB *quarter wavelength transformer*, based on the operation phases pair $(\phi_1, \phi_2) = (-\pi/2, -3\pi/2)$, are fully determined by Eqs. (5.17) and (5.18). The parameters obtained in this manner as well as the actual parameters taking into account parasitic effects and fabrication constraints are indicated in the caption of the figure. The slight discrepancies between the ideal and actual design parameters are essentially due to two factors: the availability of SMT components only in discrete values and the natural RH parasitics of the SMT chip components. These parasitics induce a small phase delay in the LH section (assumed in the analysis to exhibit pure phase advance), which results in smaller phase advance, and requires reduction of the length of the microstrip line for compensation. This is the reason why the actual microstrip line in this design is 17% shorter than the predicted length.[8]

Figs. 5.4 and 5.3 show a DB quarter-wavelength *short-circuited* stub and a DB quarter-wavelength *open-circuited stub*, respectively, based on the design

---

[7]It is important to realize that in the phase equation, we do not have $\phi_1^{RH} = \phi^{RH}(\omega = \omega_1) \neq -\pi/2$ and $\phi_2^{RH} = \phi^{RH}(\omega = \omega_2) \neq -3\pi/2$. ϕ^{RH} is designed to contribute to exactly the appropriate phase shift inducing $\phi_1 = \phi^{LH}(\omega_1) + \phi^{RH}(\omega_1) = -\pi/2$ according to Eqs. (5.15) and $\phi_2 = \phi^{LH}(\omega_2) + \phi^{RH}(\omega_2) = -3\pi/2$ according to Eqs. (5.15) and (5.16).

[8]In practice, the exact design procedure would typically be as follows: 1) Compute L_L and C_L by the last two relations of Eq. (5.17), and use the closest possible combination of chip components; 2) measure the isolated LH section including the real chips (using TRL calibration) and compare the measured phase shift ϕ^{LH} with the theoretical phase shift given by Eq. (5.18b); the measured ϕ^{LH} is smaller than the theoretical one by a value δ, which corresponds to the phase delay induced by the RH parasitics; 3) calculate the actual microstrip line length to be used, corresponding to the phase shift $|\phi^{RH}| - |\delta|$; 4) fabricate the complete line, including the LH and RH sections, and fine-tune the RH length for optimal performances.

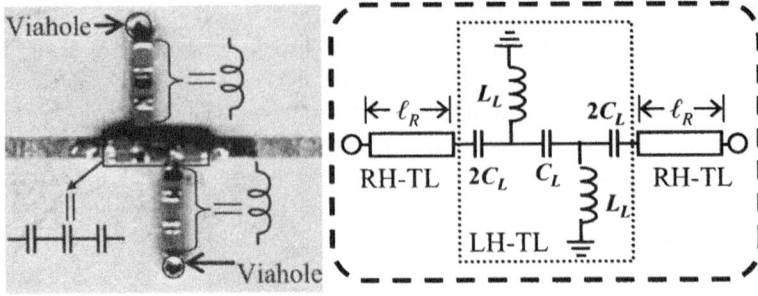

Fig. 5.2. Prototype and equivalent circuit of a DB CRLH TL implementation utilizing SMT chip components for the LH contribution and a conventional microstrip line (on each side of the LH section) for the RH contribution. The substrate used is Duroid 6010 with permittivity 10.2 and thickness 1.27 mm. The two frequencies of operation are $f_1 = 0.88$ GHz and $f_2 = 1.67$ GHz. The ideal parameters obtained from Eqs. (5.17) and (5.18) are ($Z_c = 50\ \Omega$, $N = 2$) $L_L = 14.32$ nH, $C_L = 5.73$ pF, $L_R = 12.81$ nH, $C_R = 5.13$ pF, $Z_{cR} = 50\ \Omega$, $\phi_1^{RH} = -162.38°$, leading to $w_{\mu strip} = 1.18$ mm (width) and $\ell_{\mu strip} = 2\ell_R = 58.9$ mm (length). The actual design parameters are $L_L = 14.1$ nH (3×4.7 nH, in series), $C_L = 6$ pF (2×3 pF, in parallel), $w_{\mu strip} = 1.2$ mm, and $\ell_{\mu strip} = 49.4$ mm. From [1], © IEEE; reprinted with permission.

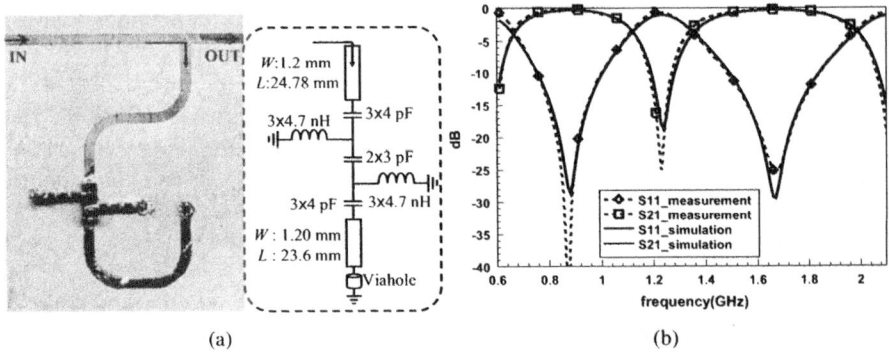

(a)   (b)

Fig. 5.3. DB quarter-wavelength short-circuited stub corresponding to the design of Fig 5.2. (a) Prototype and equivalent circuit. (b) Simulated and measured S-parameters. From [1], © IEEE; reprinted with permission.

parameters of Fig 5.2 for the frequencies $f_1 = 0.88$ GHz and $f_2 = 1.67$ GHz. The input impedances of short and open TLs are given by [2]

$$Z_{in}^{short} = jZ_c \tan(\beta\ell) \overset{d=\lambda/4, 3\lambda/4}{=} \infty, \tag{5.19a}$$

$$Z_{in}^{open} = -jZ_c \cot(\beta\ell) \overset{d=\lambda/4, 3\lambda/4}{=} 0, \tag{5.19b}$$

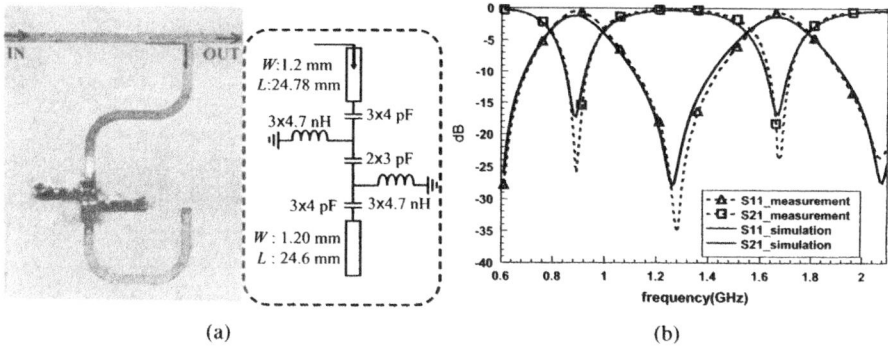

Fig. 5.4. DB quarter-wavelength open-circuited stub corresponding to the design of Fig 5.2. (a) Prototype and equivalent circuit. (b) Simulated and measured S-parameters. From [1], © IEEE; reprinted with permission.

and are thus infinite and zero, respectively, at the impedance-inversion frequencies, that is, at the frequencies f_1 (where $\ell = \lambda/4$) and f_2 (where $\ell = 3\lambda/4$). Consequently, the shorted and opened stubs on a transmission line exhibit the characteristics $|S_{21}| \approx 0$ dB and $|S_{11}| \approx 0$ dB, respectively, at both f_1 and f_2. This is seen in Figs. 5.4 and 5.3, where excellent agreement between simulated and measured results can be observed.

5.1.3 Passive Component Examples: Quadrature Hybrid and Wilkinson Power Divider

The DB quarter-wavelength TL described in the previous section can be applied to provide DB functionality in any TL-based microwave component. The design procedure essentially consists in replacing conventional PRH TL branches by CRLH TL branches designed to exhibit the required phase shifts at the desired frequencies for a given operation (Fig 5.1). This section presents two examples of passive DB components: the quadrature hybrid (QH) [1] and the Wilkinson power divider (WPD).

5.1.3.1 Quadrature Hybrid   The conventional QH, also known as a the branch-line hybrid, is a 3-dB directional coupler with a $90°$ phase difference between the outputs of the through and coupled arms [2]. A DB quadrature hybrid is obtained by replacing the $-90°$ RH branches by CRLH branches exhibiting $-90°$ and $-270°$ phase shifts at f_1 and f_2, respectively, as shown in the schematics of Fig 5.5.[9] Therefore, the DB quarter-wavelength CRLH TL theory presented in Section 5.1.2 is directly applicable here.

---

[9]The only difference between the two circuits of Fig 5.5(a) and 5.5(b) is the sign in the phase difference between the output signals of ports 2 and 3.

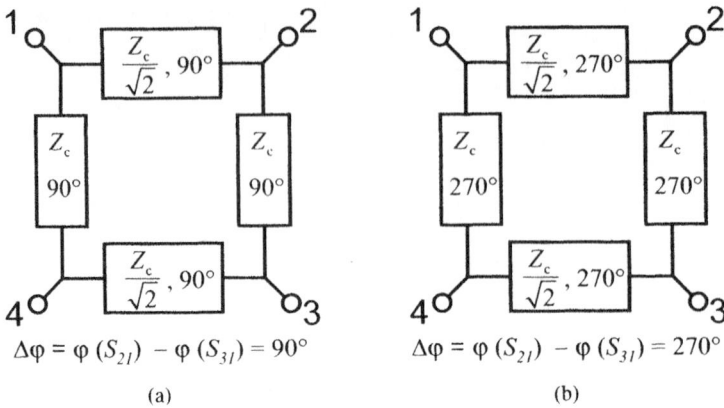

$$\Delta\varphi = \varphi(S_{21}) - \varphi(S_{31}) = 90°$$

(a)

$$\Delta\varphi = \varphi(S_{21}) - \varphi(S_{31}) = 270°$$

(b)

Fig. 5.5. Conceptual schematics of the DB QH. (a) At the first frequency, f_1. (b) At the second frequency, f_2. The labels 1, 2, 3, and 4 correspond to the input, through output, coupled output, and isolated ports, respectively, and Z_c represents the characteristic impedance of the input and output lines. From [1], © IEEE; reprinted with permission.

Fig 5.6 shows a DB LH-SMT/RH-microstrip QH prototype[10] operating at the two frequencies ($f_1 = 0.93$ GHz and $f_2 = 1.78$ GHz) which are in the ratio of $f_2/f_1 = 1.59$.[11] The magnitude characteristics for this prototype are shown in Figs. 5.7, and Fig 5.8 shows the phase differences between the two output ports.

In both simulation and measurement, the phase differences between S_{31} and S_{21} are $\pm 90°$, and quadrature phase differences are obtained with errors less than $1.5°$ at f_1 and f_2 (Fig 5.8). The amplitude imbalance between S_{31} and S_{21} is less than 0.5 dB in both passbands, which means that the incident power is evenly divided between port 2 and port 3. The 1-dB bandwidth, defined as the frequency range in which the amplitude imbalance is less than 1 dB, S_{21} and S_{31} are larger than -5 dB, isolation/return loss is larger than 10 dB, and quadrature phase error is less than 11.1%, is larger than 3.4%.

A DB rat-race (RR) coupler, or ring hybrid, based on the same principle as this CRLH DB QH is also demonstrated in [1]. The operating frequencies of this RR are $f_1 = 1.5$ GHz and $f_2 = 3.0$ GHz ($f_2/f_1 = 2$). The three $-90°$ RH branches of the conventional RR are replaced by three CRLH branches of $-90°$ at f_1 and $-270°$ at f_2, whereas the $-270°$ branch of the conventional RR is replaced by the series connection of three similar CRLH TL sections.

5.1.3.2 Wilkinson Power Divider
The conventional Wilkinson power divider (WPD) is a three-port network used for power division (or power combination)

---

[10]The tuning stub was added (also included in the simulation) to compensate for the effects of amplitude imbalance between the through and coupled ports caused by the frequency variations of the reactance of the chip inductors and susceptance of the chip capacitors.

[11]These two frequencies are chosen so as to correspond to the first and second bands of the global system for mobile (GSM) communications (900 base station and 1800 mobile phone).

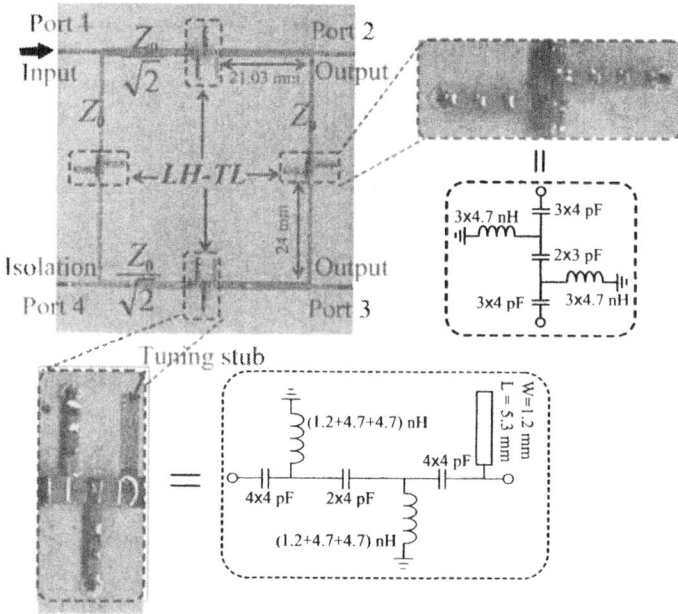

Fig. 5.6. Prototype and equivalent circuit of a CRLH DB QH using SMT chip components for the LH contribution and a conventional microstrip line contribution for the RH contribution, as the TLs of Section 5.1.2. The characteristic impedance is $Z_c = 50 \ \Omega$ ($Z_c/\sqrt{2} \approx 35.4 \ \Omega$). The overall dimension of the circuit is 81.8 mm × 71.3 mm. From [1], © IEEE; reprinted with permission.

[2]. The interest of this component lies in the fact that it can have all three ports matched and isolation between the output ports thanks to the presence of a resistance between the two output (or input) branches.[12] Arbitrary power division is achievable, but equal-split (3 dB) is the most common case. Fig 5.9 shows the schematic of a DB WPD replacing the $-90°$ branches of the conventional WPD by CRLH branches inducing a phase of $+90°$ (LH range, phase advance) at the first frequency and the phase shift of $-90°$ (RH range, phase delay) at the second frequency,

$$\phi_1 = \phi(\omega_1) = +\frac{\pi}{2}, \tag{5.20a}$$

$$\phi_2 = \phi(\omega_2) = -\frac{\pi}{2}, \tag{5.20b}$$

[12]It is well-known that a loss-less (non-dissipative) three-port network cannot be matched simultaneously at all of its ports [3]. However, the introduction of a resistance makes matching at all ports possible and is exploited in the Wilkinson divider/combiner.

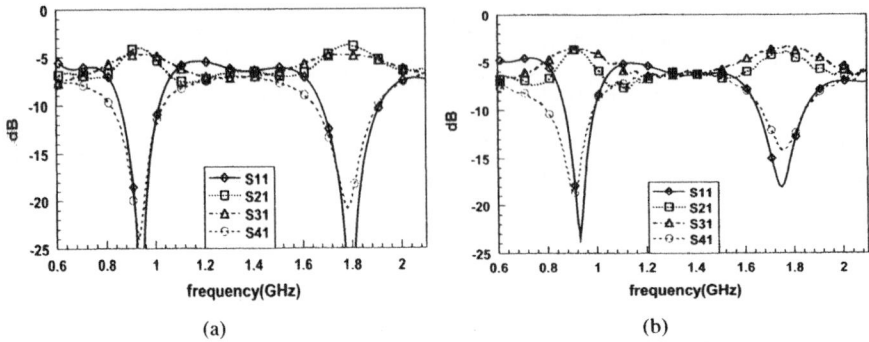

Fig. 5.7. S-parameters for the CRLH DB QH of Fig 5.6. (a) Simulation. (b) Measurement. From [1], © IEEE; reprinted with permission.

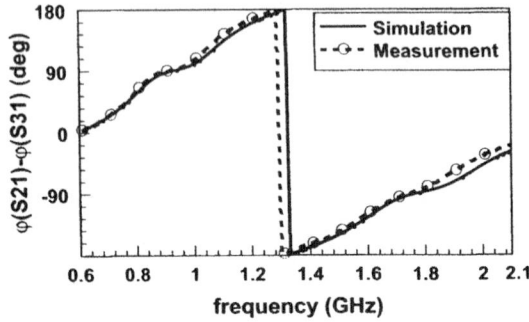

Fig. 5.8. Phases difference $\Delta\varphi = \varphi(S_{21}) - \varphi(S_{31})$ for the CRLH DB QH of Fig 5.6. From [1], © IEEE; reprinted with permission.

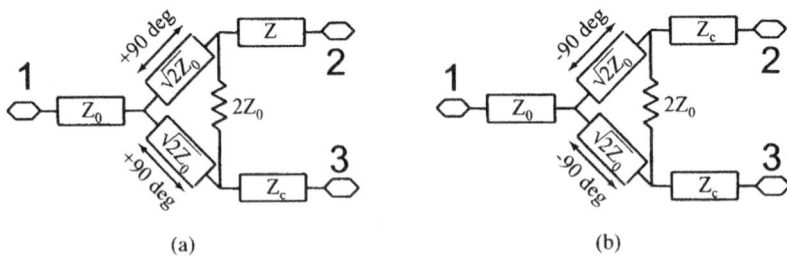

Fig. 5.9. Conceptual schematics of the CRLH DB WPD. (a) At the first frequency, f_1. (b) At the second frequency, f_2. Z_c represents the characteristic impedance of the input and output lines.

leading, using Eq. (5.14), to the design parameters

$$L_R = \frac{Z_t \pi \left[(\omega_1/\omega_2) + 1\right]}{2N\omega_2 \left[1 - (\omega_1/\omega_2)^2\right]},$$

(5.21a)

$$C_R = \frac{\pi \left[(\omega_1/\omega_2) + 1\right]}{2N\omega_2 Z_t \left[1 - (\omega_1/\omega_2)^2\right]},$$

(5.21b)

$$L_L = \frac{2N Z_t \left[1 - (\omega_1/\omega_2)^2\right]}{\pi\omega_1 \left[1 + (\omega_1/\omega_2)\right]},$$

(5.21c)

$$C_L = \frac{2N \left[1 - (\omega_1/\omega_2)^2\right]}{\pi\omega_1 Z_t \left[1 + (\omega_1/\omega_2)\right]}.$$

(5.21d)

Fig 5.10 shows a DB LH-SMT / RH microstrip WPD prototype operating at the two frequencies $f_1 = 1.0$ GHz and $f_2 = 3.1$ GHz, and the corresponding results are presented in Fig 5.11.

5.1.4 Nonlinear Component Example: Quadrature Subharmonically Pumped Mixer

After presenting examples of CRLH DB *passive* components applications in Section 5.1.3, namely, a DB QH and a DB WPD, we propose now a more

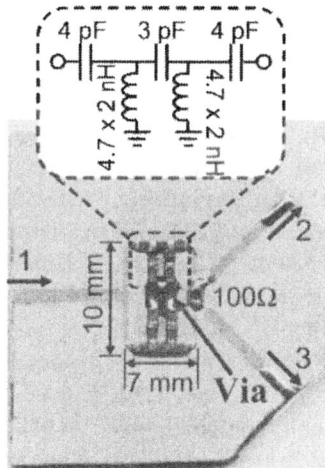

Fig. 5.10. Prototype and equivalent circuit of a CRLH DB WPD prototype using SMT chip components for the LH contribution and a conventional microstrip line for the LH contribution, as the TLs of Section 5.1.2. The frequencies of operation are $f_1 = 1.0$ GHz and $f_2 = 3.1$ GHz. The characteristic impedance is $Z_c = 50$ Ω ($\sqrt{2}Z_c \approx 70.7$ Ω). The overall dimension of the circuit is 7 mm × 10 mm.

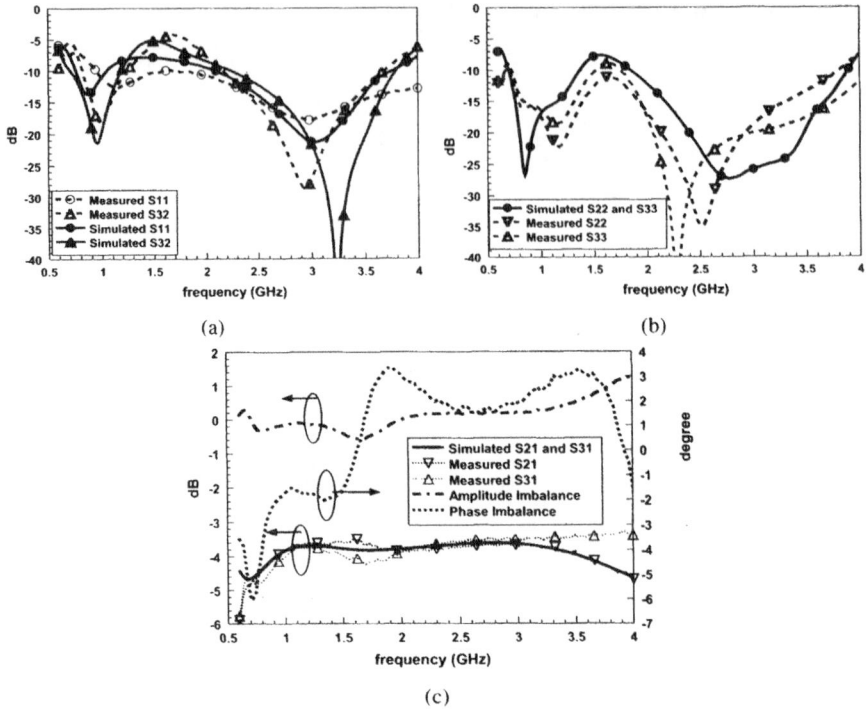

Fig. 5.11. Characteristics of the CRLH DB WPD of Fig 5.10. (a) Input return loss and isolation. (b) Output return loss. (c) Output amplitude and phase imbalance.

advanced CRLH DB *nonlinear* application, including several CRLH DB components, thereby illustrating the generality of the CRLH DB concept: a *dual-band quadrature subharmonically pumped mixer* (DB QSHPM).[13] Subharmonic mixers are particularly useful at high (millimeter-wave) frequencies where reliable and stable LO sources are often unavailable, inefficient, or prohibitively expensive [4]. Quadrature mixers are utilized in direct conversion systems, where the in-phase/quadrature (I/Q) nature of IF ports is beneficial to broad-band digital baseband modulation and demodulation schemes [5].

A circuit diagram of the DB QSHPM and a photograph of corresponding prototype are shown in Figs. 5.12 and 5.13, respectively. The complete mixer includes two CRLH DB WPDs, one CRLH ±45° DB phase advance/delay TL, and two subharmonically pumped mixers (SHPMs) containing each two CRLH *dual-passband dual-stopband* (DPDS) *shunt terminations*, representing a total of seven CRLH DB components. The DB-WPDs, which were described in Section 5.1.3.2, are used to split the RF and LO powers toward the I and Q branches; two different DB-WPD designs are required since the RF and LO

---

[13]The mixer to be presented here is passive as exhibits conversion loss. However, extension to *active* mixers, with conversion gain, is straightforward.

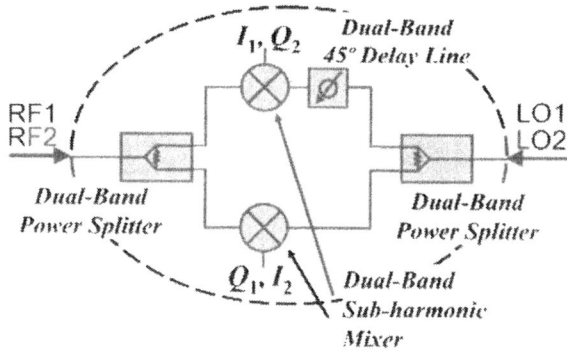

Fig. 5.12. Circuit diagram of a DB QSHPM.

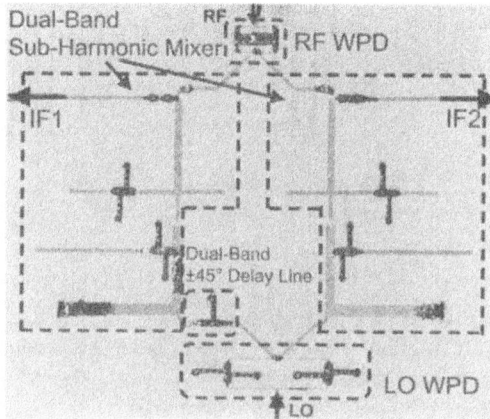

Fig. 5.13. Prototype a DB QSHPM. The DB RF and LO frequencies are $(f_{RF_1}, f_{LO_1}) =$ (1.0, 0.51) GHz for the first pair and $(f_{RF_2}, f_{LO_2}) = (3.06, 1.56)$ GHz for the second pair. The overall dimension of the circuit is 89.3 mm × 103.8 mm. A lowpass filter (simple series-inductor shunt-capacitor filter, not shown in this picture) with cutoff frequency of 50 MHz is used in the IF lines to suppress f_{LO}.

frequencies are different. The DB phase advance/delay TL is identical to the TL described in Section 5.1.2 with $(\phi_1, \phi_2) = (-45°, +45°)$, $\pm45°$ corresponding to $\pm90°$ at the utilized second harmonic of the LO, $2f_{LO}$. The two SHPMs are identical to each other; we will describe them now.

Fig 5.14 shows a DB SHPM of the type used in the DB QSHPM of Figs. 5.12 and 5.13. The core of this mixer is an antiparallel diode pair. This diode arrangement is known to automatically suppress fundamental mixing products as well

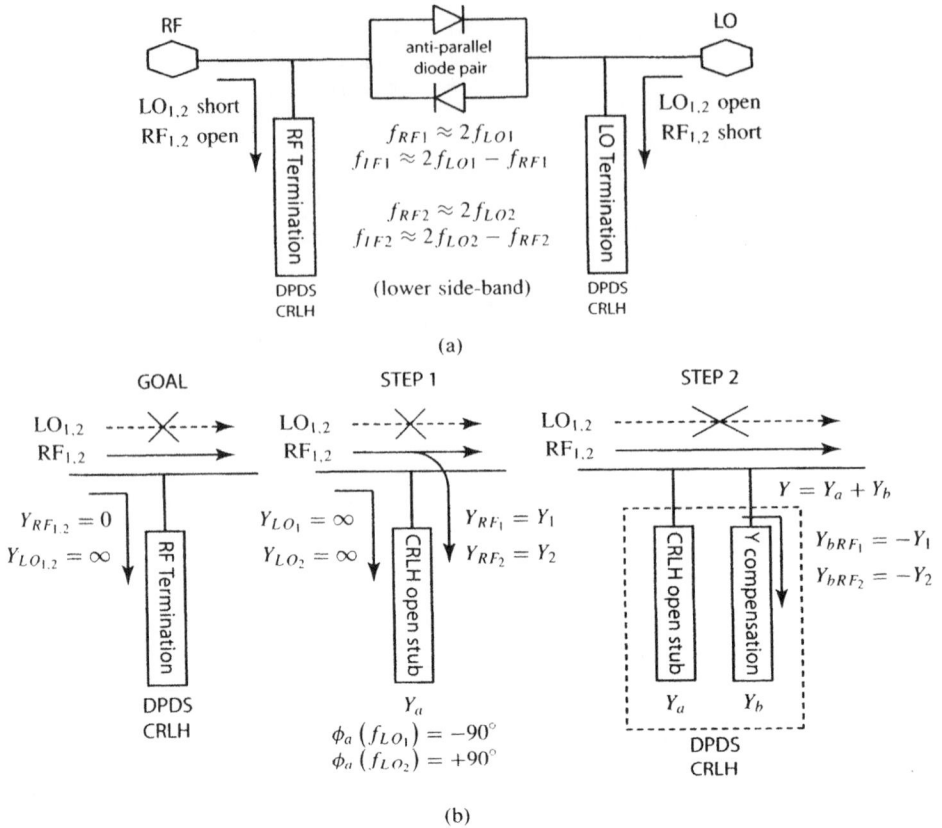

Fig. 5.14. DB SHPM based on CRLH dual-passband dual-stopband (DPDS) shunt stub terminations. (a) Circuit diagram of the DB-SHPM (2nd LO harmonic) with DPDS stubs. (b) Design procedure of the DPDS RF termination in (a). The design of the DPDS LO termination in (a) is simply obtained by the substitution RF↔LO.

as odd harmonics of the LO [6][14] and is for this reason particularly suited for second harmonic mixing.

A DB SHPM is obtained from the antiparallel diodes pair by terminating LO at the RF port and RF at the LO port while presenting matched conditions to RF at the RF port and to LO at the LO port for both band-pairs of RF/LO frequencies $\left(f_{RF_1}, f_{LO_1}\right)$ and $\left(f_{RF_2}, f_{LO_2}\right)$, as shown in Fig 5.14(a). This complex function, involving four different frequencies, is accomplished by the CRLH DPDS terminations following the procedure presented in Fig 5.14(b) for the case of the RF termination, the principle being transposable to the LO termination

---

[14] As shown in [6], only frequencies $m f_{LO} + n f_{RF}$ where $m + n$ is an *odd* integer (in particular the desired frequency $2 f_{LO} - f_{RF}$) can flow outside the diode loop, with the other products and harmonics (including DC and the undesired frequencies $2 f_{LO}, f_{LO} \pm f_{RF}$) being "captured" (flowing) inside the loop. The LO frequency f_{LO} is suppressed by a simple lowpass filter.

by simple exchange of RF and LO. First, a CRLH DB inverter stub of admittance $Y_a(f)$, similar to those presented in Section 5.1.2 (open or short) using for instance $(\phi_1, \phi_2) = (+90°, -90°)$, is introduced to terminate the two LO frequencies, f_{LO_1} and f_{LO_2}, by presenting infinite admittances $Y_{LO_1} = \infty$ and $Y_{LO_2} = \infty$. This stub provides the appropriate termination at the LO frequencies but does not induce the required zero admittance at the RF frequencies in general, $Y_{RF_1} = Y_1 \neq 0$ and $Y_{RF_2} = Y_2 \neq 0$, which produces unacceptable RF power loss. To solve this problem, a second compensating stub of admittance $Y_b(f)$ is added to the first stub, so that the global admittance is now $Y(f) = Y_a(f) + Y_b(f)$, as shown in Fig 5.14(b). If this admittance is adjusted so that $Y_{bRF_1} = -Y_1$ and $Y_{bRF_2} = -Y_2$[15], the RF conditions $Y_{RF_1} = 0$ and Y_{RF_2} are also satisfied, and dual-band mixing is achieved. The admittances values of the CRLH DPDS RF termination are listed in Table 5.1 for easier understanding.

The experimental results of the DB QSHPM of Fig 5.10 are shown in Fig 5.15. LO to RF isolation [Fig 5.15(b)] has its maxima at $f_{LO_1} = 0.51$ and $f_{LO_2} = 1.55$ GHz. Conversion loss [Figs. 5.15(c) and 5.15(d)] is defined as the power ratio between the RF power at the mixer input to a single channel output (I or Q). Conversion loss is observed to exhibit minima at $f_{RF_1} = 1.0$ GHz and $f_{RF_2} = 3.06$ GHz. These results conform the DB operation with frequencies $(f_{RF_1}, f_{LO_1}) = (1.0, 0.51)$ GHz and $(f_{RF_2}, f_{LO_2}) = (3.06, 1.56)$ GHz.

In the scheme of the DB QSHPM of Figs. 5.12 and 5.13, the phase difference between the IF output with the delay line and the other IF output in Fig 5.13 is $+90°$ at (f_{RF_1}, f_{LO_1}) and is $-90°$ at (f_{RF_2}, f_{LO_2}).[16] Consequently, DB frequency mixing at the same IF can be achieved by utilizing a $90°$ hybrid to separate the two outputs. The design of Fig 5.12 corresponds to this case, since $f_{IF_1} = f_{IF_2} = 200$ MHz.

TABLE 5.1. Admittance of the CRLH DPDS RF Termination of Fig 5.14(b) at the Different Frequencies of Interest.

f	Y_a	Y_b	$Y = Y_a + Y_b$
f_{LO_1}	∞	Y_x	∞
f_{LO_2}	∞	Y_y	∞
f_{RF_1}	Y_1	$-Y_1$	0
f_{RF_2}	Y_2	$-Y_2$	0

[15]In principle, a DB stub would be required to satisfy this requirement. However, if Y_1 and Y_2 are very small, as is the case in the prototype of Fig 5.10, the design can be simplified by using a single-band stub presenting a small susceptance of opposite sign, sufficient to reduce Y_1 and Y_2 to negligible values.

[16]The delay line, which is $+45°$ at f_{LO_1}, induces $+90°$ phase shift at $2f_{LO_1}$, while the same delay line, which is $-45°$ at f_{LO_2}, induces $-90°$ phase shift at $2f_{LO_2}$. Therefore, after mixing, the difference between the IF outputs with and without the delay line are $+90°$ for $f_{IF_1} = 2f_{LO_1} - f_{RF_1}$ and $-90°$ for $f_{IF_2} = 2f_{LO_2} - f_{RF_2}$.

(a)   (b)

(c)   (d)

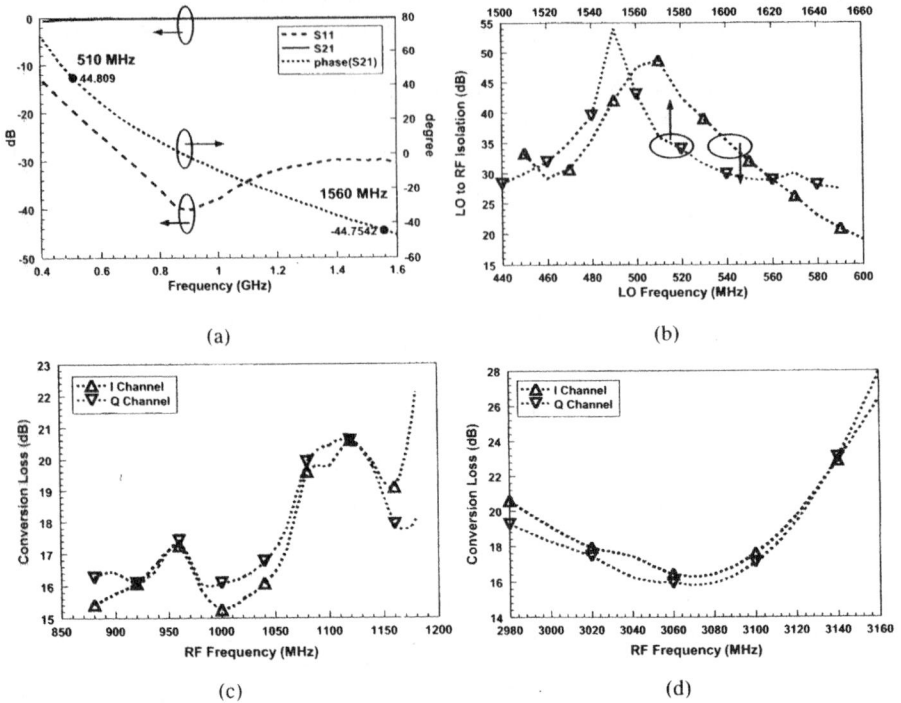

Fig. 5.15. Measured results for the DB QSHPM of Fig 5.13. (a) Frequency response of the isolated DB $\pm 45°$ delay line used in one of the two LO branches for orthogonal outputs generation at $2f_{LO}$. (b) LO to RF isolation versus LO frequency with $P_{RF} = -20$ dBm, $P_{LO} = +4$ dBm. (c) Conversion loss versus RF frequency around the first LO band ($f_{LO,1} = 1.0$ GHz) with $f_{IF} = 20$ MHz (fixed) (LO is adjusted at each frequency point following $f_{LO} = (f_{RF} + f_{IF})/2$. (d) Conversion loss versus RF frequency around the second LO band ($f_{LO,2} = 3.1$ GHz) in the same conditions as in (c).

It should be noted that the operating frequencies of the prototypes, including SMT chip components are limited by the self-resonance of the chips, which is typically around 5 GHz to 10 GHz, depending on the values of the components. In contrast, monolithic microwave integrated circuit (MMIC) implementations of the described QH and RR, and other similar components, may be useful in many DB applications of modern mobile communication, wireless local area network (WLAN) multistandards, and uplink/downlink satellite systems, since MMIC passive components typically exhibit much higher self-resonant frequencies.

5.2 ENHANCED-BANDWIDTH COMPONENTS

The CRLH principles exposed in Chapter 3 can also lead to *bandwidth enhancement* in various microwave components, such as for instance couplers, baluns, and phase shifters. An essential difference between *enhanced-bandwidth* (EB) components, which will be presented in this section, and DB components, which were presented in the previous section, is that in the case of EB components only *some* of the conventional TL branches in a given component are replaced by CRLH branches, whereas in the case of DB components this replacement pertains to *all* of the branches. The key idea in EB components lies in a judicious combination of the hyperbolic-linear[17] dispersion characteristic of the CRLH TL with the linear dispersion characteristic of conventional (PRH) TL sections.

5.2.1 Principle of Bandwidth Enhancement

The problem of bandwidth enhancement can be generally formulated in the following manner. Consider an arbitrary component operating at a frequency ω_s and including two sets of PRH TL sections with phase shifts of $\phi_{1,s}$ and $\phi_{2,s}$, respectively, at ω_s.[18] We will show that it is possible to maximize the bandwidth of this component by replacing one of these two sets of PRH TL sections by CRLH TL sections and keeping the other set unchanged.

The phase shift induced by a PRH TL of length ℓ can be written

$$\phi_{PRH} = -\beta\ell = -\frac{n\ell}{c}\omega, \quad \text{with} \quad n = \sqrt{\mu_{eff}\,\varepsilon_{eff}}, \tag{5.22}$$

where μ_{eff} and ε_{eff} are the effective permeability and permittivity of the line, respectively. If this TL is required to exhibit a phase shift of $\phi_{PRH,s}$ at ω_s

$$\phi_{PRH}(\omega = \omega_s) = -\frac{n\ell}{c}\omega_s = \phi_{PRH,s}, \tag{5.23}$$

then Eq. (5.22) can be rewritten as

$$\phi_{PRH}(\omega) = \frac{\phi_{PRH,s}}{\omega_s}\omega. \tag{5.24}$$

On the other hand, the phase shift induced by an N-cell (balanced) CRLH TL can be written as [Eq. (3.85)]

$$\phi_{CRLH}(\omega) = -N\left[\omega\sqrt{L_R C_R} - \frac{1}{\omega\sqrt{L_L C_L}}\right] \tag{5.25}$$

---

[17]In the balanced-resonances case, the real CRLH dispersion curve is, according to Eqs. (3.34) or (3.85), precisely the superposition (sum) of the *LH hyperbolic* dispersion function (dominant at lower frequencies) and of the *RH linear* dispersion function (dominant at higher frequencies).

[18]For instance, in the case of the rat-race coupler, which will be presented in Section 5.2.2, we will have $\phi_{1,s} = -90°$ and $\phi_{2,s} = -270°$.

in the frequency range around the transition frequency f_0 or, rigorously and for any frequency, by

$$\phi_{CRLH}(\omega) = N \cdot \Delta\phi(\omega), \tag{5.26}$$

where $\Delta\phi(\omega)$ is the exact phase shift induced by a CRLH unit cell as given by Eq. (5.9). Here, the CRLH TL will be required to exhibit a phase shift of $\phi_{CRLH,s}$ at ω_s

$$\phi_{CRLH}(\omega = \omega_s) = -N \left[\omega_s \sqrt{L_R C_R} - \frac{1}{\omega_s \sqrt{L_L C_L}} \right] = \phi_{CPRH,s}, \tag{5.27}$$

and, as the PRH TL, to be matched to termination ports of impedance Z_c

$$\sqrt{\frac{L_R}{C_R}} = Z_c, \tag{5.28a}$$

$$\sqrt{\frac{L_L}{C_L}} = Z_c. \tag{5.28b}$$

Finally, we need to establish a relation providing bandwidth maximization around ω_s for the considered device including both PRH and CRLH TL sections. Maximum bandwidth is achieved when the CRLH TL section phase curve exhibits the same slope as that of the PRH TL sections. This implies that the phase difference

$$
\begin{aligned}
\phi_{diff}(\omega) &= \phi_{PRH}(\omega) - \phi_{CRLH}(\omega) \\
&= \frac{\phi_{PRH,s}}{\omega_s}\omega + N \left[\omega\sqrt{L_R C_R} - \frac{1}{\omega\sqrt{L_L C_L}} \right]
\end{aligned}
\tag{5.29}
$$

is minimum at ω_s or $d\phi_{diff}/d\omega]_{\omega=\omega_s} = 0$ with $d^2\phi_{diff}/d\omega^2]_{\omega=\omega_s} < 0$; that is, with Eqs. (5.24) and (5.25),

$$\frac{\phi_{CRLH,s}}{\omega_s} + N\sqrt{L_R C_R} + \frac{N}{\omega_s^2 \sqrt{L_L C_L}} = 0. \tag{5.30}$$

The second derivative condition, ensuring that the extremum corresponding to the zero of the first derivative is a minimum, is automatically satisfied, since we have $d^2\phi_{diff}/d\omega^2 < 0]_{\omega=\omega_s} = -2N/\left(\omega_s^3 \sqrt{L_L C_L}\right) < 0$ for any set of parameters. Eqs. (5.27), (5.28a), (5.28b), and (5.30) build a linear system of four equations with the four unknowns L_R, C_R, L_L, and C_L. The solutions to this system are found to be

$$L_R = -Z_c \frac{\phi_{PRH,s} + \phi_{CRLH,s}}{2N\omega_s}, \tag{5.31a}$$

$$C_R = -\frac{\phi_{PRH,s} + \phi_{CRLH,s}}{2N Z_c \omega_s}, \tag{5.31b}$$

$$L_L = -\frac{2N Z_c}{\omega_s \left(\phi_{PRH,s} - \phi_{CRLH,s} \right)}, \tag{5.31c}$$

$$C_L = -\frac{2N}{Z_c \omega_s \left(\phi_{PRH,s} - \phi_{CRLH,s} \right)}. \tag{5.31d}$$

Examination of these expression reveals limitations in the described bandwidth maximization scheme. From the necessary condition that the LC values of the CRLH must all be positive,[19] with $\phi_{PRH,s} < 0$, we find that application of Eqs. (5.31) is restricted to the range of phases[20]

$$|\phi_{CRLH,s}| \leq |\phi_{PRH,s}|. \tag{5.32}$$

Let us now consider two examples to illustrate this BE technique. In the first example, presented in Fig 5.16, we consider that a phase difference of 90° is required between the two different sets of TLs in the component. Such a difference is typically achieved in a conventional component by using PRH TL sections with $\phi_{1,s} = -90°$ and $\phi_{2,s} = -180°$, corresponding to the phase responses shown in Fig 5.16(a). The resulting component necessarily exhibits narrow bandwidth because $\phi_{diff} = \phi_1 - \phi_2$ varies linearly with frequency around the operation frequency ω_s [Fig 5.16(a)] (and everywhere else). By replacing the second PRH TL with a CRLH TL designed according to Eq. (5.31) with $\phi_{PRH} = \phi_1$ ($\phi_{PRH,s} = -90°$) and $\phi_{CRLH,s} = 0°$, as shown in Fig 5.16(b), the bandwidth of the component is dramatically increased because the new phase difference $\phi_{diff} = \phi_{PRH} - \phi_{CRLH}$ exhibits a zero slope at ω_s, which means even an important change in frequency only induces a negligible variation in the phase difference. Defining bandwidth as the frequency range in which the phase difference variation is less than 10°, we find for the case of Fig 5.16 that fractional bandwidth has been increased from 2.8% in the conventional case [Fig 5.16(a)] to 21.7% in the EB case [Fig 5.16(b)].

---

[19]Negative values of L of C represent inversion in the reactance sign, that is, transformation from L to C if $L < 0$ or from C to L if $C < 0$, which leads to a TL different from the standard CRLH prototype. The resulting modified prototype may be useful in the cases where the CRLH cannot produce maximal bandwidth, but the difference in frequency dependence between inductive and capacitive reactances must be taken into account to obtain a physically realizable structure.

[20]$L_R, C_R > 0$ is always satisfied if $\phi_{CRLH,s} < 0$, but only satisfied when $\phi_{CRLH,s} < |\phi_{PRH,s}|$ if $\phi_{CRLH,s} > 0$; $L_L, C_L > 0$ is always satisfied if $\phi_{CRLH,s} > 0$, but only satisfied when $|\phi_{CRLH,s}| < |\phi_{PRH,s}|$ if $\phi_{CRLH,s} < 0$. The condition of Eq. (5.32) encompasses all of the cases.

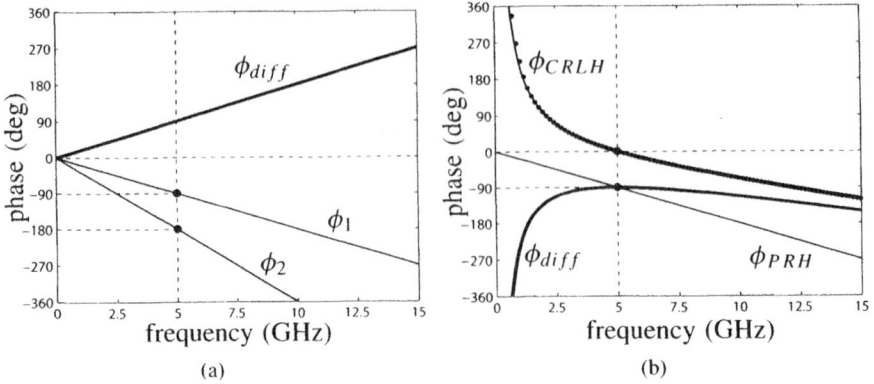

Fig. 5.16. Phase characteristics of the two sets of TL sections in a typical component operating at $f_s = 5$ GHz and requiring a phase difference of $|\phi_{diff}| = 90°$ at f_s. (a) Conventional case, using two sets of PRH TLs with $(\phi_{1,s}, \phi_{2,s}) = (-90°, -180°)$. (b) EB case, including a set of PRH TL sections and a set of CRLH TL sections, with $(\phi_{PRH,s}, \phi_{CRLH,s}) = (-90°, 0°)$, and the CRLH parameters $L_R = 0.42$ nH, $C_R = 0.17$ pF, $L_L = 6.01$ nH, $C_L = 2.43$ pF computed by Eq. (5.31) with $N = 3$. The dots on the CRLH curve are computed with the exact relation of Eq. (5.26). The transition frequency of the CRLH TLs is $f_0 = f_s = 5.0$ GHz.

In the second example, illustrated in Fig 5.17, we consider that a phase difference of 180° is required between the two different sets of TLs in the component. Such a difference is typically achieved in a conventional component by using PRH TL sections with $\phi_{1,s} = -90°$ and $\phi_{2,s} = -270°$, which corresponds to the phase responses shown in Fig 5.17(a). The results obtained by replacing the second PRH TL with a CRLH TL designed according to Eq. (5.31) with $\phi_{PRH} = \phi_1$ ($\phi_{PRH,s} = -90°$) and $\phi_{CRLH,s} = +90°$ are shown in Fig 5.17(b), offering a bandwidth improvement from 1.4% in the conventional case [Fig 5.17(a)] to 16.7% in the EB case [Fig 5.17(b)]. It should be noted that phase parameters in this second example are at the limit of the validity range, given by Eq. (5.32), of the formulas of Eq. (5.31), since $\phi_{CRLH,s} = -\phi_{PRH,s}$. In this limit, $L_R = C_R = 0$ [Eqs. (5.31a) and (5.31b)] and $f_0 = \infty$ [Eq. (3.82)], which means that the CRLH TL degenerates into a PLH TL. However, a perfectly PLH TL cannot be obtained in practice. In a practical design, the values of L_R and C_R, although possibly much smaller than those of L_L and C_L, will still be nonzero. Consequently, the attainable bandwidth will be smaller than that predicted by Eq. (5.31). However, it can still be very appreciable, as it will be shown in Section 5.2.2.

Before concluding this subsection on the principle of BE, let us discuss the variation of bandwidth as a function of the difference between the absolute values of the imposed phases at ω_s, $\phi_{diff,abs} = ||\phi_{CRLH,s}| - |\phi_{PRH,s}||$. It is well-known in the context of phase shifters that bandwidth decreases with the value of the phase shift. The same applies here. In the limit case where $\phi_{diff,abs} = 0$ (e.g., $\phi_{CRLH,s} = \phi_{PRH,s} = -\pi/2$), the optimal curve for the CRLH TL is the

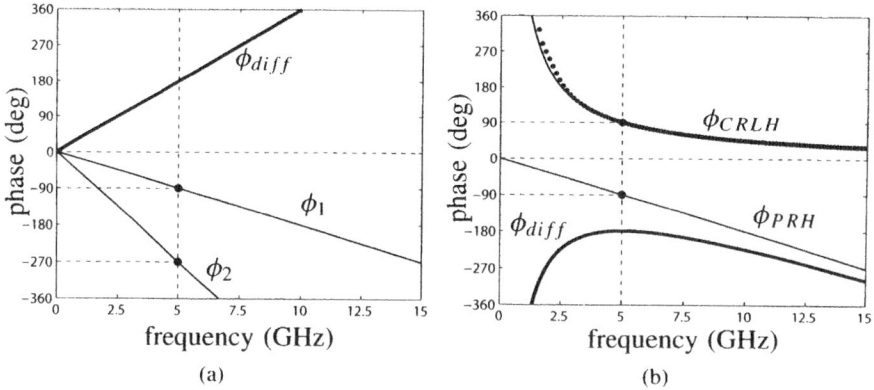

Fig. 5.17. Phase characteristics of the two sets of TL sections in a typical component operating at $f_s = 5$ GHz and requiring a phase difference of $|\phi_{diff}| = 180°$ at f_s. (a) Conventional case, using two sets of PRH TLs with $(\phi_{1,s}, \phi_{2,s}) = (-90°, -270°)$. (b) EB case including a set of PRH TL sections and a set of CRLH TL sections, with $(\phi_{PRH,s}, \phi_{CRLH,s}) = (-90°, +90°)$, and the CRLH parameters $L_R = C_R = 0$ (PLH limit), $L_L = 3.05$ nH, $C_L = 1.22$ pF computed by Eq. (5.31) with $N = 3$. The dots on the CRLH curve are computed with the exact relation of Eq. (5.26). The transition frequency of the CRLH TLs is $f_0 = \infty$ (PLH).

same as that of the PRH TL; thus, the CRLH TL degenerates into a PRH TL, with $L_L = C_L = 0$, and bandwidth, trivially, becomes infinite. In contrast, when $\phi_{diff,abs}$ increases (e.g., $\phi_{CRLH,s} = 0$ and $\phi_{PRH,s} \to -\infty$), Eqs. (5.31) reveal that L_L and C_L become smaller. Consequently, the second derivative of the phase difference ϕ_{diff} in Eq. (5.29), $d^2\phi_{diff}/d\omega^2 < 0]_{\omega=\omega_s} = -2N/\left(\omega_s^3 \sqrt{L_L C_L}\right)$, which represents the rate of variation of ϕ_{diff}, becomes more negative, which means that the ϕ_{diff} peak (bandwidth) becomes narrower. This trend is clearly apparent in the transition from Fig 5.16(b) to Fig 5.17(b), where the ($10°$-variation) bandwidth is decreasing from 21.7% to 16.7%.

5.2.2   Rat-Race Coupler Example

We will now illustrate the BE principles exposed in Section 5.2.1 with the example of an EB rat-race coupler (RRC) [7]. A RRC, or hybrid ring, is a four-port network with a $0°$ phase shift or $180°$ phase shift between its two output ports [2], as illustrated in Fig 5.18 for both the conventional RRC and the EB RRC. This figure shows that if port 1, called the Σ port, is used as the input, the signal is evenly split into two in-phase components at ports 2 and 3 and port 4 is isolated, whereas if port 4, called the Δ port, is used as the input, the signal is evenly split into two anti-phase components at ports 2 and 3 and port 1 is isolated. This component is widely used in microwave circuits, for instance in balanced mixers and power amplifiers.

The conventional RRC is made of three $-90°$ PRH TL sections and one $-270°$ PRH TL section. In the EB RRC, the $-270°$ PRH TL section is replaced

(a)

(b)   (c)

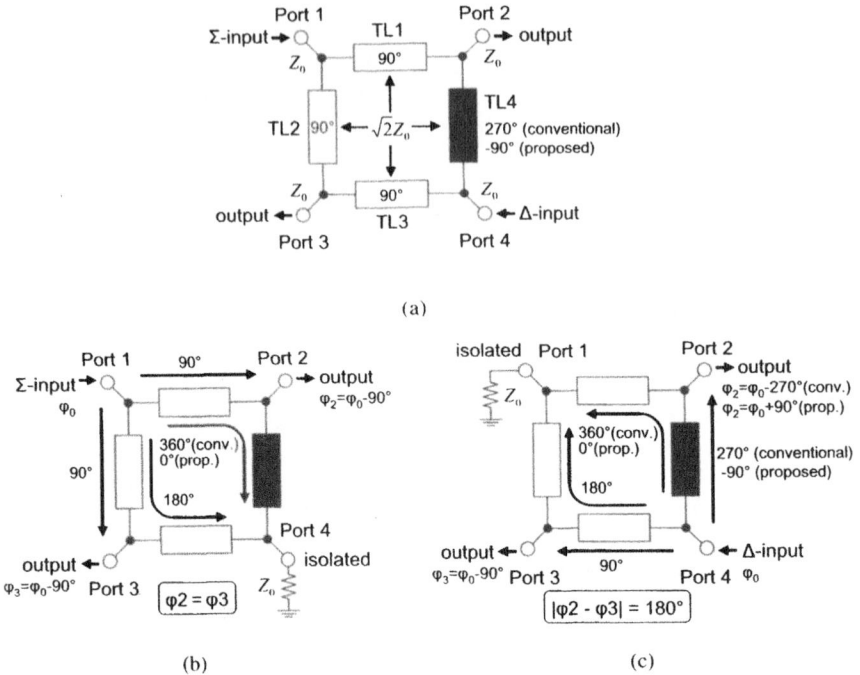

Fig. 5.18. Schematics of conventional and EB RRCs. (a) Equivalent circuit. (b) In-phase operation. (c) Anti-phase operation. From [7], © IEEE; reprinted with permission.

by a $+90°$ CRLH TL section. This substitution does not change the output phase difference, as shown in Fig 5.18, and the resulting component is therefore equivalent to the conventional RRC *at the design frequency.* The difference between the conventional RRC and the new component will only appear in the *bandwidth*, which is expected to be significantly enhanced in the EB RRC, if based on the design guidelines of Eqs. (5.31), according the explanations given in Section 5.2.1.

Fig 5.19 shows an EB RRC prototype, where the $+90°$ CRLH TL section has been implemented in SMT chip components. The TL sections appearing in the CRLH TL account for the parasitic RH effects produced by the soldering pads. Series-L/shunt-C configurations, allowing smaller LC values, are used to avoid limitations caused by self-resonance of the SMT chip components, smaller LC values having larger self-resonances. Replacement of the $-270°$ PRH TL section by the $+90°$ CRLH TL section results in a 67% size reduction compared with the conventional RRC. However, this benefit is only due to the SMT implementation, which could also be used in the PRH case and not to the CRLH TL. The really original and useful property of interest here is bandwidth enhancement.

The performances of the EB RRC of Fig 5.19 compared with those of a conventional RRC on the same substrate are shown in Fig 5.20 and summarized

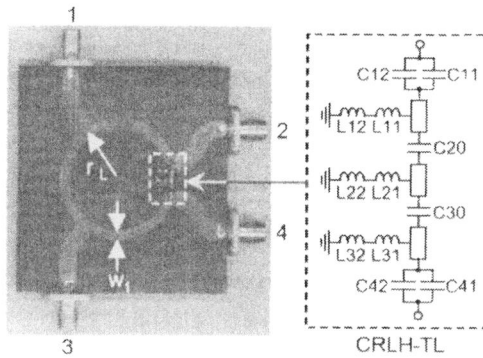

Fig. 5.19. Prototype of an SMT-CRLH/microstrip-PRH enhanced-bandwidth rat-race coupler (EB RRC), utilizing a $+90°$ CRLH TL section, with circuit shown on the right-hand side, in replacement of the conventional PRH $-270°$ section. The substrate used is Duroid 6510 with permittivity 2.2 and thickness 1.57 mm. The value of all of the inductances is 4.7 nH, and the value of all of the capacitors is 1.2 pF. The frequency of operation is $f_s = 2.0$ GHz. From [7], © IEEE; reprinted with permission.

in Table 5.2. It can be seen that the EB RRC exhibits the important 58% and 49% bandwidth enhancement at 2 GHz in the anti-phase and in-phase modes, respectively. The bandwidth enhancement with respect to the design frequency is mostly toward the higher frequencies, as expected from the phase characteristics of Figs. 5.16 and 5.17.

5.3 SUPER-COMPACT MULTILAYER "VERTICAL" TL

The DB and EB components presented in Sections 5.1 and 5.2, respectively, are implemented with discrete SMT chip components, essentially providing the series capacitance and shunt inductance required for left-handedness. As pointed out in Section 5.1, chip component implementations are appropriate for low-frequency applications, but they are limited in frequency by the self-resonance of the chips. In addition, the fact that chip components are available only in discrete values represents an important design limitation. Another limitation of chips is that they cannot be easily integrated into complex structures, such as multilayer or/and 3D structures. Finally, chip components tend to be inappropriate for radiative applications (Chapter 6) because the geometry and constitution of the chips cannot be modified to provide desired radiation characteristics.

In this section, we switch to a purely *distributed implementation* of a CRLH TL, that is, a configuration that includes only dielectric slabs and patterned metallizations and no commercial chip components. It should be noted that the DB

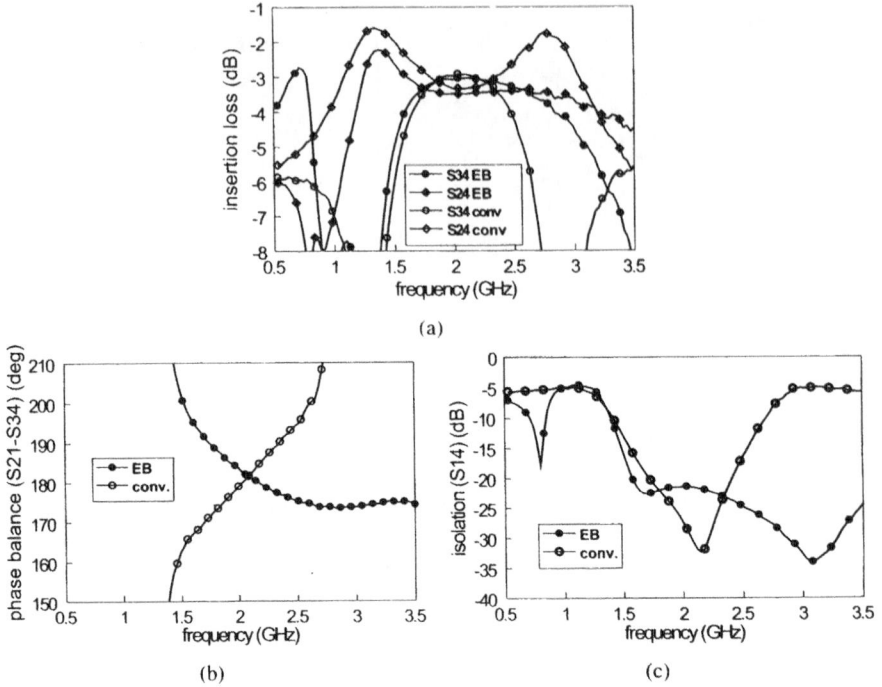

(a)

(b)   (c)

Fig. 5.20. Measured performances of the EB RRC of Fig 5.19 compared with those of a conventional RRC on the same substrate. (a) Insertion loss. (b) Phase balance. (c) Isolation. From [7], © IEEE; reprinted with permission.

TABLE 5.2. Summary of the Measured Performances of the EB RRC of Fig 5.19 Compared with those of a Conventional RRC on the Same Substrate.

	Conventional		Proposed	
	Σ Input	Δ Input	Σ Input	Δ Input
Output [dB]	-3.14 ± 0.25	-3.15 ± 0.25	-3.19 ± 0.25	-3.28 ± 0.25
Frequency range [GHz]	1.73–2.32	1.73–2.32	1.67–2.59	1.65–2.62
Bandwidth [%]	29	29	43	46
Phase balance [deg]	0 ± 10	-180 ± 10	0 ± 10	180 ± 10
Frequency range [GHz]	1.68–2.40	1.67–2.33	1.36–(3.5)	1.68–(3.5)
Bandwidth [%]	35	33	>88	>70
Isolation [dB]	< −20		< −20	
Frequency range [GHz]	1.69–2.38		1.54–(3.5)	
Bandwidth [%]	34		>78	
Return loss [dB]	< −15		< −15	
Frequency range [GHz]	1.53–2.48		1.72–2.54	
Bandwidth [%]	47		39	

and EB principle developed in Sections 5.1 and 5.2, respectively, are universal in the sense that they do not depend on the specific implementation. Therefore, DB and EB components can be readily designed in purely distributed form as well.

The structure described in this section is a *multilayered "vertical"* (MLV) CRLH TL. The first interest of this structure is that it provides an *alternative to the interdigital-capacitor/stub-inductor configuration* introduced in Section 3.3 (which will be applied to couplers in Section 5.4). The MLV CRLH TL is very different from planar CRLH TLs such as the interdigital-capacitor/stub-inductor TL. Whereas in the latter propagation occurs in the plane of the substrate, the former has a direction of propagation that is *"vertical" or perpendicular to the plane of the substrate*. The interest of the MLV CRLH TL, in addition to representing an alternative architecture for a CRLH TL, is that it constitutes a *super-compact TL*, as will be shown next.

5.3.1 "Vertical" TL Architecture

Achieving small size in a LH structure seems a priori a challenging task. A LH structure requires a series capacitance C_L and a shunt inductance L_L, whereas natural (RH) materials provide a series inductance L_R and a shunt capacitance C_R, which globally results in the CRLH TL structure with the circuit model of Fig 3.14. As a consequence, the reactances C_L and L_L must be produced artificially under the form of structured components, such as interdigital capacitors and stub inductors (Section 3.3). Unfortunately, attempts to reduce the size of such *planar* components typically result in reduction of the values of the reactances associated with $C_L - L_L$ and $L_R - C_R$, shifting the dispersion characteristics to higher frequencies, and also in reduction of the ratio between the LH and the RH reactances, eventually leading to complete disappearance of left-handedness if the size of the unit cell becomes too small.

By replacing the planar capacitor C_L by parallel-plate capacitors perpendicular to the direction of propagation, large series capacitance values can be obtained over a very small physical length with a moderate footprint, and miniaturized LH structures can thereby be obtained. The LH shunt inductance can then still be generated by a stub inductor. In addition, the RH series-L/shunt-C can be reduced to extremely small values if necessary, which increases the bandwidth of the LH range. An immediate benefit of such an architecture for MTMs is that the reduced size of the unit cell provides *improved homogeneity*, due to reduction of the electrical size of the unit cell (p/λ_g) in comparison with that achieved in planar configurations, while requiring more unit cells for a given electrical length.

The above considerations suggest a ML CRLH TL structure of the type shown in Fig 5.21. Such a structure can be easily fabricated with today's printed circuit board (PCB) or low-temperature cofired ceramics (LTCC) processes.[21] The ML

---

[21]In current LTCC technology, the distance between the plates can be accurately controlled with a resolution of up to 1 μm.

(a)

(b)   (c)

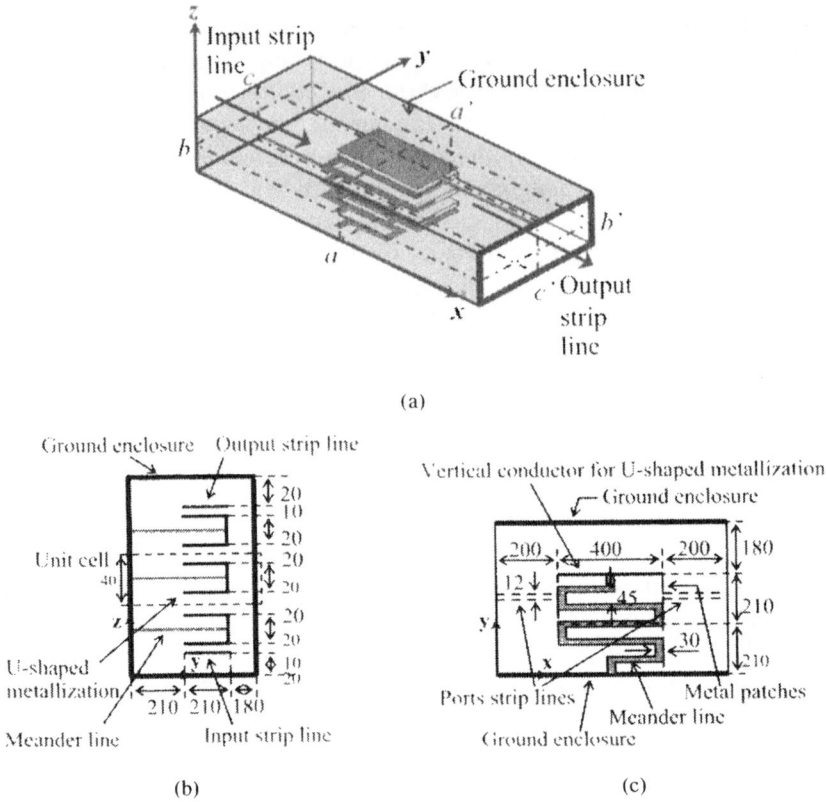

Fig. 5.21. Geometry of the MLV CRLH TL. Here the structure is periodic, with period $p = 1.57$ mm and number of unit cells $N = 3$, so that the total length $\ell = Np = 4.72$ mm. (a) 3D-view. (b) yz-plane cross section $a - a'$. (c) xy-plane cross section $b - b'$. All the dimensions are in mil. From [8], © IEEE; reprinted with permission.

CRLH TL is constituted of (periodically or not) stacked LH unit cells, each of which is constituted by a pairs of U-shaped parallel plates connected to a ground enclosure by meander lines. The ports of this TL are located on either side of the structure, one at the top of it and the other one at the bottom of it. The gaps between the U-shaped metallizations provide the LH series capacitances C_L, while the meander lines provide the LH shunt inductances L_L. The vertical conductors linking to two plates of the U-shaped structure introduce a small RH series inductance L_R and the spacing between these conductors and the metallic enclosure introduce a RH shunt capacitance C_R, which can be made extremely small with a large spacing. An impedance transformation, which is here simply a stepped impedance transformation, is necessary for the transition between the 50 Ω termination strip lines and the patches to which they are connected and which need to provide the capacitance value of $2C_L$ for matching.

5.3.2   TL Performances

The performances of the MLV CRLH TL of Fig 5.21 are shown in Fig 5.22. The agreement between full-wave FEM/FDTD results and LC network results based on extracted parameters[22] is seen to be very good.

It should be noted that the size of the unit cell p referred to the wavelength of the embedding material λ_d is extremely small: $p/\lambda_d \approx 1/150$ ($p/\lambda_0 \approx 1/488$) at the center frequency (400 MHz)! This is exceptionally small in comparison with the ratios typically obtained in most of the MTMs, which are usually of the order 1/10 to 1/15. This small size ensures perfectly homogeneous behavior of the structure.

It can be seen in Fig 5.22(a) that the TL exhibits a -10 dB bandwidth of 21.7%, extending from 260 to 660 MHz. The dispersion diagram of the structure, shown in Fig 5.22(b), reveals that this frequency band is the LH band and that the structure is unbalanced (gap above the LH band),[23] the RH range extending beyond the frequency range shown. The bandwidth predicted from the dispersion diagram, extending from 260 to 815 MHz (50.0%), is more than twice larger than the -10 dB bandwidth of the actual structure. This discrepancy is explained as follows. The lower cutoff frequencies (f_{cL}) of the scattering parameters and dispersion diagram agree very well with each other because the lower cutoff in the actual structure is very sharp and therefore well defined. In contrast, the higher cutoff frequencies [$f_{\Gamma 1} = f_{se} = 1/(2\pi\sqrt{L_R C_L})$] are very different. This is because this cutoff in the actual structure is very smooth and therefore not clearly defined, whereas the dispersion diagram, in terms of β, provides a binary information on the existence or not existence of a passband.[24] If we consider the -20 dB bandwidth instead of the -10 dB bandwidth, the scattering parameters and dispersion diagram bandwidths compare very well.

---

[22] The extraction procedure presented in Section 3.3.3 can be used. A less elegant but sometimes simpler procedure could be the following: 1) estimate C_L by the static approximation $C = \varepsilon A/h$; 2) full-wave compute the dispersion diagram with PBCs of the real TL structure to obtain the values of the LH cutoff frequency f_{cL} and of the $\beta = 0$ resonance frequencies $f_{\Gamma 1}$ and $f_{\Gamma 2}$. The quantity f_{cL} is given analytically in Eq. (3.94a) (general CRLH TL), [Eq. (3.96a)] (balanced CRLH TL) or [Eq. (3.97a)] (PLH approximation), while the quantities $f_{\Gamma 1}$ and $f_{\Gamma 2}$ are given by the minimum and the maximum, respectively, of Eqs. (3.54a) and (3.54b); 3) solve the resulting system of four equations to determine the four unknowns L_R, C_R, L_L, and C_L. In case one of these parameters is not available from full-wave simulations (e.g., $f_{\Gamma 2}$, which may be too high in frequency as in Fig 5.22(b), curve fitting with the S_{11} and S_{21} curves using ABCD-to-LC relation of Eq. (3.66) can be used.

[23] As shown in Chapter 3, an unbalanced CRLH TL can only be matched over a restricted frequency band because its characteristic impedance Z_c is frequency dependant, as seen in Eq. (3.14). This expression also shows that the frequency dependence of Z_c decreases as the amount of unbalance is decreased, i.e., $f_{sh}/f_{se} \to 1$. In the case of Fig 5.22, the TL is strongly unbalanced since $f_{sh}/f_{se} = \sqrt{(L_L C_R)/(L_R C_L)} = Z_L/Z_R \approx 7$. The line is matched here to the impedance $Z_L = 31$ Ω while $Z_R = 217$ Ω. This indicates that the insertion loss could be reduced by decreasing the unbalance ratio f_{sh}/f_{se}.

[24] Quantitative information in the dispersion diagram, correlated with the scattering parameters of the actual structure, can be obtained by taking into account the attenuation factor α (e.g., Fig 3.3(b) and Section 3.1.4).

(a)

(b)

(c)

(d)

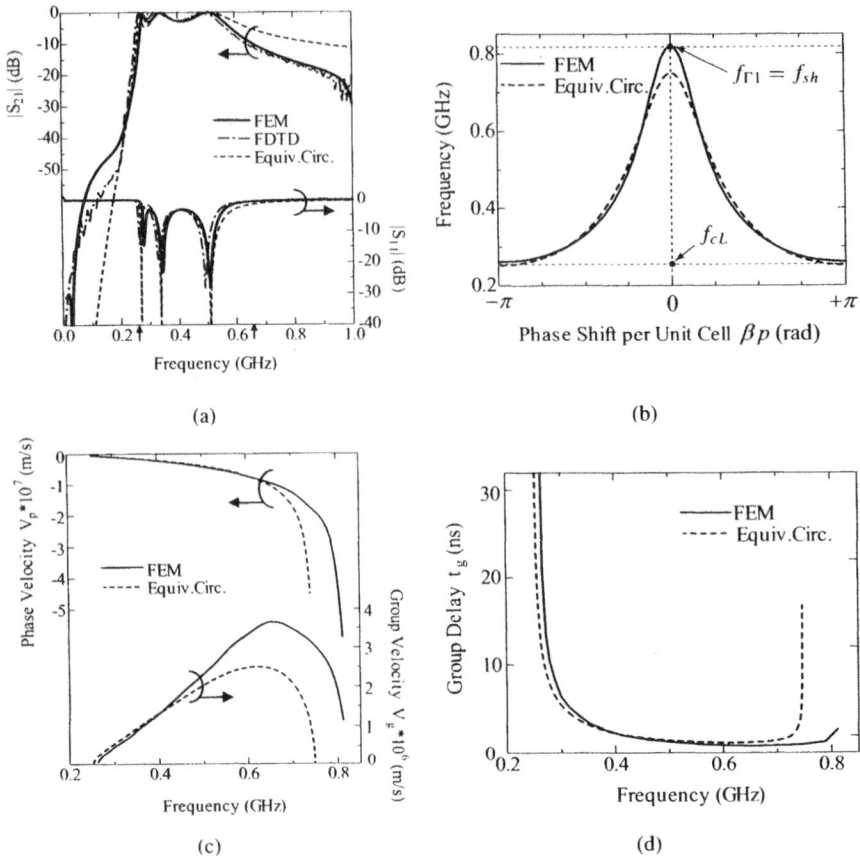

Fig. 5.22. Simulated frequency characteristics for the MLV CRLH TL of Fig 5.21 with embedding dielectric of permittivity $\varepsilon_r = 10.2$. (a) Insertion loss and return loss. (b) Dispersion diagram, computed by Eq. (3.116b) for the circuit analysis. (c) Phase and group velocities, computed by Eqs. (3.121) and (3.122), respectively, for the circuit analysis. (d) Group delay, computed by $t_g = \ell/v_g$ for the circuit analysis. For the equivalent circuit model, the extracted CRLH TL parameters are $L_R = 4.7$ nH, $C_R = 0.1$ pF, $L_L = 9.2$ nH, $C_L = 9.6$ pF, yielding the resonances $f_{\Gamma 1} = f_{se} = 1/(2\pi\sqrt{L_R C_L}) = 750$ MHz and $f_{\Gamma 2} = f_{sh} = 1/(2\pi\sqrt{L_L C_R}) = 5242$ MHz. This corresponds to an unbalanced design, with $Z_R = \sqrt{L_R/C_R} = 217\ \Omega$ and $Z_R = \sqrt{L_R/C_R} = 31\ \Omega$. From [8], © IEEE; reprinted with permission.

Considering the phase and group velocities [Fig 5.22(c)], we verify that the phase velocity is negative, with a resonance pole corresponding to the cutoff $f_{\Gamma 1} = f_{se}$ at around 0.8 GHz, and that the group velocity is zero in the stop bands (LH highpass stopband from 0 to 0.26 GHz and unbalanced CRLH stopband above 0.8 GHz). The group delay [Fig 5.22(d)], although strongly varying at the edge of the passband, becomes relatively flat around the center frequency of the

LH passband, which means that in this frequency range dispersion is moderate and that a modulated signal can therefore be transmitted along this line with small distortion.

The LH backward-wave propagation phenomenon, along the vertical direction of propagation of the MLV CRLH TL structure, is verified by FDTD analysis in Fig 5.23.

The experimental prototype corresponding to the dimensions in Fig 5.21 is shown in Fig 5.24, and the corresponding measured performances are presented

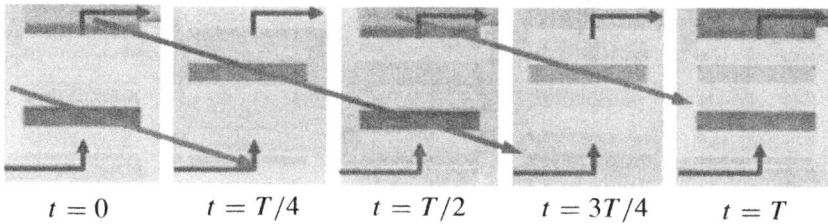

$$t = 0 \qquad t = T/4 \qquad t = T/2 \qquad t = 3T/4 \qquad t = T$$

Fig. 5.23. Backward-wave effect observed by FDTD snapshots within one time period at 0.4 GHz (time period $T = 2.5$ ns) for the MLV CRLH TL of Fig 5.21 in the xy-plane cross section $c - c'$. The input and output lines are located at the bottom and top of the structure, respectively. From [8], © IEEE; reprinted with permission.

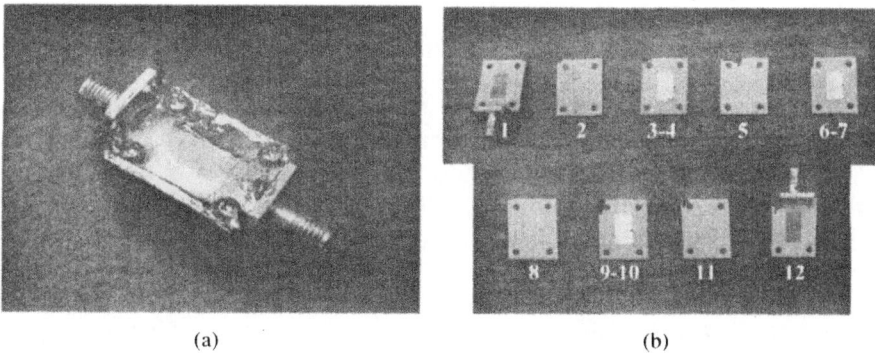

(a)   (b)

Fig. 5.24. Prototype of the MLV CRLH TL of Fig 5.21. The frequency of operation is centered at 400 MHz, with the extremely small size of $0.08\lambda_d \times 0.06\lambda_d \times 0.016\lambda_d$ along the directions $x - y - z$ ($\lambda_d = \lambda_0/\sqrt{\varepsilon_r} = 235$ mm at the center of the passband, 400 MHz). The substrate is Duroid 6510 with permittivity $\varepsilon_r = 10.2$ and thicknesses $h = 10$ and $h = 20$ mil. (a) Assembled prototype. (b) Layer parts before assembling. Parts #1 and #12: patch with stepped-impedance transition to a port (input/output); Parts #2 and #11: separator for generation of $2C_L$ ($h = 10$ mil); Parts #3 and #4, #6 and #7, #9 and #10: assembled U-shaped metallization with meander line is sandwiched inside the U-shape; Parts #5 and #8: separator for generation of C_L ($h = 20$ mil). From [8], © IEEE; reprinted with permission.

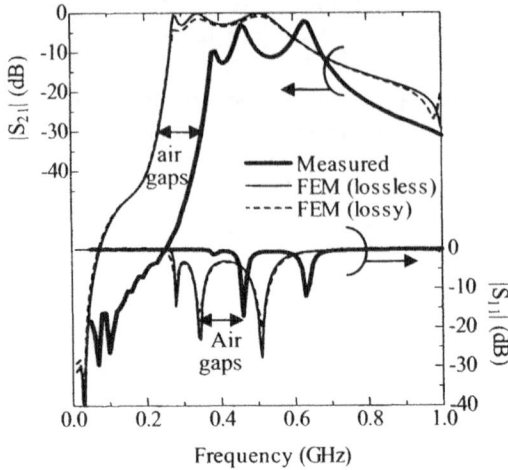

Fig. 5.25. Experimental insertion and return losses, compared with full-wave FEM results, for the MLV CRLH TL of Fig 5.21. In simulation, "loss-less" means that all metallizations are treated as perfect electric conductors, whereas "lossy" means that the conductivity of copper (5.8×10^7 S/m) is introduced for metal. The frequency shift between the simulation and measurement results is clearly explained by the fact that the effective permittivity of the dielectric is lowered by the unavoidable air films existing between the layers in the hand-made prototype of Fig 5.24. From [8], © IEEE; reprinted with permission.

in Fig 5.25. Due to the unavailability of LTCC technology, the prototype is a very rough hand-made structure built just for the proof of concept. Nevertheless, the measured results are in reasonable agreement with full-wave predictions, except for a systematic shift of the measured curves by around 100 MHz to higher frequencies, easily explainable by the decreased effective permittivity of the dielectric medium due to the presence of air films in the hand-made prototype. The larger insertion loss observed in the experiment is probably due to misalignment between the 12 different layers in the prototype.

This MLV CRLH TL exhibits the remarkably small size of $0.08\lambda_d \times 0.06\lambda_d \times 0.016\lambda_d$ ($\lambda_d = \lambda_0/\sqrt{\varepsilon_r}$) along the directions $x - y - z$ at the center of the pass-band, 400 MHz, where a phase shift of $+67°$ (phase advance) is achieved. The MLV CRLH TL represents thus a very interesting miniaturized slow-wave structure. Comparison between the loss-less and lossy cases in full-wave simulation shows that the size reduction has not been done at the expense of insertion loss, as revealed by FEM lossy analysis, whereas losses typically increase dramatically if size is reduced too much in a given structure, due to high current concentrations at the discontinuities of the metallizations.

It should be emphasized that the prototype described here represents more an illustration of the ML concept than a realistic implementation for real-life applications. The return loss is much too high, as seen in Fig 5.25. The reason for such high loss is that the present design is unbalanced, as pointed out in the caption of Fig 5.22, which results in frequency-dependant characteristic impedance preventing broadband matching [Eqs. (3.14)/(3.32), Eq. 3.86, and Section 3.2.3]. The observed passband in Fig 5.25 is the LH band only, extending from $f_{c,LH} =$ [Eq. (3.94a)] and f_{se}. The resonance $f_{se} = 1/(2\pi \sqrt{L_R C_L}) = 750$ MHz is much smaller than the resonance $f_{sh} = 1/(2\pi \sqrt{L_L C_R}) = 5,242$ MHz, so that a huge gap of 150% fractional bandwidth exists between these two resonances, which corresponds to a very frequency dependent characteristic impedance, according to Eq. (3.14), and hence poor matching ($Z_R/Z_L = 7$, whereas the balanced or matched condition requires $Z_R/Z_L = 1$). This situation $f_{se} \ll f_{sh}$ is due to the very high series capacitance C_L provided by the MIM configuration utilized. The matching condition $Z_R = Z_L$ can be obtained in different manners, as for instance by keeping f_{se} constant (unchanged C_L and L_R) and bringing the metal enclosure (ground plane) much closer to the parallel-plate structure [see Fig 5.21(b)] on the side of the bottoms of the "Us" to increase C_R by the appropriate amount (which would represent for the present parameters an increase by a factor $7^2 = 49$).

5.3.3 Diplexer Example

In this section, we present a more advanced application of the MLV CRLH structure than the simple miniature slow-wave TL described above, a *MLV CRLH diplexer*. Conventional diplexers [9, 10] often use lowpass and highpass filters for design simplicity, low insertion loss, and low cost. However, in applications where higher selectivity is required, the utilization of two narrow bandpass filters is necessary. This typically implies increasing the size in a monopackaged device to avoid spurious coupling effects between the integrated capacitors and inductors of the two channels. There is thus a trade-off to find between the size of a diplexer and its isolation in conventional RH implementations. The miniature ML CRLH TL presents then a potential interest in highly selective compact diplexers. This TL can be designed as a narrow-band filter by setting the frequencies f_{cL} and $f_{\Gamma l}$ close to each other. Very high selectivity despite a very small footprint is achievable in this structure by increasing the number of the vertically stacked unit cells.

The geometry and dimensions of the MLV CRLH diplexer are shown in Fig 5.26. The diplexer consists of two different MLV CRLH TLs operating each as a band-pass filter centered at the desired frequency of the corresponding channel. The input line is branched out into two lines to feed the two channels; these interconnecting lines are much shorter than the guided wavelength and exhibit therefore negligible distributed effects. In addition, the output lines are set in different opposite planes (top and bottom) so that propagations through the two filters occur in opposite directions, resulting in excellent isolation between the two channels due to cancelation of mutual coupling.

Fig 5.27 shows a picture of the hand-made MLV CRLH diplexer. The overall size of this prototype is of $0.163\lambda_d \times 0.096\lambda_d \times 0.045\lambda_d$ along the directions $x - y - z$ ($\lambda_d = \lambda_0/\sqrt{\varepsilon_r}$) at the center of the first channel.

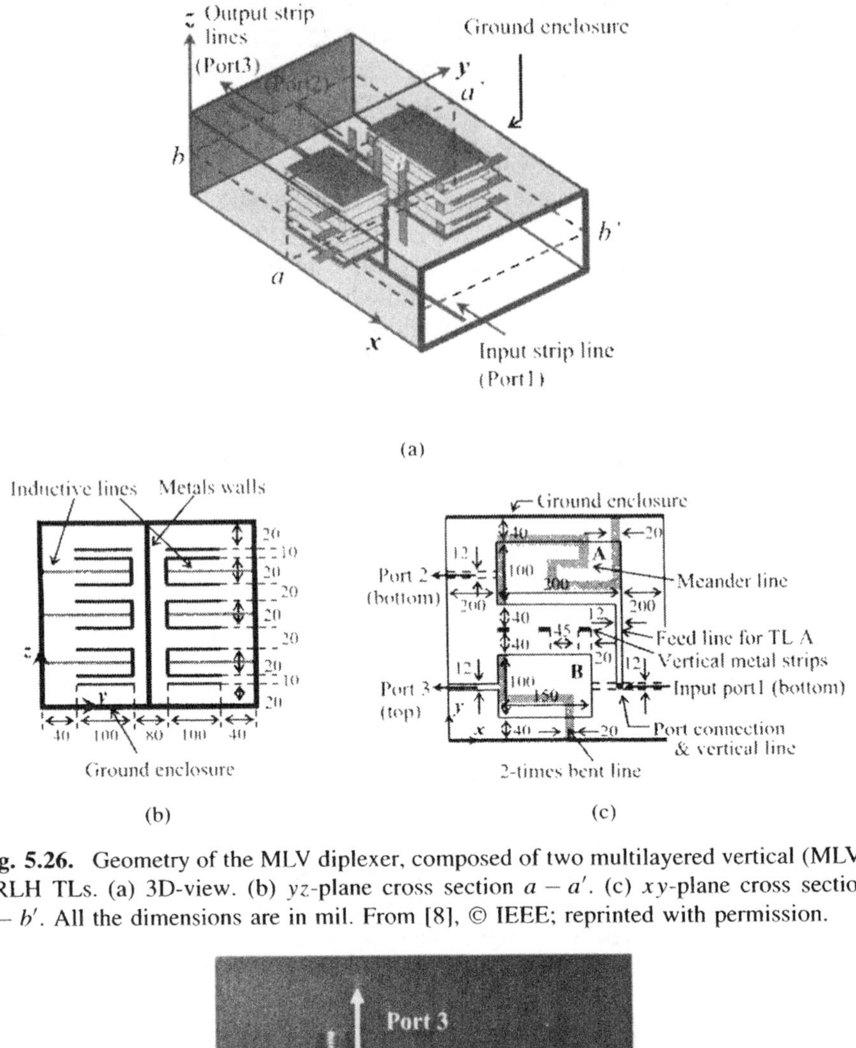

(a)

(b)

(c)

Fig. 5.26. Geometry of the MLV diplexer, composed of two multilayered vertical (MLV) CRLH TLs. (a) 3D-view. (b) yz-plane cross section $a - a'$. (c) xy-plane cross section $b - b'$. All the dimensions are in mil. From [8], © IEEE; reprinted with permission.

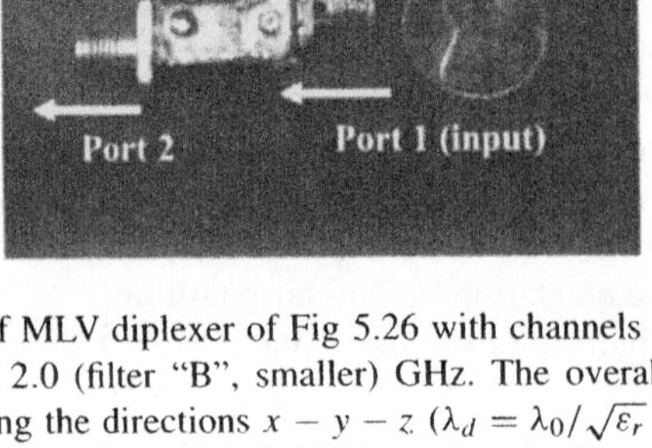

Fig. 5.27. Prototype of MLV diplexer of Fig 5.26 with channels at frequencies 1.0 GHz (filter "A", larger) and 2.0 (filter "B", smaller) GHz. The overall size is of $0.163\lambda_d \times 0.096\lambda_d \times 0.045\lambda_d$ along the directions $x - y - z$ ($\lambda_d = \lambda_0/\sqrt{\varepsilon_r} = 94$ mm at the center of the first channel, 1 GHz). From [8], © IEEE; reprinted with permission.

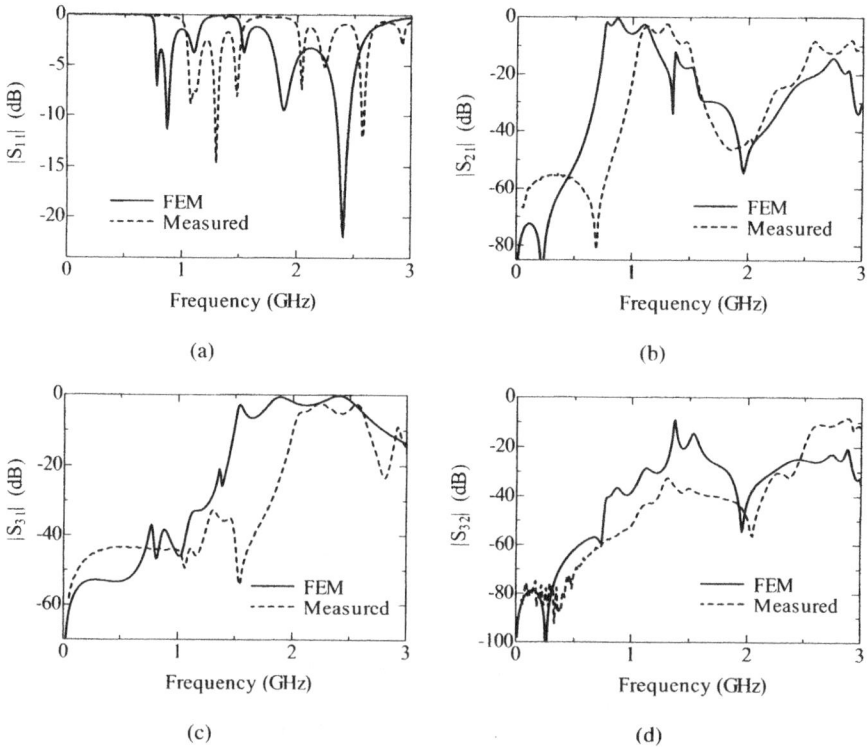

Fig. 5.28. FEM-simulated and measured scattering characteristics (magnitude) of the MLV diplexer of Fig 5.26. (a) Return loss, S_{11}. (b) Transmission coefficient from port 1 to port 2, S_{21}. (c) Transmission coefficient from port 1 to port 3, S_{31}. (d) Isolation between ports 2 and 3, S_{32}. From [8], © IEEE; reprinted with permission.

The FEM-simulated and measured scattering characteristics are shown in Fig 5.28. Except for the frequency shift, again explained by the existence of air films in the experimental prototype, good agreement is observed between the simulated and measured curves. The channels at 1.0 and 2.0 GHz are clearly apparent, and excellent isolation is obtained (38.8 GHz at 1.0 GHz and 48.9 GHz at 2.0 GHz). The return loss is high due to the fact that the two MLV TLs are unbalanced but could be reduced by balancing the TLs, as explained in Section 5.3.2.

5.4 TIGHT EDGE-COUPLED COUPLED-LINE COUPLERS (CLCs)

MTM coupled-line couplers (CLCs) are obtained by replacing one or both of the two lines constituting conventional CLCs by CRLH TLs. In the case of edge-coupled CLCs, this substitution results in unique *tight edge-coupled CLCs*, providing arbitrarily high (up to almost 0 dB) coupling level, in addition to the broad-bandwidth characteristics of typical conventional CLCs. These MTM CLCs present thus a distinct advantage over conventional edge-coupled CLCs, which suffer severe coupling-level limitations. Two types of CRLH edge-coupled CLCs will be presented. The first one (Section 5.4.2) is an *impedance coupler* (IC), so termed because its coupling mechanism is based on the difference between the *characteristic impedances* of the even and odd modes; this coupler is symmetric because it is constituted of two identical CRLH TLs. The second one (Section 5.4.3) is a *phase coupler* (PC), so termed because its coupling mechanism is based on the difference between the *phase velocities* of the c and π modes; this coupler is asymmetric as it is constituted of two different TLs, one conventional PRH TL and one CRLH TL. Although based on fundamentally different principles, these two CLCs will be shown to exhibit comparable performances. We start in Section 5.4.1 with a synthetic overview of CLCs, which should ease the understanding of the relatively involved mechanisms of the couplers presented in Sections 5.4.2 and 5.4.3.

5.4.1 Generalities on Coupled-Line Couplers

A CLC is a four-port network constituted by the combination of two unshielded TLs in close proximity to each other, as depicted in Fig 5.29.[25] Due to this proximity, the electromagnetic fields of each line interact with each other, which results in power exchange between the two lines, or coupling. Essentially, two different coupling mechanisms are possible, depending on the nature of the TLs constituting the coupled-line structure and the interspacing s between these lines: impedance coupling (IC) and phase coupling (PC).[26] Conventional IC couplers are called backward-wave couplers, and conventional PC couplers are called forward-wave couplers, due to the location of their coupled port with the respect to the input port, as indicated in Fig 5.29. The reader is referred to comprehensive texts such as [11] for the detailed derivations leading to the formulas presented in this subsection.

5.4.1.1 TEM and Quasi-TEM Symmetric Coupled-Line Structures with Small Interspacing: Impedance Coupling (IC) A (perfectly) TEM TL (e.g., coaxial, parallel-plate, and strip-line TLs) is a TL in which the component of the wave vector transverse to the direction of propagation z is zero, $k_\perp = 0$, so that the electric and magnetic fields are both in the plane perpendicular to z, in a plane said *transverse* to z. Therefore, the propagation constant of a TEM TL is identical to that of the dielectric medium (ε_r, μ_r) surrounding the metallizations of the line, $\beta = nk_0$ ($n = \sqrt{\varepsilon_r \mu_r}$), associated with the phase velocity $v_p = c/n$. When two *identical* TEM TLs are combined so as to form a *symmetric* coupled-line

---

[25] Although we will consider mainly edge-coupled CLCs, the theory of the section fully applies to broadside-coupled CLCs, and any types of other CLCs.

[26] This terminology (IC and PC) is not conventional. It is introduced here to avoid later confusion, as it will appear in Sections 5.4.2 and 5.4.3.

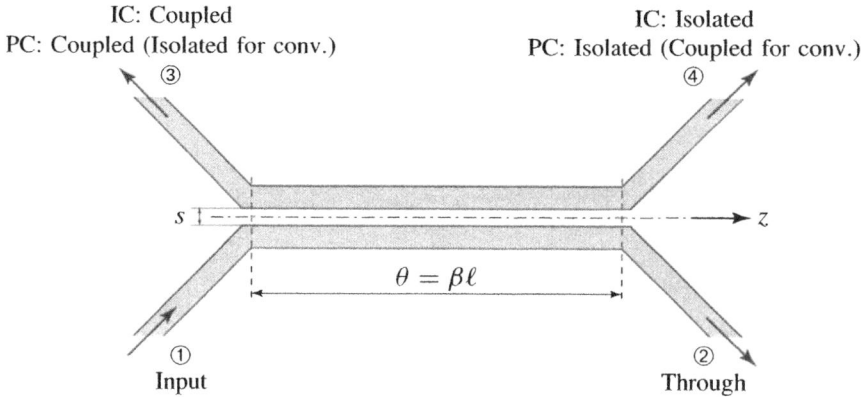

Fig. 5.29. Geometry and ports designation of a general (edge-coupled) CLC. In the case of the IC, the port locations are identical for the conventional [2] and MTM couplers. In the case of the PC, the coupled and isolated ports are exchanged between the cases of the conventional [12] and MTM couplers.

structure, the TEM nature of propagation is preserved due to symmetry. Two fundamental modes of propagation can exist in the resulting coupled-line structure, the *even mode* and the *odd mode*, represented in Fig 5.30, and the physical fields existing in the structure can always be expressed as the superposition of these two modes. Because the even and odd modes are associated with a magnetic wall or open-circuit and with an electric wall or short circuit, respectively,[27] the even/odd four-port networks are reduced to even/odd two-port networks (ports $1 \rightarrow 2$ or $3 \rightarrow 4$ in Fig 5.29), which can be treated separately as simple fictitious even/odd TLs. Because even and odd modes are necessarily TEM (by symmetry), the even and odd TLs exhibit the *same propagation constant*, $\beta_e = \beta_o = \beta = nk_0$. On the other hand, they have *different characteristic impedances* Z_{ce} and Z_{co} due to their different per-unit-length effective capacitances $C'_e = C'_{PP}$ and $C'_o = C'_{PP} + 2C'_{edge}$, as shown in Fig 5.30: $Z_{ce} = \sqrt{L'/C'_e} = 1/(cC'_e) = 1/(cC'_{PP})$ and $Z_{co} = \sqrt{L'/C'_o} = 1/(cC'_o) = 1/[c(C'_{PP} + 2C'_{edge})]$ ($Z_{ce} > Z_{co}$).

By writing the voltages and currents at each port k ($k = 1, 2, 3, 4$) $V_k = V_{ke} + V_{ko}$ and $I_k = I_{ke} + I_{ko}$, we can calculate the input impedance of the coupled-line structure as $Z_{in} = V_1/I_1 = (V_{1e} + V_{1o})/(I_{1e} + I_{1o})$, and we find thereby that $Z_{in} = Z_c$ (input matching condition to terminations of impedance Z_c) is achieved under the condition

$$Z_c = \sqrt{Z_{ce} Z_{co}}, \tag{5.33}$$

---

[27] In the case of CLCs, the even mode is associated with an open circuit and the odd mode with a short circuit. But other couplers, such as for instance the E-plane waveguide branch line coupler, have even/odd modes characterized by the opposite association: even-short/odd-open.

H-wall
open-circuit (virtual open)

(a)

E-wall
short-circuit (virtual ground)

(b)

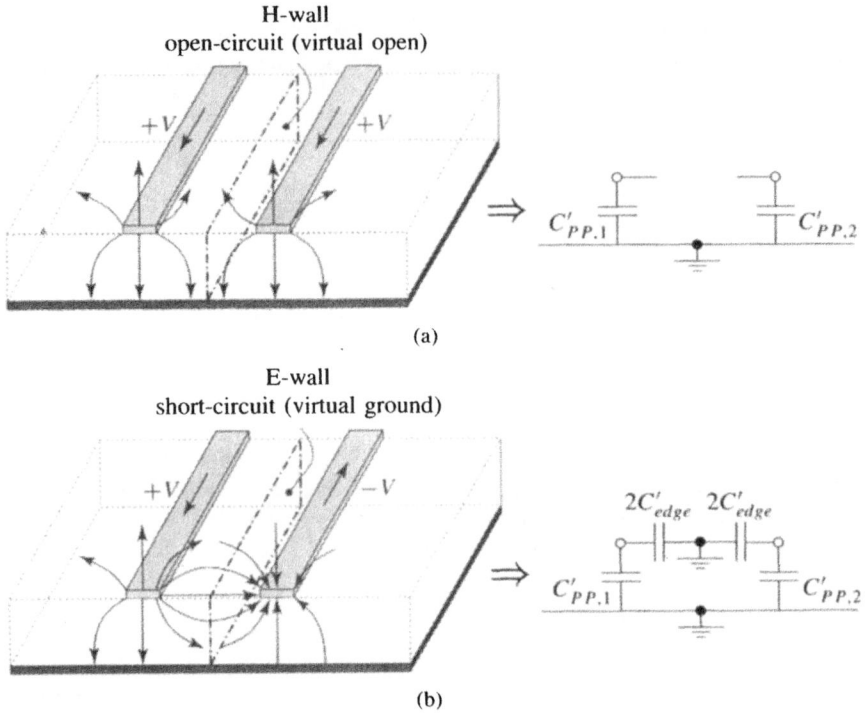

Fig. 5.30. Even and odd mode field distributions in a coupled-line structure and corresponding equivalent circuits. (a) Even-mode. (b) Odd-mode. If the structure is symmetric, the parallel-plate capacitances are equal, $C'_{PP,1} = C'_{PP,2} = C'_{PP}$. Primes indicate per-unit-length quantities.

from which all the scattering parameters of the coupled-line structure are found as

$$S_{11} = 0, \tag{5.34a}$$

$$S_{21} = \frac{\sqrt{1 - k^2}}{\sqrt{1 - k^2}\cos\theta + j\sin\theta}, \tag{5.34b}$$

$$S_{31} = \frac{jk\sin\theta}{\sqrt{1 - k^2}\cos\theta + j\sin\theta} = C_Z, \tag{5.34c}$$

$$S_{41} = 0, \tag{5.34d}$$

where $\theta = \beta\ell$ is the electrical length of the coupler and the quantity

$$k = \frac{Z_{ce} - Z_{co}}{Z_{ce} + Z_{co}}, \tag{5.35}$$

is called the *coupling factor*, for a reason that will become obvious shortly. S_{11} is the return loss, which has been mentioned to be zero from Eq. (5.33), and S_{41} is related to the isolation, $I_{dB} = -20 \log_{10}(S_{41})$, which is infinite for a perfectly TEM structure due to perfect anti-phase between the even and odd voltages at port 4. It follows that directivity, $D_{dB} = I_{dB} - C_{dB} = 20 \log(|S_{13}|/|S_{14}|)$, is also infinite.[28] Thus, power is split between ports 2 and 3, with 2 being the through port, and 3 the coupled port. Comparison of Eqs. (5.34b) and (5.34c) reveals that the through and coupled ports are in quadrature with respect to each other, irrespectively to the electrical length θ of the structure, a feature that is useful in several applications. Because the coupled port is located on the same side of the structure as the input port and power is subsequently coupled backward to the direction of the source $(-z)$, this coupler is conventionally called a *backward-wave coupler*. It is clearly seen in Eq. (5.34) that coupling (S_{31}) depends, via k, on the *difference between the even and odd characteristic impedances*, which justifies the more general terminology *impedance coupler* (IC) that is used here. We note therefore $S_{31} = C_Z$, where C_Z stands for IC coupling coefficient, and IC coupling in dB is $C_{Z,dB} = -20 \log_{10}(S_{31})$. It can be seen from Eq. (5.34c) that maximum IC coupling (with the shortest electrical length) occurs when $\theta = \pi/2$, or $\ell = \lambda/4$, and that the maximum coupling coefficient is then exactly equal to the coupling coefficient

$$C_{Z,max} = C_Z(\theta = \pi/2) = k, \tag{5.36}$$

which justifies its name. This formula reveals the drawback of conventional edge-coupled backward-wave couplers: high coupling levels require strong contrast between Z_{ce} and Z_{co} from Eq. (5.35),[29] which is difficult and often impossible to obtain in practical structures due to fabrication constraints. The main limitation is the *minimum attainable spacing* s between the two coupled lines. For a given substrate and technology, reduction of this parameter is the only available degree of freedom to increase the ratio $Z_{ce}/Z_{co} = (C'_{PP} + C'_{edge})/C'_{PP}$ for increase in k via increase in C'_{edge}. But the edge capacitance C'_{edge} tends to be so weak in comparison with the parallel-plate capacitance C'_{PP} that unpractically small spacing s/λ would be required for high levels of coupling. In practice, the coupling levels of these couplers are typically lower than 10 dB. This limitation prevents their utilization in many applications such as balanced mixers and amplifiers where 3 dB coupling is usually required, despite their attractive feature of broad bandwidth ($>25\%$), not shared by intrinsically tight-coupling couplers such as the quadrature hybrid or the rate race coupler.

A dominantly but not perfectly TEM TL is called a quasi-TEM TL (e.g., microstrip TL). In such a line, there exists a nonzero but small transverse component of the wavenumber ($k_\perp \neq 0$ and $k_\perp/\beta \ll 1$), and the even/odd mode

---

[28]The directivity is a measure of the coupler's ability to isolate forward and backward waves and represents thus an essential figure of merit in couplers.
[29]The limit case of complete coupling, $k = 1$, would require $Z_{ce}/Z_{co} = \infty$!

velocities are therefore slightly different, $\beta_e \neq \beta_o$. However, for weak coupling, the approximation $\beta_e \approx \beta_o$ can be made and the initial design can be determined by still using Eqs. (5.33) through (5.36). As coupling becomes tighter, these equations become less valid, and the modified conditions

$$Z_c = \left(\frac{Z_{ce} \sin \theta_e + Z_{co} \sin \theta_o}{Z_{ce} \sin \theta_o + Z_{co} \sin \theta_e} \right) \sqrt{Z_{ce} Z_{co}}, \qquad (5.37a)$$

$$\theta = \frac{1}{2} (\theta_e + \theta_o) = \frac{2\pi}{\lambda_0} \frac{\sqrt{\varepsilon_{ee}} + \sqrt{\varepsilon_{eo}}}{2} \ell, \qquad (5.37b)$$

where $\theta_e = \beta_e \ell$ and $\theta_o = \beta_o \ell$, must be used for input matching and electrical length at the center frequency, respectively [9]. Thus, in a non perfectly TEM structure, matching ($S_{11} = 0$) is still achievable, ensuring Eq. (5.37a), but isolation cannot be perfect ($S_{41} \neq 0$) due to the unbalance in the even/odd modes velocities, as will appear clearly below. As the difference between even/odd phase velocities increases, $|S_{41}|$ increases and directivity becomes poorer.

5.4.1.2 Non-TEM Symmetric Coupled-Line Structures with Relatively Large Spacing: Phase Coupling (PC)

It has been pointed out that a very small spacing between the two lines (Fig 5.29) is required for significant impedance coupling $C_Z = S_{31}$. If a *relatively* large spacing s/λ exists between the two lines, the edge capacitance C'_{edge} is negligible in comparison with the parallel-plate capacitance C'_{PP} (Fig 5.30); therefore, the even and odd *characteristic impedances are equal*, $Z_{ce} = Z_{co}$. This results in suppression of impedance coupling according to Eq. (5.35), $k = C_{Z.max} = 0$. If the structure is in addition TEM, all the power is transferred to the through port according to Eq. (5.34b), which represents a trivial non interesting case. If however some degree of non-TEM nature is allowed, $k_\perp/\beta \neq 0$, and thus *different phase velocities* for the even and odd modes, a very different phenomenon occurs, despite the fact that the coupled-line structure under consideration is still the same (Fig 5.29). Considering still, for simplicity, a quasi-TEM structure, we obtain the following scattering parameters

$$S_{11} = 0, \qquad (5.38a)$$

$$S_{21} = e^{\frac{-j(\beta_e + \beta_o)\ell}{2}} \cos \left[\frac{(\beta_e - \beta_o)\ell}{2} \right], \qquad (5.38b)$$

$$S_{31} = 0, \qquad (5.38c)$$

$$S_{41} = -j e^{\frac{-j(\beta_e + \beta_o)\ell}{2}} \sin \left[\frac{(\beta_e - \beta_o)\ell}{2} \right] = C_\phi, \qquad (5.38d)$$

where input matching is again ensured by Eq. (5.33). It appears from these relations that port 3, which was the coupled port in the case of small-s/TEM-propagation impedance coupler, has become the isolated port in the present case of large-s/non-TEM-propagation and that port 4, which was the isolated port in the impedance coupler, has become the coupled port here. The present structure is therefore conventionally called a *forward-wave coupler*. Because, as seen in Eq. (5.38), coupling depends now on the *difference between the even/odd phase velocities*, we will use here the more general terminology of *phase coupler* (PC), and we call $S_{41} = C_\phi$ the PC coupling coefficient.

It appears in Eq. (5.38d) that *complete power transfer* from the input to the coupled ports ($|C_\phi| = 1$) can be achieved if the length of the coupler is chosen as (shortest length)

$$\ell = \frac{\pi}{|\beta_e - \beta_o|} = \frac{\lambda_0}{2\left|\sqrt{\varepsilon_{ee}} - \sqrt{\varepsilon_{eo}}\right|}, \tag{5.39}$$

which is referred to as the *coherence length*. In addition, comparing Eqs. (5.38b) and (5.38d) shows that the through and coupled ports are always in phase quadrature, as in the case of the impedance coupler. Moreover, relation Eq. (5.38d) immediately shows, as announced above, why and how directivity is reduced in a non perfectly TEM impedance coupler when the difference between the even/odd phase velocities increases.

In practical microwave structures, the difference between the even and odd mode velocities are relatively small, which would require prohibitively long structures of several tens of wavelengths. For this reason, phase couplers (conventionally forward-wave) are rarely used in microwave engineering. In contrast, several wavelengths still represent small dimensions in optics engineering, whereas the quarter-wavelength dimension of impedance couplers (conventionally backward-wave) tends to be unpractically small. Thus, phase couplers are the CLCs almost exclusively used optics, whereas impedance couplers are the CLCs almost exclusively used in microwaves.

5.4.1.3 Summary on Symmetric Coupled-Line Structures

In summary, a symmetric coupled-line structure, such as the one represented in Fig 5.29, exhibits the following characteristics:

- If s/λ is very small (Z_{ce}/Z_{co} large) and the structure is perfectly TEM ($\beta_e = \beta_o$), a (backward-wave) impedance coupler (IC) is obtained, with coupling $C_Z = S_{31}$ infinite isolation ($S_{41} = 0$).

- If s/λ is very small (Z_{ce}/Z_{co} large) but the structure is only quasi-TEM ($\beta_e \approx \beta_o$), we still have a (backward-wave) impedance coupler, but isolation and therefore directivity are not infinite; as the degree of non-TEM nature increases, directivity decreases.

- If s/λ is relatively large ($Z_{ce}/Z_{co} \approx 1$) and the structure is not perfectly TEM ($\beta_e \neq \beta_o$), but either quasi-TEM ($\beta_e \approx \beta_o$) or strongly non-TEM ($\beta_e \neq \beta_o$), a (forward-wave) phase coupler is obtained, with $C_\phi = S_{41}$, and theoretically possible infinite isolation ($S_{31} = 0$) and directivity.

- A practical symmetric CLC is strictly speaking both an impedance and a phase coupler since both types of couplers share a qualitatively identical configuration. What determines whether the structure is more of the former or of the latter category is only quantitative, depending on the interspacing s/λ and degree of TEM nature.

5.4.1.4 Asymmetric Coupled-Line Structures

When the two lines constituting a CLC are *different*, the structure is *asymmetric*.[30] In this case, decomposition in even and odd modes is not possible anymore. The analysis becomes more difficult, and the even/odd modes have to be replaced by the more general c/π modes, which are two fundamental independent modes, described in [11]. The most general possible CLC, which corresponds to the case of a non-TEM asymmetric structure, can be fully characterized by six quantities: the two complex propagation constants of the c and π modes, γ_c and γ_π, the two characteristic impedances of one of the two lines in the presence of the other one for the c and π modes, $Z_{c,c1}$ (or $Z_{c,c2}$) and $Z_{c,\pi1}$ (or $Z_{c,\pi2}$), and the two voltage ratios between one line and the other one for the c and π modes, R_c and R_π, related to the impedances by

$$\frac{Z_{c,c1}}{Z_{c,c2}} = \frac{Z_{c,\pi1}}{Z_{c,\pi2}} = -R_c R_\pi. \qquad (5.40)$$

As in the case of symmetric coupled-line structures, impedance-coupled and phase-coupled asymmetric couplers are possible [11]. In this section, only the latter category needs to be considered. The through and coupled scattering parameters of an asymmetric PC CLC are given by

$$S_{21} = \left\{ \cos\left[\frac{(\beta_c - \beta_\pi)\ell}{2}\right] - j\frac{1-p}{1+p}\sin\left[\frac{(\beta_c - \beta_\pi)\ell}{2}\right] \right\} e^{\frac{-j(\beta_c+\beta_\pi)\ell}{2}}, \qquad (5.41a)$$

$$S_{41} = -2j\frac{\sqrt{p}}{1+p}\sin\left[\frac{(\beta_c - \beta_\pi)\ell}{2}\right] e^{\frac{-j(\beta_c+\beta_\pi)\ell}{2}} = C_\phi, \qquad (5.41b)$$

with

$$p = -\frac{R_c}{R_\pi}. \qquad (5.41c)$$

It can be seen in Eq. (5.41b) that the coherence length ($|S_{41}| = |S_{41}|_{max}$) in the asymmetric coupler takes a form similar to that of the symmetric coupler [Eq. (5.39)] with the simple exchanges $e \leftrightarrow c$ and $o \leftrightarrow \pi$,

$$\ell = \frac{\pi}{|\beta_c - \beta_\pi|}. \qquad (5.42)$$

This formula provides maximum coupling but not necessarily complete coupling. Complete coupling ($|S_{41}| = 1$) requires in addition $p = 1$, which is equal

---

[30]One possible interest of asymmetric CLCs is that they provide broader bandwidth than their symmetric counterparts.

voltage ratios for the c and π modes. It can be shown that in the limit case where the two lines become identical (symmetric structure), we have $\beta_c = \beta_e$, $\beta_\pi = \beta_o$, $Z_{c,c1} = Z_{c,c2} = Z_{ce}$, $Z_{c,\pi 1} = Z_{c,\pi 2} = Z_{co}$, $R_c = 1$ and $R_\pi = -1$, so that Eqs. (5.41a) and (5.41b) are reduced to Eqs. (5.38b) and (5.38d), respectively. In this case, $p = -1$ is automatically satisfied, consistently with Eq. (5.39).

5.4.1.5 *Advantages of MTM Couplers*

As already pointed out above, conventional CLCs exhibit the advantage of broad bandwidth (typically more than 25%) but can achieve only loose coupling levels (typically less than 10 dB) in the case of edge-coupled configurations.[31] In contrast, branch line couplers, such as the quadrature hybrid or the rat-race, naturally achieve tight (3 dB) coupling but suffer of poor bandwidth (typically less than 10%). A widely used planar coupler providing both broad bandwidth and tight coupling is the Lange coupler [13]. But this coupler has the disadvantage of requiring cumbersome bonding wires with subsequent parasitic effects at high frequencies. The CRLH couplers presented in this section, both the IC coupler of Section 5.4.2 and the PC coupler of Section 5.4.3, represent novel and unique alternatives to the Lange coupler, because they are capable of arbitrarily tight coupling (up to virtually 0 dB!) over a broad bandwidth (more than 30%), while still being planar and without requiring bonding wires [14, 15], as will be shown.

5.4.2 Symmetric Impedance Coupler

An interdigital/stub prototype of CRLH IC edge-coupled CLC is shown in Fig 5.31 [14]. A CRLH IC coupler is a symmetric coupled-line structure constituted of two identical CRLH TLs. Because the CRLH TLs are quasi-TEM, the overall coupled-line structure is also quasi-TEM by symmetry, and the even/odd propagation constants can therefore be considered approximately equal, $\beta_e \approx \beta_o$. Consequently, phase coupling will be negligible, as seen in Eq. (5.38d), $C_\phi \approx 0$, and the only possible coupling mechanism will be an impedance coupling mechanism of the type expressed in Eq. (5.34c), with C_Z large. We will see however that the MTM IC mechanism is significantly different from the IC mechanism in conventional couplers.

The equivalent circuit model for the IC CRLH coupler is shown in Fig 5.32, with decomposition into the corresponding even and odd equivalent circuits. Because the topologies of both of these circuits are identical to those of the standard CRLH TL under the substitutions

$$L_R \rightarrow L_R + 2L_m = L_{Re} \quad \text{(even mode)}, \tag{5.43a}$$

$$C_R \rightarrow C_R + 2C_m = C_{Ro} \quad \text{(odd mode)}, \tag{5.43b}$$

where L_m and C_m are the mutual inductance and coupling capacitance, respectively, the even/odd (fictitious) TLs can be completely characterized by using the theory developed in Chapter 3.

---

[31] Broadside configurations naturally provide much higher coupling levels, but such nonplanar configurations tend to be more expensive and may be impractical in some applications.

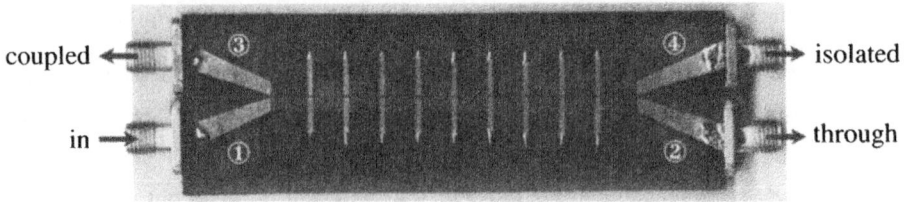

Fig. 5.31. Prototype of a 0-dB (9-cell) impedance coupling (IC) edge-coupled directional coupler constituted of two interdigital/stub CRLH TLs. The two microstrip lines of the conventional coupler have been replaced by two CRLH TLs identical to that of Fig 3.42, with the LC parameters given in Fig 3.43. The spacing between the two lines is of $s = 0.3$ mm. From [14], © IEEE; reprinted with permission.

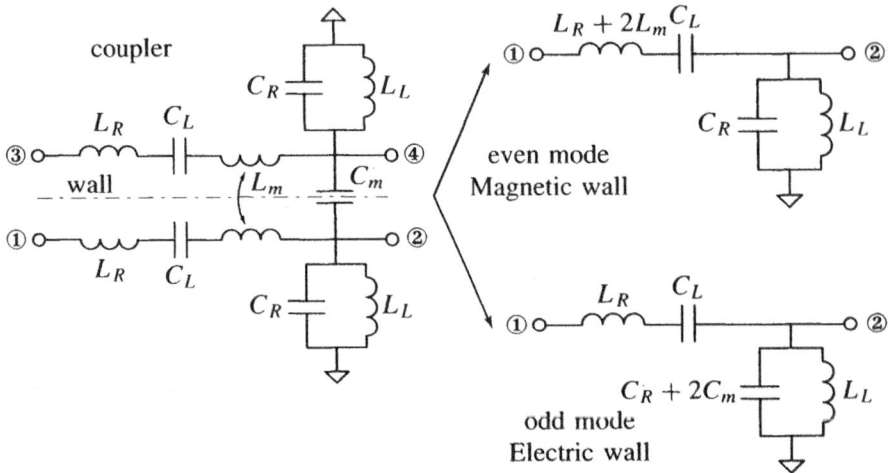

Fig. 5.32. Equivalent circuit model for the unit cell of an IC CRLH coupler, as the one of Fig 5.31, and corresponding even/odd modes TL models. The topologies of the models of both the even and the odd TLs are identical to those of a simple CRLH TL (Fig 3.14), where L_R has been replaced by $L_R + 2L_m$ and C_L has been replaced by $C_L + 2C_m$, L_m and C_m being the mutual inductance and coupling capacitance, respectively. From [14], © IEEE; reprinted with permission.

Thus, the even/odd characteristic impedances Z_{ci} $(i = e, o)$ are given from Eq. (3.124) by

$$Z_{ce} = \sqrt{\frac{L_L}{C_L}} \sqrt{\frac{1 - \omega^2 L_{Re} C_L}{1 - \omega^2 L_L C_R}}, \tag{5.44a}$$

$$Z_{co} = \sqrt{\frac{L_L}{C_L}} \sqrt{\frac{1 - \omega^2 L_R C_L}{1 - \omega^2 L_L C_{Ro}}}, \tag{5.44b}$$

and the even/odd complex propagation constants $\gamma_i = \alpha_i + j\beta_i$ Z_{ci} $(i = e, o)$ (dispersion/attenuation diagram) are, rigorously given from Eq. (3.115) by

$$\gamma_e = \frac{1}{p}\cosh^{-1}\left\{1 - \frac{1}{2}\left[\omega^2 L_{Re}C_R + \frac{1}{\omega^2 L_L C_L} - \frac{L_{Re}C_L + L_L C_R}{L_L C_L}\right]\right\}, \quad (5.45a)$$

$$\gamma_o = \frac{1}{p}\cosh^{-1}\left\{1 - \frac{1}{2}\left[\omega^2 L_R C_{ro} + \frac{1}{\omega^2 L_L C_L} - \frac{L_R C_L + L_L C_{Ro}}{L_L C_L}\right]\right\}, \quad (5.45b)$$

which may be anticipated to be approximately equal ($\gamma_e \approx \gamma_o$) under the quasi-TEM propagation assumption. In addition, the even/odd series/shunt resonance frequencies are given from Eq. (3.54) by

$$\omega_{se,e} = \frac{1}{\sqrt{L_{Re}C_L}} \quad \text{and} \quad \omega_{sh,e} = \frac{1}{\sqrt{L_L C_R}}, \quad (5.46a)$$

$$\omega_{se,o} = \frac{1}{\sqrt{L_R C_L}} \quad \text{and} \quad \omega_{sh,o} = \frac{1}{\sqrt{L_L C_{Ro}}}, \quad (5.46b)$$

whereas the even/odd maximum attenuation frequencies, relevant in the unbalanced case that will be seen to be prevailing here, is given from Eq. (3.13) by

$$\omega_{oe} = \frac{1}{\sqrt[4]{L_{Re}C_R L_L C_L}}, \quad (5.47a)$$

$$\omega_{0o} = \frac{1}{\sqrt[4]{L_R C_{Ro} L_L C_L}}. \quad (5.47b)$$

The even/odd scattering parameters for the IC coupler of Fig 5.31 are shown in Fig 5.33. These are naturally qualitatively identical to the scattering parameters of a simple CRLH TL (Fig 3.23(a)), as they correspond to similar equations. Because each of the CRLH TL of the coupler, isolated from the other one, is characterized by the balanced parameters (L_R, C_R, L_L, C_L) (same line as in Fig 3.43(b)), the even/odd TLs with modified parameters $(L_{Re}, C_R, L_L, C_L)/(L_R, C_{Ro}, L_L, C_L)$ are necessarily unbalanced, which gives rise to the observed gaps, characterized by Eqs. (5.46) and (5.47).

The even/odd characteristic impedances for the IC coupler of Fig 5.31 are shown in Fig 5.34, exhibiting as expected imaginary values in the stop bands.[32]

In general, input matching ($S_{11} = 0$) of a coupled-line structure, as pointed out above, is achieved if the condition of Eq. (5.33) is satisfied. For this reason, Fig 5.35 examines the magnitudes of the even/odd characteristic impedances as well as their product for the structure of Fig 5.31. The results show that the quantity $\sqrt{Z_{ce}Z_{co}}$ is fairly constant and equal to Z_0 over a broad frequency range,

---

[32]The even/odd impedances can be computed either directly by Eq. (5.44) or from the even/odd S-parameters by the formulas $Z_{ci} = \sqrt{(\Pi_i - 1)/(\Pi_i + 1)}$ $(i = e, o)$ with $\Pi_i = (S_{21i}^2 - S_{11i}^2 - 1)/(2S_{11i})$ [14].

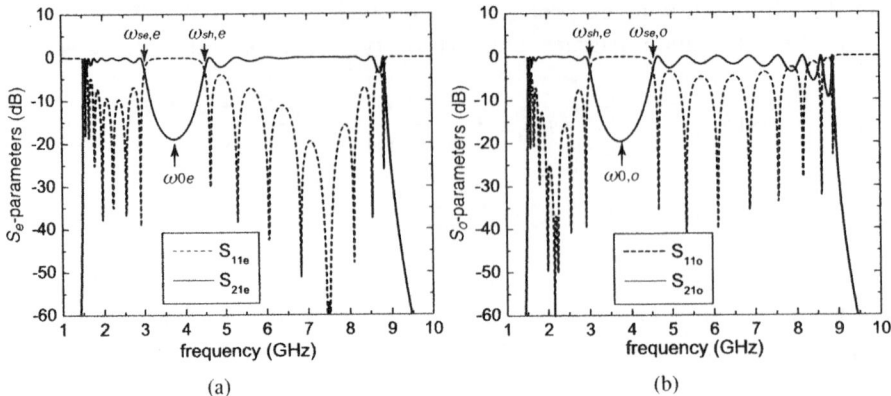

Fig. 5.33. Circuit simulated S-parameters for the even and odd modes of the coupler of Fig 5.31 with the unit cell shown in Fig 5.32 using the extracted parameters $L_m = 1.0$ nH and $C_m = 0.8$ pF. (a) Even mode: $f_{se,e} = 3.14$ GHz, $f_{0e} = 3.67$ GHz, and $f_{sh,e} = 4.30$ GHz. (b) Odd mode: $f_{se,o} = 3.13$ GHz, $f_{0o} = 3.68$ GHz, and $f_{sh,o} = 4.33$ GHz. From [14], © IEEE; reprinted with permission.

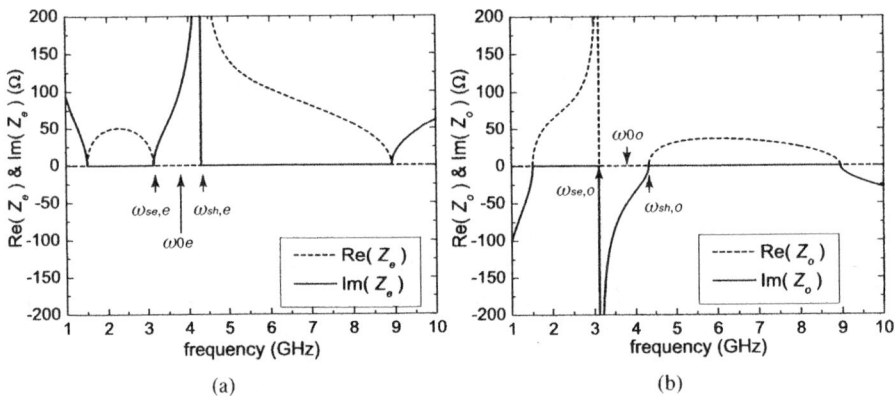

Fig. 5.34. Circuit simulated even/odd characteristic impedances (real/imaginary parts) for the coupler of Fig 5.31 computed by Eq. (5.44) with the extracted parameters of Fig 5.33. (a) Even mode. (b) Odd mode. From [14], © IEEE; reprinted with permission.

which indicates that a broad matching frequency range will be obtained. The frequency ω_{eq} where the magnitudes of the even/odd characteristic impedances are equal (not identical to ω_0), $|Z_{ce}| = |Z_{co}|$, can be easily derived from Eq. (5.44)

$$\omega_{eq} = \sqrt{\frac{L_m C_L + L_L C_m}{L_L C_L (L_R C_m + L_m C_R + L_m C_m)}}, \quad (5.48)$$

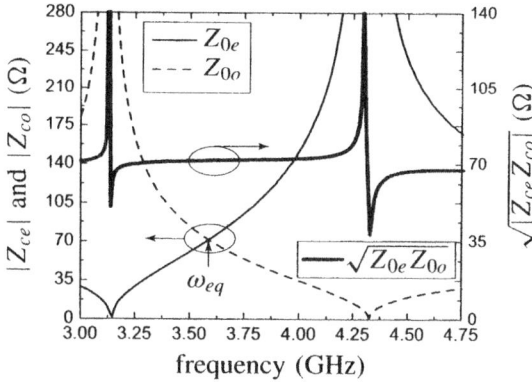

Fig. 5.35. Circuit simulated even/odd characteristic impedances (magnitude) for the coupler of Fig 5.31. Left-hand axis: magnitudes of the impedances. Right-hand axis: square root of product of the impedances appearing in Eq. (5.33). The Z_{ce}/Z_{co} crossing frequency is 3.59 GHz and $Z_c = 50 \, \Omega$. From [14], © IEEE; reprinted with permission.

from which we have

$$\omega \left\{ \begin{array}{c} < \\ > \end{array} \right\} \omega_{eq} \Rightarrow Z_{ce} \left\{ \begin{array}{c} < \\ > \end{array} \right\} Z_{co}, \tag{5.49}$$

as verified in Fig 5.35.

The most noteworthy point to be observed in Figs. 5.33 and 5.34 is that, very unusually, the matching condition of the coupler Eq. (5.33) is satisfied *in the gap of the even/odd modes TLs*, whereas in the conventional case such gaps cannot even exist because the lines are of PRH nature. Since matching is obtained within the even/odd TL gaps, we need to generalize the expression of the IC coupling coefficient in Eq. (5.34c) by changing $\theta = \beta\ell$ into $\theta = \gamma\ell = (\alpha + j\beta)\ell$, where $\alpha_e \approx \alpha_o \approx \alpha$ and $\beta_e \approx \beta_o \approx \beta$ under the quasi-TEM assumption. The IC coefficient of Eq. (5.34c) becomes then

$$C_Z = S_{31} = \frac{(Z_{ce} - Z_{co}) \tanh\left[(\alpha + j\beta)\ell\right]}{2Z_c + (Z_{ce} + Z_{co}) \tanh\left[(\alpha + j\beta)\ell\right]}. \tag{5.50}$$

Since the even/odd TLs are operated in their gap, we have $|\beta| \ll \alpha$, so that $\alpha + j\beta \approx \alpha$. In addition, we can consider that $\alpha\ell > 1$ in the middle of the gap, as verified in Fig 5.36 for the coupler of Fig 5.31, so that $\tanh(\alpha\ell) \approx 1$. Under these two conditions, we obtain

$$C_Z \approx \frac{\left(\dfrac{Z_{ce}}{Z_c} - \dfrac{Z_{co}}{Z_c}\right)}{2 + \left(\dfrac{Z_{ce}}{Z_c} + \dfrac{Z_{co}}{Z_c}\right)}. \tag{5.51}$$

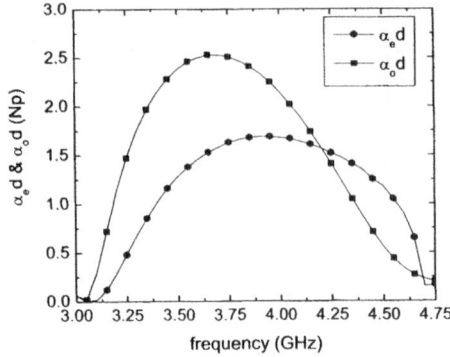

Fig. 5.36. Full-wave (MoM) simulated even/odd modes attenuation lengths $\alpha_e d / \alpha_o d$. From [14], © IEEE; reprinted with permission.

By using Eq. (5.33) to eliminate either Z_{ce} or Z_{co} and also taking into account the fact that $Z_{ci} = j\text{Im}(Z_{ci})$ ($i = e, o$) (Fig 5.35), this expression is further transformed into

$$C_Z \approx \frac{\xi + \xi^{-1}}{2j + (\xi - \xi^{-1})}, \quad \text{with} \quad \xi = \frac{Z_c}{\text{Im}(Z_{co})} \quad \text{or} \quad \xi = \frac{\text{Im}(Z_{ce})}{Z_c}, \qquad (5.52)$$

with magnitude

$$|C_Z| = |S_{31}| \approx \left| \frac{\xi + \xi^{-1}}{2 + j(\xi - \xi^{-1})} \right| = \frac{\xi + \xi^{-1}}{\sqrt{4 + (\xi - \xi^{-1})^2}} = 1.^{33} \qquad (5.53)$$

We have just shown that *complete IC backward coupling* can be achieved in a symmetric CRLH CLC if the condition $\alpha \ell > 1$, which requires CRLH unbalanced even/odd modes (obtainable from individual balanced CRLH TLs). It is clear that if complete coupling can be achieved, *any level of coupling can be achieved* by reducing $\alpha \ell$, that is by reducing the number of cells of the CRLH structure or increasing the lines interspacing s. This result indicates that the IC CRLH is a unique novel type of IC CLC, since coupling in conventional IC CLCs is severely limited by the maximum attainable ratio Z_{ce}/Z_{co}, as seen in Eq. (5.35). We will also see shortly that this interesting property of arbitrary coupling level is not obtained at the expense of bandwidth reduction in comparison with conventional couplers.

---

[32]The last equality is easily verified by developing the denominator as $\sqrt{4 + (\xi - \xi^{-1})^2} = \sqrt{\xi^2 + \xi^{-2} + 2} = \sqrt{(\xi + \xi^{-1})^2} = \xi + \xi^{-1}$, which is equal to the numerator.

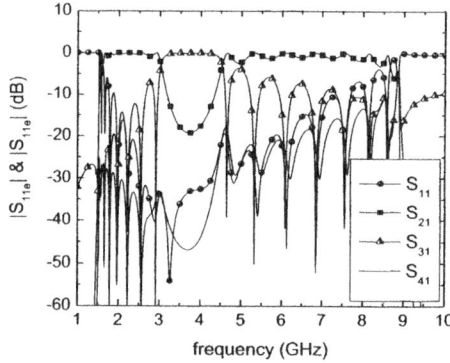

Fig. 5.37. *S*-parameters for the coupler of Fig 5.31 reconstructed by Eq. (5.54) with the full-wave even and odd parameters of Fig 5.33 (from circuit model of Fig 5.32). From [14], © IEEE; reprinted with permission.

Fig 5.37 shows the circuit model scattering parameters for the CRLH IC coupler of Fig 5.31 reconstructed by the general coupler formulas [11]

$$S_{11} = \frac{S_{11e} + S_{11o}}{2}, \tag{5.54a}$$

$$S_{21} = \frac{S_{21e} + S_{21o}}{2}, \tag{5.54b}$$

$$S_{31} = \frac{S_{11e} - S_{11o}}{2} = C_Z, \tag{5.54c}$$

$$S_{41} = \frac{S_{21e} - S_{21o}}{2}, \tag{5.54d}$$

with the even/odd scattering parameters of Fig 5.33. Fig 5.38 shows the same scattering parameters obtained by the full-wave simulation. Finally, Fig 5.39 presents the experimental results compared with the circuit results of Fig 5.37. Excellent agreement between LC circuit analysis, full-wave simulation, and measurement can be observed. The simulation results show that, in addition to complete backward coupling in the range from 3.2 to 4.5 GHz, there is a range of through coupling from 1.5 to 3.1 GHz, suggesting possible dual-band/dual-mode operation. In the experimental results, around 0.5 dB coupling (despite the relatively large interspacing of $s = 0.3$ mm) is obtained in the range from 3.2 to 4.6 GHz, which corresponds to a bandwidth of 36%. The measured directivity ($D_{dB} = I_{dB} - C_{dB} = S_{41,dB} - S_{31,dB}$) is approximately 25 dB. The length of the coupler is approximately $1.05\lambda_e$, where λ_e is the effective conventional microstrip wavelength. This length is relatively large, but still one order of magnitude smaller than the length required in a coupled-line forward-wave directional coupler for complete power transfer.

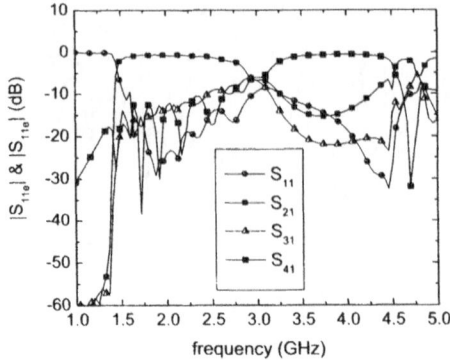

Fig. 5.38. Full-wave simulated (MoM) S-parameters for the coupler of Fig 5.31. The coupling bandwidth corresponding to more than -2.5 dB coupling extends from 3.2 to 4.5 GHz. From [14], © IEEE; reprinted with permission.

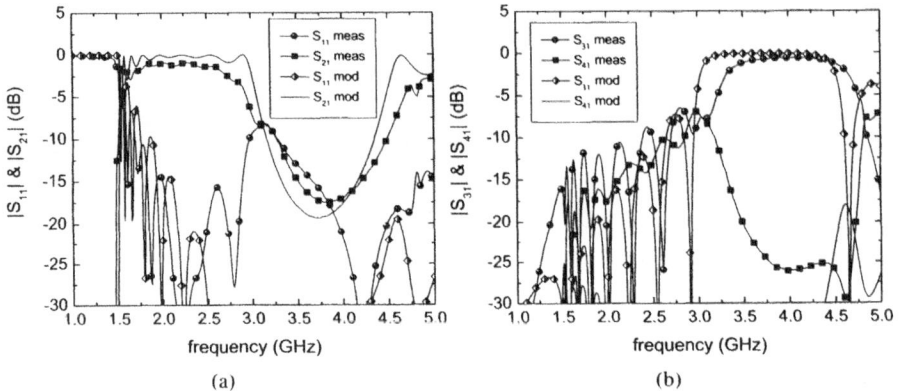

Fig. 5.39. Comparison between the circuit-model results of Fig 5.37 and the measured results. (a) S_{11} and S_{21}. (b) S_{31} and S_{41}. From [14], © IEEE; reprinted with permission.

Inspection of Eq. (5.54c) reveals that complete IC coupling requires fulfillment of the double condition

$$|S_{11e}| = |S_{11o}| = 1 \quad \text{and} \quad \phi(S_{11e}) - \phi(S_{11o}) = 180°. \tag{5.55}$$

The first of these conditions is verified in Fig 5.33, whereas the second one is verified in Fig 5.40.

The main characteristics of the IC CRLH coupler can be summarized as follows:

- The CRLH IC coupler can achieve *any arbitrary level of coupling*, up to almost complete coupling (0 dB), even with a relatively large spacing between the lines.

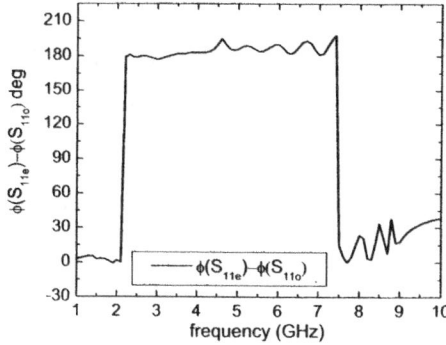

Fig. 5.40. Phase difference between the even and odd reflection parameters plotted in Fig 5.33. From [14], © IEEE; reprinted with permission.

- The bandwidth of the CRLH IC corresponds to the *LH bandwidth of the isolated CRLH TLs*. This can be seen in the following manner. The isolated CRLH TLs are balanced, $\omega_{se} = \omega_{sh} = \omega_0$. By observing Eq. (5.46), we notice that $\omega_{sh,e} = \omega_{sh}$ and $\omega_{se,o} = \omega_{se}$, which means that the upper edges of the even/odd gaps correspond to the transition frequency ω_0 of each of the individual CRLH TL forming the coupler. Thus, the frequency range of the coupler lies entirely in the LH range of the individual CRLH TLs.

- The *electrical length* of the CRLH coupler is *not* $\lambda_g/4$ as for the conventional IC coupler [see Eq. (5.35)]. In fact, the electrical length of the coupler at the transition frequency f_0 of each CRLH TL (upper edge of the coupling range) is known to be 0 and to increase progressively toward lower frequencies (LH range). So, the electrical length of the CRLH IC coupler ranges from 0 to larger values as frequency is decreased from the upper to the lower edges of the coupling bandwidth.

- *The even/odd modes characteristic impedances of the CRLH IC coupler are purely imaginary* in the coupling range. This means that the even/odd Tls operate in a stopband. Consequently coupling is not related to the electrical length of the even/odd lines $\theta = \beta \ell$, but to the *attenuation length $\alpha \ell$ of the even/odd modes*. This length is clearly directly proportional to the physical length ℓ. It is also directly proportional to the even/odd attenuation factor α, which depends on the width of the even/odd gaps. The larger the even/odd gaps, the larger the value of α at a given frequency, all the parameters being rigorously designable with Eqs. (5.43) to (5.53).

A 0-dB coupler is not very interesting per se, except possibly for DC blocks [11]. The motivation for describing a 0-dB coupler was to demonstrate that any coupling level is achievable. A practically more useful CRLH IC coupler, a 3-dB coupler, is shown in Fig 5.41, and the corresponding scattering parameters are shown in Fig 5.42.

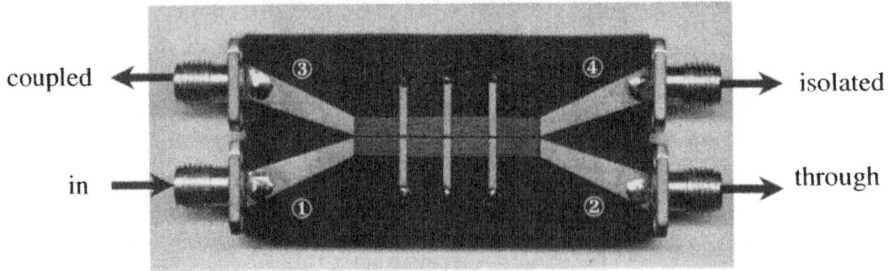

Fig. 5.41. 3-dB (3-cell) backward "impedance" edge-coupled directional coupler. The spacing between the lines is $s = 0.3$ mm. From [14], © IEEE; reprinted with permission.

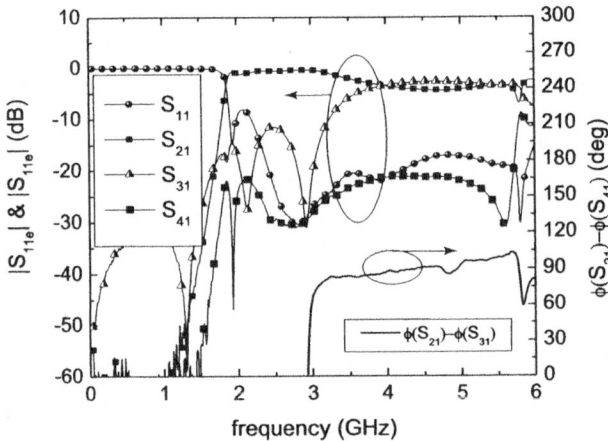

Fig. 5.42. Measured S-parameters of the CRLH coupler of Fig 5.41. From [14], © IEEE; reprinted with permission.

The performances of this coupler are as follows. An amplitude balance of 2 dB is achieved over the huge fractional bandwidth of around 50%, from 3.5 to 5.8 GHz. The quadrature phase balance is $90° \pm 5°$ from 3.0 to 4.0 GHz. The average directivity is 20 dB. The length of the coupler is around $\lambda_e/3$. No attempt has been made to try to reduce that length, but it is believed that this length can be reduced by the spacing s, eliminating one-cell and increasing the LC loadings to obtain exactly 3 dB coupling.

Finally, Fig 5.43 represents a comparative plot for the coupling level achieved by the conventional and CRLH IC couplers versus the spacing between the lines. It can be observed that the coupling enhancement obtained with the CRLH coupler varies from around 10 dB to 25 dB as the line spacing is increased.

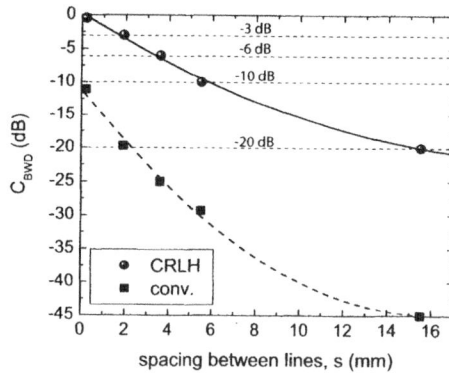

Fig. 5.43. Comparisons of coupling levels for conventional and CRLH couplers on the same substrate (RT/Duroid 5880, $\varepsilon_r = 2.2$ and $h = 62$ mils) as a function of the spacing s between the two lines. From [14], © IEEE; reprinted with permission.

5.4.3 Asymmetric Phase Coupler

In Section 5.4.2, we presented a symmetric IC CRLH coupler, constituted of two identical CRLH TLs operated in their LH range. In this section, we will explore an asymmetric PC coupler, constituted by the combination of a conventional (PRH) TL and of a CRLH TL, operated in its LH range [14]. Fig 5.44 shows a microstrip-interdigital/stub CRLH PC coupler with its circuit model.

This PC coupler can be intuitively understood in the following manner. As a signal is injected at port 1, power (Poynting vector, $\vec{S}$) propagates toward port 2, but phase (propagation constant, β) propagates backward toward port 1 because the CRLH TL is operated in its LH range. Therefore, because coupling to the other line (3 − 4) occurs through transverse evanescent waves following the phase direction, phase and power propagate in the same direction toward port 3 in the PRH microstrip line. Consequently, *backward coupling* is functionally achieved, although the coupling mechanism is of PC nature, normally associated with forward coupling in conventional PC couplers. Therefore, in the PC coupling coefficient formula of Eq. (5.38d), the scattering parameter S_{31} has to be replaced by S_{41} because of the reversal of phase direction in the CRLH TL ($C_{\phi,\text{FWD,conv}} \rightarrow C_{\phi,\text{BWD}}$).

We have seen that in Section 5.4.1.4 that in an asymmetric CLC maximum PC coupling corresponds to the coherence length of

$$\ell_{max} = \frac{\pi}{|\beta_c - \beta_\pi|}. \tag{5.56}$$

(a)

(b)

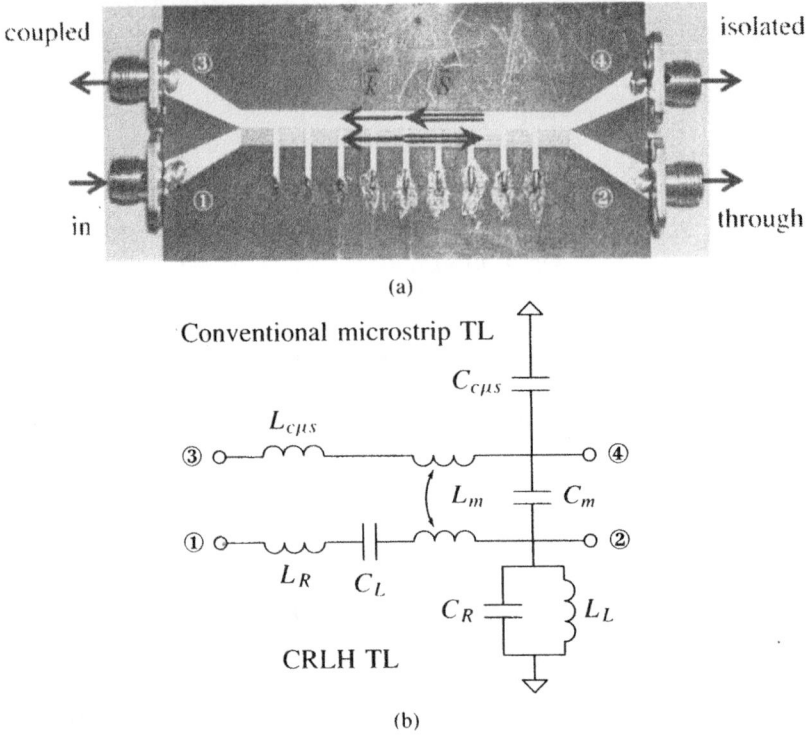

Fig. 5.44. Phase coupling (PC) edge-coupled directional coupler. (a) 0-dB coupling prototype constituted of one conventional microstrip line (top line) and one interdigital/stub CRLH TL (bottom line). (b) Equivalent circuit model for the unit cell. The design parameters of the (9-cell) CRLH TL (bottom line) are the same as in Fig 3.42, with the extracted circuit parameters L_R, C_R, L_L and C_L given in Fig 3.43(a). The microstrip line ($c\mu s$) is modeled by the parameters $L_{c\mu s}$ and $C_{c\mu s}$ representing the equivalent series inductance and shunt capacitance, respectively, for one (lumped-implemented) unit cell. The parameters C_m and L_m represent the coupling capacitance and mutual inductance, respectively. The spacing between the lines is $s = 0.3$ mm and the length of the coupler is $\ell = 62$ mm. From [15], © IEEE; reprinted with permission.

involving the c and π modes. Because polarities in an *individual* RH TL and in an *individual* LH TL are opposite ($\beta_{LH} - \|\beta_{RH}$), the isolated microstrip (RH) and CRLH (LH) TLs may be considered approximation of the c/π equivalent TLs: $\beta_{\mu sp} \to \beta_c \to \beta_e$ and $\beta_{CRLH} \to \beta_\pi \to \beta_o$. Although the small difference $|\beta_e - \beta_o|$ leads to poor coupling in the conventional case, we have here $\beta_o \to -|\beta_{CRLH}|$, so that the difference in the denominator of Eq. (5.56) is turned into a *sum*

$$\ell_{max} = \frac{\pi}{|\beta_{CRLH}| + \beta_{c\mu s}}. \qquad (5.57)$$

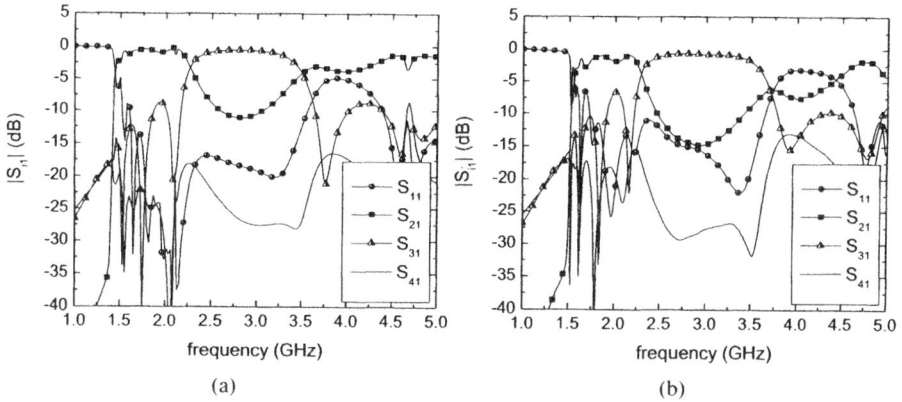

Fig. 5.45. *S*-parameters for the coupler of Fig 5.44(a). (a) Full-Wave (MoM) simulation. (b) Measurement. From [15], © IEEE; reprinted with permission.

Thus, the negative sign in the denominator of Eq. (5.56) is replaced by a *positive sign*, which suggests that a significantly smaller ℓ_{max} can be attained in the microstrip/CRLH coupler. This expression is based on an excessively crude assumption in practice, as will be confirmed in Fig 5.46(b), but it provides a useful insight into the possible capability of high coupling in an asymmetric RH/LH coupler.

Fig 5.45 shows the full-wave simulated and measured scattering parameters for the coupler of Fig 5.44(a). Excellent agreement can be observed between simulated and experimental results. Quasi 0-dB (around −0.7 dB) backward coupling is achieved over the range from 2.2 to 3.8 GHz (−10 dB bandwidth in measurement), which represents the broad fractional bandwidth of 53%. An excellent directivity of $\ell = 30$ dB is achieved in the measured prototype. The asymmetric coupler presented here was built on the same substrate, has the same length and uses two CRLH lines identical to those of the symmetric CRLH-CRLH coupler presented in Section 5.4.2. However, its operating frequency range is lower (3.3 to 4.7 GHz for the symmetric coupler of Section 5.4.2), which means that, after rescaling, the asymmetric PC PRH-CRLH coupler is *more compact* than its symmetric counterpart. On the other hand, due to its asymmetry, the PRH-CRLH *does not exhibit* 90 *degrees phase balance* in contrast to the symmetric CRLH-CRLH coupler.

It should be noted that, as in the case of the IC coupler, if complete coupling is possible, any arbitrary coupling level can be achieved, by reducing the length of the structure or increasing the spacing between the two lines.

Fig 5.46(a) shows the scattering parameters of the coupler obtained from the circuit model depicted in Fig 5.44(b). The agreement with full-wave simulations and measurements is remarkable, and the circuit model can therefore be safely used to describe and design such a type of coupler. The parameters of the microstrip line were computed from the formulas $L_{c\mu s} = L'(\ell/N)$ and $C_{c\mu s} = C'(\ell/N)$, where $C' = \sqrt{\varepsilon_{eff}}/(c_0 Z_c)$ and $L' = Z_c^2 C'$ [2], and the coupling

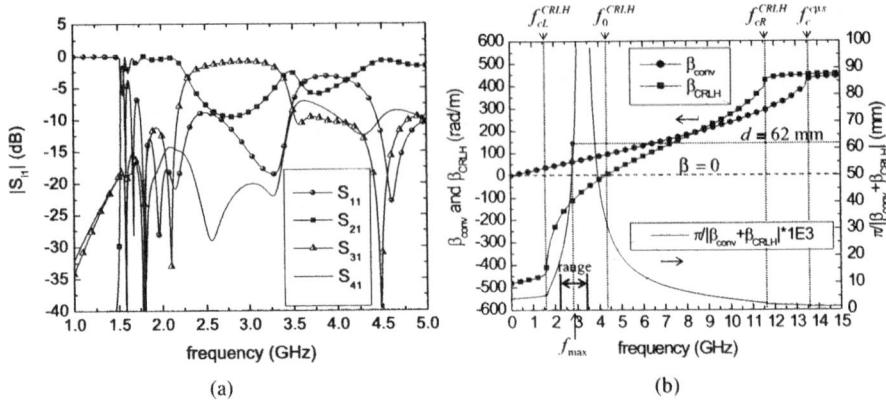

(a)   (b)

Fig. 5.46. Results obtained from the circuit model of Fig 5.44(b) with the extracted parameters $L_{c\mu s} = 1.64$ nH, $C_{c\mu s} = 0.33$ pF, $L_R = 2.21$ nH, $C_R = 0.45$ pF, $L_L = 3.04$ nH, $C_L = 0.61$ pF, $L_m = 0.27$ nH, $C_m = 0.33$ pF. (a) S-parameters. (b) Phase constants in the isolated conventional and CRLH TLs (left-hand axis) and right-hand term of Eq. (5.57) (right-hand axis). The values of the reference frequencies computed from Eqs. (5.58), (5.59), and (5.60) are $f_{c\mu s} = 13.6$ GHz, $f_{cL}^{CRLH} = 1.8$ GHz, $f_0^{CRLH} = 4.3$ GHz, $f_{cR}^{CRLH} = 11.9$ GHz. The frequency computed from the coherence condition Eq. (5.57) is $f_{max} = 2.8$ GHz. From [15], © IEEE; reprinted with permission.

capacitance C_m and mutual inductance L_m were estimated from curve fitting with the full-wave and measurement results of Fig 5.45.

Fig 5.46(b) (left-hand axis) shows the propagation constants of the isolated microstrip and CRLH TLs constituting the coupler [Fig 5.44(b)]. The lowpass cutoff frequency of the microstrip lumped model of Fig 5.44(b)] is given by

$$f_c^{c\mu s} = \frac{1}{\pi \sqrt{L_{c\mu s} C_{c\mu s}}}. \tag{5.58}$$

For the CRLH TL, the LH highpass and RH lowpass cutoff frequencies are rigorously given by Eq. (3.94) and approximately given by Eq. (3.98) (balanced case), the latter reading

$$f_{cL}^{CRLH} \approx \frac{1}{4\pi \sqrt{L_L C_L}}, \tag{5.59a}$$

$$f_{cR}^{CRLH} \approx \frac{1}{\pi \sqrt{L_R C_R}} + f_{cL}, \tag{5.59b}$$

respectively, whereas the LH-to-RH transition frequency is given by Eq. (3.82)

$$f_0^{CRLH} = \frac{1}{2\pi \sqrt[4]{L_R C_R L_L C_L}}. \tag{5.60}$$

Fig 5.46(b) (right-hand axis) also shows the right-hand term of Eq. (5.57). Maximum coupling is expected when this quantity is equal to the length of the coupler ℓ. The coherence frequency obtained from Eq. (5.57), $f_{max} = 2.8$ GHz, is consistent with the fact that 2.8 GHz lies in the center of the frequency range of the coupler [Figs. 5.45 and 5.46(a)]. However, Eq. (5.57) also predicts a maximum coupling at the frequency $f = 3.8$ GHz, which is not observed in the experiment. This suggests that the "weak" coupling assumption in Eq. (5.57) (c and π modes βs are equal to the isolated CRLH and microstrip βs) is too crude in reality. Note that a pole of the right-handed term of Eq. (5.57) appears at the frequency where the microstrip and CRLH line have same magnitude (3.2 GHz) but opposite signs. This pole would suggest that an infinite length is required for complete coupling at this frequency, which contradicts the results. Consequently, the weak coupling assumption giving [Eq. (5.57)] is clearly inappropriate at this particular frequency.

5.5 NEGATIVE AND ZEROTH-ORDER RESONATOR

In the applications of the previous sections (dual-band, enhanced-bandwidth, multilayer reduced-size and coupler components), the CRLH TL sections are always designed to be matched to the elements that they are connected to and therefore to support the propagation of traveling waves. They are thus treated as TLs terminated by a matched load. If instead of being matched to the external elements, a CRLH TL is *open-ended* or *short-ended*, it produces standing waves due the open/short boundary conditions and becomes a *resonator*. Although this is also true for a PRH TL, we will discover that a CRLH TL resonator exhibits some unusual characteristics, such as allowing negative and zeroth-order resonances, with interesting perspectives for dual-band and size-independent resonators.

5.5.1 Principle

In a conventional (PRH) distributed resonator, the resonance frequencies ω_m correspond to the frequencies where the physical length ℓ of the structure is a multiple of half a wavelength or, equivalently, the electrical length $\theta = \beta \ell$ is a multiple of π,

$$\ell = m\frac{\lambda}{2} \quad \text{or} \quad \theta_m = \beta_m \ell = \left(\frac{2\pi}{\lambda}\right) \cdot \left(\frac{m\lambda}{2}\right) = m\pi, \qquad (5.61a)$$

with

$$m = +1, +2, \ldots + \infty, \qquad (5.61b)$$

as illustrated in Fig 5.47.

The resonance frequencies are obtained by sampling the dispersion curve $\omega = \omega(\beta_{PRH})$, where $\beta_{PRH} = \omega\sqrt{L'_R C'_R}$ from Eq. (3.36), with a sampling rate of π/ℓ

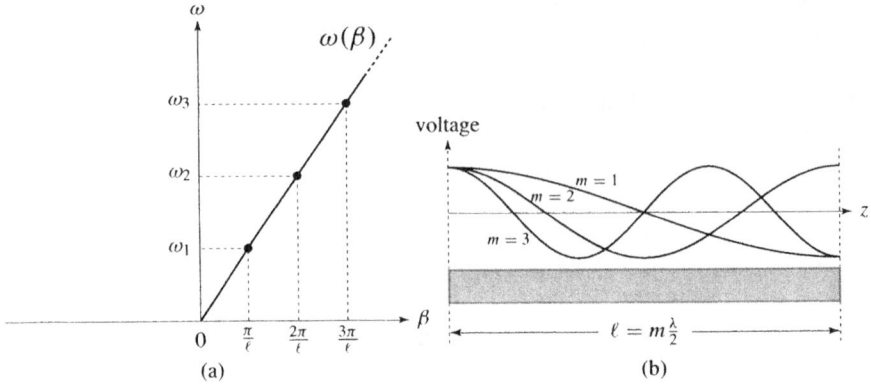

Fig. 5.47. PRH (conventional) TL resonator. (a) Dispersion relation of the TL and resonance frequencies ω_m of the corresponding resonator. (b) Typical field distributions of the resonance modes; more precisely, voltage distribution for the case of an open-circuited TL.

in the β variable, as it appears graphically in Fig 5.47(a). An *infinite number of resonances* exist in an ideal continuous distributed TL, since such a line has an infinite bandwidth, extending from $\omega = 0$ to $\omega = \infty$.

Because the electrical length of a PRH TL can only be strictly positive, only *nonzero and positive* resonances [Eq. (5.61b)] are allowed in PRH TL resonators and because the dispersion curve of a PRH TL is linear, the PRH resonances are *harmonics* of the fundamental ω_1, that is $\omega_m = m\omega_1$.

Fig 5.48 shows the characteristics of a CRLH TL resonator, which is based on an ideal homogeneous balanced CRLH TL (Section 3.1) for simplicity. The case of the practical LC network implementation will be treated in Section 5.5.2. In contrast to the PRH TL, the CRLH TL can have $\beta = 0$ (transition frequency) and $\beta < 0$ (LH range). Therefore, the electrical length $\theta = \beta\ell$ can be zero and negative, which means that the resonance index m becomes symmetrically defined around $m = 0$. We have thus

$$\ell = |m|\frac{\lambda}{2} \quad \text{or} \quad \theta_m = \beta_m\ell = \left(\frac{2\pi}{\lambda}\right)\cdot\left(\frac{m\lambda}{2}\right) = m\pi, \tag{5.62a}$$

with

$$m = 0, \pm1, \pm2, \ldots \pm\infty. \tag{5.62b}$$

The resonance frequencies are obtained by sampling the dispersion curve $\omega = \omega(\beta_{CRLH})$, where $\beta_{CRLH} = \omega\sqrt{L'_R C'_R} - 1/\left(\omega\sqrt{L'_L C'_L}\right)$ from Eq. (3.34), with a sampling rate of π/ℓ in the β variable, as it appears graphically in Fig 5.48(a). An *infinite number of resonances* is also obtained because the line considered is also an ideal continuous distributed TL structure.

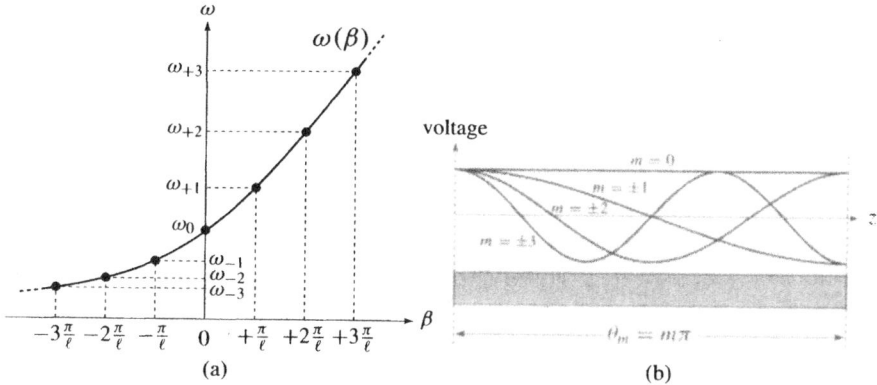

Fig. 5.48. CRLH TL resonator (ideal homogeneous TL case). (a) Dispersion relation of the TL and resonance frequencies ω_m of the corresponding resonator. (b) Typical field distributions of the resonance modes; more precisely, voltage distribution for the case of an open-circuited TL.

The CRLH TL resonator has several differences compared with the PRH TL resonator:

- *Negative resonances* ($m < 0$) and ($m = 0$) exist in addition to the conventional positive resonances ($m > 0$).
- Except for the mode $m = 0$, *each mode (m) has a corresponding mode with the opposite sign (−m) with identical field distribution*. Therefore, each pair $(-m, +m)$ should have identical impedance.[33] This property has been exploited to realize a dual-band CRLH antenna (Section 6.7).
- Due to the nonlinear nature of the CRLH dispersion curve, especially in the LH range, the *resonant frequencies are not in harmonic ratios*, that is, $\omega_m \neq \omega_{-1}/m$ for $m < 0$ and $\omega_m \neq m\omega_1$ for $m > 0$. A strong compression of the resonance spectrum is obtained at low frequencies.
- The $m = 0$ mode is particularly interesting: This mode corresponds to a flat field distribution (no voltage gradient) and is therefore *not related to the physical length of the TL*, which suggests that it can be used to realize, theoretically, arbitrarily small resonators.

5.5.2 LC Network Implementation

Because no continuous distributed CRLH is readily available in nature, the CRLH TL resonator has to be implemented in the form of an LC network structure (Section 3.2). For the sake of the argument, we will first consider a periodic (and

---

[33]The impedances are expected to be identical *at* the resonance frequency, but the dissymmetry in the CRLH dispersion curve indicates that the bandwidths will be different for $+m$ and $-m$.

still balanced) CRLH configuration, with period p, as in Section 3.2.6. We also start with a loss-less structure, which is sufficient for our initial purpose of only determining the position of the resonances. Losses will be introduced later on in correlation with the quality factor.

An essential difference between an ideal homogeneous CRLH TL and an LC network CRLH TL is that bandwidth is infinite in the former case Figs. 3.3/3.9(a) whereas it is limited in the latter case (Fig 3.35). As a consequence, an LC network CRLH TL resonator exhibits a *finite number of resonance frequencies*. This is best understood by considering Fig 5.49. The LC network dispersion relation is given by Eq. (3.118)

$$\beta(\omega) = \frac{s(\omega)}{p} \sqrt{\omega^2 L_R C_R + \frac{1}{\omega^2 L_L C_L} - \frac{L_R C_L + L_L C_R}{L_L C_L}}, \qquad (5.63)$$

and is bandwidth-limited by the LH highpass and RH lowpass cutoff frequencies given by Eq. (3.94) [Eq. (3.96) in the balanced case]. Therefore, only a finite number of resonances can "fit" within the passband as the dispersion curve $\omega = \omega(\beta_{CRLH})$ is sampled with a sampling rate of π/ℓ in the β variable. The apparent (infinite number of) space harmonic resonances appearing by periodicity outside the Brillouin zone are degenerated with the fundamental resonances (same fields distributions) and do not correspond to additional frequencies. For a resonator of length ℓ constituted of N unit cells with period p (i.e., $\ell = Np$), the Brillouin zone is divided into N regions of width π/ℓ on each side of the axis $\beta = 0$. There is thus $N - 1$ positive resonances and $N - 1$ negative resonances plus the zeroth-order resonance, which represents a total of $2N - 1$ resonances. It

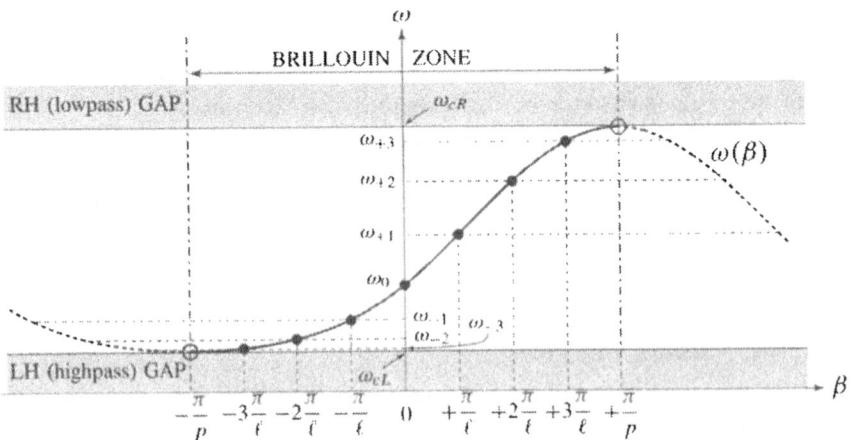

Fig. 5.49. Resonances of a (balanced) periodic LC network CRLH TL resonator constituted of N unit cells (Here $N = 4$). The fields distributions are similar to those shown in Fig 5.48(b). The length of the resonator ℓ and the period p are related by $\ell = Np$, which results in Brillouin zone edges of $\pm N\pi/\ell = \pm\pi/p$.

should be noted that the frequencies located at the edges of the Brillouin zone, represented by an empty circle in Fig 5.49, do not provide resonance frequencies for the resonator because they correspond to Bragg frequencies where the *unit cell* is resonating (size p) instead of the full structure (size ℓ).

5.5.3 Zeroth-Order Resonator Characteristics

In Section 5.5.1, we considered the case of a balanced [$\omega_{se} = \omega_{sh}$ in Eq. (3.9) or Eq. (3.54)] CRLH structure, for which no gap exists between the LH and the RH ranges (Fig 5.49). Whereas the balanced condition is necessary for good matching in the propagating TL operation [Eq. (3.14)], it does not need to be satisfied in resonator operation. If the CRLH TL is unbalanced, it exhibits two resonance frequencies, ω_{se} and ω_{sh}, *when it is terminated by matched load* (i.e., used as a propagating transmission line) [Fig 3.35(b)]. If the line is open-ended or short-ended to form a resonator, what will happen? Will ω_{se} and ω_{sh} be both resonances of the resulting resonator or will only one of them or neither of them? To address these questions, we will consider the open-ended and short-ended cases separately.

In the *open-ended* case, represented in Fig 5.50(a), the input impedance Z_{in} seen from one end of the resonator toward the other end is given by

$$Z_{in}^{\text{open}} = -jZ_c \cot(\beta\ell) \overset{\beta \to 0}{\approx} -jZ_c \frac{1}{\beta\ell}$$
$$= -j\sqrt{\frac{Z'}{Y'}} \left(\frac{1}{-j\sqrt{Z'Y'}} \right) \frac{1}{\ell} = \frac{1}{Y'\ell} = \frac{1}{Y'(Np)} = \frac{1}{NY}, \tag{5.64}$$

where Y is the admittance of the CRLH unit cell, given as $Y = j[\omega C_R - 1/(\omega L_L)]$ by Eq. (3.53b). This result shows that the input impedance of an open-ended is $1/N$ times the impedance of the anti-resonant L_L/C_R shunt tank, $Z_{in} = 1/(NY)$. Because N is a simple integer, not affecting the susceptance, the resonance of the whole resonator is the same as the resonance of the admittance Y [2]. Therefore, there is a single resonance frequency,

$$\omega_{res}^{\text{open}} = \omega_{sh} = \frac{1}{\sqrt{L_L C_R}}. \tag{5.65}$$

In contrast, no resonance occurs at $\omega_{se} = 1/\sqrt{L_R C_L}$.

In the *short-ended* case, represented in Fig 5.50(b), the input impedance Z_{in} seen from one end of the resonator toward the other end is given by

$$Z_{in}^{\text{short}} = jZ_c \tan(\beta\ell) \overset{\beta \to 0}{\approx} -jZ_c\beta\ell$$
$$= -j\sqrt{\frac{Z'}{Y'}} \left(j\sqrt{Z'Y'} \right) \ell = Z'\ell = Z'(Np) = NZ, \tag{5.66}$$

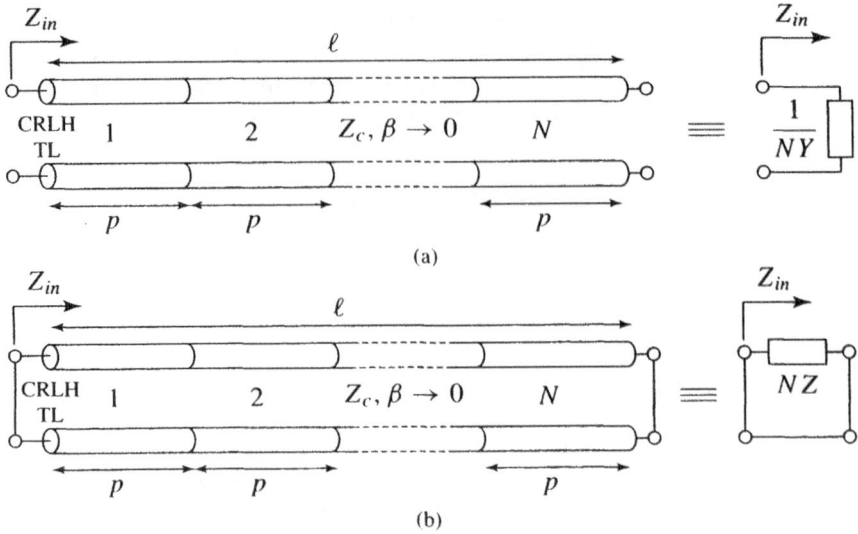

Fig. 5.50. CRLH TL resonator and equivalent circuits in terms of unit cell immittances, Z and Y. Each unit cell is identical to the LC cell shown in Fig 3.14(a). (a) Open-ended case. (b) Short-ended case.

where Z is the admittance of the CRLH unit cell, given as $Z = j[\omega L_R - 1/(\omega C_L)]$ by Eq. (3.53a). This result shows that the input impedance of an open-ended is N times the impedance of the resonant L_R/C_L shunt tank, $Z_{in} = NZ$. Since N is a simple integer, not affecting the reactance, the resonance of the whole resonator is the same as the resonance of the series impedance Z [2]. Therefore, there is a single resonance frequency,

$$\omega_{res}^{short} = \omega_{se} = \frac{1}{\sqrt{L_R C_L}}. \tag{5.67}$$

In contrast, no resonance occurs at $\omega_{sh} = \sqrt{L_L C_R}$.

In the particular case of balanced resonances, the zeroth-order resonance described in Section 5.5.2 occurs for both the open-ended and the short-ended resonators, since $\omega_0^{bal} = \omega_{se} = \omega_{sh}$.

The remarkable feature of the zeroth-order resonator is that, either for an open-ended or a short-ended configuration, *the resonance frequency depends only on the circuit elements L_R/C_L or L_L/C_R of the unit cell and not on the physical length ℓ of the resonator.* This suggests that a zeroth-order resonator could be made arbitrarily small[34], the limitation in size reduction being the minimum footprint required by the LC elements for the required LC values in a given

---

[34]It could naturally also be made arbitrarily large. Although this is not of interest in microwave applications, it may be interesting at higher frequencies where the size of a conventional resonator may be too small in some cases.

technology. Consider a resonator constituted of N cells occupying each a length p for a total length of $\ell = Np$ with unit cell values (L_R, C_R, L_L, C_L). For a given configuration (open-/short-ended), this resonator exhibits one zeroth-order resonance, which depends only on the LC values according to Eqs. (5.65) and (5.67). If we can reduce the footprint p of the unit cell, for instance by switching from an "in-plane" technology such as the microstrip one (Section 3.3.2) to a "vertical" technology such as the multilayer one (Section 5.3), the overall length ℓ of the resonator can be reduced without affecting the resonances. The limit in size reduction is only determined by the limit in the capacity of engineering small LC components with sufficiently high inductance/capacitance values.

The loss mechanism in the CRLH TL resonator at the zeroth-order resonant state is also different from that of conventional resonators due to infinite wavelength. Let us consider first the open-ended case. Since the input impedance of the open-ended resonator is equal to $1/N$ times the impedance of the shunt tank circuit of the unit cell [Eq. (5.64)], the equivalent LCG values are $[L_L/N, NC_R, 1/NG]$, as indicated in Fig 5.51(a). The unloaded Q-factor is then obtained as

$$
\begin{aligned}
Q_0^{\text{open}} &= \frac{1/NG}{\omega_{sh}(L_L/N)} = \frac{1/G}{\omega_{sh}L_L} \\
&= \omega_{sh}(1/NG) \cdot NC_R = \omega_{sh}(1/G)C_R \\
&= \frac{1}{G}\sqrt{\frac{C_L}{L_R}}.
\end{aligned}
\tag{5.68}
$$

It can be seen that the unloaded-Q depends only on the loss G in the shunt tank circuit and *not on the loss R in the series tank circuit* of the unit cell. This can be understood by the following consideration: At the zeroth-order resonance, ω_{sh}, there is no voltage gradient (neither in magnitude nor in phase) between the different nodes of the resonator due to the infinite-wavelength wave; therefore, no current flows along the series resistor R. Consequently, no power is dissipated by

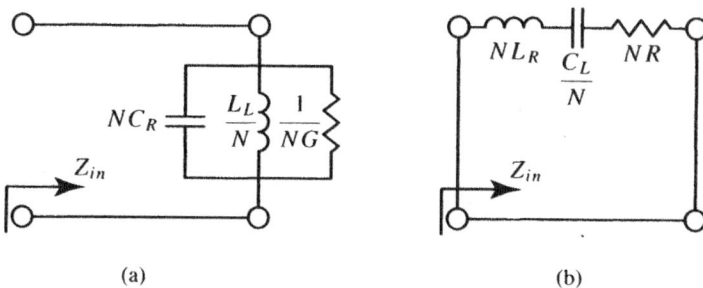

(a)   (b)

Fig. 5.51. Equivalent circuit and immittances for the CRLH TL resonator of Fig 5.50 for the $n = 0$ mode (and only for this mode). (a) Open-ended case. (b) Short-ended case.

the series resistance R. In contrast, currents can flow from each nonzero voltage nodes to the ground plane, which explains the dependence of Q_0 on the shunt admittance G. This appears clearly in Fig 5.51(a), where it is seen that the equivalent total series branch is a simple short connection (R "invisible"), whereas the equivalent total shunt branch includes the resistance $1/G$ (dissipating power).

The unloaded Q-factor of the short-ended case is obtained in a similar manner. Since the input impedance of the short-ended resonator is equal to N times the impedance of the series tank circuit of the unit cell [Eq. (5.66)], the equivalent LCR values are $[NL_R, C_L/N, NR]$, as indicated in Fig 5.51(b). The unloaded Q-factor is then obtained as

$$Q_0^{\text{short}} = \frac{NR}{\omega_{se} NL_R} = \frac{R}{\omega_{se} L_R}$$

$$= \omega_{se} NR \cdot C_L/N = \omega_{se} RC_L \tag{5.69}$$

$$= R\sqrt{\frac{C_R}{L_L}}.$$

In this case, the unloaded-Q depends only on the loss R in the series tank circuit and *not on the loss G in the shunt tank circuit* of the unit cell. This appears clearly in Fig 5.51(b), where it is seen that the equivalent total shunt branch is a simple short connection ($1/G$ "invisible") whereas the equivalent total series branch includes the resistance R (dissipating power).

A remarkable and essential feature of the zeroth-order resonator, pertaining to both the open-ended and short-ended configurations, is that *the unloaded Q-factor is independent on the number of unit cells* (and therefore length ℓ, of course) of the resonator, according to Eqs. (5.68) and (5.69). A zeroth-order resonator seems therefore to have the potential to exhibit a higher Q than that of conventional resonators, since only one of the two possible (series and shunt) loss contributions is present.

5.5.4 Circuit Theory Verification

The main characteristics of the CRLH TL resonator are verified by standard circuit analysis in Fig 5.52. Fig 5.52(a) shows the resonances of the CRLH TL, distributed along the dispersion curve of the TL, for a given set of parameters $(N; L_R, C_R, L_L, C_L)$, in the general case of an unbalanced design.

Fig 5.52(b) shows the corresponding transmission resonance frequencies of the resonator excited via coupling capacitance/inductance: In the open-ended case, the resonator is coupled to the external ports by a very small series capacitance C_e ($|Z_{C_e}| = 1/\omega C_e$ negligible in comparison with Z_L, and Z_R), whereas in the short-ended case, the resonator is coupled to the external ports by a very small shunt inductance L_e ($|Y_{L_e}| = 1/\omega L_e$ negligible in comparison with $1/Z_L$ and $1/Z_R$). As expected, $2N - 1$ resonances (i.e., here, for $N = 7$, 13 resonances)

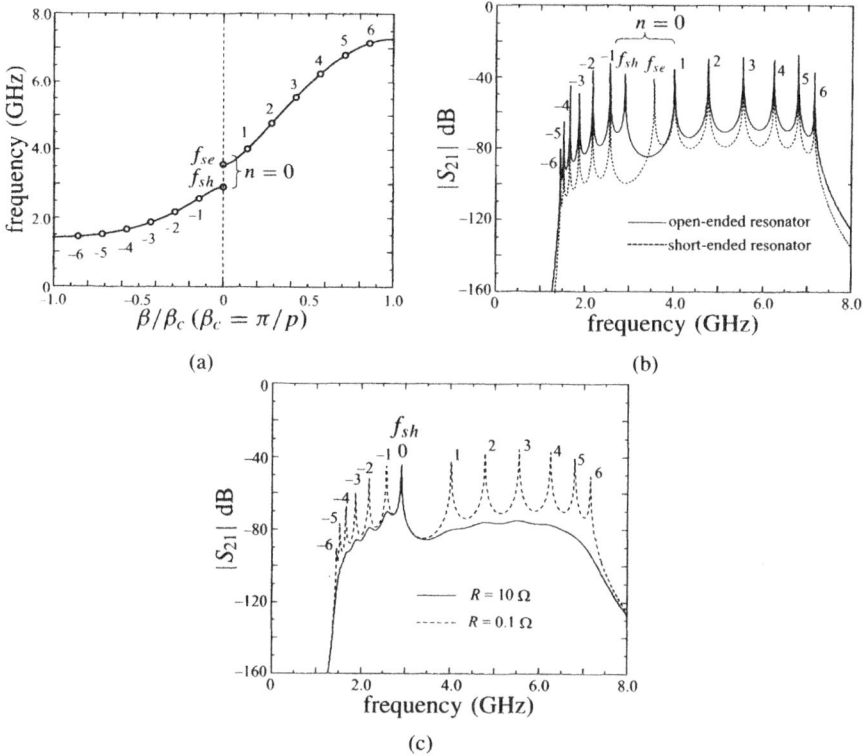

(a)

(b)

(c)

Fig. 5.52. Circuit theory verification of the negative/zeroth-order (unbalanced) CRLH TL resonator for the parameters $N = 7$, $L_R = 2.0$ nH, $C_R = 1.5$ pF, $L_L = 2.0$ nH, $C_L = 1.0$ pF, $C_e = 0.01$ pF (input/output series coupling capacitance for the open-ended configuration), and $L_e = 0.01$ nF (input/output shunt coupling inductance for the short-ended configuration); $f_{sh} = 2.91$ GHz and $f_{se} = 3.56$ GHz. (a) Resonance frequencies along the dispersion curve (the dispersion curve is computed by PBCs assuming an infinitely periodic structure, but the resonances are obtained by circuit theory for $N = 7$). (b) Corresponding resonances appearing in the transmission parameter S_{21}. (c) Effect of series resistance increase in the case of the open-ended configuration with $G = 0.1$ S.

are observed and there are $N - 1$ negative resonances, $N - 1$ positive resonances, and one zeroth-order resonance. As predicted in Eqs. (5.65) and (5.67), the zeroth-order resonance is at ω_{sh} for the open-ended configuration and at ω_{se} for the short-ended configuration.

Finally, Fig 5.52(c) shows the effect of loss increase in the series resonant tank of the unit cell for the open-ended resonator. Although increasing the series resistance (from $R = 0.1$ Ω to $R = 10$ Ω) decreases the Q-factor for all the $n \neq 0$ modes, as in conventional resonators, the Q-factor of the $n = 0$ mode is unaffected by the increase of R, as expected from Eq. (5.68).

5.5.5 Microstrip Realization

To demonstrate the concept of zeroth-order resonance, an open-ended CRLH TL resonator is implemented in microstrip technology [16], as shown in Fig 5.53.

Fig 5.54 shows full-wave (MoM) simulations of this resonator. The simulated resonances (Fig 5.54(a)) are seen to agree well with the circuit theoretical values for the negative and zero resonances, whereas some discrepancies, which might be due to higher order mode effects and frequency dependence of the LC components, are observed for the positive resonances. The electric field distributions [Fig 5.54(b)] demonstrate the expected uniform field distribution of the zeroth-order mode.

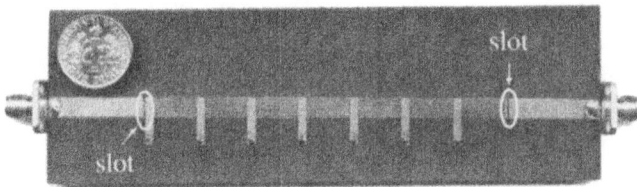

Fig. 5.53. Prototype of a 7-cell microstrip CRLH TL (Section 3.3.2) open-ended resonator. Coupling is achieved via two weakly capacitive slots of width 200 μm on the input/output microstrip traces at the edges of the structure. The substrate used is Duroid 6510 with permittivity 2.2 and thickness 1.57 mm.

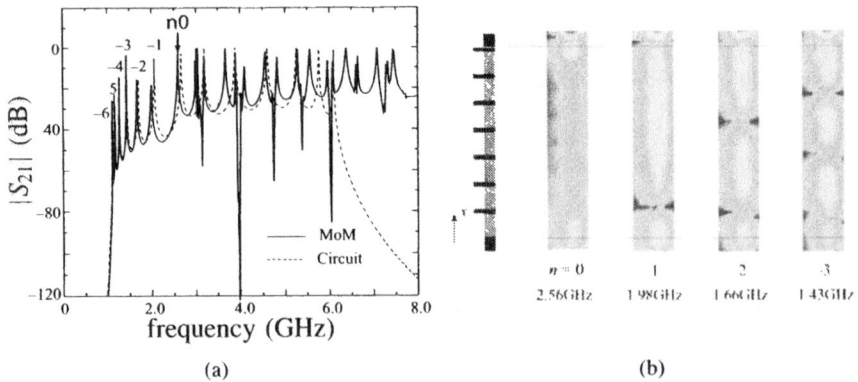

Fig. 5.54. Full-wave (MoM) analysis of the microstrip CRLH TL resonator of Fig 5.53. (a) Transmission parameter S_{21}, compared with the curve obtained by circuit theory. The $n = 0$ resonance is obtained at $f_{sh} = 2.56$ GHz. A metal shield is surrounding the structure. (b) Electric field distribution (x-component) at 1.5 mm ($= 0.013\lambda_0$) above the structure for the $n = 0$ mode and a few $n < 0$ modes. The lighter areas correspond to the higher intensities.

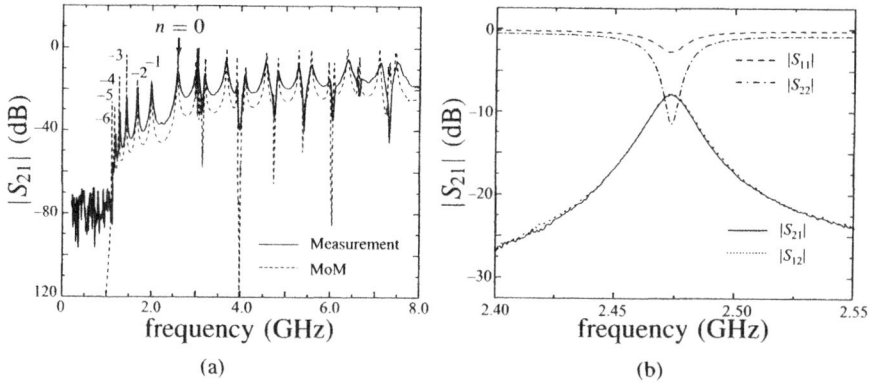

Fig. 5.55. Measured frequency characteristics of the microstrip CRLH TL resonator of Fig 5.53. The $n = 0$ resonance is obtained at $f_{sh} = 2.48$ GHz. No shielding enclosure is used, neither for the simulation nor for the measurement. (a) Transmission parameter over a broad frequency range including all the resonances. (b) Transmission and reflection parameters over a small frequency range zoomed around the $n = 0$ mode.

The experimental results are presented in Fig 5.55. Good agreement is observed between the measured and predicted results. It should be noted that, since the CRLH mode $n = 0$ lies in the fast-wave region of the dispersion diagram, it is leaky in nature and can therefore radiate (Section 6.1). Although this property may be exploited in potentially subwavelength antenna applications, it represents an undesirable effect for guided-wave applications, such as filters, where a packaging shield would be required for optimal performances.

The unloaded Q-factor, Q_0 at the zeroth-order resonance can be computed from the transmission and reflection scattering parameters [Fig 5.55(b)]. The Q-factor obtained in this manner from measurement is $Q_0 = 290$, whereas the Q-factor of a conventional half-wavelength microstrip resonator with same frequency is around $Q_0 = 430$. However, no optimization has been carried out at this point, and it is expected that the performances of the CRLH resonator can be significantly improved, for instance, by increasing the ratio C_L/L_R in the unit cell according to Eq. (5.68). A smaller CRLH with approximately identical zeroth-order resonance has been experimented.

REFERENCES

1. I.-H. Lin, M. De Vincentis, C. Caloz, and T. Itoh. "Arbitrary dual-band components using composite right/left-handed transmission lines," *IEEE Trans. Microwave Theory Tech.*, vol. 52, no. 4, pp. 1142–1149, April 2004.

2. M. D. Pozar. *Microwave Engineering*, Third Edition, John Wiley & Sons, 2004.

3. F. E. Gardiol. *Introduction to Microwaves*, Artech House, 1983.

4. S. A. Maas. *Microwave Mixers*, Second Edition, Artech House, 1993.

5. B. P. Lathi. *Modern Digital and Analog Communication Systems*, Third Edition, Oxford University Press, 1998.

6. M. Cohn, J. E. Degenford, and B. A. Newman. "Harmonic mixing with an antiparallel diode pair," *IEEE Trans. Microwave Theory Tech.*, vol. 52, no. 3, pp. 667–673, March 2004.

7. H. Okabe, C. Caloz, and T. Itoh. "A compact enhanced-bandwidth hybrid ring using an artificial lumped-element left-handed transmission-line section," *IEEE Trans. Microwave Theory Tech.*, vol. 52, no. 3, pp. 798–804, March 2004.

8. Y. Horii, C. Caloz, and T. Itoh. "Super-compact multi-layered left-handed transmission line and diplexer application," *IEEE Trans. Microwave Theory Tech.*, vol. 53, no. 4, pp. 1527–1534, April 2005.

9. I. Bahl and P. Bhartia. *Microwave Solid State Circuit Design*, Wiley Interscience, 1988.

10. A. R. Brown and G. M. Rebeiz. "A high-performance integrated K-band diplexer," *IEEE Trans. Microwave Theory Tech.*, vol. 47, no. 8, pp. 1477–1481, Aug. 1999.

11. R. Mongia, I. Bahl and P. Bhartia. *RF and Microwave Coupled-Line Circuits*, Artech House, 1999.

12. B. E. A. Saleh and M. C. Teich. *Fundamentals of Photonics*, Wiley-Interscience, 1991.

13. J. Lange. "Interdigital stripline quadrature hybrid," *IEEE Trans. Microwave Theory Tech.*, vol. 17, no. 12, pp. 1150–1151, Dec. 1969.

14. C. Caloz, A. Sanada, and T. Itoh. "A novel composite right/left-handed coupled-line directional coupler with arbitrary coupling level and broad bandwidth," *IEEE Trans. Microwave Theory Tech.*, vol. 52, no. 3, pp. 980–992, March 2004.

15. C. Caloz and T. Itoh. "A novel mixed conventional microstrip and composite right/left-handed backward-wave directional coupler with broadband and tight coupling characteristics," *IEEE Microwave Wireless Compon. Lett.*, vol. 14, no. 1, pp. 31–33, Jan. 2004.

16. A. Sanada, C. Caloz, and T. Itoh, "Zeroth-order resonance in composite right/left-handed transmission line resonators," *Asia-Pacific Microwave Conference*, vol. 3, pp. 1588–1592, Seoul, Korea, Nov. 2003.

6

RADIATED-WAVE APPLICATIONS

In Chapter 5 we presented a number of novel CRLH MTM *guided-wave* concepts and applications, where energy remains essentially confined within the MTM structures. However, if a MTM structure is open to free space and supports a fast-wave mode, called a *leaky-wave* (LW) mode, it radiates and can therefore be used as an antenna or, by reciprocity, as a reflector[1]. Unique *radiation effects* may be obtained with CRLH structures due to their rich and unusual *propagation properties*. In addition, novel resonant-type MTM antennas, based on the MTM resonators presented in Section 5.5, can also be designed. In Chapter 6, we present several novel LW and resonant-type CRLH MTM *radiated-wave* concepts and related antenna-reflector applications.

Section 6.1 presents the fundamental elements of LW structures needed in the following sections. Section 6.2 describes the phenomenon of CRLH LW backfire-to-endfire (including broadside) radiation with a frequency-scanned antenna application. A fixed-frequency electronically scanned version of this antenna is demonstrated in Section 6.3, and derivative "smart" reflectors are presented in Section 6.4. Whereas the previous sections mainly regard 1D structures, 2D LW structures are described in Section 6.5, with the example of a conical-beam bidimensional LW antenna. The next two sections deal with resonant-type (as opposed to leaky) antennas; Section 6.6 presents a zeroth-order

---

[1]MTM LW radiation is fundamentally similar to Vavilov-Čerenkov radiation (Section 2.5), the mere distinction between the two phenomena being that in the latter case the radiating particle or (beam of particles) propagates in a bulk medium, whereas in the former case the radiating wave typically propagates along a transmission line structure.

Electromagnetic Metamaterials: Transmission Line Theory and Microwave Applications,
By Christophe Caloz and Tatsuo Itoh
Copyright © 2006 John Wiley & Sons, Inc.

(size-independent) resonant CRLH antenna, whereas Section 6.7 demonstrates a dual-frequency CRLH resonating antenna. Finally, Section 6.8 demonstrates how focusing can be achieved by a straight interface, or "meta-interface" by using the principle of phase conjugation.

6.1 FUNDAMENTAL ASPECTS OF LEAKY-WAVE STRUCTURES

Sections 6.1.1 and 6.1.2 summarize the fundamental aspects of conventional leaky-wave structures. A comprehensive text on this topic is found in [1]. Section 6.1.3 points out the similarities and differences between MTMs and conventional LW structures.

6.1.1 Principle of Leakage Radiation

A *leaky wave* is a traveling wave progressively leaking out power as it propagates along a waveguiding structure. *Leaky-wave (LW) structures* are structures supporting one or several leaky waves. They are typically used as antennas, where the leakage phenomenon is generally associated with high directivity. LW antennas are fundamentally different from resonating antennas, in the sense that they are based on a *traveling-wave* as opposed to a *resonating-wave* mechanism. Therefore, their size is not related to the operation frequency, but to directivity.

A LW structure is schematically represented in Fig 6.1. The leaky wave in free space above the structure exhibits the general wave form

$$\psi(x, z) = \psi_0 e^{-\gamma z} e^{-jk_y y} = \left(e^{-j\beta z} e^{-\alpha z}\right) e^{-jk_y y}, \qquad (6.1)$$

where $\gamma = \alpha + j\beta$ is the complex propagation constant of the wave in the direction of the waveguide (z), and k_y is the propagation constant perpendicular to this direction, related to β by

$$k_y = \sqrt{k_0^2 - \beta^2}, \qquad (6.2)$$

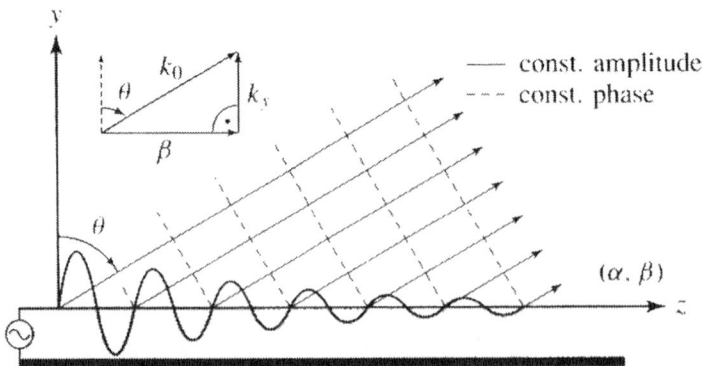

Fig. 6.1. Schematic representation of a general LW structure. The solid and dashed lines above the structure represent the constant amplitude and constant phase planes, respectively, which are perpendicular to each other [3].

where $k_0 = \omega/c$ is the free space wavenumber. This relation suggests the two following possible situations. If the wave is slower than the velocity of light ($v_p < c$, *slow wave*) or $\beta > k_0$, the perpendicular propagation constant is imaginary, $k_y = j\,\mathrm{Im}(k_y)$, and there is exponential decay along y away from the interface, $e^{+\mathrm{Im}(k_y)y}$ with $\mathrm{Im}(k_y) < 0$ ensuring decay (as opposed to unphysical growth). If in contrast the wave is faster than the velocity of light ($v_p > c$, *fast wave*) or $\beta < k_0$, the perpendicular propagation constant is real, $k_y = \mathrm{Re}(k_y) = q$ and there is propagation along y, $e^{-j\mathrm{Re}(k_y)y} = e^{-jqy}$, which means that leakage radiation occurs. Thus, a slow wave, characterized by $\beta > k_0$, is a guided wave, whereas a fast-wave, characterized by $\beta < k_0$, is a leaky wave. The region of the dispersion diagram where the condition $\beta < k_0$ (or $\omega > \beta c$) is called the *radiation region or radiation cone* and is represented in Fig 6.2. Any waveguiding structure with dispersion curves $\omega(\beta)$ penetrating into the radiation cone is a LW structure and may be used as an antenna at frequencies within the radiation cone.

The complex propagation constant γ contains the two fundamental parameters of a leaky-wave structure. The quantity β is the *propagation constant* of the waveguide. As illustrated in the inset of Fig 6.1, β *determines the angle* θ_{MB} *of radiation of the main beam* following the simple relation

$$\theta_{MB} = \sin^{-1}(\beta/k_0) \qquad (6.3)$$

and, via this relation, the width of the main beam by [1]

$$\Delta\theta \approx \frac{1}{(\ell/\lambda_0)\cos(\theta_{MB})} = \frac{1}{(\ell/\lambda_0)\cos\left[\sin^{-1}(\beta/k_0)\right]}, \qquad (6.4)$$

where ℓ is the length of the structure and λ_0 the free space wavelength ($\lambda_0 = 2\pi/k_0$). If the waveguide is dispersive (i.e., β is a nonlinear function of ω),

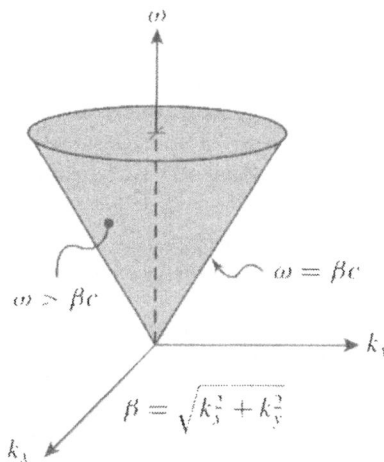

Fig. 6.2. Radiation cone for a 2D structure.

the quantity β/k_0 depends on frequency and therefore the main beam angle is changed as a function of frequency, $\theta_{MB} = \theta_{MB}(\omega)$, according to Eq. (6.3).[2] This phenomenon is called *frequency scanning*. Eq. (6.3) also reveals that radiation in any angle from *backfire* ($\theta_{MB} = -90°$) to *endfire* ($\theta_{MB} = +90°$) can be potentially achieved if the propagation constant of the wave varies across the range $\beta \in [-k_0, +k_0]$. However, when frequency is changed, according to Eq. (6.4), the width of the main beam is also changed and increases when the angle increases from broadside ($\theta_{MB} = 0$) toward backfire or endfire.

The other fundamental quantity of a LW structure is the *leakage factor, α*. Whereas in a purely guiding structure the real part of the complex propagation constant usually represents an undesirable loss coefficient (e.g., ohmic or dielectric losses), the leakage factor in a LW antenna is clearly a necessary parameter, as it represents radiation, and more precisely the amount of radiation per unit length.[3] If the α is small, which is usually desired in LW antennas, the structure can be made long before all the power has leaked out of it; therefore, the radiation aperture [4] is large and high-directivity is obtained. If in contrast α is large, all the power has leaked out after a short length and the beam is fat due to the resulting small radiation aperture. Most LW structures tend to have relatively small leakage factors (typically $\alpha/k_0 < 0.05$) and are thus highly directive. In practice, a LW antenna is usually designed to radiate about 90% of the power, corresponding the electrical length of

$$\frac{\ell}{\lambda_0} \approx \frac{0.18}{\alpha/k_0}, \qquad (6.5)$$

while the remaining 10% is absorbed by a matched load [1]. This formula indicates that, for instance, a typical ratio of $\alpha/k_0 = 0.02$ corresponds to an electrical length of $\ell/\lambda_0 \approx 9$; this corresponds to a much larger radiation aperture than that of a resonant-type antenna (e.g., patch antenna) and explains the superior directivity of LW structures compared with that of resonant structures. It should be noted however that the aperture size in the direction transverse to the direction of propagation is usually very small. Consequently, a 1D leaky-wave structure produces a *fan beam*, narrow in the scan plane and fat in the transverse plane. Combining several 1D LW antennas in an array configuration naturally leads to pencil beams and provides in addition 2D scanning capability, using frequency

---

[2]Otherwise, if β is a linear function of frequency, $\beta = A\omega$ (where A is a constant), $\beta/k_0 = (A\omega)/(\omega/c) = Ac$, which is a constant. Therefore, according to Eq. (6.3), the angle of the main beam θ_{MB} does not vary with frequency.

[3]If the waveguide or TL is loss-less, the only contribution to $\alpha = \mathrm{Re}(\gamma)$ is the leakage factor. Then this factor can be simply determined from the magnitude of the scattering parameter S_{21} by Eq. (3.69b), $\alpha = -\ln|S_{21}|/\ell$. However, if ohmic losses (α_c) or/and dielectric losses (α_d) are present in the line, this formula corresponds to the *total* attenuation factor, $-\ln|S_{21}|/\ell = \alpha_{tot}$, including all the transmission losses, $\alpha_{tot} = \alpha + \alpha_c + \alpha_d$, and the amount of leakage contribution α in the resulting factor α_{tot} cannot be discriminated from the amount of undesired/real losses α_c and α_d. By using high-quality commercial substrates, we usually have $\alpha_c, \alpha_d \ll \alpha$, so that the formula $\alpha \approx -\ln|S_{21}|/\ell$ provides a good (slightly overestimated) approximation of the leakage factor.

scanning in the plane of the LW structures and electronic scanning (typically used in conventional arrays phase shifters) in the transverse plane.

Before concluding this subsection, let us consider a theoretical aspect regarding the mathematical nature of a LW mode. A leaky wave is called a *complex wave* because it has wavenumber components with distinct natures in the different directions (e.g., $k_z = \beta$ for propagation and k_y for radiation in Fig 6.1) [3]. In the discussion of guidance/leakage associated with examination of the perpendicular component of the wavenumber k_y in Eq. (6.2), we have implicitly assumed that k_z was either purely imaginary (in the slow-wave region) or purely real (in the fast-wave region), which is a sufficient assumption for the understanding of the leakage mechanism and design of LW antennas. However, in general, k_y has both a real and an imaginary part, $k_y = \mathrm{Re}(k_y) + j\mathrm{Im}(k_y)$. Whereas, as mentioned above, $\mathrm{Im}(k_y) < 0$ (exponential decay) in the slow-wave region, it can be shown that a branch point occurs at $\beta = k_0$ so that $\mathrm{Im}(k_y) > 0$ (*improper wave*) in the fast-wave region, corresponding to exponential growth as one moves away from the structure. Such a physical paradox can be understood by examining Fig 6.1. At a given distance from the origin, $z = z_0$, we observe that we indeed meet planes of apparently increasing amplitudes as we move away from the structure, which explains the reason why $\mathrm{Im}(k_y) > 0$. However, it is obvious that no decrease of fields due to radiation has been taken into account when we initially introduced k_y in Eq. (6.1). In reality, there is, of course, radiation free-space attenuation, so that the "constant amplitude planes" are not really constant but have amplitudes decreasing along the direction of k_0. Thus, physically, it is true that the radiated wave increases over a short distance along y; however, after the increase along y due to the improper nature of the leaky wave has been exactly compensated by the decrease along k_0 due to free space radiation, the wave decreases monotonically, in conformity with the Sommerfeld radiation condition.

It should be noted that the complex nature of the complex propagation constant, $\gamma = \alpha + j\beta$, in a MTM LW structure induces a complex eigenfrequency, $\omega = \omega_{\mathrm{real}} + j\omega_{\mathrm{imag}}$, when this eigenfrequency is sought from the propagation constant β. This fact may be understood as follows. From the point of view of transmission along the waveguiding structure, leakage is perceived as loss. In the TL approach, losses are represented by a per-unit-length resistance R' and a per-unit-length conductance G', which results for a CRLH TL into the general (lossy) dispersion relation [Eq. (3.45)]

$$\gamma = \alpha + j\beta = \sqrt{[R'G' - X'(\omega)B'(\omega)] - j[R'B'(\omega) + G'X'(\omega)]}, \qquad (6.6)$$

where

$$X'(\omega) = \omega L'_R - \frac{1}{\omega C'_L}, \qquad (6.7a)$$

$$B'(\omega) = \omega C'_R - \frac{1}{\omega L'_L}. \qquad (6.7b)$$

The dispersion diagram corresponding to Eq. (6.6) can be computed either by solving $\gamma = \gamma(\omega)$ (γ solution sought for each value of ω), which provides the propagation constant $\beta(\omega)$ and the leakage factor $\alpha(\omega)$ (including also ohmic/dielectric losses, method used in Sections 3.2.6 and 4.1) or by solving the inverse problem $\omega = \omega(\gamma)$ (ω solution sought for each value of γ). The latter method is more difficult because the solutions have to be sought in the complex γ plane. In practice, when no closed-form solution is available and one has to resort to full-wave analysis with PBCs, the problem $\omega = \omega(\beta)$ is solved instead. As long as the mode lies in the guided (slow-wave) region, $\gamma = j\beta$ ($\alpha = 0$) and therefore real solutions, corresponding to a purely imaginary square root in Eq. (6.6),[4] are found for ω. When in contrast the mode penetrates in the leaky (fast-wave) region, it becomes complex, $\gamma = \alpha + j\beta$. This means that the square root in Eq. (6.6) has both an imaginary and a real part both depending on ω, and the resulting complex function of ω is "incorrectly" set equal to $j\beta$. As a consequence, the eigenfrequencies also become complex, $\omega = \omega_{real} + j\omega_{imag}$. The imaginary part of the eigenfrequencies is related to the leakage losses ($\omega_{imag} = 0$ if $R' = G' = 0$ in a balanced CRLH TL, as shown in Section 3.1.3) and could be related to R' (G' can be considered zero if only leakage loss is involved) [5] by inverting Eq. (6.6). This phenomenon of complex eigenfrequency can be associated with the *relaxation time* considered in resonant cavities [6].

6.1.2   Uniform and Periodic Leaky-Wave Structures

LW structures can be classified in two categories with distinct characteristics: uniform and periodic LW structures.

6.1.2.1   Uniform LW Structures   A *uniform LW* structure is a LW waveguide with a *cross section invariant along the direction of propagation* or with only a small, smooth, and continuous variation (taper) of the cross section along the direction of propagation. In a uniform LW antenna, only the *dominant mode* of the structure[5] propagates and therefore this mode has to be *fast* for radiation to occur, according to Eq. (6.2). If the structure is PRH, as it is the case in conventional uniform LW antennas, it follows from Eq. (6.3) that *only forward angles can be obtained* since $\beta > 0$ at all frequencies. In addition, *broadside radiation is not possible* because it would require $\beta = 0$, which would occur at DC! These angle-range limitations severely restrict the interest of conventional uniform LW antennas despite their attractive high-directivity feature. Another typical problem in uniform LW antennas is that they often require a complex and inefficient feeding structure, to selectively excite the mode of interest and suppress the lower-frequency modes, as will be illustrated in an example below.

---

[4]The principle is the same if no closed form solution is available.

[5]Here "dominant" refers to the lowest-frequency *fast-wave* (leaky) mode. In some case, as for instance the microstrip LW antenna (Fig 6.3), there may be a *slow-wave* (guided) mode in a lower frequency range; this slow-wave mode is then eliminated by a mode suppressing mechanism, so that the lowest-frequency fast-wave mode may be regarded as the dominant mode of the resulting structure.

A perfectly uniform LW structure has a complex propagation constant that does not vary along the direction of the structure, $\gamma \neq \gamma(z)$. Whereas a perfectly uniform propagation constant, $\beta \neq \beta(z)$, is desirable for maximum directivity with a given structure length, a constant leakage factor, $\alpha \neq \alpha(z)$ does not produce optimal radiation performances, the exponential decay of power along the line leading to high sidelobe level when the length of the structure is finite.[6] Therefore, smooth-tapering techniques, strongly depending on the specific LW structure, have been developed [7] to design a leakage factor profile $\alpha(z)$ minimizing the sidelobe level while keeping a relatively invariant phase constant profile $\beta \neq \beta(z)$.[7] A closed-form formula, based on the fact that the radiation pattern corresponds to the Fourier transform of the aperture distribution, $A(z)$, has been established to reduce the sidelobes and to obtain a desired radiation pattern with a structure of length ℓ. This formula reads [1]

$$
\alpha(z) = \frac{|A(z)|^2}{\dfrac{2P(0)}{P(0) - P(\ell)} \displaystyle\int_0^\ell |A(\xi)|^2 \, d\xi - \displaystyle\int_0^z |A(\xi)|^2 \, d\xi},
\tag{6.8}
$$

where $P(z)$ is the power along the line.

To illustrate the characteristics of uniform LW structures, we consider the *microstrip LW antenna* of Fig 6.3(a) [8, 9]. In this antenna, the dominant leaky mode is an odd mode, excited by a balun and corresponding to the first higher-order mode of the conventional microstrip line, the even/guided mode of which is fully suppressed by the periodic slots etched in the center of the line.[8] Due to its odd-mode configuration, this antenna exhibits a polarization perpendicular to the direction of the line. The complexity, and subsequent size inefficiency, of the feeding mechanism, requiring both a balun and an even-mode suppressor, is obvious in this configuration.

Fig 6.3(b) shows the $\beta - \alpha$ diagram of this antenna. It can be seen that $\beta_{min}/k_0 \approx 0.1$, which corresponds from Eq. (6.3) to a minimum angle of radiation of $\theta_{MB} \approx 6°$. Clearly, broadside radiation cannot be achieved. Moreover, α becomes very large and therefore directivity very poor for small angles in this specific structure. An obvious problem is the fact that the narrow-bandwidth nature of the excitation balun limits the usable bandwidth, which further restricts the available scanning range.

---

[6]This fact can be easily understood by considering the LW structure as a continuous array fed at each incremental length Δz by a virtual source with a magnitude corresponding to the exponential decay of the leakage factor $|I| = e^{-\alpha z}$ and phase corresponding to the accumulated phase of the propagating wave $\varphi(I) = e^{-j\beta z}$ [5].

[7]A difficulty in LW antennas is that α and β cannot always be controlled independently, which means that it is sometimes not possible to obtain simultaneously the desired radiation patterns, mostly determined by $\alpha(z)$, and the desired frequency scanning relation, mostly determined by $\beta(\omega)$.

[8]Another solution to suppress the microstrip line mode is to use short-circuiting vias to the ground plane.

(a)

(b)

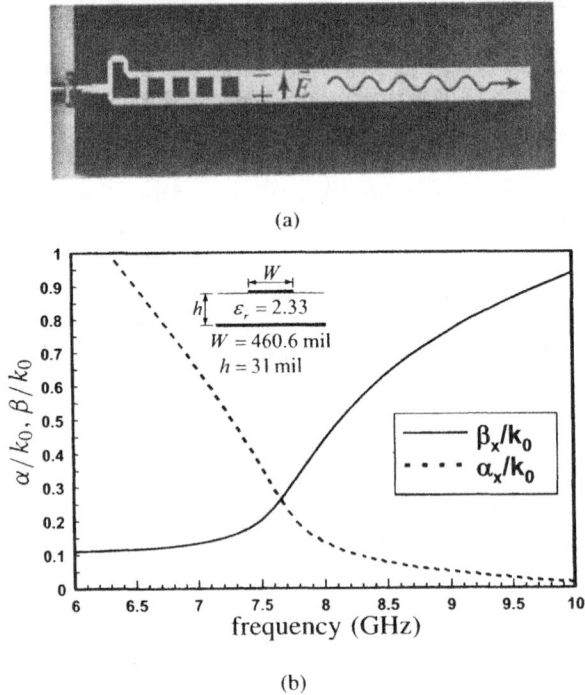

Fig. 6.3. Microstrip LW antenna. (a) Prototype. A balun is used to excite the odd/fast (leaky) higher-order mode of the structure, and periodic slots at the center of the line (input) are used to completely suppress the lower-frequency even/slow (guided) mode of the conventional microstrip line. As a consequence of suppression of this mode, the leaky mode becomes the dominant mode of the resulting structure. Due to the odd-mode configuration, the polarization of the antenna is perpendicular to the direction of the line. (b) $\beta/\alpha - \omega$ diagram. From [2], © IEEE; reprinted with permission.

6.1.2.2 Periodic LW Structures A *periodic LW* structure is a LW waveguide with a *cross section periodically modulated along the direction of propagation.* Due to periodicity, the wave can be represented in the form of a Floquet expansion [Eq. (3.103)] and is thus constituted by the superposition of an infinite number of space harmonics [Eq. (3.106)]

$$\beta_n = \beta_0 + \frac{2n\pi}{p}, \tag{6.9}$$

which can be straightforwardly inserted into Eqs. (6.3) and (6.4) to determine the radiation angle of the main beam and its width of any space harmonic n. In contrast to uniform LW antennas, periodic LW structures have a *slow-wave/guided dominant mode* and their radiation can therefore be obtained only from the contribution of one or several *fast-wave space harmonics* to the total field, as suggested in the dispersion diagrams shown in Figs. 3.33(b) and 3.33(c), where the radiation cone corresponds to the two lines $n = 0_-$ and $n = 0_+$. Another difference with uniform LW antennas is that backward radiation ($\theta_{MB} < 0$) can be achieved by using a negative space harmonic (typically the space harmonic $n = -1$) in addition to forward radiation using a positive space harmonic (Fig 3.33). However, the systematic presence of a gap at $\beta = 0$ in conventional periodic structures prevents broadside radiation because the gap edges correspond to a Bragg resonances ($v_g = 0$) associated with standing waves, whereas only traveling waves can be leaky, according to Section 6.1.1. The presence of this gap also prevents a continuous radiation from backward to forward angles because no radiation is possible over the extent of the gap. Finally, even if the gap is relatively limited small, switching from backward to forward angles requires switching from the excitation of a negative space harmonic to a positive space harmonic, which may have very different modal field distributions and input impedances, causing problems of feeding. For these reasons, in practice, a periodic LW antenna can radiate only in a restricted range of backward angles or in a restricted range of forward angles, with the exclusion of broadside.[9] Examples of periodic LW structures include dielectric rectangular rods with grooves or metal strips, stepped-impedance resonating microstrip lines, and asymmetrical rough waveguides [1].

6.1.3 Metamaterial Leaky-Wave Structures

A *MTM LW* structure does not really represent a third category of LW structures in addition to the uniform and periodic categories. A MTM LW structure is usually structurally periodic because periodicity is a computational and fabrication convenience. However, it does not need to be periodic to operate, because it uses only the fundamental mode when it has a periodic configuration (Fig 3.28), whereas periodicity is essential to generate leaky space harmonics in a periodic LW structure. A MTM LW structure essentially behaves as a uniform LW structure due to its effective-homogeneity nature. So, MTM LW structures may be considered to electromagnetically belong to the category of uniform LW structures even if they are structurally periodic.[10]

---

[9]Broadside radiation can be achieved in an indirect manner by using a relatively short structure and omitting to terminate it by a matched load. In this case, the reflected wave radiates a beam on the other side of the normal to the antenna than the incident-wave beam (the resulting pattern cannot be completely symmetric due to the fact that the reflected wave has a smaller magnitude). If the angle is small (i.e., β/k_0 is small), the superposition of these two beams can generate a maximum at broadside. But when the antenna is scanned, there is then always an undesired symmetric beam due to the reflected wave. Another solution would be to excite the antenna in its center of symmetry, so that two leaky waves of identical power may contribute to symmetric broadside radiation at frequency where β is close to zero, by superposition; but this approach also produces an undesired pair of beams at other frequencies.

[10]The following terminology comment is appropriate here. The radiating mode of a conventional uniform LW antenna is called the *dominant leaky* mode because it lies below the cutoff of higher order modes. At the same time, a MTM, if periodic, is known to operate in the *fundamental* mode. So, the radiating mode could be called either "fundamental" (material point of view) or dominant (antenna point of view). Because periodicity is not an essential attribute of MTMs, we prefer to use the term "dominant" to designate the radiating mode in the context of LW antennas.

6.2   BACKFIRE-TO-ENDFIRE (BE) LEAKY-WAVE (LW) ANTENNA

In principle, any open CRLH TL structure can operate as a LW antenna since the CRLH dispersion curve always penetrates in the radiation region, the transition frequency ω_0 begin necessarily a fast-wave point since $v_p(\omega_0) = \infty$ from $\beta(\omega_0) = 0$. The most general dispersion diagram of a *balanced* CRLH TL structure [Eq. (3.119)],

$$\beta(\omega) = \frac{1}{p} \left(\omega \sqrt{L_R C_R} - \frac{1}{\omega \sqrt{L_L C_L}} \right), \qquad (6.10)$$

is represented in Fig 6.4 with the air lines $\omega = \pm kc$ delimiting the radiation region. This $\omega - \beta$ diagram exhibits the following four distinct regions: the LH-guided region, the LH-leaky region, the RH-leaky region, and the RH-guided region, which are characterized in Table 6.1. If the CRLH structure is unbalanced, there is an additional gap region, from $\min(\omega_{se}, \omega_{sh})$ to $\max(\omega_{se}, \omega_{sh})$ between the LH-leaky and RH-leaky regions, but this region is generally unfavorable for antennas because it prevents broadside radiation and introduces a gap in the scanning range.

Fig 6.5 illustrates the operation principle of the CRLH LW antenna. According to Eq. (6.3), *backfire radiation* ($\theta = -90°$) is achieved at the frequency ω_{BF}

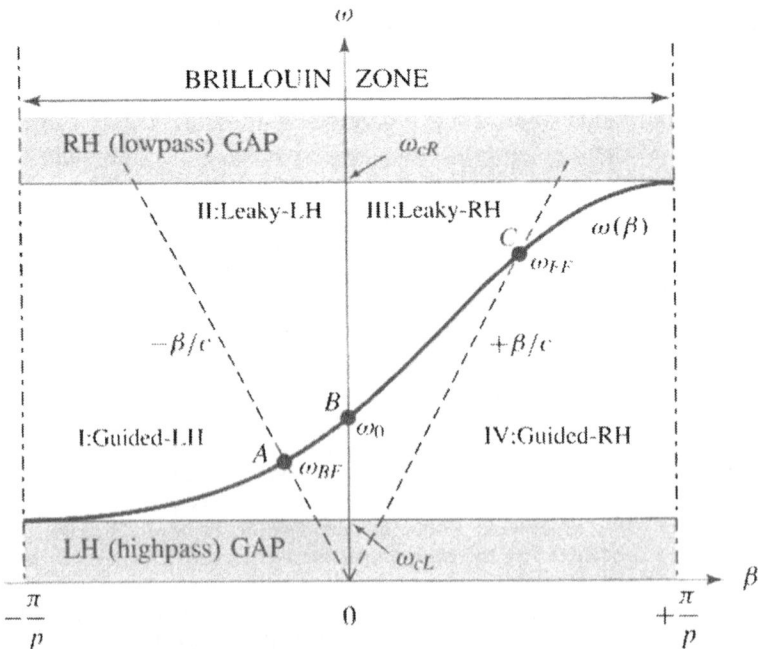

Fig. 6.4. Dispersion diagram of a network balanced CRLH TL structure with its four distinct regions, I: Guided ($v_p < c$) LH ($\beta < 0$), II: Leaky ($v_p > c$) LH ($\beta < 0$), III: Leaky ($v_p > c$) RH ($\beta > 0$), IV: Guided ($v_p < c$) RH ($\beta > 0$).

TABLE 6.1. Different Dispersion Regions of a Balanced CRLH TL Structure.

Region	Frequency range	β	v_p
LH-guided	$\omega_{cL} - \omega_{BF}$	< 0 (LH)	< c (slow-wave)
LH-leaky	$\omega_{BF} - \omega_0$	< 0 (LH)	> c (fast-wave)
RH-leaky	$\omega_0 - \omega_{EF}$	> 0 (RH)	> c (fast-wave)
RH-guided	$\omega_{EF} - \omega_{cR}$	> 0 (RH)	< c (slow-wave)

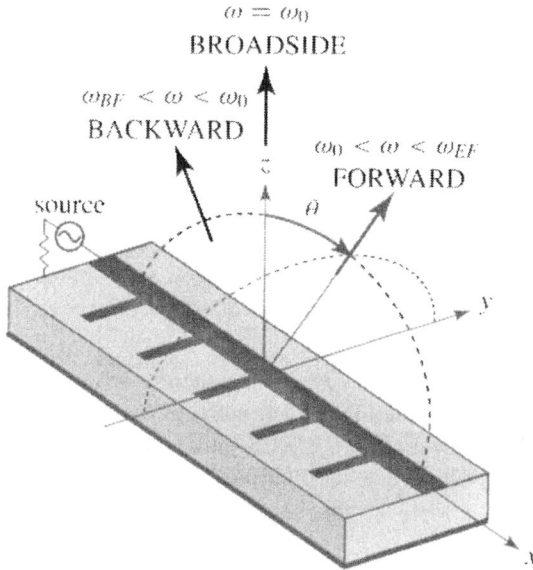

Fig. 6.5. Sketch of a CRLH TL LW antenna illustrating the three radiation regions: backward ($\beta < 0$, LH range), broadside ($\beta = 0$, transition frequency, f_0), and forward ($\beta > 0$, RH range).

where $\beta = -k_0$ (point A in Fig 6.4), *broadside radiation* ($\theta = 0°$) is achieved at the frequency ω_0 where $\beta = 0$ (point B) and *endfire radiation* ($\theta = +90°$) is achieved at the frequency ω_{EF} where $\beta = +k_0$ (point C). Therefore, *backfire-to-endfire frequency-scanning* capability is provided by an open balanced CRLH structure. This backfire-to-endfire capability, first demonstrated experimentally in [10] and explained by the CRLH concept in [11], is a very unique feature for a LW antenna, which cannot be obtained in conventional (uniform or periodic) LW structures, as explained in Section 6.1. In addition to this continuous backward-to-forward angle scanning characteristic, we should emphasize the exceptional *broadside radiation* capability of this antenna. Broadside radiation is a consequence of the fact that the usual gap is suppressed (by mutual cancelation of the series/shunt branches of the CLRH structure under the balanced condition). Because there is no gap, group velocity is different from zero at the origin,

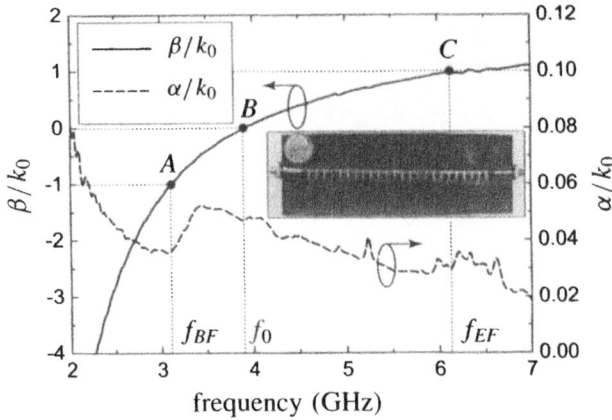

Fig. 6.6. Measured dispersion (β) and attenuation (α) diagrams for the microstrip interdigital-C/stub-L CRLH LW antenna shown in the inset (same structure as that of Fig 3.42). The β and α curves are computed from the measured scattering parameter S_{21} with Eqs. (3.69a) and (3.69b), respectively. The points A, B, and C correspond to the points shown in Fig 6.4: $f = f_{BF}$: backfire radiation ($\theta = -90°$), $f = f_0$: broadside radiation ($\theta = 0°$), $f = f_{EF}$: endfire radiation ($\theta = +90°$).

$v_g(\omega_0) \neq 0$ [Eqs. (3.41) and (3.123)]; consequently, a traveling wave (with infinite v_p) exists along the structure and broadside radiation can occur. Finally, a further advantage of the present CRLH LW antenna is that, operating in a fundamental mode, it can be fed by a very simple and efficient (small and broadband) mechanism, such as a simple line, in contrast to typical LW structures (e.g., Fig 6.3).

Experimental demonstration of the described characteristics of a CRLH LW antenna is provided in Figs. 6.6 and 6.7 for the microstrip structure of Fig 3.42.[11] Fig 6.6 shows dispersion/attenuation diagrams $\beta(\omega)/\alpha(\omega)$ of the structure. In this design, the full-space scanning frequency range extends from around 3.2 GHz to 6.2 GHz, with the broadside transition at 3.9 GHz. Measured radiation patterns for the same structure are shown in Fig 6.7 at three frequencies corresponding to backward, broadside, and forward radiation angles. Finally the frequency-scanning law $\theta_{MB}(f)$ of the antenna is shown in Fig 6.8.

As explained in Section 6.1.1, the directivity of a LW antenna increases with its physical length until the length where most of the power has leaked out. In the case of the CRLH antenna presented here, this trend is apparent in Fig 6.9(a), which shows that for the length corresponding to $N = 24$ cells (prototype demonstrated above), the gain (≈ 7 dB) could be significantly increased (up to 12 dB) by

---

[11] If an open CRLH structure, such as this microstrip structure, is intended to operate as a transmission line instead of as an antenna, a shielding cover may have to be added on the top of it to limit radiation losses. It should be noted however that, because the radiation loss is extremely small for lengths smaller than λ_{eff}, such a precaution may not be necessary in most applications. For instance, in the backward couplers of Section 5.4, quasi complete coupling is achieved despite the fact that the structures are open to air.

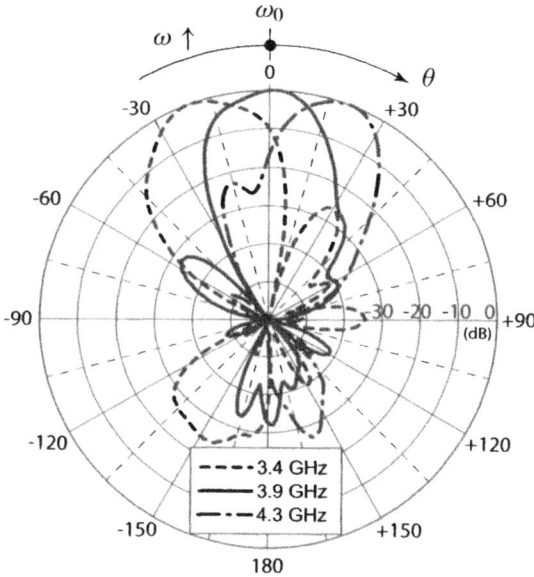

Fig. 6.7. Measured radiation patterns for the CRLH LW antenna of Fig 6.6 at three different frequencies: $f = 3.3$ GHz ($\beta < 0$, backward radiation), $f = f_0 = 3.9$ GHz ($\beta = 0$, broadside radiation), and $f = 4.3$ GHz ($\beta > 0$, forward radiation).

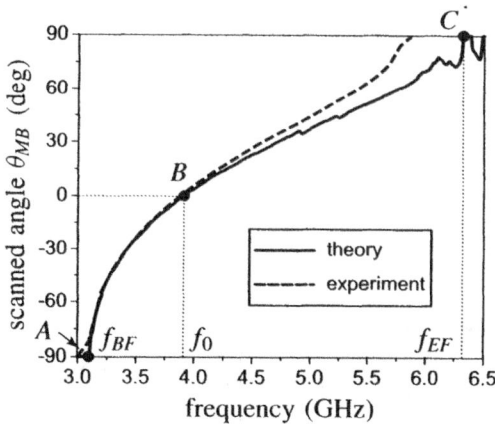

Fig. 6.8. Scanned angle versus frequency, $\theta_{MB}(\omega)$, for the CRLH LW antenna of Fig 6.6. The theoretical curve is computed from the analytical expression of β Eq. (6.10) by Eq. (6.3) with the extracted parameters (Section 3.3.3) $L_R = 1.38$ nH, $C_R = 0.45$ pF, $L_L = 3.75$ nH, $C_L = 1.23$ pF. The experimental curve is obtained by measuring the angles of the main beam.

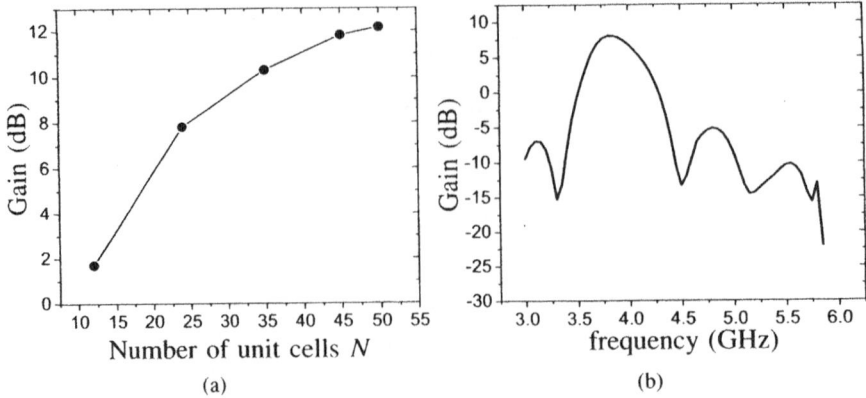

Fig. 6.9. Full-wave (MoM) simulated gain for the CRLH LW antenna of Fig 6.6. (a) Gain versus number of cells at $f = f_0 = 3.9$ GHz (broadside radiation). (b) Gain versus frequency (and therefore angle [Eq. (6.3)]) for $N = 24$.

doubling the length of the antenna. The gain versus frequency curve of Fig 6.9(b) confirms the trend predicted by Eq. (6.4): A maximum gain is obtained at broadside, and this gain decreases when the beam moves toward grazing angles.

A useful array factor approach has been developed [5] for fast and convenient design of MTM LW antennas. In this approach, the LW structure is treated as an array antenna in which the radiating elements are the unit cells of the MTM. The radiation pattern $R(\theta)$ is then well approximated by the array factor $AF(\theta)$ of the equivalent array,

$$R(\theta) = AF(\theta) = \sum_{n=1}^{N} I_n e^{j(n-1)k_0 p \sin\theta + j\xi_n}, \tag{6.11a}$$

with

$$\xi_n = (n-1)k_0 p \sin\theta_{MB} \quad \text{and} \quad I_n = I_0 e^{-\alpha(n-1)p}, \tag{6.11b}$$

where N and p represent the number of unit cells and period of the structure, respectively, I_n is an exponentially decaying function determined by the leakage factor $\alpha = -\ln|S_{21}|/\ell$ [Eq. (3.69b)], and θ_{MB} represents the angle of the main beam, which is determined by $\theta_{MB} = \arcsin(\beta/k_0)$ [Eq. (6.3)] from the dispersion relation $\beta(\omega)$ [Eq. (6.10)]. This formula may be applied to any LW structure once α and β have been determined and provides excellent results in condition that the electrical length of the structure is sufficiently large ($\ell/\lambda_0 > 1$), as it is generally the case in LW structures.[12] Typical results are shown in Fig 6.10.

[12]This fact can be understood in the following manner. The radiation pattern $R(\theta)$ in an array is the product of the element pattern $E(\theta)$ and the array factor $AF(\theta)$, $R(\theta) = E(\theta) \cdot AF(\theta)$. Consider two limiting cases. If $\ell/\lambda_0 \to 0$, then $AF(\theta) \to 1$ (omnidirectional); therefore, $R(\theta) \to E(\theta)$, which

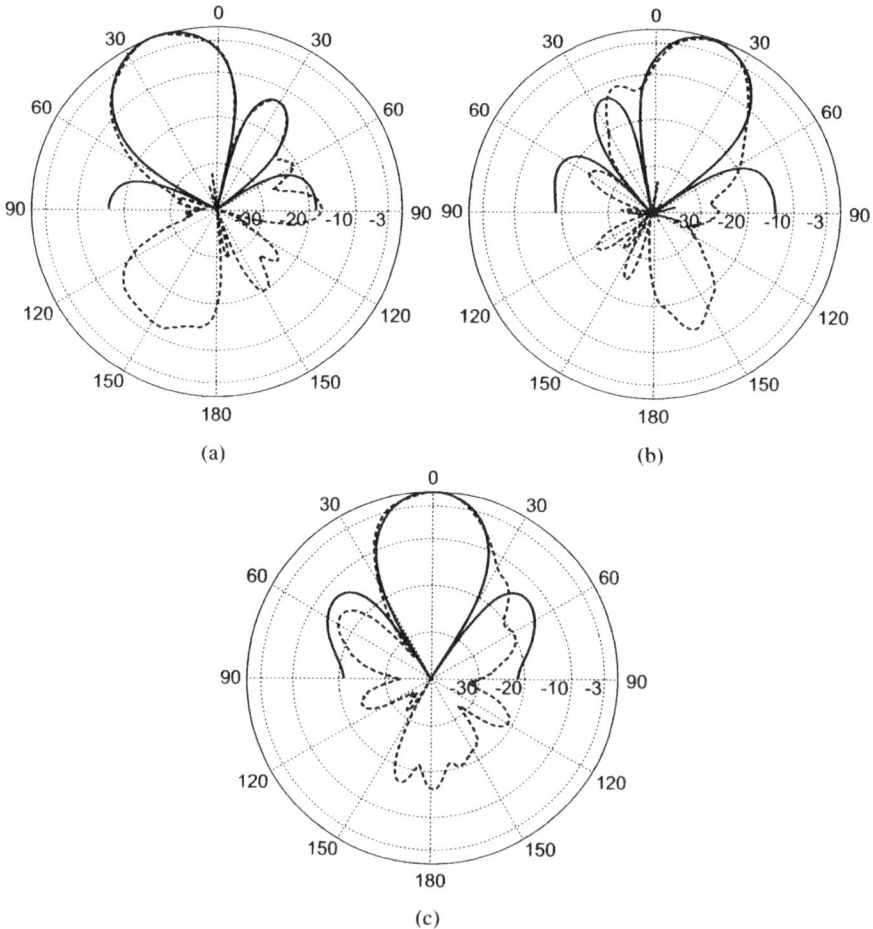

(a)

(b)

(c)

Fig. 6.10. Radiation patterns predicted by the array factor approach [Eq. (6.11)] (solid lines) compared with measurement (Fig 6.7) (dashed lines). (a) Backward radiation, $f = 3.4$ GHz. (b) Forward radiation, $f = 4.3$ GHz. (c) Broadside radiation, $f = 3.9$ GHz. The parameters used in the AF formulas are $L_R = 2.45$ nH, $C_R = 0.5$ pF, $L_L = 3.38$ nH, $C_L = 0.68$ pF, $R = 1\ \Omega$, $G = 0$ S, $N = 24$, $p = 0.61$ mm.

6.3 ELECTRONICALLY SCANNED BE LW ANTENNA

The frequency-scanned nature of conventional LW antennas is a disadvantage, which, in addition to the absence of broadside radiation and continuous scanning capability, has limited their applications in modern communication systems, generally requiring fixed frequency operation for effective channelizing.

is very different from $AF(\theta)$ if $E(\theta)$ is very directive. In contrast, if $\ell/\lambda_0 \to \infty$, then $AF(\theta) \to \delta(\theta - \theta_{MB})$ (infinite directivity); therefore, $R(\theta) \to E(\theta_{MB})$, which corresponds to a sample of the element pattern at θ_{MB}, where θ_{MB} is the radiation angle predicted by the array factor. In practical cases, where ℓ/λ is relatively large (but not infinite), $AF(\theta)$ provides thus a good estimate of $R(\theta)$.

In the past, significant efforts have been directed toward developing frequency-independent LW antennas with another scanning mechanism than frequency tuning [12, 13, 14, 15]. Horn et al. [12] used PIN diodes as switches to electronically vary the radiation angle by controlling the guided wavelength. However, only two discrete radiation angles were available in this approach because diodes have only two states (biased/ unbiased). Maheri et al. [13] reported a magnetically scannable LW antenna built on a ferrite slab structure, in which the radiation angle is scanned by tuning a DC magnetic field. However, a DC magnetic field supply is not practical for most microwave applications. In [14], PIN diodes were applied as switches to control the period of the structure, but this was again limited to two discrete radiation angles.

To overcome these drawbacks while taking full advantage of the unique features of the BE LW antenna presented in Section 6.2, a novel electronically controlled BE LW antenna incorporating varactor diodes was proposed in [16]. This antenna is capable of *continuous beam steering* by tuning the reverse bias voltages of the varactors at a fixed frequency. In addition, it provides the functionality of *beam-width control* by using a nonuniform biasing distribution of the diodes, whereas beamwidth is conventionally controlled by changing the geometrical parameters of the structure or with a phased array of antennas [17, 18]. This antenna represents an attractive alternative to conventional phased arrays with its advantages of requiring only one radiating element, utilizing a very simple and compact feeding mechanism and not necessitating any phase shifters.

6.3.1 Electronic Scanning Principle

The main beam angle θ_{MB} of a LW antenna is a function of the propagation constant along the structure, $\theta_{MB} = \theta_{MB}(\beta)$, according to Eq. (6.3), and the propagation constant in a CRLH TL depends both on frequency and on the LC parameters, $\beta = \beta(\omega; L_R, C_R, L_L, C_L)$, as seen for instance in Eq. (6.10). Therefore, the angle of a leaky CRLH structure can be steered either by frequency tuning, as in the frequency-scanned antenna of Section 6.2 [$\theta_{MB} = \theta_{MB}(\omega)$], or by LC parameters tuning at a fixed operation frequency. In the latter case, varactor diodes can be integrated along the structure in each cell to provide continuously variable capacitances via the control of their reverse bias voltage V [$\theta_{MB} = \theta_{MB}(V)$].[13] A prototype of electronically scanned CRLH BE LW antenna is shown in Fig 6.11 [16].

As the value of the varactors capacitances (for C_L, C_R or both) are varied by variation of the bias reverse voltage, the dispersion curve of Fig 6.4 is shifted in the $\omega - \beta$ plane (and also slightly altered in shape). A "continuous family" of curves $\beta(V)$ is obtained. Therefore, the radiation angle is varied by Eq. (6.3). This principle is illustrated in Fig 6.12, where three bias conditions are represented.

---

[13]It would also be possible to utilize ferrite components to obtain a continuously tuned inductance, but this option is less attractive in planar circuit technologies because ferrite components tend to have high losses and to be too bulky; this problem may however be alleviated with the advent of ferrite thin film technology. Ferroelectric capacitors may also be an option, especially when higher power handling and higher frequency operation are required.

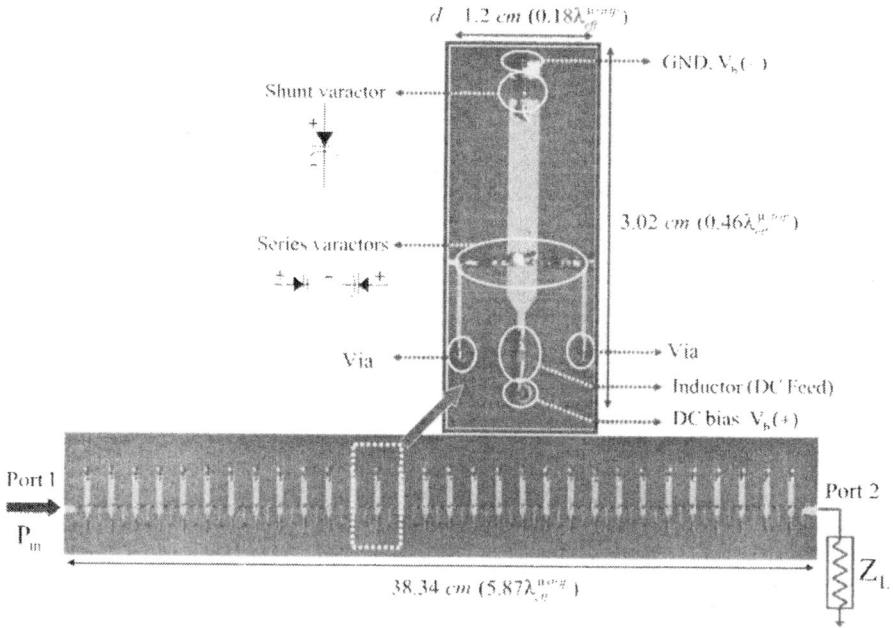

Fig. 6.11. Prototype of a 3.3-GHz 30-cell electronically scanned CRLH BE LW antenna implemented in microstrip technology (Section 3.3.2). The structure is built on Duroid5880 with permittivity 2.2 and thickness 62 mil, and the varactors are Si abrupt varactor diodes. The dimension of one cell is 1.2 × 1.2 cm, and the length of the overall structure is 38.34 cm ($5.87\lambda_{eff}^{\mu strip}$). From [16], © IEEE; reprinted with permission.

If the value of β can be varied from $\beta_{BF} = -\omega/c$ to $\beta_{EF} = +\omega/c$ in a voltage range $[V_{BF}, V_{EF}]$, backfire-to-endfire electronic scanning is achieved. It should be noted that only the dispersion curve crossing the frequency axis at ω_0, $\beta(V_0)$, needs to be balanced, since ω_0 does not correspond to the transition frequency for the other curves. This fact relaxes the design constraints in the design of the antenna.

6.3.2 Electronic Beamwidth Control Principle

In the description of the electronically scanned antenna of Section 6.3.1, it was implicitly assumed that the bias distribution was *uniform*, which means that bias voltage is the same for all the cells of the structure, as illustrated in Fig 6.13(a). This results in a truly periodic structure, since all the cells have the same capacitance parameters. In addition, this structure can be considered uniform [$\beta \neq \beta(z)$] since its unit cell is much smaller than wavelength, and therefore all the incremental lengths (e.g., period p) of the structure radiate toward the same direction, which leads to maximum directivity. In contrast, if a *nonuniform* bias voltage distribution is applied, as shown in Fig 6.13(b), the structure becomes nonuniform [$\beta = \beta(z)$]. Therefore, the different cells radiate toward different angles,

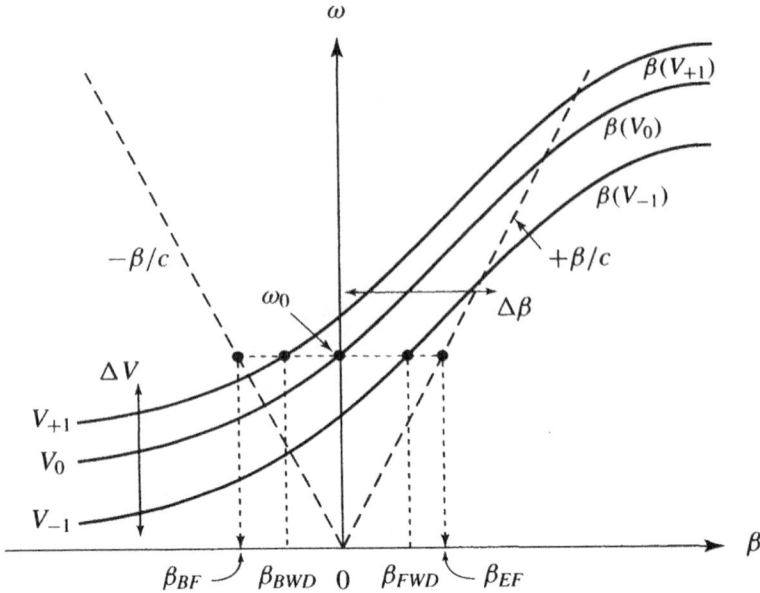

Fig. 6.12. Principle of the electronically scanned BE LW antenna. The operation frequency is fixed at f_0, where different values of β [Eq. (6.10)] and hence of θ_{MB} [Eq. (6.3)] are obtained for bias different voltages.

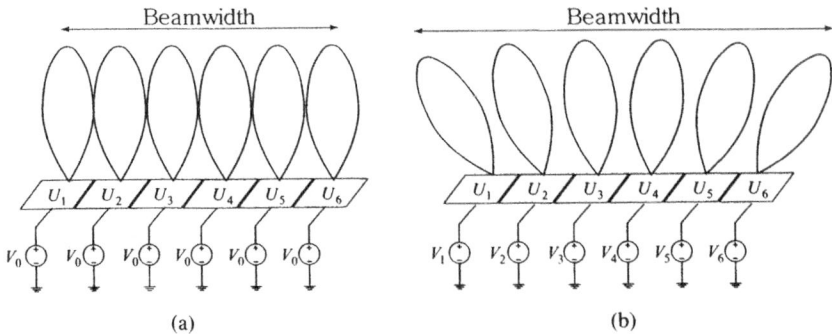

Fig. 6.13. Principle of electronic beamwidth control in BE LW antenna. (a) Uniform bias voltage distribution (maximum directivity). (b) Nonuniform bias voltage distribution (increased beamwidth). From [16], © IEEE; reprinted with permission.

and, as a result of superposition of all the radiation contributions, the beamwidth is increased. Thus, nonuniform biasing provides a mechanism for electronic control of the beamwidth of the antenna. This beamwidth control feature could be exploited to *equalize the gain-frequency characteristic* of the structure [Eq. (6.4) and Fig 6.9(b)]. For this purpose, nonuniform biasing would have to be applied at broadside (maximum gain under uniform biasing), and the biasing distribution would have to be progressively uniformized as the angle $|\theta_{MB}|$ increases toward backfire or endfire.

The array-factor approach described in Section 6.2 [Eq. 6.11)] can be conveniently used to analyze and optimize nonuniform LW antennas, since it can deal with an array of an arbitrary number of elements N which can be all different from each other. In the electronically scanned case, the angle θ_{MB} of the main beam is determined by Eq. (6.3) from the dispersion relation $\beta(V)$ to be established in Eq. (6.12).

6.3.3   Analysis of the Structure and Results

Fig 6.14 shows the unit cell layout and equivalent circuit for the electronically scanned CRLH BE LW antenna prototype of Fig 6.11. In this design, the unit cell contains two series and one shunt varactors in order to maintain a fairly constant characteristic impedance and to provide enough degrees of freedom to achieve a wide scanning range. Due to the head-to-head orientation of the three varactor diodes, with their anodes grounded through the shorted stubs, a relatively simple biasing circuit is possible and the nonlinear performances of the diodes are optimal. The resulting structure, due to the incorporation of the varactors and their bias circuits, is significantly different from the standard CRLH TL (Fig 3.18). However, it will be shown to provide similar performances

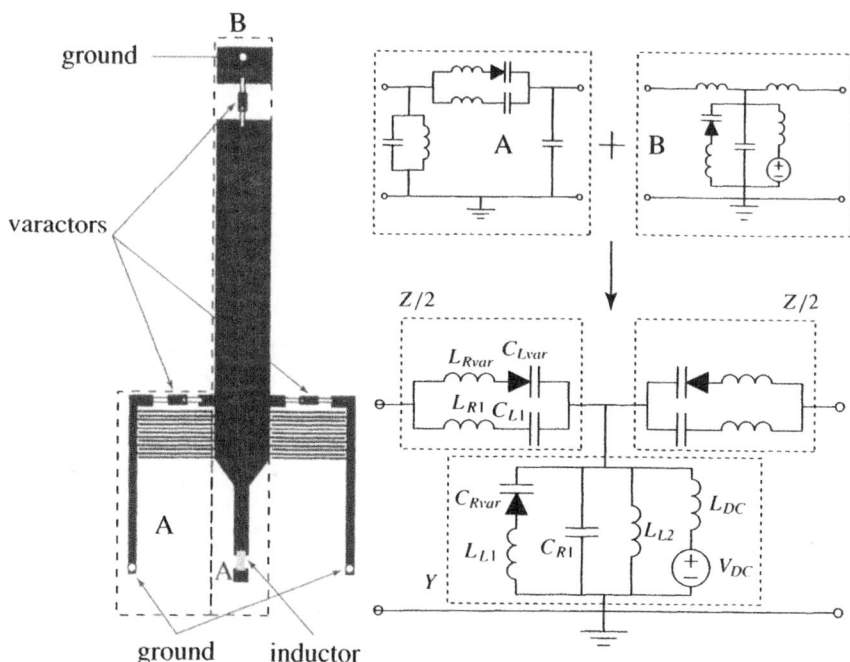

Fig. 6.14. Unit cell layout and equivalent circuit for the electronically scanned CRLH BE LW antenna prototype of Fig 6.11. From [16], © IEEE; reprinted with permission.

with continuous angle scanning from backward to forward angles including broadside.

The dispersion relation of the modified CRLH TL of Fig 6.14 can be computed by Eq. (3.115),

$$\beta(V) = \frac{1}{p} \cos^{-1}\left[1 + \frac{Z(V)Y(V)}{2} \right],\qquad(6.12)$$

with the modified immittances

$$
\begin{aligned}
Z(V) &= \left[j\omega L_{Rvar} + \frac{1}{j\omega C_{Lvar}(V)} \right] \parallel \left[j\omega L_{R1} + \frac{1}{j\omega C_{L1}} \right] \\
&= j\frac{1 - \omega^2 \left[L_{Rvar} C_{Lvar}(V) + L_{R1} C_{L1} \right] + \omega^4 L_{Rvar} C_{Lvar}(V) L_{R1} C_{L1}}{\omega^3 C_{Lvar}(V) C_{L1} \left[L_{Rvar} + L_{R1} \right] - \omega \left[C_{L1} + C_{Lvar}(V) \right]}
\end{aligned}
\qquad(6.13)
$$

and

$$
\begin{aligned}
Y(V) &= \left[\frac{1}{j\omega L_{L1}} \parallel j\omega C_{Rvar}(V) \right] + j\omega C_{R1} + \frac{1}{j\omega L_{L2}} \\
&= \frac{j\omega C_{Rvar}(V)}{1 - \omega^2 L_{L1} C_{Rvar}(V)} + j\omega C_{R1} + \frac{1}{j\omega L_{L2}}.
\end{aligned}
\qquad(6.14)
$$

The capacitance value (C_{Lvar} and C_{Rvar}) of the varactor diodes decreases exponentially as the reverse bias voltage increases. A careful examination of Eqs. (6.13) and (6.14) would reveal that, as a result, at relatively low voltages, both $Z(V)$ and $Y(V)$ decrease with increasing voltage for typical design parameters. Therefore, the magnitude of the product $Z(V)Y(V)$ in Eq. (6.12) decreases. Because this product is negative (otherwise, β would be complex and no propagation would occur), $\cos(\beta p) = 1 + Z(V)Y(V)$ increases toward 1 with increasing V, which implies that $|\beta|$ decreases toward 0. Consequently, from Eq. (6.3), the main beam angle will be steered toward broadside as the voltage is increased from low values. In addition, it can be verified that the slope of the relation $\theta_{MB}(V)$, $d\theta_{MB}(V)/dV = (\partial\beta/\partial V)/\sqrt{1 - (\beta/k_0)^2}$ is a negative quantity. It follows that the direction of scanning will be from forward to backward scanning as the voltage is increased. This variation is verified theoretically and experimentally at Fig 6.15, where an angle scanning range of $[+50°, -50°]$ is observed in a bias voltage range of $[0\ \text{V}, 22\ \text{V}]$. Good agreement is obtained between theory and measurement, the discrepancies being mostly due to the simplified varactor model used.

The measured insertion and return losses as a function the reverse bias voltage are shown in Fig 6.16. A reasonable and relatively flat matching curve ($S_{11} = -12$ dB to -8 dB) is obtained over the voltage range of 0 V to 22 V. The large

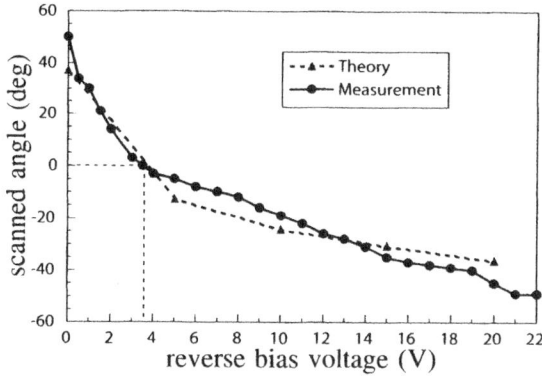

Fig. 6.15. Scanned angle versus reverse bias voltage for the electronically scanned CRLH BE LW antenna of Fig 6.11 ($f_0 = 3.3$ GHz). The theoretical curve is computed by Eq. (6.12) with LC parameters extracted from full-wave (MoM) simulation following the procedure of Section 3.3.3 at different voltages. From [16], © IEEE; reprinted with permission.

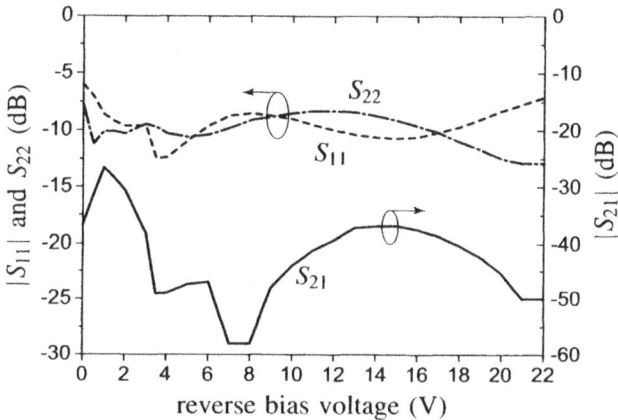

Fig. 6.16. Measured insertion and return loss for the antenna of Fig 6.11 as a function of the (uniform) bias voltage ($f_0 = 3.3$ GHz). From [16], © IEEE; reprinted with permission.

insertion loss is caused essentially by radiation leakage but also by resistive losses in the varactors.

Fig 6.17 shows the radiation patterns measured for different bias voltages, corresponding to the curve of Fig 6.15. As in any LW structure, the gain is maximum at broadside and decreases as one moves toward backfire or endfire. In this antenna, a gain of 18 dBi was obtained at broadside.

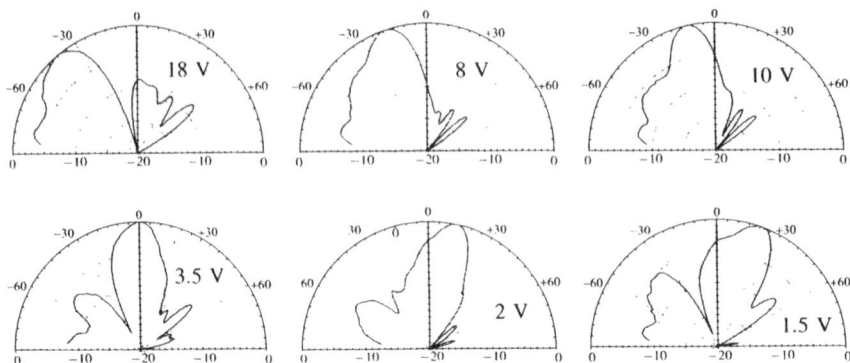

Fig. 6.17. Measured normalized radiation patterns (dB scale) for the electronically scanned CRLH BE LW antenna of Fig 6.11 for some (uniform) bias voltages (Fig 6.15) ($f_0 = 3.3$ GHz). From [16], © IEEE; reprinted with permission.

Finally, Fig 6.18 shows how nonuniform biasing can be used to modulate the beamwidth. In Fig 6.18(a), increased-beamwidth forward radiation is obtained with a nonuniform distribution of voltages extending from 0 V to 10 V. Although, as seen in Fig 6.18, the voltages above 3.5 V correspond to backward radiation in the case of uniform biasing, the net radiation angle is in the forward range because the influence of the first cells with lower voltages is higher than that of the last cells with higher voltages due to the exponential decay of the leaky wave. The last cells with higher voltages contribute to increase the beamwidth toward broadside. In Fig 6.18(b), increased-beamwidth backward radiation is obtained with a nonuniform distribution of voltages extending from 5 V to 15 V. In this case, all of the cells are biased at voltages corresponding to backward radiation in the uniform case (Fig 6.18), with the lower voltages increasing the beamwidth toward broadside and the higher voltages increasing the beamwidth toward backfire.

Additional information on this antenna regarding the effects of the varactors, such as parasitic resistive losses and intermodulation distortion, is provided in [16]. In particular, the effects of the harmonics generated by the nonlinearity of the varactor components used is shown to be sufficiently moderate to successfully recover a 10-Mbps BPSK modulated signal.

It should be noted that the nonuniformly biased LW CRLH antenna represents a nonuniform MTM structure, since this antenna is constituted of cells with different parameter values.

6.4 REFLECTO-DIRECTIVE SYSTEMS

The concept of CRLH backfire-to-endfire frequency/electronically scanned LW antennas can be extended to novel kinds of reflectors. This section presents three examples of such LW reflectors: a purely passive retrodirective reflector (Section 6.4.1), an arbitrary-angle frequency tuned reflector (Section 6.4.2), and an arbitrary-angle electronically-tuned reflector (Section 6.4.3).

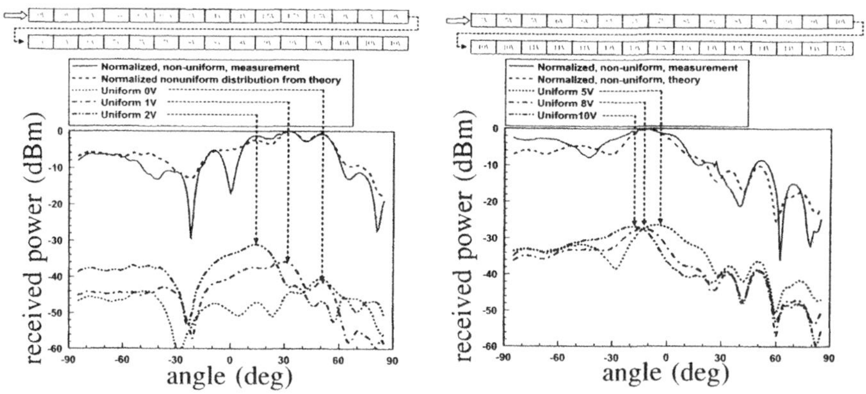

Fig. 6.18. Measured radiation patterns for the electronically scanned CRLH BE LW antenna of Fig 6.11 for some for nonuniform bias voltage distributions compared with the uniformly biased case patterns ($f_0 = 3.3$ GHz). The theoretical curves are computed by Eq. (6.11). (a) Voltage distribution range from 0 to 10 V (forward radiation). (b) Voltage distribution range from 5 to 15 V (backward radiation). The numbers in the boxes above the graphs indicate the applied reverse voltages in the 30 cells of the structure. This graph should be compared with Fig 6.15. From [16], © IEEE; reprinted with permission.

6.4.1 Passive Retro-Directive Reflector

Perhaps the simplest LW reflector that can be conceived is a purely passive retro-directive reflector terminated at one end by a short circuit and at the other end by a matched load, as depicted in Fig 6.19 [19]. When an incident wave hits the LW antenna reflector, under an angle θ, it transforms into a traveling wave, which propagates along the line toward the short-circuited end. The wave is then reflected back by the short circuit and propagates along the line in the opposite direction, producing a reflected beam at the same angle θ as the incident beam. The matched load at the other end absorbs the remaining power of the reflected wave, avoiding thereby a possible spurious beam at the opposite angle $-\theta$. This reflector is thus *retrodirective*, and may be considered as the LW counterpart of the passive Van Atta retrodirective array [20]. However, the CRLH LW reflector presents the advantage over the Van Atta reflector of providing retrodirectivity with a single element without requiring an array with complex and narrow-band interconnections.

In practice, a purely passive retrodirective reflector suffers the fundamental limitation of the reflected wave being difficult to discriminate from the transmitted wave because the two waves have the same frequency. In this case, time gating[14] is necessary at the receiver to isolate the reflected wave, which

---

[14]Time gating is a time domain filtering function available on some network analyzers. If the distance between the interrogator and the reflector is ℓ, the returned signal is "gated" at $\Delta t = 2\ell/c$ (time

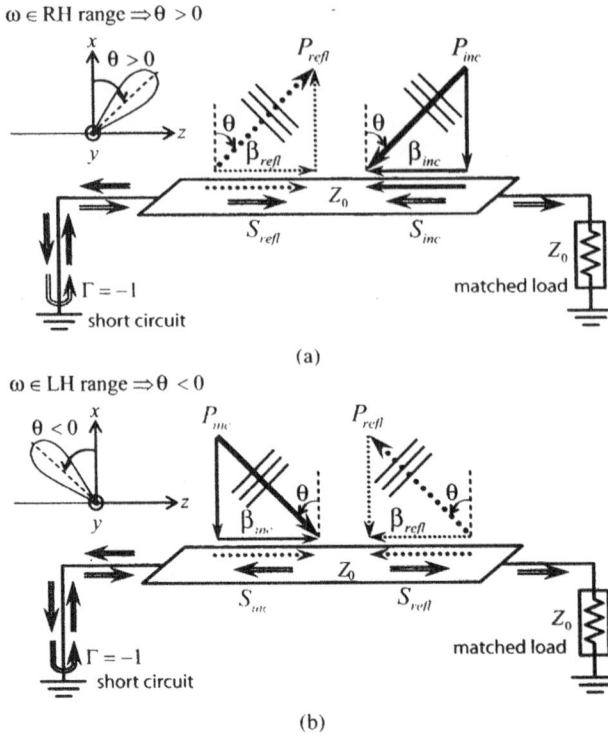

Fig. 6.19. Shorted/matched purely passive retro-directive CRLH LW reflector. The symbols β and S correspond to the propagation constant and Poynting vector, respectively. (a) RH range (forward operation). (b) LH range (backward operation).

is typically much weaker than the leakage signal between the transmitting and receiving test antennas of the interrogator in the case of monostatic operation (where the transmitting and receiving antennas are collocated) [4].

Fig 6.20 shows a prototype of the shorted/matched purely passive retro-directive CRLH LW reflector described in Fig 6.19, where several CRLH TLs have been etched in parallel on the substrate to obtain sufficient radar cross section. The monostatic radar cross section (RCS) measurements of this reflector are shown in Fig 6.21. Although RCS local peaks are observed at the expected angles, these results highlight another problem of purely passive reflectors (also shared by the Van Atta array): *specular reflection* ($\theta = 0$) produced by the ground plane of the overall structure dominates over retrodirective reflection. For this reason, the active LW reflectors of the next two subsection are of much more significant practical interest than the purely passive LW reflector described in this section.

taken by the wave for the round trip to and back from the reflector), so that the leakage signal form the transmitting and receiving antennas of the interrogator is filtered out and does not mask the signal of interest.

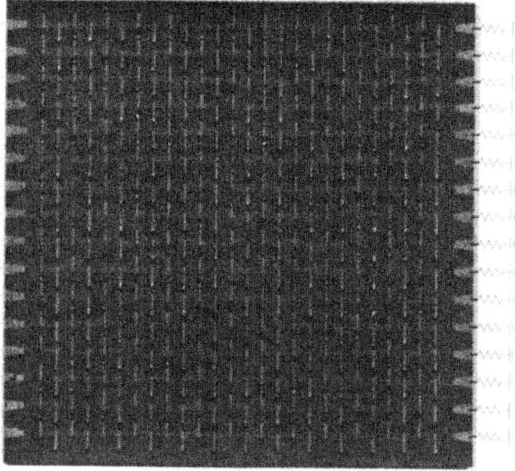

Fig. 6.20. Prototype of a shorted/matched purely passive retro-directive microstrip CRLH TL (Section 3.4) LW reflector [19] operating according to the principle described in Fig 6.19. Although only one line is functionally necessary, 16 parallel (independent) lines similar to that of Fig 3.43(a) with the angle-frequency relation of Fig 6.8 have been used to increase the radar cross section of the reflector. The lines are terminated by short circuits to the ground plane on the left-hand side and terminated by chip 50 Ω matched loads on the right-hand side.

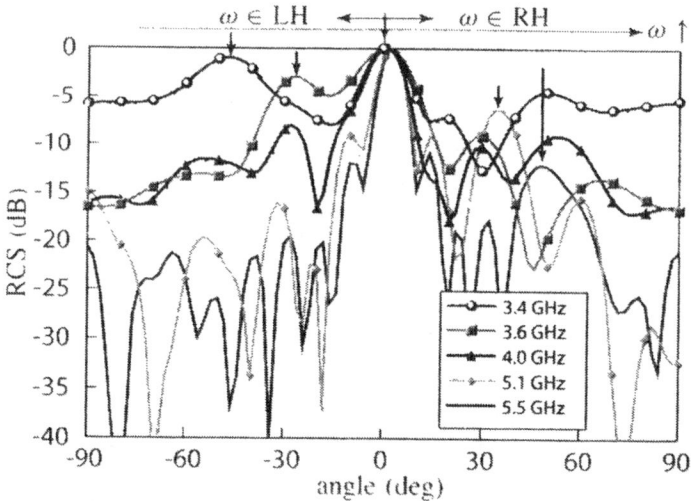

Fig. 6.21. Mono-static RCS measurements of the retrodirective reflector of Fig 6.20 (to be compared with the angle-frequency relation of Fig 6.8).

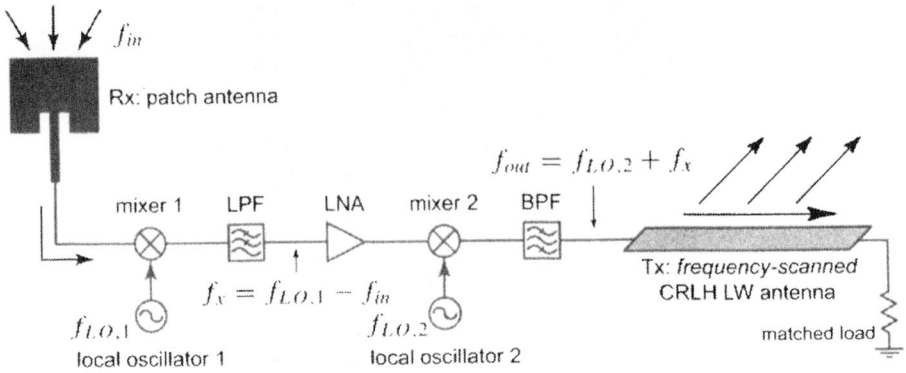

Fig. 6.22. Schematic of the arbitrary-angle frequency tuned CRLH LW reflector.

6.4.2 Arbitrary-Angle Frequency Tuned Reflector

We have seen that the CRLH retrodirective reflector of Section 6.4.1, although conceptually interesting, suffers two major drawbacks that limit its utilization in practical applications. First, monostatic detection of the reflected signal necessitates time gating to discriminate the signal reflected by the reflector from the leakage signal between the Tx and Rx antennas of the interrogator because the Tx and Rx frequencies are the equal, $f_{Tx} = f_{Rx}$; this is a serious limitation because the distance to the reflector must be known a priori to apply time gating. Second, even with appropriate time gating, the parasitic specular reflection can be stronger than the useful retrodirective reflection if the RCS of the antenna is too small, and this phenomenon cannot be avoided in a simple manner. For these reasons, a more elaborate reflector system was developed, which is the arbitrary-angle frequency tuned CRLH LW reflector [21] presented in this subsection.

The general idea of this reflector is to receive the incoming signal at a fixed frequency by a quasi-omnidirectional patch antenna (arbitrary input angle) and to retransmit or "reflect" this signal under *any* desired output angle with a CRLH BE LW antenna (Section 6.2). Because this antenna is a frequency-scanned device, the incoming signal has to be converted to the appropriate frequency according to the angle-frequency relation of the antenna (Fig 6.8). This new function is accomplished by *heterodyne mixing*.

The LW reflecto-directive reflector may be seen as a LW counterpart of the conventional heterodyne retrodirective array [22, 23]. However , the latter is only capable of returning the incident signal toward the source $\theta_{out} = \theta_{in}$, whereas the reflector presented here provides the additional functionality of returning the signal to any arbitrary angle (θ_{out} arbitrary, $\theta_{out} \neq \theta_{in}$ in general); for this reason, this reflector is not called retrodirective but *reflecto-directive*, since it "reflects" the incident signal to different desired directions. Moreover, this reflecto-directive system requires only one antenna element, whereas conventional retrodirective systems need an array of several radiating elements.

A schematic of the arbitrary-angle frequency tuned CRLH LW reflector is shown in Fig 6.22. The incoming signal of frequency f_{in} is first downconverted by a first mixer, with fixed frequency $f_{LO,1}$, to the intermediate frequency $f_x = f_{LO,1} - f_{in}$. The resulting signal is lowpass filtered and enhanced by an LNA and upconverted by a second mixer to the frequency range of the LW antenna,

$$f_{out} = f_{LO,2} + f_x = f_{LO,2} + f_{LO,1} - f_{in}, \tag{6.15}$$

where $f_{LO,2}$ is tuned to yield the frequency providing the desired output angle, according to angle-frequency law of the antenna (e.g., Fig 6.8). A bandpass filter centered at the output RF frequency f_{out} rejects unwanted frequencies before the Tx antenna. In this scheme, two mixers are required if f_{in} lies in the band of the LW antenna (as it is typically the case) because, if only one mixer were used, the input signal could not be filtered out and would therefore generate a spurious angle corresponding to the frequency f_{in} (except in the case where the desired output angle corresponds to the frequency equal to the input frequency, $f_{out} = f_{in}$). This problem is solved by the proposed two-mixers circuit, where f_{in} is suppressed by the lowpass filter before the second mixer. According to Eq. (6.15), the tuning frequency $f_{LO,2}$ is controlled according to the formula

$$f_{LO,2} = f_{out} + f_{in} - f_{LO,1}, \tag{6.16}$$

where f_{in} and $f_{LO,1}$ are fixed, and f_{out} is dictated by the angle-frequency law of the CRLH LW antenna.

Fig 6.23 shows a prototype of the arbitrary-angle frequency tuned CRLH LW reflector. The input frequency is fixed ($f_{in} = 5.0$ GHz). The intermediate frequency ($f_x = 1.0$ GHz) is chosen low enough to allow efficient amplification by way of LNA and far enough from the range of the output frequencies ($f_{out} = 3.2 \ldots 6.0$ GHz) so that the signal at the input frequency is suppressed by way of lowpass filtering for all angles θ_{out}.

The experimental results for this reflecto-directive system are shown in Fig 6.24. The source test antenna sends a signal at the frequency 5.0 GHz at three different fixed angles: $\theta_{in} = -30°$, $0°$, and $+30°$. For each of these incidence angles, the bistatic RCS [24] of the system is measured at three different output frequencies: $f_{out} = 3.5$ GHz, 4.5 GHz, and 5.5 GHz, corresponding to $f_{LO,2} = 2.5$ GHz, 3.5 GHz, and 4.5 GHz, respectively, according to Eq. (6.16). For all the incidence angles tested, clear RCS peaks appear at the angles $\theta_{out} = -20°$, $+20°$, and $+50°$, for the frequencies $f_{out} = 3.5$ GHz, 4.5 GHz, and 5.5 GHz, respectively, in agreement with the LW antenna angle-frequency relation of Fig 6.8.

6.4.3 Arbitrary-Angle Electronically Tuned Reflector

As in the BE LW antenna, fixed frequency operation in the LW reflector may be desired in practical applications for effective channelization, whereas frequency tuning is required in the reflecto-directive system of Section 6.4.2 to steer the reflection angle θ_{out}. The frequency-scanned LW antenna (Section 6.2) utilized

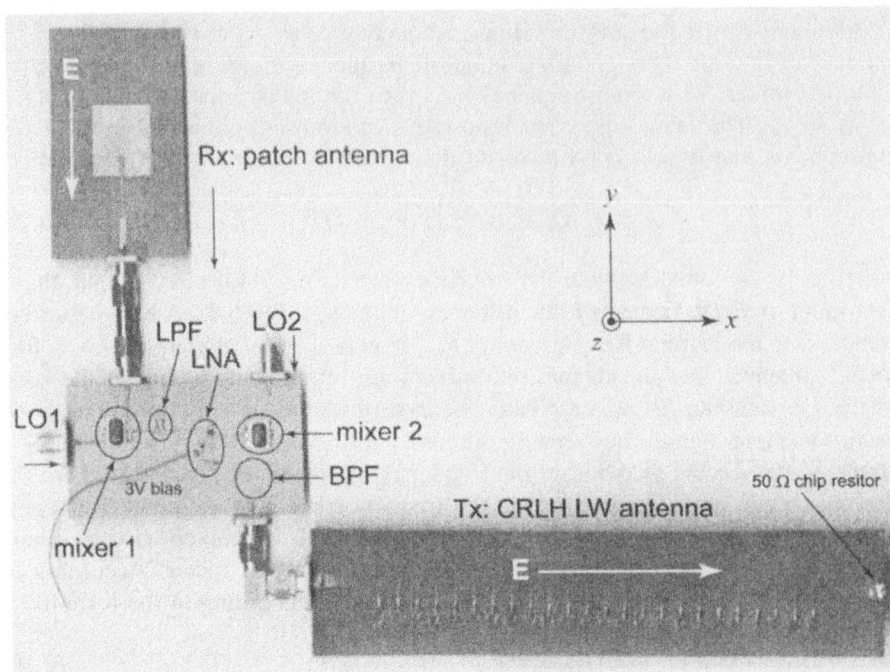

Fig. 6.23. Prototype of the arbitrary-angle frequency tuned CRLH LW reflector of Fig 6.22 including the CRLH LW antenna of Fig 3.43(a) with the angle-frequency relation of Fig 6.8. The patch antenna operates at $f_{in} = 5.0$ GHz and the intermediate frequency is chosen to be $f_x = 1.0$ GHz. The polarization of the Rx and Tx antennas are along the y and z directions, respectively.

in the reflector can then be replaced by its electronically scanned counterpart (Section 6.3), so that the reflection angle is controlled electronically by the bias voltage of integrated varactors at a fixed frequency. In fact, the resulting system is even simpler than that of the frequency tuned reflector, because heterodyning is not required any more since both the Rx and Tx antennas are independent of frequency. The electronically tuned reflector is thus simpler and cheaper for integration into wireless packages and does not consume any LO power.

The schematic of the proposed system is shown in Fig 6.25. It consists of four components: a receiving (Rx) antenna, a low noise amplifier (LNA), a band-pass filter (BPF), and a transmitting (Tx) antenna. A patch antenna is used for Rx because its quasi-omnidirectionality allows detection of the incoming signal from any arbitrary angle at a fixed frequency. An LNA is employed to compensate for free space and passive device attenuation. The undesired harmonics from the LNA are suppressed by the BPF. The electronically scanned LWA is used for Tx. Therefore, the output frequency f_{out} can be kept constant and chosen to be identical to the input frequency f_{in}. Since the scanned angle of this antenna is a function of the bias voltage, according to Eqs. (6.3) and (6.12), the desired reflected angle can be obtained by electronic tuning at a fixed frequency.

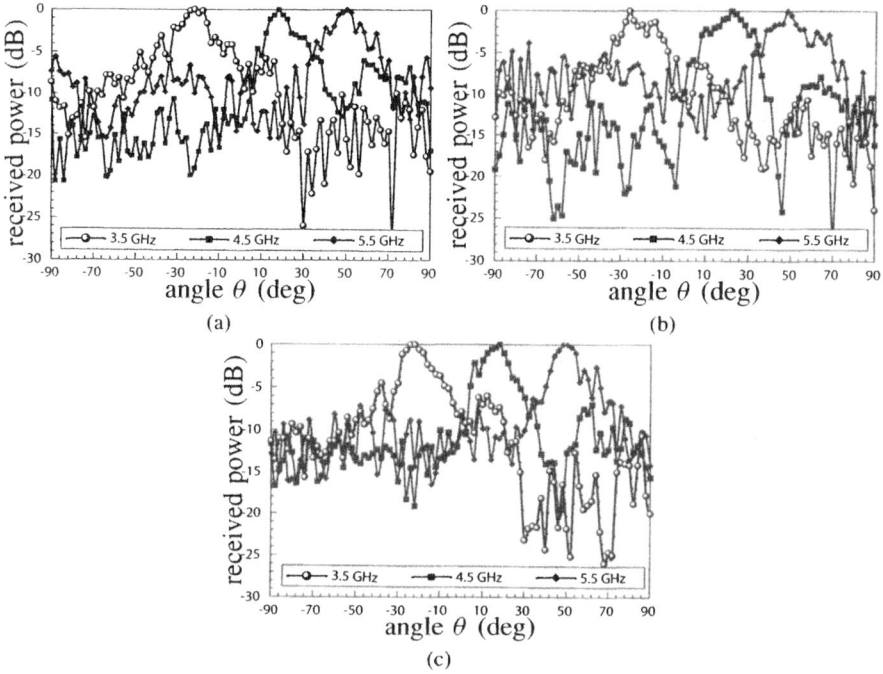

Fig. 6.24. Measured bistatic (normalized) RCS of the arbitrary-angle frequency tuned CRLH LW reflector of Fig 6.23 for the input frequency $f_{in} = 5$ GHz for different positions of the source. (a) $\theta_{in} = -30°$. (b) $\theta_{in} = 0°$. (c) $\theta_{in} = +30°$. Each graph shows the normalized RCS for three different output frequencies: $f_{out} = 3.5, 4.5,$ and 5.5 GHz corresponding to the output angles $\theta_{out} = 3.5°, 4.5°,$ and $5.5°$ according to Fig 6.8. From [21], © IEEE; reprinted with permission.

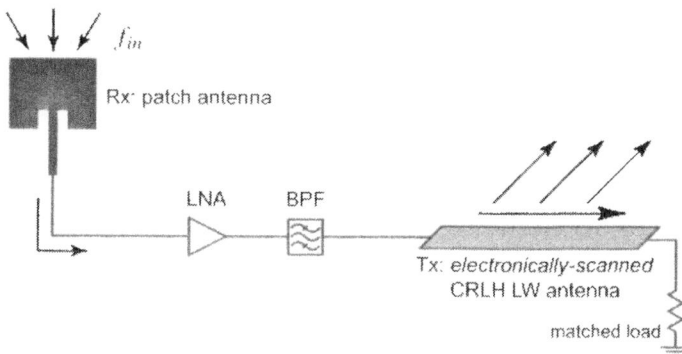

Fig. 6.25. Schematic of the arbitrary-angle electronically tuned CRLH LW reflector.

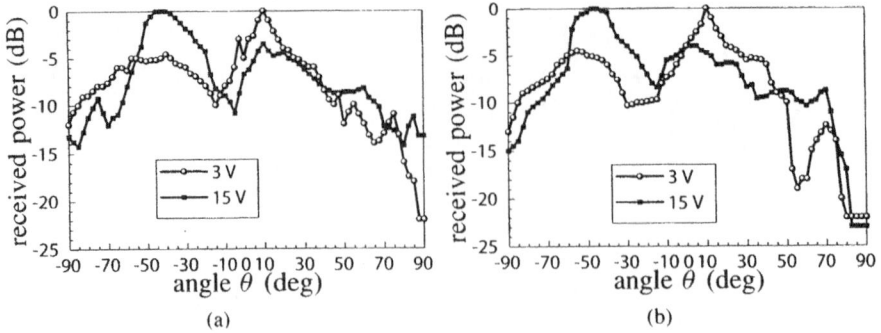

Fig. 6.26. Measured bistatic (normalized) RCS of the arbitrary-angle electronically tuned CRLH LW reflector corresponding to Fig 6.25 with the LW antenna of Fig 6.11 (with angle-voltage law shown in Fig 6.15) at the fixed input/output frequency $f_{in} = f_{out} = 3.3$ GHz for two arbitrary source positions. (a) $\theta_{in} = 0°$. (b) $\theta_{in} = -20°$. For both source positions, the reflected angles of $\theta_{out} = 10°$ and $-40°$ are detected at the reverse bias voltages of 3 V and 15 V, respectively, in agreement with Fig 6.15.

Fig 6.26 presents experimental results for the arbitrary-angle electronically tuned CRLH LW reflector, utilizing the electronically scanned LW antenna of Fig 6.11 with the angle-voltage law shown in Fig 6.15. Because detection is difficult when the transmitting and receiving frequencies are the same, due to mutual coupling between the Tx and Tx test antennas, a frequency converter slightly changing the output frequency was used in this measurement. The expected reflection angles corresponding to the angle-voltage law of the electronically scanned LW antenna are detected.

The backfire-to-endfire capability of the MTM CRLH LW antenna thus enables very novel and versatile reflector systems. The electronically tuned CRLH LW reflector presented in this subsection is particularly remarkable because, with the requirement of only one LNA, one BPF, and two antennas (Rx/Tx) easily integrable on single substrate/package, it provides more functionality (arbitrary angle) than complex and expensive conventional retrodirective arrays.

6.5 TWO-DIMENSIONAL STRUCTURES

As shown in Chapter 4, 1D MTM structures (Chapter 3) can be straightforwardly extended to 2D MTM structures due to their effective-homogeneity. It is therefore natural to wonder what benefits 2D MTM may possibly offer in terms of radiation properties for antenna and reflector applications. In particular, can 2D MTMs or metasurfaces operate as 2D LW antennas, and, if this is the case, what functionalities can they provide in addition to those of 1D structures? It is the purpose of this section to address these questions.

6.5.1 Two-Dimensional LW Radiation

Possible 2D CRLH MTM structures have been described in Section 4.5. These structures, and their derivatives, are essentially mushroom-type structures

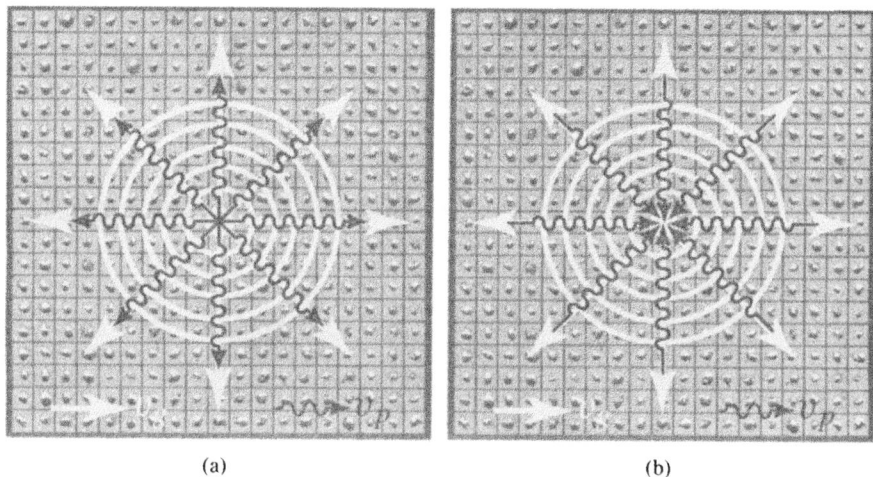

(a)   (b)

Fig. 6.27. Top view of a 19 × 19-cell (square lattice) open mushroom structure prototype (see Figs. 4.30, 4.32, and 4.34) with illustration of an homogeneous/isotropic (circular) wave launched by a source placed at the center. The substrate used is Rogers RT/duroid 5880 with dielectric constant $\varepsilon_r = 2.2$ and thickness $h = 50$ mil, and the unit cell size is 5 mm. The extracted CRLH parameters are $L_R = 0.66$ nH, $C_R = 0.44$ pF, $L_L = 1.66$ nH, and $C_L = 0.09$ pF. (a) RH range $(v_p \| v_g)$. (b) LH range $(v_p - \| v_g)$.

(Figs. 4.30 or 4.31), a prototype of which is shown in Fig 6.27, and are either *open* [Figs. 4.32 and 4.34(b)] like microstrip structures in 1D or *closed* [Fig 4.34(c)] like strip-line structures in 1D. Obviously, open structures are required to produce free space radiation. However, closed structures including radiating slots can also be envisioned. As pointed out in Section 4.5.3, the open 2D MTM structure generally couples to the TM air mode [Fig 4.35(a)], which results in a dominant mixed RH-LH mode, whereas the closed structure supports a purely LH dominant mode.

The CRLH parameters (L_R, C_R, L_L, C_L) of 2D MTM structures can be extracted from full-wave analysis of the unit cell by using the 2D Bloch impedance concept following the method developed in Section 4.5.3. Once these parameters are determined, the TL network approaches of Section 4.1, 4.2, and 4.3 can be used to determine in a fast and insightful manner the characteristics of the metastructure: the dispersion diagram is computed by the relation (4.10), solution of the general 2D LC network eigenvalue problem (Section 4.1), whereas the propagation characteristics and edges matching conditions of a practical finite-size structure excited by an arbitrary[15] source can be analyzed by driven-solution circuit techniques such as the TMM presented in Section 4.2 or the TLM presented in Section 4.3.

Fig 6.28 shows the dispersion diagram and typical LW fields distributions for the open mushroom structure of Fig 6.27. The dominant mixed RH-LH mode is

---

[15]The source may be connected to any cell of the structure and may be either single or multiple.

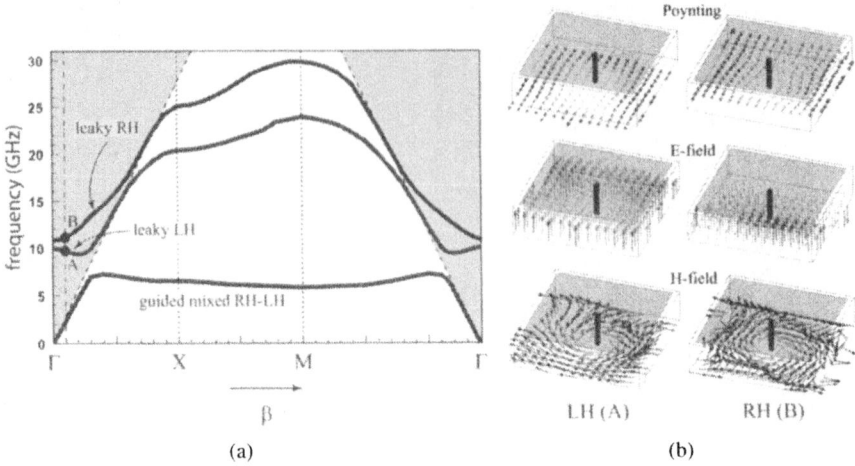

Fig. 6.28. Modal characteristics for the open mushroom structure of Fig 6.27 obtained by full-wave (FEM) analysis. (a) Dispersion diagram. The shaded areas correspond to the radiation cone or leaky-wave region (Fig 6.2). (b) Field distribution of the leaky LH and RH modes at the points A and B, respectively, of the dispersion diagram, corresponding to the spectral value $\beta = (0.1\pi/p)\,\hat{x}$.

purely guided (slow-wave) and can therefore not be used for radiation purposes. Instead, the next two modes are utilized. In fact, the LW LH mode corresponds (similar fields distribution) to the extension of the LH slow-wave part of the dominant mode, which couples to the air mode $k_0 = \beta$ when the structure is open.[16] The field distributions verify the quasi-TEM nature of the LW modes, also observed in the corresponding modes of the closed mushroom structure (Fig 4.36). In addition, the antiparallelism (at the same phase instant) between the Poynting vectors of the LH and RH modes confirms the opposite "handedness" of these modes.

Fig 6.29 shows surface and isofrequency contours representations of the LW LH and RH modes and verifies their quasi-isotropy in the complete LW region. Since the radiation cone has its symmetry axis at the spectral origin ($\beta = 0$) a high degree of isotropy is easy to obtain in CRLH LW antennas.

6.5.2 Conical-Beam Antenna

If a 2D MTM CRLH structure is *excited in its center* in a frequency range close to the Γ point of the BZ, where fast-wave regime prevails and effective-homogeneity is ensured, an homogeneous/isotropic circular [$\beta \neq \beta(\varphi)$] leaky wave is generated, as suggested in Fig 6.27. Fig 6.29(a) shows the phase fronts and power directions when the structure is excited in its RH range, while Fig 6.29(b) shows the phase fronts and power directions when the structure is excited in its LH range. In the latter case, the propagation constant is radially directed inward

---

[16]In the closed structure (Fig 4.33), such coupling does not exist and the negative gradient LW curve continuously extends to the slow-wave region along the dominant LH mode, as shown in Fig 4.35(b).

(a)

(b)

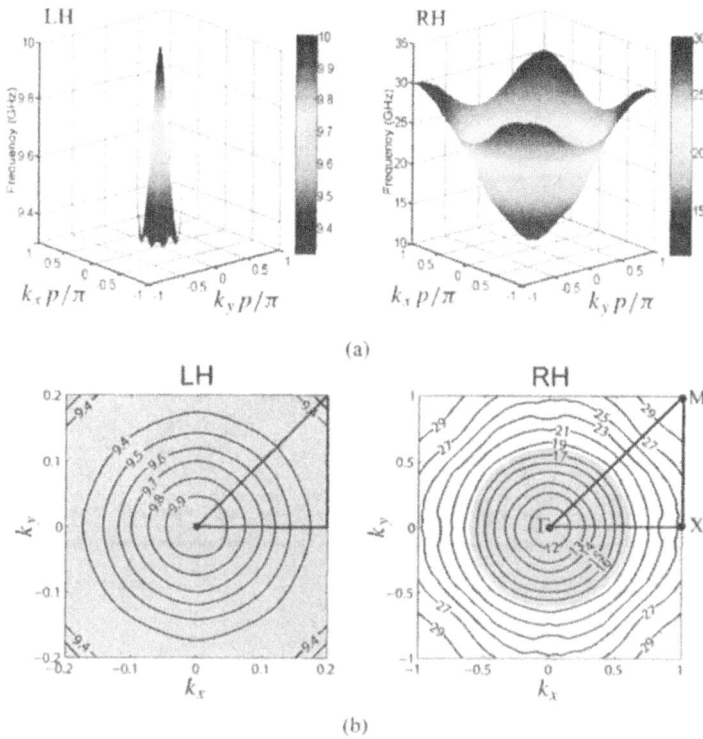

Fig. 6.29. Leaky LH and RH modes of Fig 6.28 represented over the full BZ domain, $(k_x, k_y) = [-\pi/p, +\pi/p, -\pi/p, +\pi/p]$ with $\beta = \sqrt{k_x^2 + k_y^2}$. (a) Mode surfaces $f_i(k_x, k_y)$ ($i = $ LH,RH). (b) Isofrequency contour curves $f_i(k_x, k_y) = $ const ($i = $ LH,RH). The shaded circular regions correspond to the LW region.

toward the source with the opposite direction to that of the Poynting vector.[17] As a consequence of the circular and fast wave propagation from the center of a structure, a *conical beam* is radiated, where maximum power is obtained along a circle when the antenna illuminates a plane surface, as illustrated in Fig 6.30.[18]

Because of the 2D nature of the LW, the same type of conical beam is obtained in LH and in the RH range. This is illustrated in Fig 6.30(b), where it is seen that for a given azimuthal angle of the main beam, φ_{MB}, the radiated power in $(\varphi_{MB}, \theta_{MB})$ originates from the outgoing wave at φ_{MB} in the RH range and from

[17]In a time domain animation, the wavefronts are seen to propagate from large distances to the center and collapse onto the source. This is the unusual effect that we would observe if we would throw a pebble in a pond of LH water!

[18]If the structure is excited at frequencies relatively far from the spectral origin (Γ point), its isofrequency curves degenerate into distorted circles and eventually into discontinuous curves, as seen in Figs. 4.5 or 4.6. In the former case, the conical beam degenerates into an anisotropic cone with a cross section similar to that of the isofrequency curve; in the latter case, radiation is restricted to a limited range of angles around the diagonal direction of the structure.

(a)

(b)

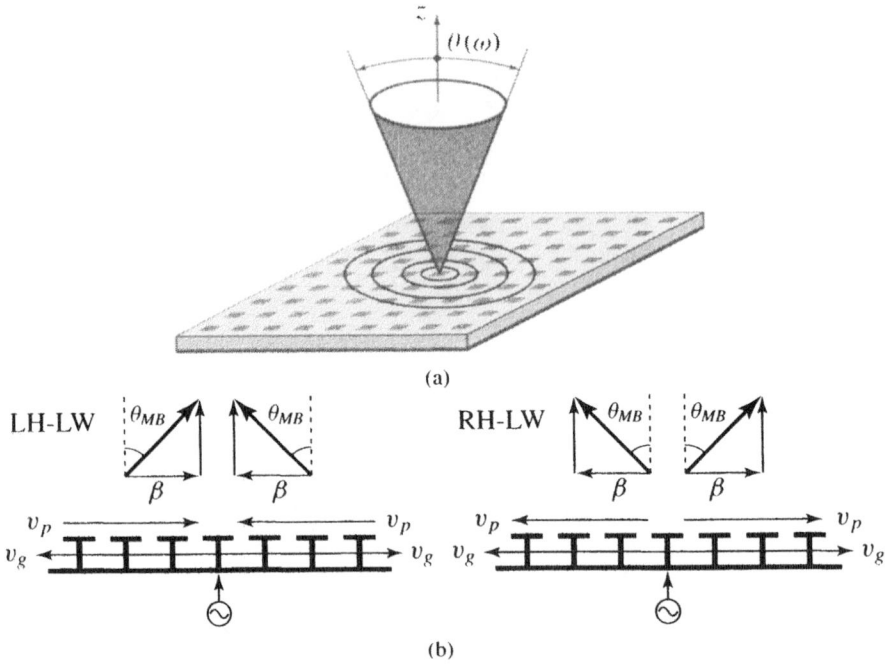

Fig. 6.30. Principle of a conical-beam 2D LW antenna. (a) Perspective view illustration. (b) Radiation phenomenon in the LH and RH ranges. A conical beam is obtained in both the LH and in the RH ranges; as frequency is increased, the solid angle of the cone decreases in the LH range and increases in the RH range.

the incoming wave at $\varphi_{MB} + \pi$ in the LH range. But, because a conical beam is observed in both the LH and RH ranges, these two ranges can be distinguished only by the *variation of the scanned angle in frequency*. It can be inferred form the dispersion diagram [Fig 6.28(a)] and the scanning angle law Eq. (6.3) that as frequency is increased the cone closes up (θ_{MB} decreases) in the LH range while it opens up (θ_{MB} increases) in the RH range.

The experimental radiation patterns and scanning angle versus frequency law are shown in Figs. 6.31 and 6.32, respectively, and are seen to confirm the theoretical predictions. For this antenna, the maximum and minimum gains are 17.3 dB at 10.1 GHz and 12.4 dB at 11.0 GHz, respectively. These values are extremely high considering that the radiated power is distributed circularly along the cone.

In the angle-frequency law shown in Fig 6.32, the theoretical curve is determined by the array factor approach [5] in 2D, corresponding to the formula

$$R(\theta, \varphi) = AF(\theta, \varphi) = \left[\sum_{m=1}^{M} I_{m1} e^{j(m-1)(k_0 p_x \sin\theta \cos\varphi + \xi_{xm})} \right]$$
$$\cdot \left[\sum_{n=1}^{N} I_{1n} e^{j(n-1)(k_0 p_y \sin\theta \sin\varphi + \xi_{ym})} \right], \qquad (6.17)$$

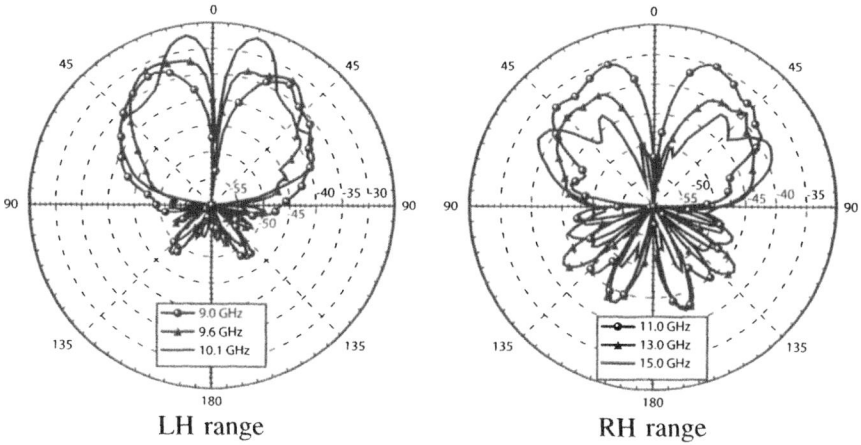

LH range RH range

Fig. 6.31. Measured radiation patterns for the antenna of Fig 6.27 corresponding to the dispersion diagram of Fig 6.28(a). As frequency is increased, the cone closes up and opens up in the LH and RH ranges, respectively.

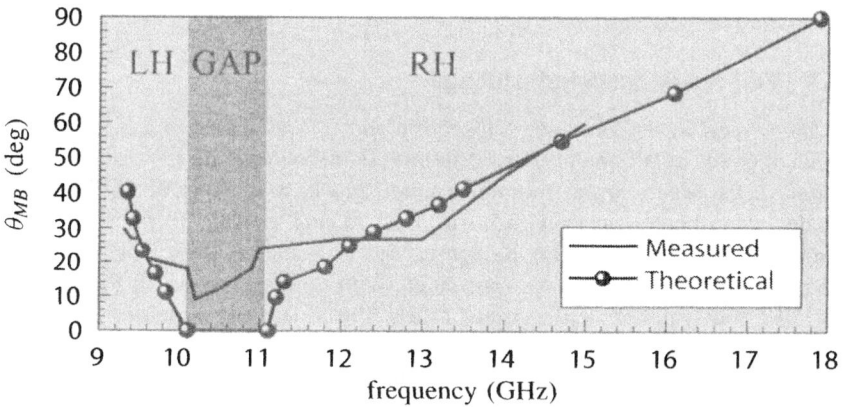

Fig. 6.32. Theoretical and measured scanning angle versus frequency law for the antenna of Fig 6.27.

with

$$\xi_{xm} = -(m-1)k_0 p_x \sin\theta_{MB} \cos\varphi_{MB} \quad \text{and} \quad I_m = I_0 e^{-\alpha_x(m-1)p_x}, \quad (6.18\text{a})$$

$$\xi_{yn} = -(n-1)k_0 p_y \sin\theta_{MB} \sin\varphi_{MB} \quad \text{and} \quad I_n = I_0 e^{-\alpha_y(n-1)p_y}, \quad (6.18\text{b})$$

where p_x and p_y are the periods, M and N are the number of cells, and α_x and α_y are the leakage factors along the x and y directions, respectively. In the prototype of Fig 6.27, we have $p_x = p_y = 5$ mm, $M = N = 19$, $\alpha_x = \alpha_y = \alpha$. The direction $(\theta_{MB}, \varphi_{MB})$ of the main beam is determined with this formula once

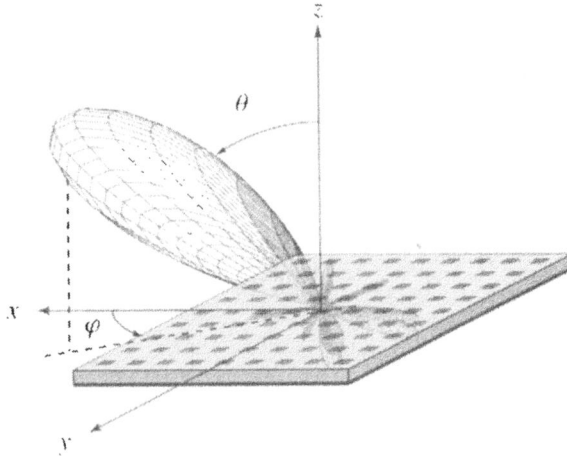

Fig. 6.33. Illustration of a full-space scanning meta-surface.

β [yielding θ_{MB} by Eq. (6.3)] and α (yielding I_{m1}, I_{1n}) have been computed, the angle φ_{MB} being set here to any arbitrary value due to azimuthal symmetry.

6.5.3   Full-Space Scanning Antenna

We have seen in the previous subsection that a LW metasurface excited in its center operates as a conical-beam antenna. If in contrast the LW meta-surface is *excited at its edges*, it can radiate a *pencil-beam*, as a 1D LW MTM structure radiates a fan-beam (Sections 6.2 and 6.3). In this manner, full-space scanning, as suggested in Fig 6.33, may be achieved. For instance, if the surface is square and is excited along one of its four sides, with constant phase all the way along this side, the structure will operate much like a 1D LW structure (with higher RCS) and will exhibit backfire-to-endfire scanning capability in the plane considered. This means that elevation scanning $\theta = \theta(\omega)$[19] is obtained at a fixed azimuthal angle $\varphi = \varphi_0$. Because the structure is azimuthally isotropic and effectively homogenenous, it may be shaped in an arbitrary manner, as for instance cut out into the hexagonal shape shown in Fig 6.34, allowing switched azimuthal scanning along with frequency or electronic elevation scanning.

Although some important issues, such as development of an efficient feeding structure and out-of-angle leakage of the leaky-waves, are still to be resolved, it seems that this antenna structure presents an interesting potential for full-space scanning applications, where it might represent a simple and inexpensive alternative to conventional phased arrays.

---

[19]Electronic scanning $\theta = \theta(V)$ utilizing varactors is also possible in a 2D surface [25]. Another even more promising option would consist of the use of a distributed ferroelectric thin film below the structure to steer the angle by capacitance modulation; a simpler fabrication process, a simplified bias scheme, higher frequency capability, and larger power handling would result in this approach.

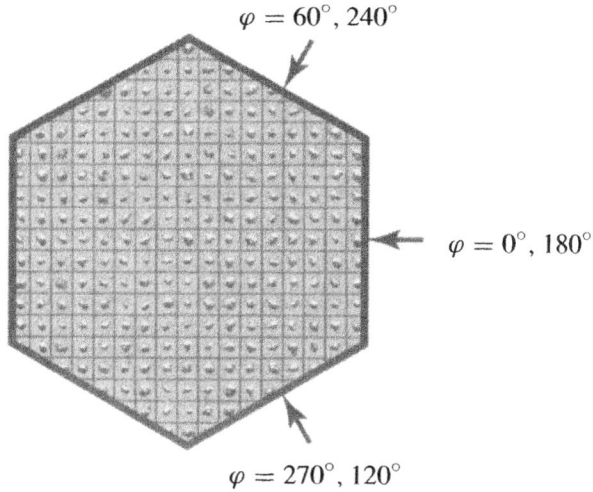

Fig. 6.34. Example of full-space scanning CRLH LW meta-surface using frequency (or electronic) scanning for elevation, $\theta(\omega)$ [or $\theta(V)$], and switched scanning for azimuth. In this hexagonally shaped structure, an azimuthal resolution of $\Delta\varphi = 60°$ is obtained.

Realizing a balanced design to close up gap between the LH to RH range, thereby allowing broadside LW radiation, represents a more difficult task in a 2D than in a 1D structure. A special balancing technique, using two TLs of different sections as the inductances L_R between MIM-type capacitors C_L, is demonstrated in [26].

6.6 ZEROTH ORDER RESONATING ANTENNA

In the previous sections of this chapter, MTM radiation was based on a *leaky-wave mechanism*, where the operating wave is a (fast) *traveling wave* progressively leaking out energy along the structure to produce a beam in the direction of phase coherence. However, since a MTM structure is effectively homogeneous, or uniform,[20] it can also be open-ended or short-ended, as any uniform structure, to produce radiation based on a *resonance mechanism*, where the operating wave is a *standing wave* exhibiting constant-phase field distribution along the structure, which leads to broadside radiation, and constructive field interference above the structure to allow this radiation. This section and the next section present examples of resonant-type MTM antennas, showing how MTM concepts can be beneficial to resonant antennas and provide unique characteristics not shared by conventional structures. It is important to realize that in these antennas what resonates is the overall structure, due to open/short boundary conditions and not its structural elements or unit cells.

---

[20]Electromagnetically, it is a uniform transmission line, surface or volume, in 1D, 2D, and 3D, respectively.

The MTM antenna considered in this section is a *zeroth-order resonating antenna*, based on the zeroth-order resonator presented in Section 5.5 [27]. This resonator was shown to be essentially a CRLH TL structure with short or open coupling terminations operated at an arbitrarily-designed transition frequency ω_0, where constant-phase and *constant-magnitude* field distribution are achieved along the structure (Fig 5.48), which corresponds to infinite propagation constant and therefore zeroth-order mode ($m = 0$) according to the resonance condition of Eq. (5.61b), $\ell = |m|\lambda/2$. Due to the unique nature of this zeroth-order mode, the size of the resonator does not depend on its physical length but only on the amount of reactance provided by its unit cells. In terms of antenna, this represents a potential interest for miniaturization. Reduction in size of an antenna, by whatever technique, necessarily results in the decrease of its directivity D and thereby generally of its gain G, assuming a constant efficiency factor $k = G/D$ [4]. A potential advantage of the zeroth-order antenna lies in its constant-magnitude field distribution. The corresponding perfectly even repartition of energy along the structure leads to smaller ohmic losses than in other small antennas where high current concentrations near discontinuities produce high losses. Consequently, higher efficiency factor k can be obtained in the zeroth-order antenna, which may mitigate, to some extent, the reduction in gain subsequent to size reduction. Another zeroth-order effect helping to achieve an optimal gain for a given physical size of the antenna is the fact that the effective aperture [24, 4] is increased and becomes larger than that of a typical antenna resonating in its first mode. Comparative gain analysis with conventional small antennas is not provided here, but the operation principle and experimental demonstration of the zeroth-order resonating antenna are presented as a starting point for further developments.

Any CRLH architecture can be used to implement this antenna. As a microstrip alternative to the structure of Figs. 3.39 and 3.42, the *via-less* configuration shown in Fig 6.35 [28] is used here. Such a via-less configuration may be advantageous in processes where implementation of vias is problematic.

The direct connection of the inductor L_L to the ground is replaced in this structure by capacitive coupling through the *virtual-ground capacitance* C_g, which is a high-value capacitance obtained by a large patch area. Consequently, the LH-inductive branch of the standard CRLH model [Fig 3.14(a)] is transformed into the impedance $Z_{L_C + C_g} = j\omega L_L - j/(\omega C_g)$. If C_g is sufficiently large, we have $Z_{L_C + C_g} \approx j\omega L_L$ and the structure becomes essentially equivalent to the standard CRLH prototype of Fig 3.14(a). Rigorously, the dispersion relation of Eq. (3.116b) is transformed into

$$\beta = \frac{1}{p}\cos^{-1}\left\{1 - \frac{1}{2}\left[\frac{\omega_L^2}{\omega^2}\zeta + \frac{\omega^2}{\omega_R^2} - \left(\frac{\omega_{sh}^2}{\omega_R^2}\zeta + \frac{\omega_{se}^2}{\omega_R^2}\right)\right]\right\}, \qquad (6.19a)$$

with

$$\zeta = \frac{\omega^2}{\omega^2 - \omega_g^2}, \qquad (6.19b)$$

interdigital capacitors

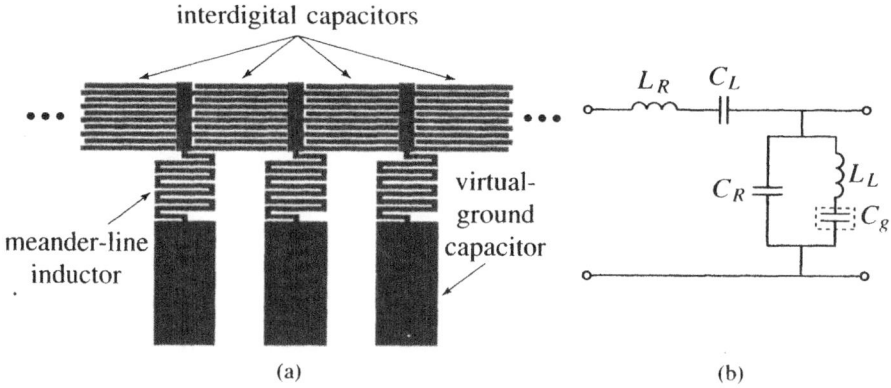

(a)   (b)

Fig. 6.35. Vialess microstrip implementation of the CRLH zeroth resonator, where the shunt inductance L_L is provided through the virtual-ground (very large) capacitance C_g: $Z_{L_C+C_g} = j\omega L_L + 1/(j\omega C_g) \approx j\omega L_L$. (a) Layout of the TL resonator. (b) Equivalent circuit of the unit cell.

where $\omega_L = 1/\sqrt{L_L C_L}$, $\omega_R = 1/\sqrt{L_R C_R}$, $\omega_{se} = 1/\sqrt{L_R C_L}$, $\omega_{sh} = 1/\sqrt{L_L C_R}$, and $\omega_g = 1/\sqrt{L_L C_g}$. As a consequence of the introduction of the capacitance C_g, the open-ended resonance of Eq. (5.65) is modified from ω_{sh} to

$$\omega_{res}^{open} = \sqrt{\omega_{sh}^2 + \omega_g^2}, \qquad (6.20)$$

while the short-ended resonance of Eq. (5.67) remains unchanged. Fig 6.36 shows dispersion diagrams of this structure. It can be seen in Fig 6.36(a) that the typical low-frequency LH and high-frequency RH characteristics of CRLH TLs are achieved for $C_g > C_{g,lim} = 1/(\omega^2 L_L)$, while Fig 6.36(b) illustrates how left-handedness in the first band is lost when this condition is not satisfied. At the limit $C_g = C_{g,lim}$, the TL is short circuited to the ground ($Z_{L_L+C_g} = 0$); no propagation occurs and a flat curve is observed in the dispersion diagram.

The resonance characteristics of a $N = 4$-cell implementation of the zeroth-order resonator structure of Fig 6.35 are shown in Fig 6.37. As explained in Section 5.5.2, $2N - 1 = 7$ resonances are obtained (i.e., here $n = 0, \pm1, \pm2, \pm3$). Since the structure is open-ended [see inset of Fig 6.37(b)], zeroth-order resonance occurs at ω_{res}^{open}, given by Eq. (6.20).

Fig 6.38 compares the size of the zeroth-order antenna with that of a conventional patch antenna of identical resonance frequency. About 75% footprint reduction is achieve in comparison with that of the $\lambda/2$ patch antenna. A much more drastic size reduction can be obtained by using vias instead of virtual-ground capacitors.

Radiation patterns of the antenna are shown in Fig 6.39. It clearly appears in Fig 6.39(b) that directivity increases as the size of the structure is increased, without any change in frequency. The angle offset in the 30-cell prototype is believed to be due to an imperfect feeding structure.

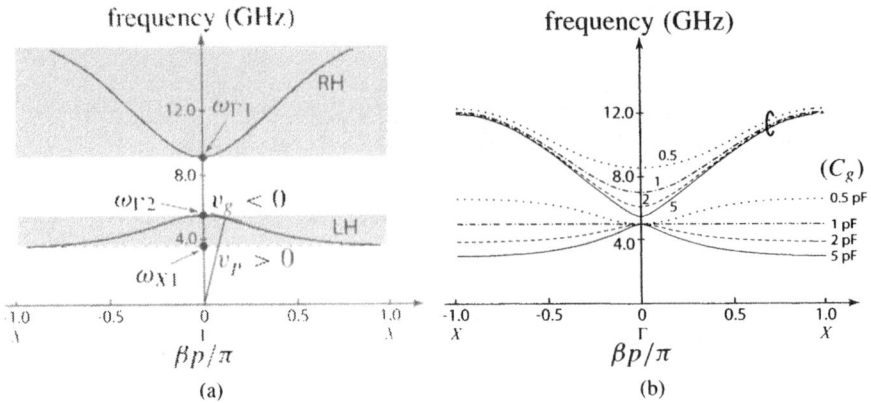

Fig. 6.36. Dispersion diagram for the MTM TL of Fig 6.35. (a) $L_R = 0.6$ nH, $C_R = 0.5$ pF, $L_L = 1.0$ nH, $C_L = 1.0$ pF, and $C_g = 5.0$ pF. (b) $L_R = L_L = 1.0$ nH, $C_R = C_L = 1$ pF, and $C_g = 0.5, 1.0, 2.0, 5.0$ pF. The limit capacitance $C_{g,lim}$ where $Z_{LC+C_g} = j\omega L_L - j/(\omega C_g) = 0$ is, for $f_{se} = f_{sh} = f_0 = 5.03$ GHz, $C_{g,lim} = 1.0$ pF.

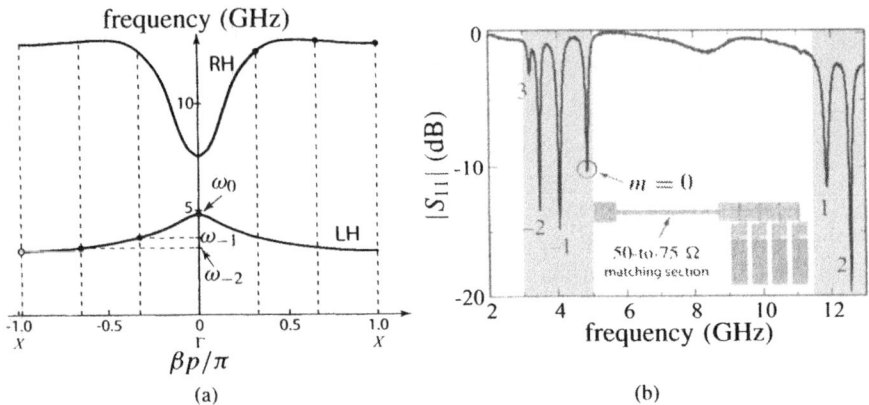

Fig. 6.37. Resonance characteristics of a 4-cell zeroth-order resonator structure [see Figs. 5.49 and 5.52(a)] with period $p = 2.5$ mm built on a substrate of permittivity 2.68 and thickness 0.787 mm. (a) FEM simulated dispersion diagram and resonances for $N = 4$. (b) Measured resonances for the prototype are shown in the inset. The zeroth-order resonance is $f_{res}^{open} = 4.88$ GHz.

6.7   DUAL-BAND CRLH-TL RESONATING RING ANTENNA

The resonating antenna of the previous section is based on the mode $m = 0$. But any mode $\pm m$ shown in Fig 5.49 can be exploited for resonant radiation, and different modes exhibit different properties. An interesting choice of modes is that of a LH/RH pair $(-m, +m)$ (same $|m|$ but negative sign for LH and positive sign for RH). As pointed out in Section 5.5.1, the modes of a pair $\pm m$

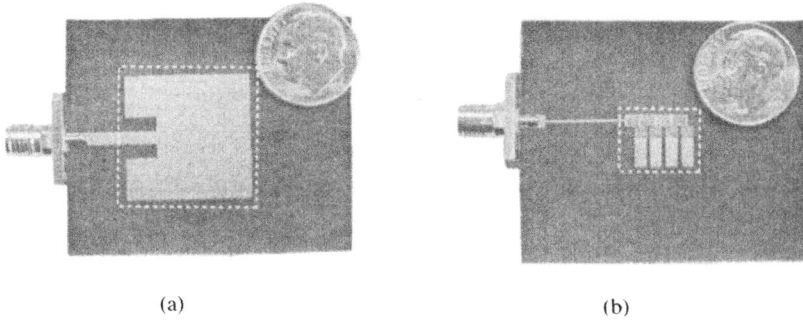

(a)   (b)

Fig. 6.38. Size comparison between for antennas resonating at 4.9 GHz. (a) Conventional patch antenna ($f_0 = 4.9$ GHz). (b) Zeroth order CRLH resonator antenna ($f_0 = 4.88$ GHz).

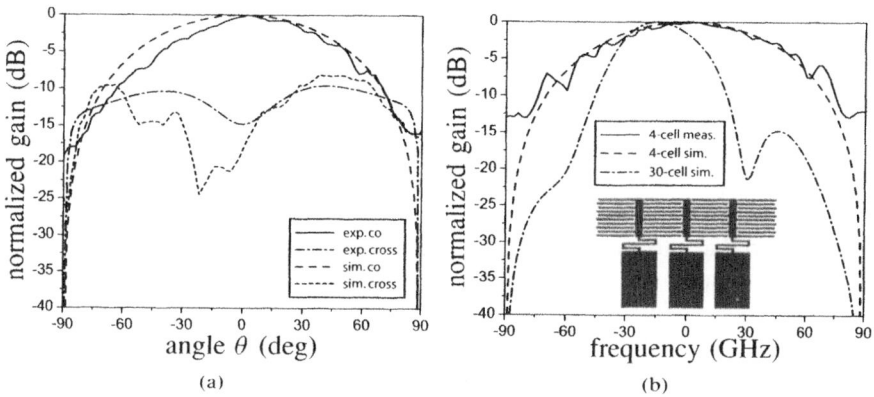

(a)   (b)

Fig. 6.39. Radiation patterns for the zeroth-order CRLH resonator antenna. (a) MoM-simulated and measured co/cross [co: direction of the line (E-plane), cross: direction of the stubs (H-plane)] polarization for the 4-cell prototype shown in the inset of Fig 6.37(b) (4-turn meander line L_L, as in Fig 6.35(a), $f_0 = 4.88$ GHz). (b) Comparison (co-pol.) between MoM-simulation and measurement results for a 4-cell and a 30-cell prototype with the layout shown in the inset (1-turn meander line L_L, $f_{4cells}^{sim} = 7.45$ GHz, $f_{4cells}^{meas} = 7.61$ GHz, $f_{30cells}^{sim} = 7.75$ GHz).

are associated with propagation constants of identical magnitude, $|\beta_{+m}| = |\beta_{-m}|$ (same effective wavelength) and consequently exhibit similar fields distributions and impedance characteristics. Therefore, they can provide *dual-band* resonant operation with a *single common feeding structure*. In addition, the resulting resonator can be designed for *any arbitrary pair of frequencies* by using CRLH design principles, as done for the dual-band components of Section 5.1. This $\pm m$ dual-band resonance, is exemplified here by the *ring antenna* of Fig 6.40, which exhibits some interesting properties specifically related to the loop configuration of the ring.

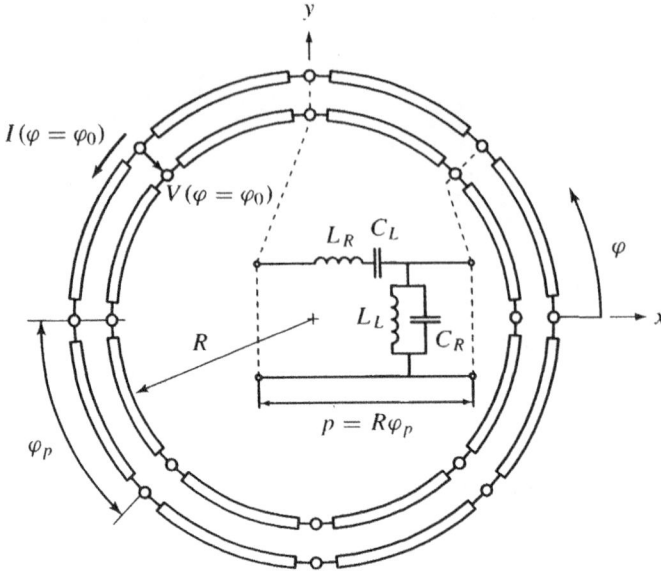

Fig. 6.40. CRLH TL ring resonator antenna configuration (composed here of $N = 8$ unit cells).

Similar to the antenna structure of the previous section, the ring of Fig 6.40 is a CRLH structure, but its closed loop configuration yields the modified boundary condition

$$\xi(\varphi_0) = \xi(\varphi_0 + 2\pi), \quad \forall \varphi_0 \tag{6.21a}$$

or

$$\xi(\zeta) = \xi(\zeta + \ell), \quad \forall \zeta = \phi R, \tag{6.21b}$$

where $\ell = Np$ denotes the circumference of the ring and ζ represents any field quantity along it. Consequently, in reference to the resonance formula of Eq. (5.61b)

$$\ell = |m|\frac{\lambda}{2}, \tag{6.22}$$

only m-even resonance modes can exist along this structure, that is modes for which the circumference of the ring is an integer multiple of wavelength.[21]

An 8-cell dual-band CRLH TL ring resonator antenna, with inductive stubs interconnected at the center, is shown in Fig 6.41. In this structure, we have for m-even modes

$$V(\varphi_0 + \pi) = \mp V(\varphi_0), \quad \forall \varphi_0, \tag{6.23}$$

---

[21] m-odd modes would correspond to the boundary condition $\xi(\zeta) = -\xi(\zeta + \ell)$, which is possible in the non-loop resonator of Section 6.6 (ℓ being the length of the structure) but excluded in a loop configuration such as the ring under consideration here.

(a)   (b)

(c)

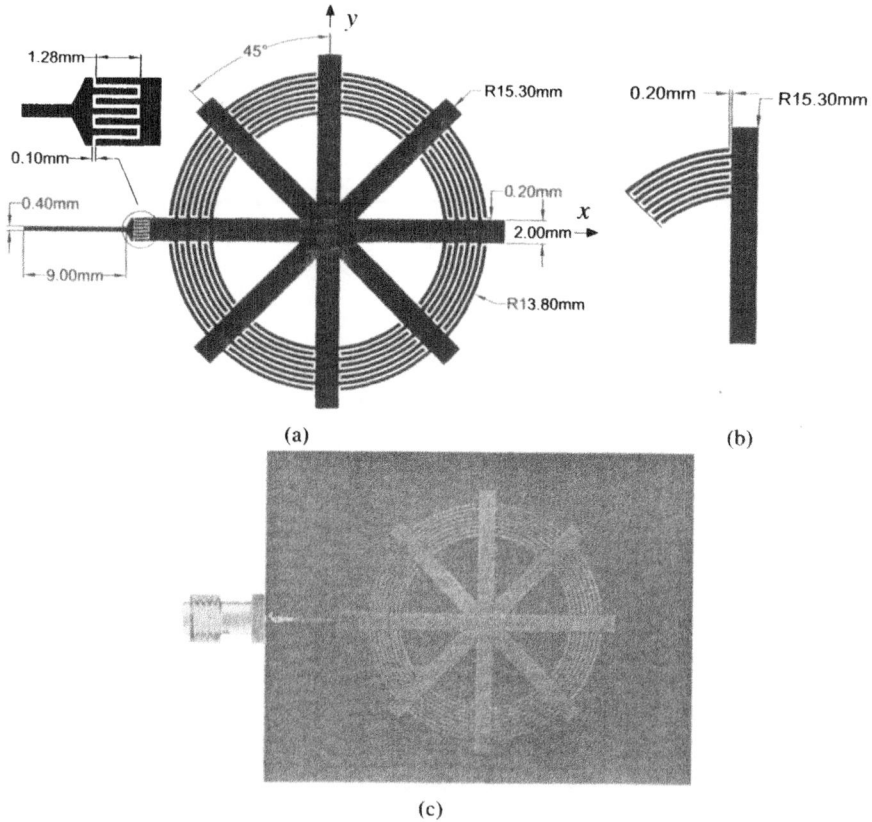

Fig. 6.41. Microstrip implementation of the dual-band m-even CRLH TL ring resonator antenna of Fig 6.40. The center of the structure is a virtual ground and therefore does not require any via (for $m \neq 0$). (a) Layout of the ring with open-ended (capacitive) coupling structure. (b) Unit cell of the ring. (c) Fabricated prototype.

where the sign "$-$" holds for $|m|/2$ odd and "$+$" holds for $|m|/2$ even. This condition implies, as shown in Fig 6.42, that the center point of the structure is a virtual ground, eliminating the need for a via to obtain CRLH resonances. The case $m = 0$ is an exception to this rule because it does not present a virtual ground in the center point, which rather becomes a common floating potential in the absence of via; the equivalent circuit in $m = 0$ is different from the CRLH prototype, which requires grounded $L_L - C_R$ (either by a real via connection or by an equivalent virtual ground), and consequently does not exhibit any CRLH resonance. The purpose of the present antenna is to provide dual-band operation based on the resonances $m = \pm 2$ ($m \neq 0$). Because the resonance $m = 0$ does not exist in the absence of the via, it cannot represent any danger of parasitic resonance.

The dispersion diagram and resonance modes for the ring antenna of Fig 6.41(c) are presented in Fig 6.43. As shown in this graph, among the $2N -$

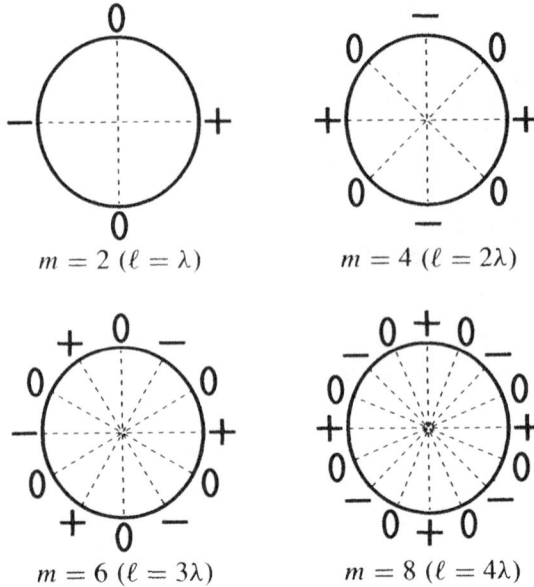

Fig. 6.42. Symmetries in a circular loop configuration (such as the ring, without considering explicitly its "spokes") for different modes. By symmetry, the voltage at the center of the structure is a virtual ground for all the modes $\pm m$, $m \neq 0$. These symmetries provide a good representation of the symmetries in the actual ring structure in the case where a relatively large number of spokes (i.e., cells N) are present.

$1 = 15$ resonances existing in a $N = 8$ cell non loop configuration (Section 6.6), only $N - 2 = 6$ resonance modes subsist in the ring. Here, the design has been optimized for radiation in the modes $m = \pm 2$.

Fig 6.44 shows the simulated and measured return losses for the prototype of Fig 6.41, and the corresponding radiation patterns are presented in Fig 6.45. It can be seen that good matching is achieved in both the modes $m = -2$ and $m = +2$ and that these modes produce relatively similar radiation patterns.

The MTM ring antenna presented in this section can be extended to a number of other applications. For instance, dual-band multipolarization ($x-$, $y-$ or $xy-$linear and RHC or LHC) operation can be obtained by feeding different "spokes" of the ring. This represents interesting potential for this structure in terms of frequency/polarization diversity for RF links in an indoor microwave communication systems (such as for instance IEEE 802.11a,g).

6.8   FOCUSING RADIATIVE "META-INTERFACES"

Using two bulk media with opposite handednesses for negative refraction (Sections 2.7 and 4.5.4) is not the only way to achieve focusing from a planar

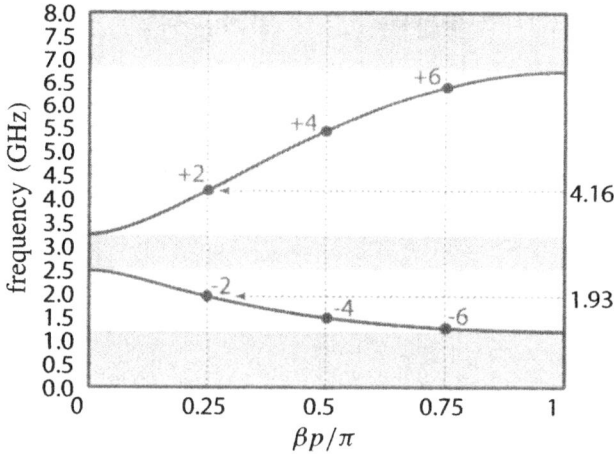

Fig. 6.43. Dispersion diagram and resonances of the CRLH TL ring of Fig 6.41, which includes $N = 8$ unit cells. Only nonzero m-even resonance modes in Eq. (6.22) meet the boundary conditions in the ring configuration. The labels on the curves correspond to the parameter m in Eq. (6.22). The antenna corresponding to the data shown is built on a substrate with permittivity of 2.2 and thickness of 1.57 mm, and exhibits the parameters $L_R = 3.44$ nH, $C_R = 0.99$ pF, $L_L = 4.20$ nH and $C_L = 0.70$ pF (i.e., $f_{se} = 2.47$ GHz and $f_{sh} = 3.24$ GHz). The $m = \pm 2$ resonances are $f_{-2} = 1.93$ GHz and $f_{+2} = 4.16$ GHz and correspond to $p = \lambda/8$ or $\ell = 8p = \lambda$.

interface. Another planar-interface focusing technique is based on the concept of *near-field phase conjugation* [29, 30, 31]. This concept is inspired from the so-called (far-field) retrodirective antenna arrays [20, 22, 23], where the phase of the incoming signal is reversed at each element (phase conjugation), which results in automatic re-radiation of the signal retro-directively toward the source. Section 6.8.1 describes a planar the focusing interface based on heterodyne mixing, while Section 6.8.2 presents a purely passive planar focusing interface based on nonuniform LW radiation. In both cases, all the cause of focusing effect is spatially concentred along a thin interface. This may represent a significant advantage over real negative refraction systems since the source and focus may exist in a same uniform and low-loss medium (typically air). Although the interface implementations of the following two sections are 1D, producing a mostly 2D focal spot in the plane of the structure, 2D implementations leading to a volumetric focal spot are also possible.

6.8.1 Heterodyne Phased Array

The principle of the implementation of the heterodyne phase array interface operating is illustrated in Fig 6.46(a) [29]. The system is designed to operate in the near-field of the array but in the far-field of its elements. The antenna array receives an incoming plane wave with incident angle θ_{inc}. This angle introduces

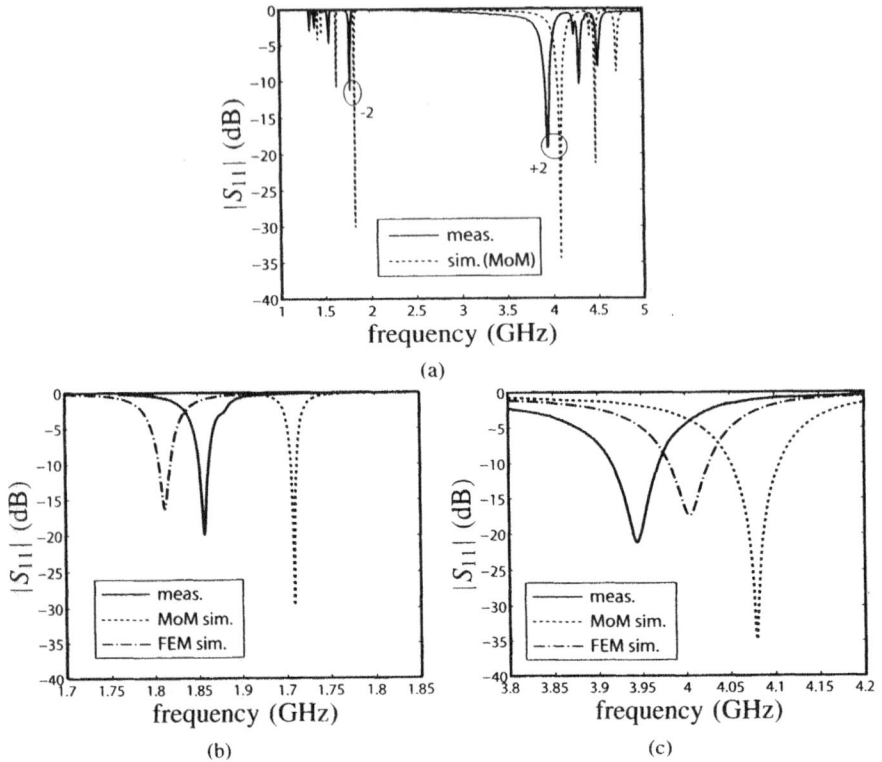

Fig. 6.44. Return loss of the ring resonator antenna of Fig 6.41. (a) Broad-band view. (b) Zoom around the resonance $m = -2$. (c) Zoom around the resonance $m = +2$.

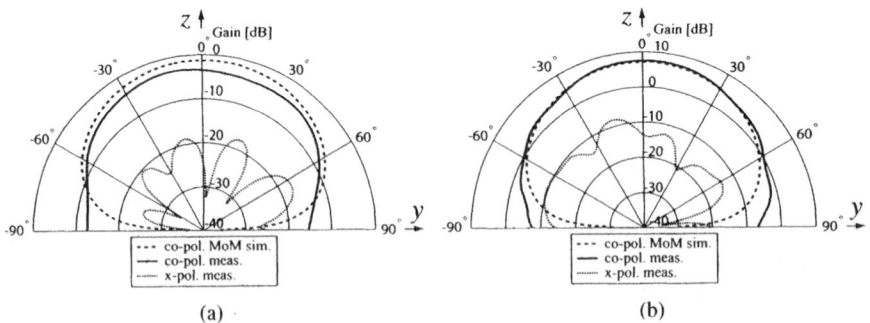

Fig. 6.45. Simulated and measured radiation patterns of the ring resonator antenna of Fig 6.41 in the $y - z$ plane ($\phi = 90°$). (a) At the resonance frequencies f_{-2}. (b) At the resonance frequencies f_{+2}.

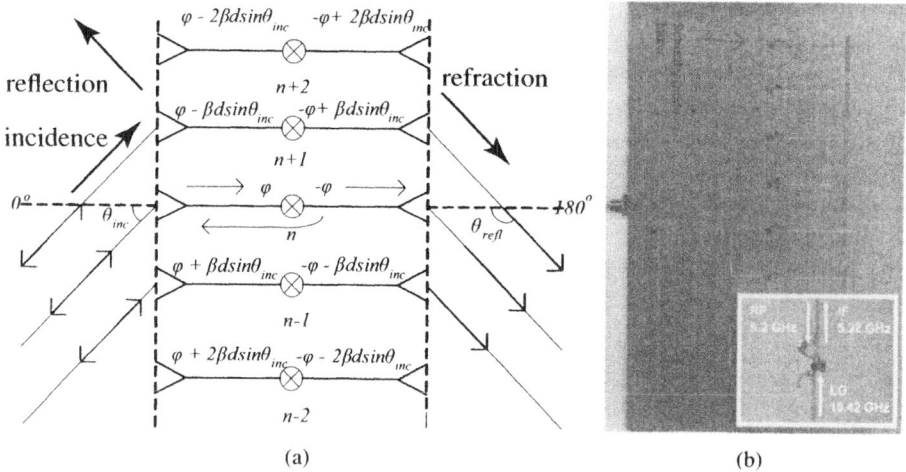

Fig. 6.46. Negative reflective/refractive "meta-interface" using a bi-directional phase-conjugating array. (a) Principle. (b) Front side of a prototype. Slots on reverse side are indicated by black rectangles. The inset shows the enlarged picture of a mixer. From [29], © IEEE; reprinted with permission.

a relative phase difference between the adjacent antenna elements causing the nth element to lead the $(n + 1)$th element by $\beta d \sin \theta_{inc}$, where d is the inter-elements spacing. After phase conjugation, the nth element lags the $(n + 1)$th element by $\beta d \sin \theta_{inc}$, causing the "reflected" signal to have an angle θ_{refl} equal to θ_{inc} and the "refracted" signal to have an angle θ_{refr} equal to $180° - \theta_{inc}$. Thus, this configuration produces an effect equivalent to the reversal of Snell's law or negative refraction. This meta-interface may also be interesting in far-field applications.

Fig 6.46(b) shows an interface prototype, where the same *bidirectional* slot array serves both the receiving and transmitting antennas. This interface is constructed of two main subcircuits, the antenna array and the mixer array. The antenna array consists of eight slot antennas with operating frequency at 5.2 GHz. The slot antenna is utilized for its omnidirectional radiation characteristic allowing for the use of a single antenna array for both the reflecting and refracting ends of the interface. The individual elements are placed a half-wavelength apart at 5.2 GHz in order to avoid grating lobes. The mixer array consists of eight Schottky diode mixers utilized for phase-conjugating heterodyne mixing. The mixer array is fed on the LO side by a corporate feed-line constructed with Wilkinson power dividers. The IF-RF side of the eight mixers is connected to the eight elements of the slot antenna array.

When operated in the far-field, the antenna array receives an incoming 5.2 GHz signal from a source located at an angle θ_{inc} from the broadside of the interface. The antenna array then feeds the RF end of the mixer. The LO end is fed by a signal of approximately twice the RF. The phase-conjugated IF signal is then radiated by the antenna array. Due to the phase-conjugation, the two main lobes of the transmitted pattern will be located at θ_{inc} and $180° - \theta_{inc}$, constituting the negative reflection and refraction of the meta-interface. Note that in measurement, while the LO is slightly greater than twice the RF in order to ease detection the IF, it can also be operated with RF equal to IF. When operated in the near-field, the received signal can no longer be considered a plane wave. Phase conjugation in the process described above results in a maximum refracted power that will mirror the location of the source.

Far-field bistatic RCS measurements are shown in Fig 6.47. A 5.2 GHz source is placed initially at $0°$ relative to the broadside of the array. The mixer side of the interface is fed with an LO of 10.42 GHz, resulting in a 5.22 GHz signal measured at the receiver. The source was then moved to $-30°$ and $+30°$ relative

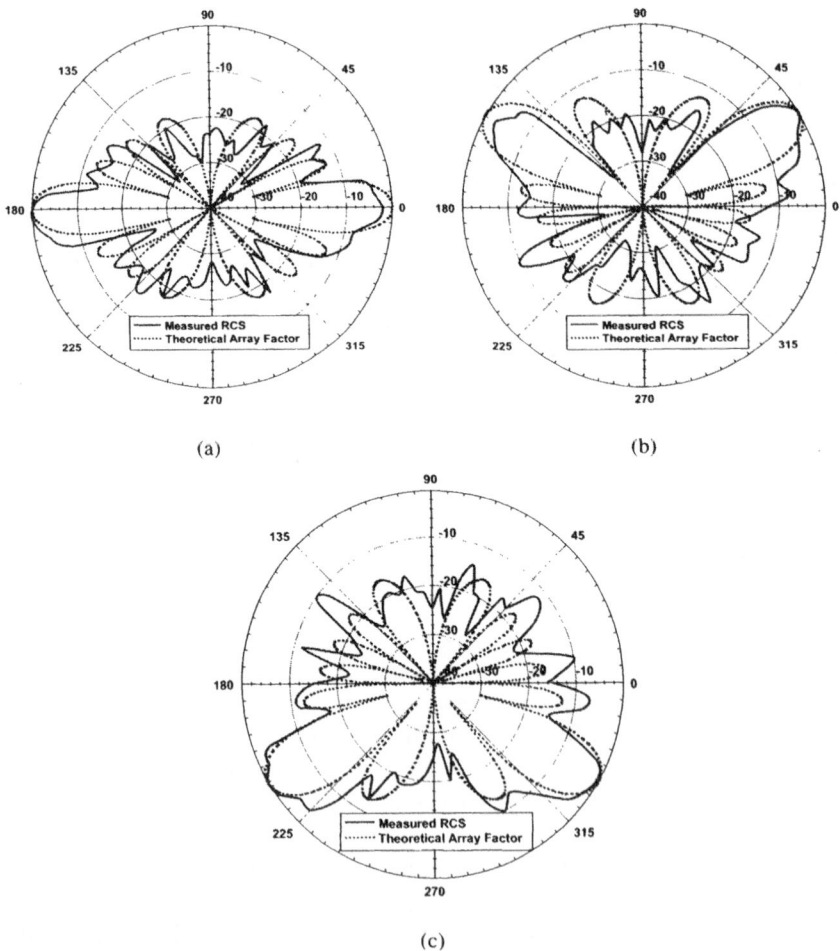

(a)   (b)

(c)

Fig. 6.47. Measured normalized RCS of meta-interface for source located at (a) $0°$, (b) $30°$ and (c) $30°$. From [29], © IEEE; reprinted with permission.

to the broadside. The array factor for an 8-element array is also included for comparison.

For the broadside case we observe that the bi-directionality of the slot antenna causes reflection at an angle of $0°$ and refraction at an angle of $180°$. When the source is moved relative to the broadside, the negative reflection and refraction become apparent. For the $-30°$ case, the main lobes are located at $-30°$ and $210°$, indicating negative reflection and refraction, respectively. Similarly, for the $+30°$ case there is reflection of angle $+30°$ and negative refraction of angle $150°$. Note that the RCS measurement closely resembles the theoretical array factor for an 8-element array. As this measurement was not performed in an anechoic chamber, the scattering due to the environment causes slightly erratic jumps in the pattern.

For the near-field measurement it was necessary for the source and the receiver to be located in the near-field of the array (less than 185 cm), but in the far-field of the slot antenna element (greater than 3 cm). The source and the receiver were placed 11 cm from the array. The set-up for this measurement is shown in Fig 6.48(a). The source was initially normal to the midpoint of the array and we will refer to this position as the origin on the source side. The receiver was moved relative to the origin on the receiver side, measuring the received power along a path parallel to the interface. As in the far-field case, the source signal was 5.2 GHz, the LO was 10.42 GHz, and the received signal was 5.22 GHz. The measurement was made with the source at 0, ±3 cm, and ±6 cm relative to the origin on the source side. Fig 6.48(b) shows the received power versus the position of the receiver for the source at the locations described above.

The figure indicates that a displacement of the source relative to the origin on the source side causes a displacement in the same direction of the peak received power relative to the origin on the receiver side. This mirroring of source and image demonstrates the converse of conventional lens behavior in which the source and image displacement are in opposite directions.

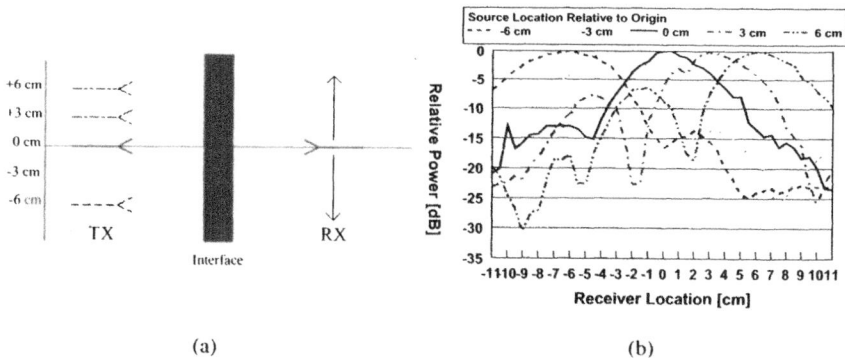

(a)   (b)

Fig. 6.48. (a) Configurations for six near-field measurements with source located at 0, 3, and 6 cm with respect to origin and (b) measured received power pattern for varying source displacements. From [29], © IEEE; reprinted with permission.

6.8.2 Nonuniform Leaky-Wave Radiator

The presence of mixers in a MTM may be undesired for several reasons, including complexity of the resulting structure, bias requirements and difficulty of homogenization (Section 7.1). For this reason, a purely passive focusing interface was proposed in [31].

This interface is a near-field planar *nonuniform LW* interface consisting of a pair of Tx/Rx nonuniform backfire-to-endfire LW antennas (Section 6.2) connected together at their center, as illustrated in Fig 6.49. The wave radiated from a point source is received by the Rx antenna, then transmitted to the Tx antenna by the interconnection, and finally radiated by the Tx antenna to form a focus at a distance f from the interface on the other side. The nonuniform profile $\beta(z)$ of the antennas is designed to provide the appropriate position-dependent radiation angle $\theta(z)$ for focus formation[22], as suggested in Fig 6.49. Although the ideal nonuniform profile is continuous, since the angles of radiation need to vary continuously, the quasi-continuous uniform profile obtained by designing MTM unit cells with appropriate different reactive parameters yields essentially the same response, since $p \ll \lambda_g$.

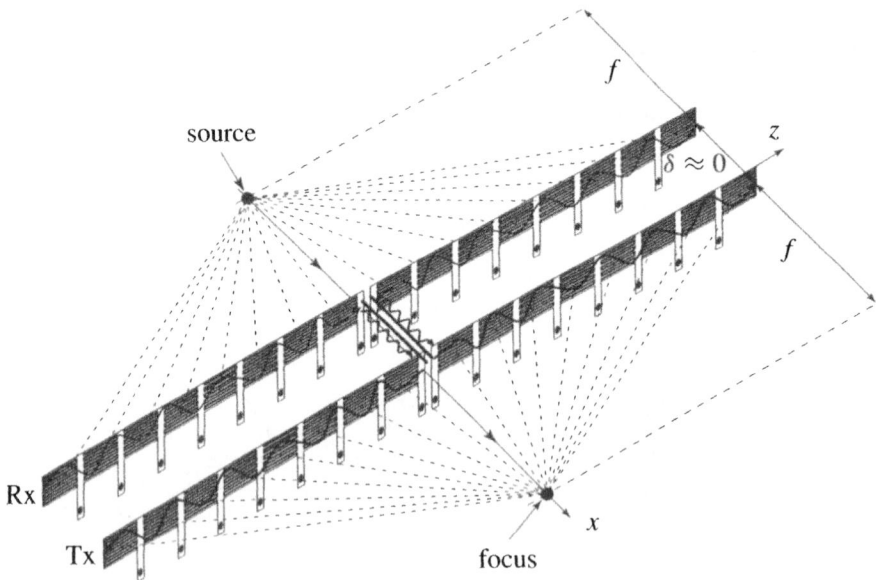

Fig. 6.49. Principle of focusing by a nonuniform LW antennas interface.

---

[22]Only the propagation constant needs to be nonuniform, $\beta = \beta(z)$, for this focusing effect. A constant characteristic impedance along the line, obtained by varying with z the inductive, and capacitive elements in the same ratios, yields the best performances. A progressive nonuniform (*tapered*) impedance profile $Z_c = Z_c(z)$ [32] is acceptable.

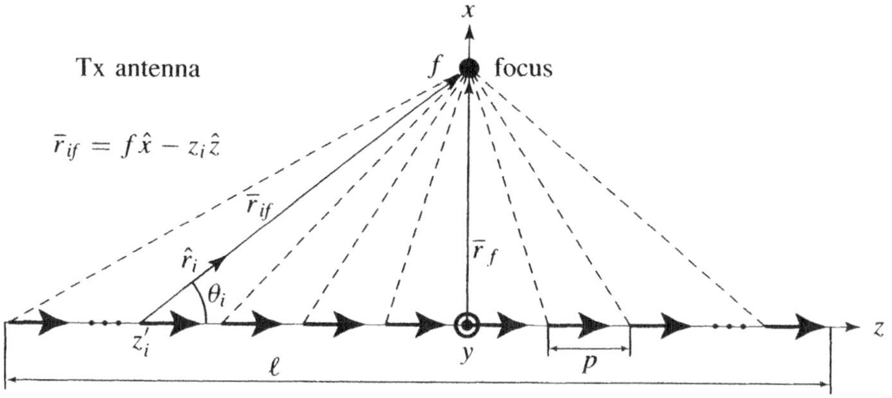

Fig. 6.50. Dipole array model for the Tx antenna of Fig 6.49.

Only the Tx antenna is considered here. The array factor approach described in Section 6.2 [5] may be straightforwardly applied to the case of a nonuniform antenna. Assuming that the electric field is mainly directed along the direction of the line (z), as in the case of the structure of Section 6.2, we model the MTM structure as an array of z-directed infinitesimal dipoles representing the contributions of the different cells. In reference with Fig 6.50, the field radiated by each dipole i at the point $\bar{r} = (x, y, z) = (f, 0, 0) = \bar{r}_f$ reads [24, 4]

$$
\begin{aligned}
\bar{E}_i(\bar{r}_{if}) &= E_0 j\omega\mu \left[1 + \frac{1}{jk_0|\bar{r}_{if}|} - \frac{1}{(k_0|\bar{r}_{if}|)^2} \right] \frac{e^{-jk_0|\bar{r}_f|}}{|\bar{r}_{if}|} \sin\theta_i \hat{\theta}_i \\
&\quad + 2E_0\eta \left[\frac{1}{|\bar{r}_{if}|} - \frac{j}{k_0|\bar{r}_{if}|^2} \right] \frac{e^{-jk_0|\bar{r}_f|}}{|\bar{r}_{if}|} \cos\theta_i \hat{r}_i \\
&= \hat{x}_i E_{xi}(\bar{r}_{if}) + \hat{y}_i E_{yi}(\bar{r}_{if}) + \hat{z}_i E_{zi}(\bar{r}_{if}),
\end{aligned}
\tag{6.24}
$$

where $\hat{\theta}_i$ and $\hat{r}_i$ are the unit vectors of the local coordinates shown in Fig 6.50.

In order to achieve focusing at the point $\bar{r}_f$, we need to determine the proper phase distribution ϕ_i for the N elements of the MTM that will maximize the magnitude of the electric field in the z direction at this point

$$
\left| E_z(\bar{r}_f) \right| = \left| \sum_{i=1}^{N} I_{i0} e^{j\phi_i} E_{zi}(\bar{r}_{if}) \right|,
\tag{6.25}
$$

where we have assumed that $|E_{xi}|, |E_{yi}| \ll |E_{zi}|$, as it is the case for the structure of Section 6.2, and where $I_{i0}e^{j\phi_i}$ and $E_{zi}(\bar{r}_{if})$ are the excitation and radiated fields of the ith element, respectively. This quantity is maximized when the contributions from all the elements are arrive in phase at $\bar{r}_f$, a condition which is met when the sum of the line phase ϕ_i and the radiation phase $\angle E_{zi}(\bar{r}_{if})$ is the same for all the elements

$$\phi_i + \angle E_{zi}(\bar{r}_{if}) = \angle E_z(\bar{r}_f), \quad \forall i. \tag{6.26}$$

By assuming that the point $\bar{r}_f$ lies in the far-field of the elements of the array (but naturally in the near-field of the array itself), we may neglect the near-field

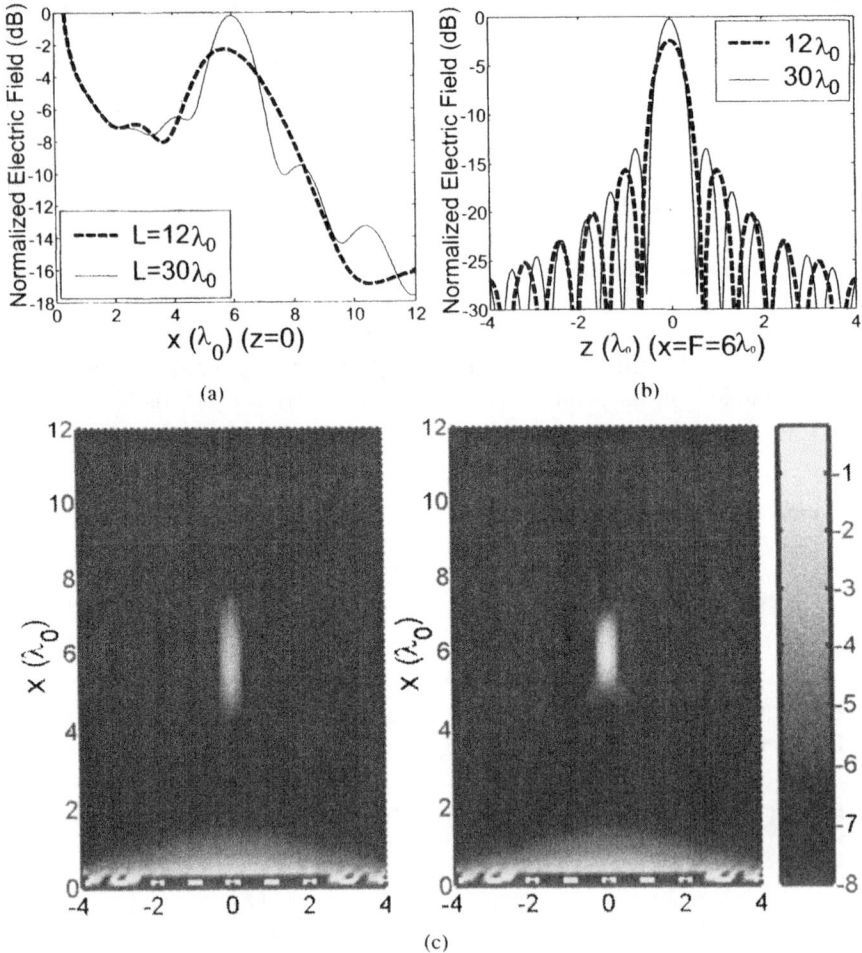

(a)   (b)

(c)

Fig. 6.51. Field distributions $|E_z|$ for different array lengths obtained from Eqs. (6.24) or (6.25) (in dB) for the phase distribution given by Eq. (6.27) with $f = 6\lambda_0$ (a) along the axis $z = 0$; (b) along $x = F = 6\lambda_0$ and $z = 0$; and over the $x - z$ plane at (c) $y = 0$ and $\ell = 12\lambda_0$ (left-hand side) and $y = 0$ and $\ell = 30\lambda_0$ (right-hand side). From [31], © IEEE; reprinted with permission.

contributions in Eq. (6.24). This reduces Eq. (6.26) to

$$\phi_i = -\angle E_{zi}(\bar{r}_{if}) + \angle E_z(\bar{r}_f) \approx k_0 |\bar{r}_{if}| + \text{constant}, \qquad (6.27)$$

which corresponds to the expected phase conjugation profile required for the generation of a focal point at $\bar{r}_f$.

Fig 6.51 demonstrates the concept of nonuniform LW interface focusing. An array of $p = \lambda_0/2$ is considered to avoid the problem of grating lobes [24]. The phase distribution is calculated using Eq. (6.27) with $f = 6\lambda_0$. A clear focus is observed at the expected location. Comparison of the two graphs in Fig 6.51(c) shows that the size of the focal spot along the x direction is reduced when the length ℓ of the interface is larger, which is due to the fact that the directivity of the array is increased when the effective aperture is increased.

In order to simplify the design of the MTM LW structure, it is possible to subdivide it into groups of different uniform LW antennas. The phase distribution of Eq. (6.27) is then approximated by a piecewise linear curve. Reasonably confined focal spots can be obtained even with a restricted number of groups [31].

REFERENCES

1. A. Oliner. "Leaky-wave antennas," in *Antenna Engineering Handbook*, Third Edition, edited by R. C. Johnson, McGraw Hill, 1993.

2. Y. Qian, B. C. C. Chang, T. Itoh, K. C. Chen, and C. K. C. Tzuang. "High efficiency and broadband excitation of leaky mode in microstrip structure," in *IEEE-MTT Int'l Symp.*, TH1B-1, Anaheim, CA, pp. 1419–1422, June 1999.

3. A. Ishimaru. *Electromagnetic Wave Propagation, Radiation, and Scattering*, Prentice Hall, 1991.

4. J. D. Kraus and R. J. Marhefka. *Antennas*, Third Edition, McGraw Hill, 2001.

5. C. Caloz and T. Itoh. "Array factor approach of leaky-wave antennas and application to 1D/2D composite right/left-handed (CRLH) structures," *IEEE Microwave Wireless Compon. Lett.*, vol. 14, no. 6, pp. 274–276, June 2004.

6. F. E. Gardiol. *Introduction to Microwaves*, Artech House, 1984.

7. F. K. Schwering and S.-T. Peng. "Design of dielectric grating antennas for millimeter wave applications," *IEEE Trans. Microwave Theory Tech.*, MTT-S. 83, no. 2, pp. 199–209, Feb. 1983.

8. W. Menzel. "A new travelling-wave antenna in microstrip," *Arch. Elektron. Uebertraegungstech.*, vol. 33, no. 4, pp. 137–140, April 1979.

9. A. A. Oliner. "Leakage from higher modes on microstrip lines with application to antennas," *Radio Science*, vol. 22, no. 6, pp. 907–912, Nov. 1987.

10. L. Liu, C. Caloz and T. Itoh. "Dominant mode (DM) leaky-wave antenna with backfire-to-endfire scanning capability," *Electron. Lett.*, vol. 38, no. 23, pp. 1414–1416, Nov. 2002.

11. C. Caloz and T. Itoh. "Novel microwave devices and structures based on the transmission line approach of meta-materials," *IEEE-MTT Int'l Symp.*, vol. 1, pp. 195–198, June. 2003.

12. R. E. Horn, H. Jacobs, E. Freibergs, and K. L. Klohn. "Electronic modulated beam steerable silicon waveguide array antenna," *IEEE Trans. Microwave Theory Tech.*, MTT-S. 28, no. 6, pp. 647–653, June 1980.

13. H. Maheri, M. Tsutsumi, and N. Kumagi. "Experimental studies of magnetically scannable leaky-wave antennas having a corrugated ferrite slab/dielectric layer structure," *IEEE Trans. Antennas Propagat.*, vol. 36, no. 7, pp. 911–917, Nov. 1988.

14. L. Huang, J.-C. Chiao, and M. P. De Lisio. "An electronically switchable leaky-wave antenna," *IEEE Trans. Antennas Propagat.*, vol. 48, no. 11, pp. 1769–1772, Nov. 2000.

15. J. U. I. Syed and A. D. Olver. "Variable beamwidth dual reflector antenna," in *Proc. IEEE AP-S Intl. Symp. Dig.*, vol. 1, pp. 92–96, April 1995.

16. S. Lim, C. Caloz and T. Itoh. "Metamaterial-based electronically-controlled transmission line structure as a novel leaky-wave antenna with tunable radiation angle and beamwidth," *IEEE Trans. Microwave Theory Tech.*,, vol. 53, no. 1, pp. 161–173, Nov. 2005.

17. K. K. Bhan, S. Ghosh, and G. P. Srivastava. "Study of control of beamwidth of radiation pattern of a waveguide using inclined slotted flanges," *IEEE Trans. Antennas Propagat.*, vol. 2, no. 3, pp. 447–450, May 1978.

18. I. A. Korisch and B Rulf. "Antenna beamwidth control using parasitic subarrays," in *Proc. IEEE AP-S Intl. Symp. Dig.*, pp. 117–120, Nov. 2000.

19. C. Caloz, S. Lim, and T. Itoh. "A novel leaky-wave retrodirective reflector using short/matched terminations," in *Proc. of European Microwave Conference*, pp. 1071–1074, Oct. 2003.

20. L. C. Van Atta. "Electromagnetic reflector," U.S. Patent 2908002, Oct. 1959.

21. S. Lim, C. Caloz, and T. Itoh. "A reflecto-directive system using a composite right/left-handed (CRLH) leaky-wave antenna and heterodyne mixing," *IEEE Microwave Wireless Compon. Lett.*, vol. 14, no. 4, pp. 183–185, April 2004.

22. C. .Y. Pon. "Retrodirective array using the heterodyne technique," *IEEE Trans. Antennas Propagat.*, vol. 22, no. 2, pp. 176–180, March 1964.

23. R. Y. Miyamoto, Y. Qian, and T. Itoh. "An active integrated retrodirective transponder for remote information retrieval-on-demand," *IEEE Trans. Microwave Theory Tech.*, vol. 49, no. 9, pp. 1910–1919, Sept. 2001.

24. C. A. Balanis. *Antenna Theory, Analysis and Design*, Second Edition, John Wiley & Sons, 1997.

25. D. F. Sievenpiper, J. H. Schaffner, H. J. Song, R. Y. Loo, and G. Tangonan. "Two-dimensional beam steering using an electrically tunable impedance surface," *IEEE Trans. Antennas Propagat.*, vol. 51, no. 10, pp. 2713–2722, Oct. 2003.

26. C. A. Allen, K. M. K. H. Leong, C. Caloz, and I. Itoh. "A two-dimensional edge excited metamaterial-based leaky-wave antenna," *IEEE-MTT Int'l Symp.*, pp. 301–304, June 2004.

27. A. Sanada, K. Murakami, I. Awai, H. Kubo, C. Caloz, and T. Itoh. "A planar zeroth-order resonator antenna using a left-handed transmission line," *34th European Microwave Conference*, Amsterdam, The Netherlands, pp. 1341-1344, Oct. 2004.

28. A. Sanada, K. Murakami, S. Aso, H. Kubo, and I. Awai. "A via-free microstrip left-handed transmission line," *IEEE-MTT Int'l Symp.*, Fort Worth, TX, pp. 301–304, June 2004.

29. C. A. Allen, K. M. K. H. Leong, and T. Itoh. "A negative reflective/refractive "meta-interface" using a bi-directional phase-conjugating array" *IEEE-MTT Int'l Symp.*, vol. 3, Philadelphia, PA, pp. 1875–1878, June 2003.

30. S. Maslovskia and S. Tretyakov. "Phase conjugation and perfect lensing," *IEEE Trans. Antennas Propagat.*, vol. 94, no. 7, pp. 4241–4243, Oct. 2003.

31. I. -H. Lin, C. Caloz and T. Itoh. "Near-field focusing by a nonuniform leaky-wave interface," *Micr. Opt. Technol. Lett.*, vol. 44, no. 5, pp. 416–418, March 2005.

32. M. D. Pozar. *Microwave Engineering*, Third Edition, John Wiley & Sons, 2004.

7

THE FUTURE OF MTMs

Although composite structures (e.g., periodic, RH artificial, chiral, and bi-anisotropic structures) have been investigated for many years, MTMs, defined as effectively homogeneous media with properties not readily available in nature (Section 1.1), really constitute a new field of science and engineering. If we consider the first demonstration of a LH structure in 2000 and 2001 as the commencement of MTMs [1, 2], this field is less than five years old in 2005! Due to their embryonic nature, MTMs are evolving at an extremely fast pace at the time of this writing. Although it is possible neither to account for all of their current developments nor to predict all of their future directions and applications, we will try in Chapter. 7 to identify a number of important challenges to be met and to anticipate potential future applications for MTMs.

7.1 "REAL-ARTIFICIAL" MATERIALS: THE CHALLENGE OF HOMOGENIZATION

The physics of most current MTMs, and in particular LH MTMs, has been well established, theoretically, numerically, and experimentally. Many novel and exciting refractive effects, such as for instance negative focusing, have been abundantly demonstrated in the literature. However, the average lattice constant p in current MTMs is still electromagnetically too large for high-quality refraction. Current MTMs are typically characterized by $p/\lambda_g = 1/5 \ldots 1/15$, so that, although refraction is dominating, diffraction/scattering effects tend to alter the

Electromagnetic Metamaterials: Transmission Line Theory and Microwave Applications,
By Christophe Caloz and Tatsuo Itoh
Copyright © 2006 John Wiley & Sons, Inc.

purity of refractive effects (e.g., diffuse focusing spot instead of highly confined focus) and increase the transmission losses. This is illustrated for instance in the distributed MTM structures of Section 4.5, where it is seen that, although negative refraction and focusing are clearly achieved (Figs. 4.45 to 4.48), refractive effects are relatively diffuse, and consequently difficult to exploit in practical applications. To obtain very pure refractive phenomena, such as those illustrated in Figs. 4.22 to 4.29,[1] it is necessary to *homogenize* the MTM, that is, to decrease the structural feature p/λ_g by one order of magnitude or more. If this challenge can be met, MTMs will behave as "real-artificial" materials, an oxymoronic term that means that the unit cells of the MTM would be so small that they would really behave as atoms in natural materials so as to produce a perfectly homogeneous macroscopic response. In natural media (such as air or teflon), the atomic lattice constant is of the order of the angström ($p \approx 10^{-10}$ m); therefore, we have in the microwave range an atomic lattice constant of the order of one thousand of millions smaller than wavelength ($p/\lambda_g = 10^{-9}$), which ensures a perfectly homogeneous response to electromagnetic waves. Such a huge ratio between λ_g and p is not necessary to obtain good refraction; a decrease of p/λ_g in MTMs from $\sim 1/10$ to $1/100$ would be perfectly sufficient, since scattering would have then become completely negligible. Is such a reduction in p/λ_g realistic or even possible at all? To address this question and identify fundamental related issues, we will consider here the paradigm of (nonresonant) CRLH MTMs.

The dispersion relation of a CRLH MTM has been extensively analyzed in this book. It is given in the balanced case from Eqs. (3.34) with Eq. (3.55) by

$$\beta(\omega) = \frac{1}{p}\left(\omega\sqrt{L_R C_R} - \frac{1}{\omega\sqrt{L_L C_L}}\right),\tag{7.1}$$

from which the refractive index is obtained as

$$n(\omega) = \frac{\beta(\omega)}{k_0},\tag{7.2}$$

where $k_0 = \omega/c$ is the free space wave number, related to the guided wavelength λ_g by

$$\lambda_g(\omega) = \frac{2\pi}{|\beta(\omega)|}.\tag{7.3}$$

Consider now a MTM with a required refractive index n_{op} at a given operating frequency ω_{op}, $n_{op} = n(\omega_{op})$ (typically negative). To homogenize the MTM, we must reduce the average lattice constant p to a significantly smaller value $p_h = p/\varsigma$ ($\varsigma \gg 1$, at least $\varsigma > 10$), *without changing* $n_{op} = n(\omega_{op})$ *and without changing the overall size of the MTM*.[2] To achieve this homogenization to p_h

---

[1] A very thin TLM mesh is used in these figures.

[2] See for instance the LH slab case in Fig 4.46 for which a focusing quality of the type shown in Fig 4.27 is targeted. The purpose of homogenization is to improve the quality of the focus, without changing its location, the operating frequency, and the overall size of the structure.

without affecting $n_{op} = n(\omega_{op})$, it is necessary to adjust the values of the inductive and capacitive reactances in the dispersion relation [Eq. (7.1)] to new values L_{Rh}, C_{Rh}, L_{Lh}, and C_{Lh}, so that $\beta(\omega)$ remains unchanged

$$
\begin{aligned}
\beta(\omega) &= \frac{1}{p_h}\left(\omega\sqrt{L_{Rh}C_{Rh}} - \frac{1}{\omega\sqrt{L_{Lh}C_{Lh}}}\right) \\
&= \frac{\varsigma}{p}\left(\omega\sqrt{L_{Rh}C_{Rh}} - \frac{1}{\omega\sqrt{L_{Lh}C_{Lh}}}\right) \\
&= \frac{1}{p}\left(\omega\sqrt{\varsigma^2 L_{Rh}C_{Rh}} - \frac{1}{\omega\sqrt{L_{Lh}C_{Lh}/\varsigma^2}}\right).
\end{aligned}
\tag{7.4}
$$

The most straightforward condition for Eq. (7.4) to be equal to Eq. (7.1) is[3]

$$
\varsigma^2 L_{Rh}C_{Rh} = L_R C_R \quad \text{and} \quad \frac{L_{Lh}C_{Lh}}{\varsigma^2} = L_L C_L,
\tag{7.5}
$$

which typically suggests $(\varsigma > 0)$[4]

$$
L_{Rh} = \frac{L_{Rh}}{\varsigma},
\tag{7.6a}
$$

$$
C_{Rh} = \frac{C_{Rh}}{\varsigma},
\tag{7.6b}
$$

$$
L_{Lh} = L_{Lh} \cdot \varsigma,
\tag{7.6c}
$$

$$
C_{Lh} = L_{Ch} \cdot \varsigma.
\tag{7.6d}
$$

Thus, the RH parameters have to be reduced by a factor ς. This fact does not seem to represent a substantial problem since the inductance and capacitance of a structure typically decrease when the size of the structure decreases (by $\varsigma > 0$), roughly by a same factor ($\varsigma > 0$). However, the LH parameters have to be *increased* by the factor ς. This seems to represent a real challenge as much more electric and magnetic energy needs to be stored in a much smaller volume!

This challenge requires new architectures and new technologies. Novel 2D configurations, including high-permittivity ceramic or ferroelectric metal-insulator-metal (MIM) capacitors as well as high-permeability ferrimagnetic spiral inductors, will have to be developed. Due to the absence of magnetic charges, the inductors are much harder to realize than the capacitors (only requiring very small spacings or very high permittivity in MIM configurations). One

---

[3]It is clear that, at a *single frequency* ω_{op}, a more general condition would be that the overall expression of Eq. (7.4) be equal to the overall expression of Eq. (7.1) without necessarily having the respective RH and LH contributions equal. However, the exact broadband characteristic of Eq. (7.1) is reproduced only if condition of Eq. (7.5) is satisfied.

[4]Of course, the following condition is not strictly necessary and can be relaxed by applying Eq. (7.5) instead.

solution to synthesize high inductance in a very small volumes could be to embed inductor wires within nano-structured (much smaller pitch than the unit cell of the MTM) materials made of nano-ferrite particles[5] in order to achieve both high permeability and low loss thanks to suppression of Foucault currents. This would give birth to a new class of MTMs, which may be termed *multiscale* MTMs, as three (or more) structural levels separated by one order of magnitude (or more) would be present: the overall MTM structure (e.g., 10 cm at 3 GHz), the unit cell of the MTM (e.g., 500 μm), and the unit cell of the effective structure embedding the inductors (e.g., 10 μm).

Another interest in being able to manipulate, as in the process of homogenization, the unit cell of CRLH MTMs is the possibility of controlling the bandwidth of the structure. This is particularly useful in 1D TLs. Assume a 1D CRLH TL structure with a given unit cell size p. If, *keeping p fixed*, one *increases* the product $L_L C_L$ and *decreases* the product $L_R C_R$, then bandwidth is increased, according to Eq. (3.80a) (lower LH highpass cutoff ω_{cL}) and Eq. (3.80b) (higher RH lowpass cutoff ω_{cR}), respectively, without any change in the transition frequency f_0 [Fig 3.22(a)]. Increase of bandwidth in the LH range, toward lower frequencies by reduction of ω_{cL}, may be beneficial for the design of super *slow-wave* LH structures, since very large values of $|\beta|$ ($\beta < 0$), and therefore highly compressed values of λ_g, can then be achieved by operating the structure just above ω_c. It should be noted however that this effect is narrow band in nature due to the very flat $\omega(\beta)$ response of the structure in this region of the dispersion diagram.

In the case of resonant-type MTMs, other configurations, with other challenges, will exist (Section 7.3). Although a bandwidth increase without prohibitive increase of losses is impossible in resonant-type MTMs (Section 1.7), homogenization of these structures will bring up a problematic of the type exposed in this section for CRLH MTMs.

7.2 QUASI-OPTICAL NRI LENSES AND DEVICES

Although the vast majority of investigations on MTMs recently reported in the literature have dealt with negative refraction in LH MTMs, very few practical applications of this phenomenon have been actually proposed. However, many applications will most likely emerge from the development of novel and high-quality MTMs. In addition, to more homogeneous dielectric-metal 2D MTMs (Section 7.1), 3D isotropic LH MTMs (Section 7.3) will probably become available soon, and there might even be avenues to develop optical LH MTMs (Section 7.4).

At microwave frequencies, a number of *quasi-optical devices* based on NRI can be envisioned and readily implemented in planar configurations such as the mushroom metastructures described in Section 4.5. In this section, as an example

---

[5]Nano-ferrites are composite ferrimagnetic or ferromagnetic materials made that include clusters of magnetic dipoles isolated from each other.

of NRI application, we present planar conical RH/LH focusing interfaces or
lenses, which exhibit unique advantages over conventional RH lenses. Although
RH microwave lenses been described in many textbooks as radomes to increase
the gain of antennas [3, 4], LH microwave lenses represent a new concept, which
paves the way for novel types of configurations with interesting characteristics.

The required interface curve to realize lenses with NRI MTM is a conic section
that can be obtained by applying the *optical path length* (OPL) technique [5] to
the generalized section shown in Fig 7.1(a). The sum of the optical path lengths
for any ray of angle θ from the origin O (focus of the conical section) to the
corresponding point P on the conical curve and from this point P to the point
Q in the plane $x = f$ must be equal to the optical path length for the ray with
$\theta = 0$ from the origin O to the point $L = r(0)$ in the same plane ($x = f$). If
this condition is met, all outgoing waves from O arrive in phase in the plane
$x = f$, so as to form an outgoing plane wave (collimation); reciprocally, all ray

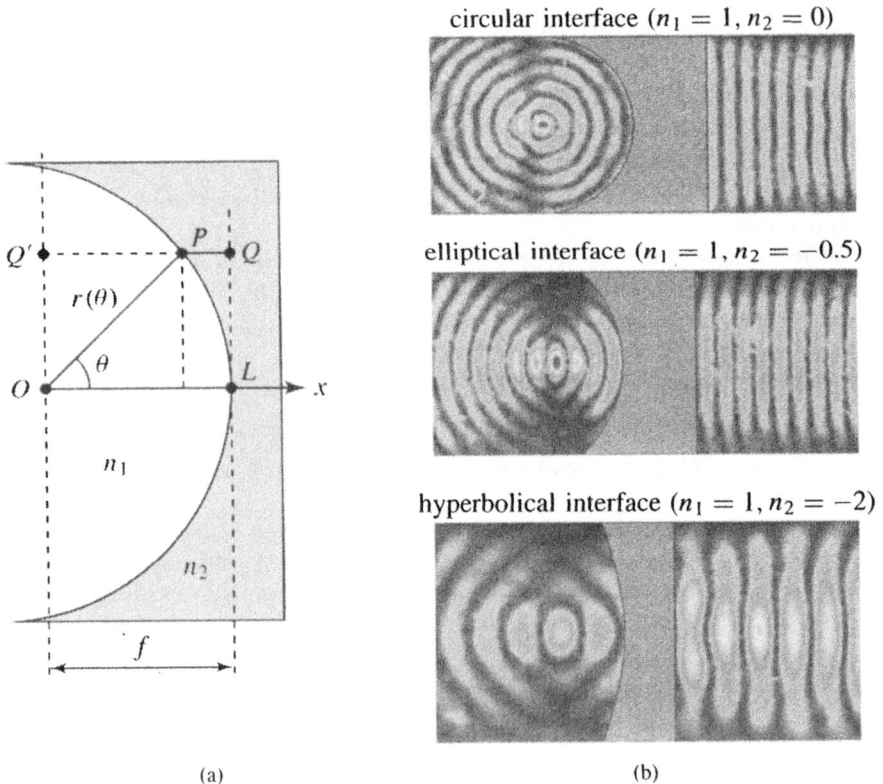

circular interface ($n_1 = 1, n_2 = 0$)

elliptical interface ($n_1 = 1, n_2 = -0.5$)

hyperbolical interface ($n_1 = 1, n_2 = -2$)

(a)   (b)

Fig. 7.1. Radiation focusing at the concave interface between a RH medium (medium 1
with refractive index n_1) and a MTM (medium 2 with refractive index n_2). (a) General
conical section interface for arbitrary ratio of refractive indexes $\Delta_n = n_2/n_1$. (b) FEM
full-wave simulated focusing effect for a spherical, an elliptical and a hyperbolical interfaces, respectively (Table 7.1).

contributions from an incident plane wave in the plane $x = f$ arrive in phase at O (focusing). Thus, circular (or spherical in a 3D configuration) wave to plane wave (or the reciprocal) is achieved.

The condition mentioned above reads

$$n_1 \overline{OP} + n_2 \overline{PQ} = n_1 \overline{OL}, \tag{7.7}$$

that is

$$n_1 r(\theta) + n_2 \left[f - r(\theta) \cos \theta \right] = n_1 f, \tag{7.8}$$

from which the following general conical function is obtained

$$r(\theta) = f \frac{n_1 - n_2}{n_1 - n_2 \cos \theta} = f \frac{1 - \Delta n}{1 - \Delta n \cos \theta}, \tag{7.9a}$$

where

$$\Delta n = \frac{n_2}{n_1} \tag{7.9b}$$

is the contrast of permittivities between the two media. By considering that medium 2 is a (CRLH) MTM, and allowing therefore negative and zero values for n_2, and that medium 1 is RH, we obtain the conical section solutions given in Table 7.1. It is seen that, depending on Δn, the required section is a circle, an ellipse, a parabola, or a hyperbola. It is worth noting (for instance by taking the limiting case $\theta \to 90°$) that, for a given contrast of electromagnetic densities Δ_n and a given focal length f, the interface between two media of different handednesses (RH and LH, $\Delta_n < 0$) yields a larger value of $r(\theta)$ and therefore a *larger radius of curvature* than an interface between two media of same handedness (RH and RH, or LH and LH). This represents an advantage for RH/LH interfaces over RH/RH interfaces, as interfaces with larger radius of curvature induce less spherical aberration due to smaller variations in the incidence angles [6] (Section 2.7). The full-wave simulation results shown in Fig . 7.1(a) confirm these predictions.

TABLE 7.1. Conical Section RH/LH Interfaces for Focusing/Collimation of Electromagnetic Waves.

$\Delta n = n_2/n_1$	Specific Equation $r(\theta)$	Type of Conical Section
$\Delta n = 0$	f	circle
$-1 \leq \Delta n \leq 0$	$f \dfrac{1 - \Delta n}{1 - \Delta n \cos \theta}$	ellipse
$\Delta n = -1$	$\dfrac{2f}{1 + \cos \theta}$	parabola
$\Delta n \leq -1$	$f \dfrac{1 - \Delta n}{1 - \Delta n \cos \theta}$	hyperbola

Conventional lenses, which are generally nonmagnetic ($\mu_{r2} = 1$), need to have a permittivity ε_{r2} different from that of their environment, in general air with $\varepsilon_{r1} = \mu_{r1} = 1$, to produce their optical effect. As a consequence, the intrinsic impedance of the lens $\eta_2 = \sqrt{\mu_2/\varepsilon_2}$ is different from that of air $\eta_1 = \sqrt{\mu_1/\varepsilon_1}$. This represents an impedance discontinuity that results in undesired reflection of part of the incident wave (see Fresnel coefficients, Section 2.8).

This drawback of conventional lenses suggested the idea of a *reflectionless parabolic (or perfectly matched) refractor*, which was first proposed and demonstrated numerically in [7] and then refined and demonstrated experimentally in [8]. The idea is to achieve focusing from an interface with two media of the same intrinsic impedance ($\eta_1 = \eta_2$), to suppress the reflections existing in conventional optical and quasi-optical lenses. This is possible by interfacing a LH medium and a RH medium with equal intrinsic impedances.[6] In this case, examination of the Fresnel coefficients [Eq. 2.59] reveals that, in the case of the parabolic interface where $\eta_1 = -\eta_2$ (Tab. 7.1), zero reflection transmission is achieved, for both polarizations under any angle of incidence, since $\theta_t = -\theta_i \forall \theta_i$ due to the absence of impedance discontinuity.[7] Consequently, the RH/LH parabolic refractor exhibits a focusing efficiency very superior to that of the RH/LH counterpart. The RH/LH *parabolic refractor* may be seen as the perfect refractive counterpart of the well-known parabolic reflector: the focusing and collimation effects are similar in both devices, except that in the case of the refractor the transformed waves are mirrored on the other side of the parabola section. Full-wave demonstration of the parabolic refractor is provided in Fig 7.2(a). A LH slab structure with same electromagnetic parameters is also shown in Fig 7.2(b) for comparison. The parabolic refractor was implemented in the 2D CRLH mushroom structure described in Section 4.5. It is shown in Fig 7.3 along with its simulated field patterns confirming the generation of plane waves from a point source.

The parabolic refractor prototype and its measured electric field are shown in Fig 7.4. Reasonable agreement can be observed between the FDTD and measured results. Although these results for the implemented refractor (Figs. 7.3 and 7.4) demonstrate the parabolic refractor concept, they are still far from those obtained with an ideal perfectly homogeneous MTM (Fig 7.2). However, much better results would be straightforwardly obtained after the structure has been homogenized, as explained in Section 7.1.

We see from these few examples that MTMs provide a new paradigm to do conventional optics and quasi-optics with nonconventional microwave structures [9], the main challenge for their successful integration into real applications being not of technological nature. An even greater diversity of novel devices will be

---

[6]It was shown in Chapters 1, 2, and 4 that the characteristic impedance $Z_c = \sqrt{L'_L/C'_L}$ or the intrinsic impedance $\eta = \sqrt{\mu_r/\varepsilon_r}$ of LH medium were positive quantities, whereas its propagation constant $\beta = -1/(\omega\sqrt{L'_L C'_L})$ or refractive index $n = -\sqrt{\mu_r \varepsilon_r}$ of such a medium were negative quantities.

[7]This is an effect comparable to the Brewster phenomenon. But this phenomenon exists exclusively at the angle $\theta_B = \tan^{-1}(n_2/n_1) > 0$ and for p-polarization at the interface between two RH media [9]. The difference here is that reflectionless transmission is obtained at all incidence angles.

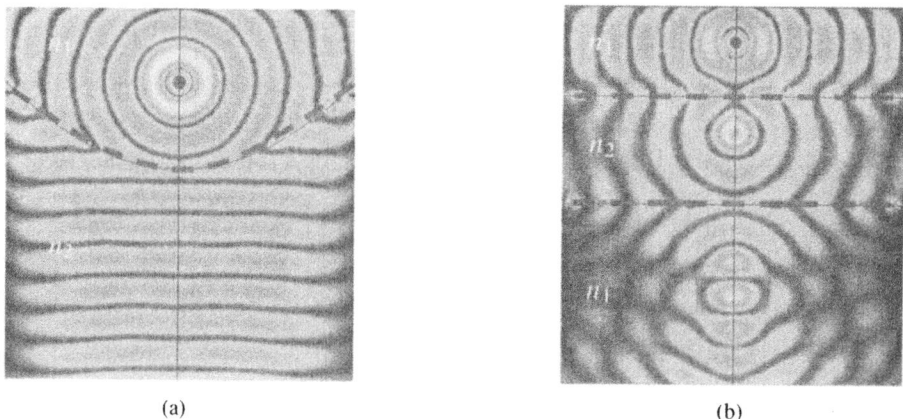

(a)   (b)

Fig. 7.2. FEM full-wave demonstration of refraction of at the interface between a RH medium with $\varepsilon_{r1} = \mu_{r1} = +1$ and a LH medium with $\varepsilon_{r2} = \mu_{r2} = -1$. Here hypothetical perfectly homogeneous LH slabs are considered. (a) Parabolic refractor. (b) LH slab. From [8], © IEEE; reprinted with permission.

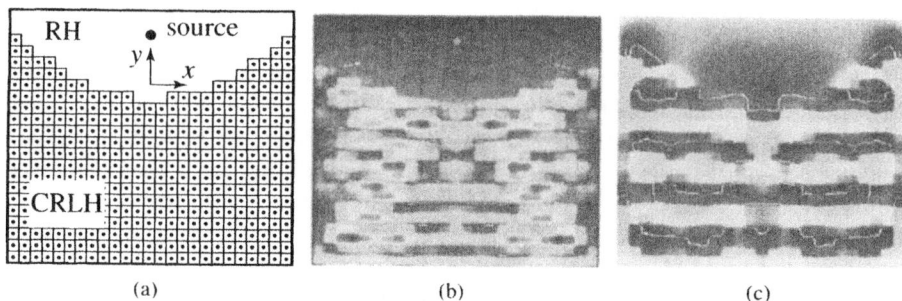

(a)   (b)   (c)

Fig. 7.3. Parabolic refractor implemented in the 2D CRLH mushroom configuration. The equation of the parabolic interface is given by $x^2 = 4fy$, with $f = 30$ mm. The entire refractor is implemented on a single piece of Rogers/RT 6010 ($h = 1.27$ mm, $\varepsilon_r = 10.2$). The required phase-matching of the RH region to the CRLH region occurs at $f = 3.80$ GHz, $|\beta p| = 72°$ (Section 4.5.4). (a) Structure. (b) FDTD-simulated E-field magnitude. (c) FDTD-simulated E-field phase. From [8], © IEEE; reprinted with permission.

available with the advent of 3D isotropic LH MTMs, the subject of the next section.[8]

7.3 THREE-DIMENSIONAL ISOTROPIC LH MTMs

Isotropic 3D effectively homogeneous structured materials, defined as structures whose "atoms" are small metal or dielectric objects large compared with atomic

---

[8]For instance, an immediate application would be the 3D implementation of the parabolic refractor described in this section.

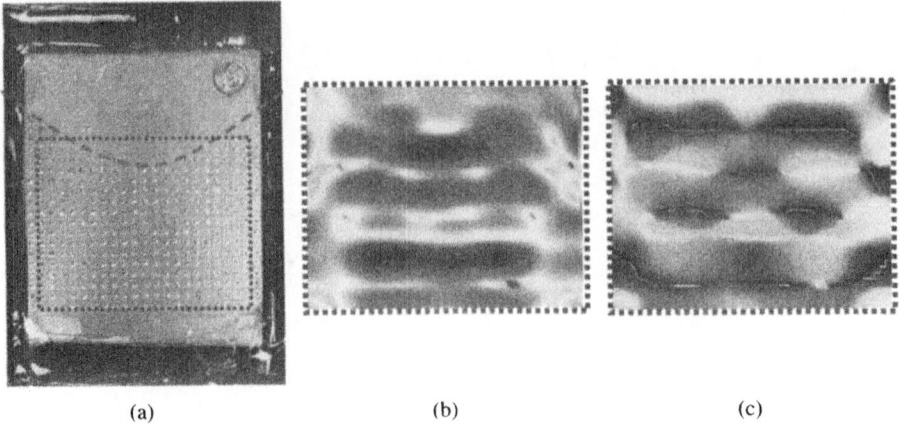

(a)   (b)   (c)

Fig. 7.4. Parabolic refractor implemented in a CRLH mushroom configuration interfaced with a PPWG structure. (a) Structure. (b) Measured E-field magnitude. (c) Measured E-field phase. Measurements are performed at 3.77 GHz at 2 mm atop the CRLH region. The CRLH region contains approximately 420 unit cells. The outlined region measures 80×100 mm^2. From [8], © IEEE; reprinted with permission.

dimensions but still small compared with the wavelength of the electromagnetic waves acting in the "host" medium in which these inclusions are embedded (e.g., [5, 10, 11, 12]),[9] and often called *artificial dielectrics*, have been studied for many years. However, it seems that no specific efforts have been dedicated toward the conception of 3D isotropic artificial structures with negative constitutive parameters, that is, 3D isotropic LH MTMs. Recently, some authors have revisited the topic of artificial dielectrics in light of recent research on MTMs and, based on the works of Lewin on the macroscopic properties of a 3D array of particles embedded in a background matrix [13], have proposed theoretical solutions for a 3D LH MTM [14, 15]. These solutions are very attractive because they represent the first specific suggestions for the realization of 3D isotropic LH MTMs.[10]

The relative effective permittivity $\varepsilon_{r,eff}$ and effective permeability $\mu_{r,eff}$ for a 3D array with average lattice constant p of spherical particles of radius r and of relative permittivity and permeability ε_p and μ_p embedded in a host

---

[9]According to the definition of Section 1.1, these well-known structures could thus be regarded as MTMs if they would exhibit properties not readily available in natural materials, but this is generally not the case.

[10]Note that the two-dimensionally LH MTM structure of Smith (Section 1.3 and Fig 1.4), in addition to being very complex and difficult to fabricate, cannot be easily extended to 3D. It seems that, *separately*, the TWs and the SRRs can be implemented in 3D architectures to yield 3D negative-ε [16] and 3D negative-μ [17] structures, respectively, but that a 3D LH MTM cannot be generated with the *combination* of the SRRs and TWs, due to the impossibility to generate in such a 3D configuration the required symmetry necessary to avoid parasitic interactions between the SRRs and the TWs, which to tend to eliminate left-handedness (Section 1.3, footnote 11).

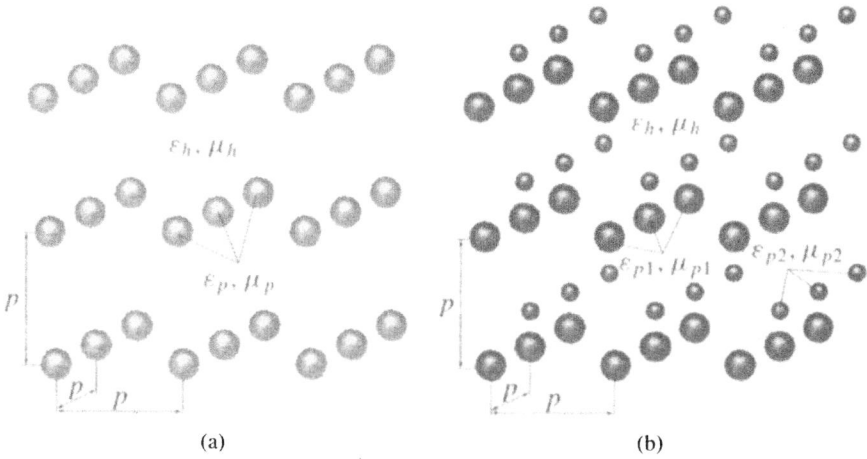

Fig. 7.5. Resonant-type 3D isotropic LH MTM ($p \ll \lambda_h$). (a) Array of magneto-dielectric spheres operating at their overlapping magnetic and electric Mie resonances [14]. (b) Double-array of dielectric spheres of two different diameters operating at the overlapping resonance frequency of TE_{011} (smaller spheres, magnetic moments) and TM_{011} (larger spheres, electric moments) [15].

medium of relative permittivity and permeability ε_h and μ_h [Fig 7.5(a)] are given by [13][11]

$$\varepsilon_{r,eff}(\omega) = \varepsilon_h \left(1 + \frac{3v_f}{\frac{F(\theta)+2K_e}{F(\theta)-K_e} - v_f} \right), \quad \text{where} \quad K_e = \frac{\varepsilon_h}{\varepsilon_p}, \qquad (7.10a)$$

$$\mu_{r,eff}(\omega) = \mu_h \left(1 + \frac{3v_f}{\frac{F(\theta)+2K_m}{F(\theta)-K_m} - v_f} \right), \quad \text{where} \quad K_m = \frac{\mu_h}{\mu_p}, \qquad (7.10b)$$

with

$$F(\theta) = \frac{2(\sin\theta - \theta\cos\theta)}{(\theta^2 - 1)\sin\theta + \theta\cos\theta}, \quad v_f = \frac{4}{3}\pi \left(\frac{r}{p}\right)^3, \quad \theta = k_0 r \sqrt{\varepsilon_p \mu_p}. \qquad (7.10c)$$

In these expression, K_e and K_m are the contrasts between the permittivities and permeabilities, respectively, of the host and particles media, and v_f is the volume filling ratio of the spherical particles with respect to the host medium. The frequency ω is included in the variable θ via $k_0 = \omega/c_0$ and is therefore present in the function $F(\theta)$. The constitutive parameters given by Eq. (7.10)

[11] These formulas were later generalized by Khizhniak for tensors for artificial material formed by an array of scatterers with arbitrary geometric shapes [11].

are clearly of *resonant* nature, just as the constitutive parameters of Pendry's plasmonic-type structures (Section 1.3): there exists a frequency ω_{0e} where $\varepsilon_{r,eff}$ exhibits a pole and a frequency ω_{0m} where $\mu_{r,eff}$ exhibits a pole. Because a pole is associated with inversion of the sign of the real part of the constitutive parameters, there are possibilities to design Lewin's type structures where the electric and magnetic resonances coincide, $\omega_{0e} = \omega_{0m}$. This is the idea exploited by Holloway et al. in [14] and by Vendik et al. in [15].

The physical principle underpinning Lewin's approach is the excitation of the *Mie resonances*[12] or resonance modes of the spheres [18]. The polarization and magnetization densities of the effective medium constituted by an array of spheres are related to the distribution of the scattering spheres and of their polarizabilities according to the formulas of Eq. (7.10). Two important conditions must be met to obtain a real effective medium based on Mie resonances. First, the electromagnetic dimension of the spheres must be of the order of the wavelength *inside them*

$$r \sim \lambda_{g,p} = \lambda_0/n_p, \quad \text{where} \quad n_p = \sqrt{\varepsilon_p \mu_p}, \tag{7.11}$$

so that they resonate. Second, the average lattice constant must be much smaller than the *effective wavelength*

$$p \ll \lambda_{g,eff} = \lambda_0/n_{eff}, \quad \text{where} \quad n_{eff} = \sqrt{\varepsilon_{eff} \mu_{eff}}, \tag{7.12}$$

to ensure that the medium is indeed effective. By combining these two relations, we obtain that

$$p \ll r\sqrt{\frac{\varepsilon_p \mu_p}{\varepsilon_{eff} \mu_{eff}}} = r\frac{n_p}{n_{eff}} < r\frac{n_p}{n_h}, \tag{7.13}$$

which shows that a Mie resonance-based MTMs requires a *high contrast of refractive indexes* between the scattering particles and the host medium, since, geometrically, we must necessarily have $2r < p$.

Another crucial consideration in this type of structure, as in any resonating structure, is the quality factor of the resonators, or their dielectric and/or magnetic losses. If these losses are too high, they damp out the resonances, so that the waves are rapidly attenuated inside the structure and no transmission occurs through it.

---

[12]In the problem of scattering by a dielectric or metallic object, three scattering regions are generally distinguished [19]. Consider for instance a spherical object of radius r. If the sphere is metallic, the region $r < 0.1\lambda_0$ is called the *Raleigh* region, where the RCS of the sphere is linearly increasing with r/λ_0; the region $r > 2\lambda_0$ is called the *optical* region, where the RCS value approaches πr^2, that is, the physical area of the cross section of the sphere; the region $0.1\lambda_0 < r < 2\lambda_0$ is called the *Mie* or resonance region, where the RCS is an oscillatory function of p/λ_0 around the value πr^2. If the sphere is dielectric, with internal wavelength $\lambda_{g,p} = \lambda_0/\sqrt{\varepsilon_p}$, where ε_p is the relative permittivity of the sphere and $\varepsilon_{g,p} \gg \varepsilon_h$ (ε_h: permittivity of the host medium) so that the sphere's surface may be approximated by a magnetic wall, the Mie resonances occur in the region $0.1\lambda_{g,p} < r < 2\lambda_{g,p}$.

Holloway et al. [14] proposed an array of *magnetodielectric* spherical particles, as shown in Fig 7.5(a), to realize a 3D isotropic LH MTM. Unfortunately, the magnetodielectric particles required for the realization of such a medium must exhibit simultaneously strongly *positive* values of ε_p and μ_p (e.g., $\varepsilon_p = 200$, $\mu_p = 200$) in order to render $\varepsilon_{r,eff}$ and $\mu_{r,eff}$ in Eq. (7.10) negative in the same frequency range, which is a combination not physically attainable at microwave frequencies. To circumvent this difficulty, the authors point out the possibility, already previously suggested by Khizhniak et al.[11], to use particles that would be composite themselves, such as particles coated with a layer of a different material (either dielectric particles coated with a magnetic materials or the opposite). They also numerically show that even relatively moderate losses totally suppress the LH range of the structure.

Due to the impractical nature of the LH MTM proposed in [14],[13] Vendik et al. suggested an interesting alternative Lewin's type LH MTM [15]. It is the structure depicted in Fig 7.5(b), which consists of a double array of *purely dielectric* spheres of two diameters $2r_1$ and $2r_2$ ($\varepsilon_{p1}, \varepsilon_{p2} \gg \varepsilon_h$, but $\mu_{p1} = \mu_{p2} = 1$).[14] The diameters of the spheres are adjusted so that the lowest Mie resonance mode, the TE_{011} (smaller spheres), and the second Mie resonance mode, the TM_{011} (larger spheres), are excited in the same frequency range. In this structure, the TE_{011} and TM_{011} resonances emulate resonating magnetic and electric dipole moments, respectively, from which left-handedness is obtained.[15] As pointed out previously, this structure should be realizable relatively easily with single crystal ferroelectric or ceramic balls.

The 3D isotropic LH MTMs described above are of resonant nature and are therefore characterized by a narrow bandwidth or by high losses. As in mono-dimensionally and bi-dimensionally LH MTMs, a three-dimensionally LH *nonresonant* TL-type MTM would thus be highly desirable. Such structures have not been demonstrated yet. Fig 7.6 shows an artistic representation of a futuristic CRLH 3D MTM structure. In this structure, the ground plane, not possible in a 3D configuration, has been replaced by interconnections between neighboring cells to provide the "shunt" inductances L'_L, whereas the series capacitances C'_L are simply provided by PPWG metal plates.

---

[13]Of course, a structure impractical today may become realistic later with the development of novel micro/nano-materials and processes. With respect to the suggestion of using composite magnetodielectric particles, we note that a 1D composite ferrimagnetic-ferroelectric LH structure that could be realized easily was already described theoretically [20].

[14]For this structure, Lewin's formulas would require an adjustment, since the field scattered by each lattice site is due to a "bi-atomic" (two spheres) basis. However, in the case of spheres with very high permittivity—which is the case of interest here ($\varepsilon_r \approx 1,000$), the interaction between the large and the small spheres is negligibly small because the fields are strongly concentrated inside the spheres as opposed to between them. This fact was confirmed by full-wave simulations, where results close to those predicted by Lewin's formulas were obtained [21].

[15]In the case of an incident plane wave, the electric dipoles are excited in the TM_{011} mode of the smaller spheres along the direction of the incident electric field, whereas the magnetic dipoles are excited in the TE_{011} mode of the smaller spheres along the direction of the incident magnetic field, so that the electric and magnetic dipoles are orthogonal to each other.

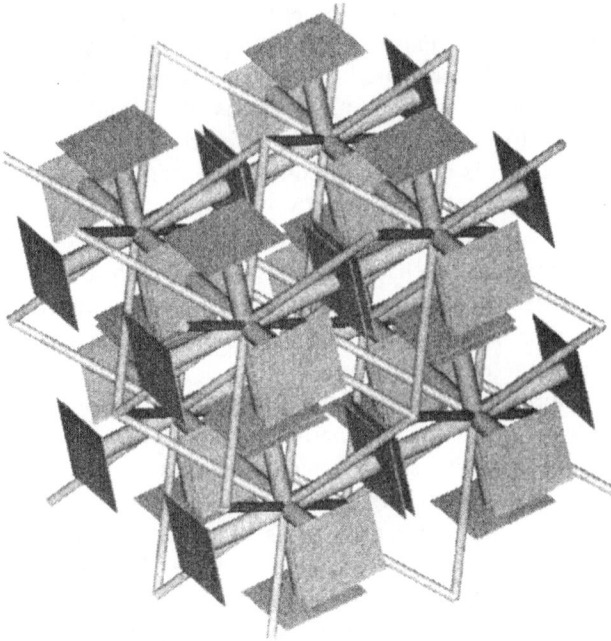

Fig. 7.6. Artistic representation of possible microwave 3D isotropic CRLH MTM. All the gray areas represent metal.

7.4   OPTICAL MTMs

Although the phenomenon of refraction is more related to optics than to microwave effects, no optical LH MTM (even 1D or 2D) seems currently available. Structures including metal, such as the SRR-TW structure or TL structures, are difficult to transpose to optical frequencies because nobel metals behave as plasmas and not as conductors at optical wavelengths, where magnetism "disappears" [22]. In addition, up-scaling of metallo-dielectric MTM structures to optical dimension represents a difficult fabrication challenge.

Unfortunately, no nonmetallic LH MTM has been proposed to date. Backward waves in dielectric Bragg gratings have been known for decades, but these backward waves are always associated with space harmonics of the global field as opposed to real fundamental-mode waves (Section 1.9), so that these waves are based on diffraction/scattering phenomena, not appropriate for refraction. Even the suggestions for 3D MTMs made of dielectric particles described in the previous sections cannot be considered for optical operation because the high permittivities required are not available in the optical range.

There have been a few suggestions to achieve LH materials in the infra-red (IR) and visible regimes. These suggestions include the possibility of using plasmonic parallel nanowires [23], modified SRRs in the near-IR region [24], closely-packed inclusions with negative permittivity and their electrostatic resonances

[25], and defects in regular photonic band gap structures [26]. Another idea, which bears some similarity with Lewin's approach described in the previous section, but with metallic instead of magnetodielectric spheres, has already been suggested [27]. This idea consists in producing negative effective permeability and permittivity by nano-inclusions made of metallic nanospheres excited in their magnetic dipole resonance and electric dipole resonances, respectively, at optical frequencies. This approach is based on the collective resonance of the array of plasmonic nanospheres arranged in a specific pattern (e.g., in a circular pattern) to form a single subwavelength "loop macro-inclusion," where it is not the conventional conduction current (as in the SRR-TW structure at microwaves) that produces the magnetic dipole moment but is the plasmonic resonant feature of every nanosphere, that induces a circulating *"displacement"* current around this loop.

7.5 "MAGNETLESS" MAGNETIC MTMs

As pointed out in Chapter 1, MTMs are not restricted to LH or CRLH structures. Among the interesting other types of MTMs, we may mention "magnetless" magnetic MTMs, which are effective structures exhibiting magnetism without including magnetic components, such as the "Swiss roll" structure [17], which is a variation of structures previously described in the literature [28, 29].

A Swiss roll is a cylinder constituted by closely packed concentric layers of metal obtained by wounding a conducting sheet in spiral around a central mandrel. The overall structure is made by a bundle of parallel rolls arranged in a closely packed array. When it is excited by a magnetic field parallel to the axis of the rolls, an electric current is induced in the spiral of the rolls, and the distributed capacitances between the turns enable transverse current flow. Due to the absence of metal connectivity in its transverse plane, the structure cuts off direct current (DC) and allows only AC flow in RF operation from the self-capacitance of the rolls, which induces a negative real part of the permeability (in the direction of the axis of the rolls) just above the resonance frequency, in a manner similar to the SRR structure [Section 1.3 and Eq. (1.4)].

The flux guiding property of the Swiss roll structure has been demonstrated experimentally [30] by enhancement of magnetic energy transmission between two coils from a roll placed between the coils and by a magnetic resonance imaging (MRI) experiment, where comparable image quality is obtained with the Swiss roll structure as with "body coils" built into the structure of the magnet. In this structure, the absence of DC magnetism is an important property to avoid perturbing the static and low-frequency magnetic fields needed to obtain image and spectral data. The authors [30] considered this concept of "magnetless" magnetic structure as a new paradigm for the manipulation of RF flux in MRI and spectroscopy systems. A related Swiss roll-based endoscope for near-field imaging, utilizing a Swiss roll medium of much greater permeability and lower loss with highly uniaxial anisotropy, has also been described [31].

7.6 TERAHERTZ MAGNETIC MTMs

Materials exhibiting a magnetic response, that is, materials with nonzero magnetic polarization [$\overline{\mathcal{M}} = \mu_0 \chi_m \overline{\mathcal{H}} \neq 0$ or $\chi_m > 0$ or $\mu_r = 1 + \chi_m > 1$ in Eq. (2.2a)], are extremely rare at terahertz (THz) and optical frequencies. There is a fundamental physics reason for this fact [22], which may be formulated as follows [32]: Magnetic polarization in materials follows indirectly either from the flow of orbital currents or from unpaired electron spins. In magnetic systems, resonant phenomena, analogous to the phonons or collective modes that lead to an enhanced electric response at infrared or higher frequencies, tend to occur at far lower frequencies, resulting in relatively little magnetic response at THz and higher frequencies.

Yen et al. [32] proposed a MTM-based avenue to redeem the electromagnetic imbalance of nature and provide magnetic materials in addition to the available electric dielectric materials at THz frequencies. Their approach consisted in generating artificial magnetic moments by exploiting the resonance of SRR particles in an SRR MTM [17]. When this structure is excited by a time-varying magnetic field with a component perpendicular to the plane of the rings, solenoidal currents are induced along the rings, giving rise to strong artificial magnetic moments at the resonance of the rings and resulting in a structure exhibiting a resonant-type macroscopic magnetic polarization. Although such a structure is resonant and is therefore characterized by a modest fractional bandwidth, such a bandwidth may be sufficient in absolute terms for THz applications. In addition, the operation frequency of the SRR MTM is tunable by proper adjustment of the structural design parameters.

An ellipsometry experiment was carried out [32] on a planar SRR MTM structure to demonstrate the magnetic response of the medium. The SRR structure tested consisted of a monolayer 2D square lattice (lattice constants used: 36, 44, and 50 µm) of SRRs deposited on substrate. In this experiment, the ratio between the reflectances of the two polarizations p ($E\|$ plane of incidence) and s ($E\perp$ plane of incidence) (Section 2.8) was measured with a Fourier transform interferometer. In the p-polarization case, the magnetic field has a component that is perpendicular to the structure (case of oblique incidence), and was therefore expected to excite the SRRs, whereas in the s-polarization case, the magnetic field is purely parallel to the plane of structure, and was therefore expected not to excite the SRRs. This ratio is therefore a measure of the magnetic response of the structure. Reflectance ratios significantly larger than 1 were measured with peaks at the frequencies corresponding to the simulated resonances of the SRRs in the range of 0.6 to 1.6 THz, which demonstrated the magnetic response of the MTM.[16]

---

[16] A broad resonance bandwidth of 30% is reported in the paper. However, the measured reflectance ratio is a relative quantity that does not account for the absolute amount of energy transferred from the source to the detector in each polarization. It is likely that this amount is very small and that a lot of energy dissipated in the strips of the rings due to their very high ohmic losses of metal at these frequencies. It is then not surprising that a broad bandwidth is observed: as pointed out in Section 1.7, bandwidth is inversely proportional to losses in a resonating structure, so that very broad bandwidth may be achieved at the expense of high losses.

More recently, a THz LH SRR-TW MTM structure (exhibiting both magnetic and electric artificial response) was demonstrated by Moser et al. [33]. This MTM, consisting of a 2D array of Ni and Au SRRs and rods embedded in an AZ P4620 photoresist matrix, was shown by a transmittance measurement to exhibit LH resonance between 1 and 2.7 THz. This represents an extension by about three orders of magnitude the spectral range accessible than in previously reported SRR-TW MTMs.

The realization of tailorable artificial electric and magnetic materials in the THz range may represent an important step toward bridging the gap between microwave/millimeter-wave and optical frequencies. It has been suggested that they may lead to the development of novel devices for biological and security imaging, biomolecular fingerprinting, remote sensing, and guidance in near-zero visibility weather conditions [32]. Such THz MTMs and derivatives may also play an important role in the development of engineering devices, such as THz sources, harmonic generators, photo-mixers, and others [34].

7.7 SURFACE PLASMONIC MTMs

Surface plasmons (SPs), also called surface polaritons, are electromagnetic waves propagating at the interface between two different media.[17] SPs have been utilized almost exclusively at optical frequencies because they require negative permittivity, as will be shown below, which is readily available only at these frequencies, from metals[18] [35]. In the microwave range, such interface waves, which are generally referred to as *Zenneck waves* [36], cannot be used in practical devices because their fields would be too loosely bound to the interface of available materials [5]. However, the advent of MTMs, with tailorable constitutive parameters, opens the way for novel SP solutions and structures *at any frequency*. This inspired to Caloz et al. the idea of producing *microwave surface plasmons*, based on MTMs with appropriate $\varepsilon - \mu$ parameters for high field confinement; these SPs could lead to the development of a novel class of microwave waveguiding and resonating structures [37].

To derive the SP dispersion relation, let us consider the canonical problem of a plane wave incident upon the interface between two arbitrary media (which may be MTMs), first in the case of perpendicular polarization, which is illustrated in Fig 2.6. The case of parallel polarization will follow by duality. From the scalar wave equation $(\nabla^2 + k_i^2)E_y = 0$ (Fig 2.6), where $k_i = (\omega/c)\sqrt{\varepsilon_{ri}\mu_{ri}}$ and $i = 1, 2$ labels the region of interest, the tangential wavenumber k_x is related to the normal wavenumber k_{iz} by

$$k_x^2 = \varepsilon_{ri}\mu_{ri}\left(\frac{\omega}{c}\right)^2 - k_{iz}^2, \tag{7.14}$$

[17]Let us recall that SPs participate in the subwavelength diffraction phenomenon in the LH slab, as pointed out in Section 2.11 and illustrated by TLM simulations in Section 4.4.4.

[18]Most metals behave as plasmas at optical frequencies, exhibiting a permittivity function of the form $\varepsilon_r(\omega) = 1 - (\omega_p/\omega)^2$ where ω_p is the plasmon frequency.

and from the boundary condition at the interface, the reflection coefficient is given by Eq. (2.59c)

$$R_\perp = \frac{\mu_{r2}k_{1z} - \mu_{r1}k_{2z}}{\mu_{r2}k_{1z} + \mu_{r1}k_{2z}}. \tag{7.15}$$

Because an SP corresponds to propagation of a guided wave along the interface, the incident energy should completely couple to this wave and no reflection should occur at frequencies where the SP exists. This leads, with Eq. (7.15), to the *Brewster condition*

$$R_\perp = 0 \quad \Leftrightarrow \quad \mu_{r2}k_{1z}^\perp = \mu_{r1}k_{2z}^\perp. \tag{7.16}$$

The perpendicular-polarization RH-LH interface dispersion relation is then obtained by inserting k_{1z} of Eq. (7.16) into Eq. (7.14) with $i = 1$ and subsequently using Eq. (7.14) with $i = 2$ to eliminate k_{2z}. The parallel-polarization relation is obtained by exact duality. We obtain thus

$$k_x^\perp(\omega) = \frac{\omega}{c} \sqrt{\frac{\mu_{r1}\mu_{r2}(\varepsilon_{r1}\mu_{r2} - \varepsilon_{r2}\mu_{r1})}{\mu_{r2}^2 - \mu_{r1}^2}}, \tag{7.17a}$$

$$k_x^\parallel(\omega) = \frac{\omega}{c} \sqrt{\frac{\varepsilon_{r1}\varepsilon_{r2}(\mu_{r1}\varepsilon_{r2} - \mu_{r2}\varepsilon_{r1})}{\varepsilon_{r2}^2 - \varepsilon_{r1}^2}}, \tag{7.17b}$$

which are related to the z component of the wavenumber in each medium from Eq. (7.14) by

$$k_{iz} = p_i - j\alpha_i = \begin{cases} \sqrt{k_i^2 - k_x^2} & \text{if } |k_x| < |k_i| \\ -j\sqrt{k_x^2 - k_i^2} & \text{if } |k_x| > |k_i| \end{cases}. \tag{7.18}$$

In the *slow wave* range $|k_x| > |k_i|$, k_{iz} is purely imaginary and therefore the wave decays exponentially away from the source as $e^{-jk_{iz}z} = e^{-a_iz}$ [sgn(a_i) = sgn(z)]; this is the condition for *nonradiative* SP formation, that is, propagation of a SP wave completely guided at the interface. The larger the value of k_x, the larger the attenuation factor a_i and therefore the more confined the energy at the interface. At the limit $k_x \to \infty$, $a_i \to \infty$ and energy is perfectly confined: this occurs at the *SP resonance frequency* ω_{SP} given by the pole $\mu_{r2}(\omega) = \pm\mu_{r1}$ of Eq. (7.17a) for $k_x^\perp$ and by the pole $\varepsilon_{r2}(\omega) = \pm\varepsilon_{r1}$ of Eq. (7.17b) for $k_x^\parallel$.

In comparison, the conventional (optical) SP only exists for parallel polarization and exhibits the dispersion relation obtained by setting $\mu_{r1} = \mu_{r2} = 1$ in Eq. (7.17b)

$$k_x(\omega) = k_x^\parallel(\omega) = \frac{\omega}{c} \sqrt{\frac{\varepsilon_{r1}\varepsilon_{r2}}{\varepsilon_{r1} + \varepsilon_{r2}}}, \tag{7.19}$$

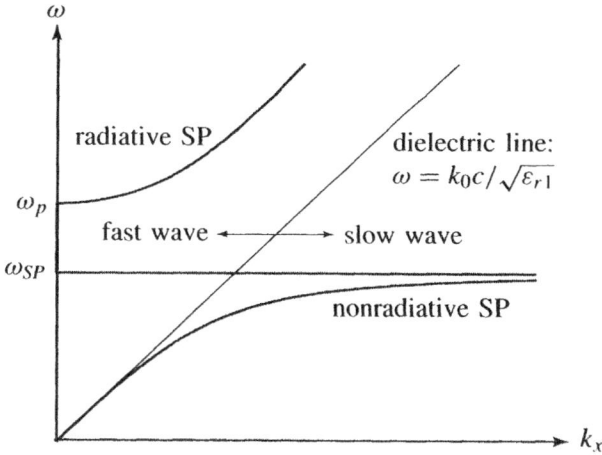

Fig. 7.7. Conventional SP computed by Eq. (7.19) (parallel polarization). From [37], © IEEE; reprinted with permission.

with the plasmonic permittivity $\varepsilon_{r2}(\omega) = 1 - (\omega_p/\omega)^2$ ($|\varepsilon_{r2}| > \varepsilon_{r1}$), plotted in Fig 7.7 [35]. This SP has a fundamental *nonradiative* (purely bound) mode from DC to ω_{SP}, since this mode is restricted to the slow-wave region of the dispersion diagram, and has a first higher-order *radiative* SP mode above ω_p.[19]

It can be observed that the generalized metamaterial SP of Eqs. (7.17a) and (7.17b) differs from the conventional SP in the following ways: 1) thanks to additional degrees of freedom in the constitutive parameters, the fundamental SP mode can be radiative because the *fast-wave* condition $|k_x| < |k_i|$ can be achieved in Eq. (7.18), yielding the z variation $e^{-jk_{iz}z} = e^{-jp_i z}$ ($p_i > 0$); 2) two SP resonances can exist due to the double pole $\mu_{r2} = \pm\mu_{r1}$ in Eq. (7.17a) and $\varepsilon_{r2} = \pm\varepsilon_{r1}$ in Eq. (7.17b), respectively; 3) an additional general condition for SP formation is that k_x be real, or that the radicands in Eqs. (7.17a) and (7.17b) be positive, so that generalized SPs may be limited to more restrictive frequency ranges than the conventional SP; and 4) a great diversity of different types of dispersion relations may be obtained depending on the parameter values.

If medium 2 is a TL CRLH MTM, the constitutive parameters of 2 are given by Eq. (3.3)

$$\varepsilon_{r2}(\omega) = \frac{C_R - 1/(\omega^2 L_L)}{p\,\varepsilon_0}, \tag{7.20a}$$

$$\mu_{r2}(\omega) = \frac{L_R - 1/(\omega^2 C_L)}{p\,\mu_0}, \tag{7.20b}$$

---

[19]The terminology "radiative" is commonly used in the SP literature [35]. Its meaning will be shortly clarified.

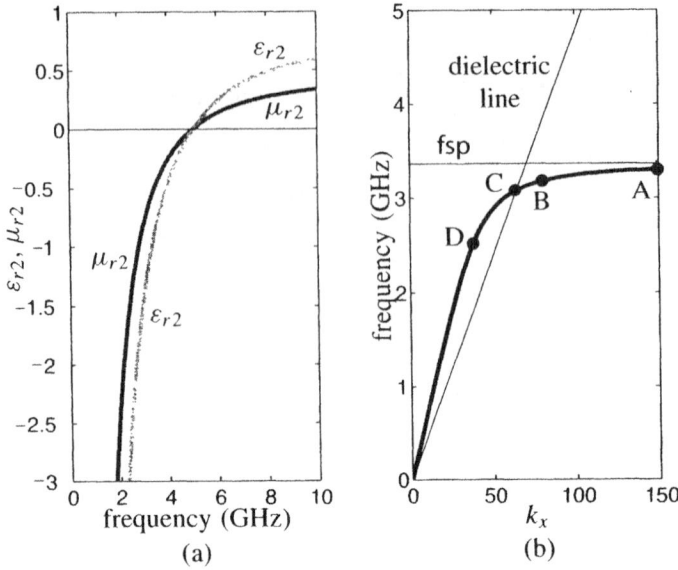

Fig. 7.8. (a) CRLH dispersive constitutive parameters computed by Eqs. (7.20a) and (7.20b) for the parameters $p = 50$ mm, $L_R = 50$ nH, $C_R = 0.2$ pF, $L_L = 5$ nH, $C_L = 0.02$ pF. For these specific design parameters, the permittivity and permeability become positive at the same frequency (balanced CRLH), 5.03 GHz. (b) Corresponding perpendicular polarization dispersion relation [Eq. (7.17a)] for the interface between air and the CRLH materials. The SP frequency, given by Eq. (7.21a), is $f_{SP} = 3.35$ GHz and the intersection with the air line occurs at 3.08 GHz. From [37], © IEEE; reprinted with permission.

where p is the period of the structure, and the corresponding SP resonances are therefore found to be

$$\omega_{SP}^{\perp} = \frac{1}{\sqrt{C_L(L_R \pm \mu_{r1}\mu_0)}}, \qquad (7.21a)$$

$$\omega_{SP}^{\parallel} = \frac{1}{\sqrt{L_L(C_R \pm \varepsilon_{r1}\varepsilon_0)}}. \qquad (7.21b)$$

We observe in these relations that *two SP resonances* (and therefore two SP branches), corresponding to the plus and minus signs in these expressions, may exist for each polarization at a RH/LH interface, whereas only one SP mode is available at the interface between two media of same handedness.

The constitutive parameters of Eqs. (7.20) and their corresponding perpendicular dispersion relation of Eq. (7.17a) are plotted in Fig. 7.8 for a particular interface. The apparently larger-than-light (if dielectric = vacuum) group velocity (slope of the dispersion curve) above the dielectric line in Fig 7.8 does not

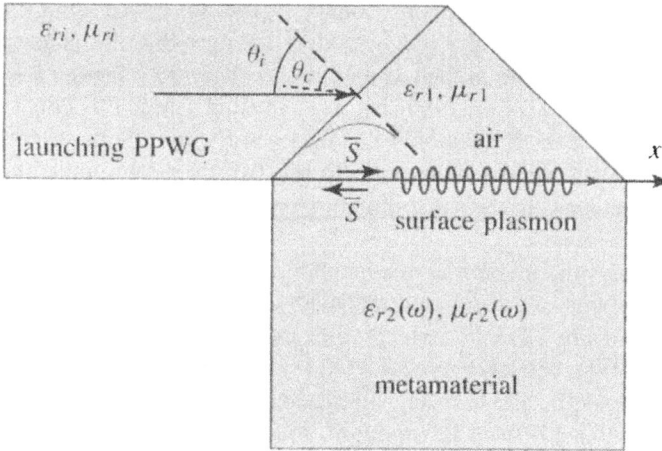

Fig. 7.9. PPWG simulation setup for the observation of the SPs. The incidence angle is $\theta_i = 45°$ and the constitutive parameters of media i and 1 are fixed to $(\varepsilon_{ri}, \mu_{ri}, n_i) = (2.5, 1, 1.58)$ and $(\varepsilon_{r1}, \mu_{r1}, n_1) = (1, 1, 1)$ (vacuum), respectively. The critical angle is consequently $\theta_c = \arcsin(n_1/n_i) = 39.2°$. The constitutive parameters of medium 2 are adjusted for each frequency to follow the CRLH frequency-dispersion of Eqs. (7.20a) and (7.20b) for medium 2, for the parameters of Fig 7.8, the data of which at the points A, B, C and D, are reported in Table 7.2. The arrows labeled $\overline{S}$ represent the Poynting vector at each side of the interface. From [37], © IEEE; reprinted with permission.

violate any physics: k_{1z} becomes real in this region and energy is simply refracted through the interface with zero reflection [Brewster condition, Eq. (7.16)]. Thus, because energy does *not* propagate along the interface, the slope of the curve in the fast-wave region does not represent the group velocity (not a mode of the interface), which solves this apparent paradox. The "radiative" part of the curve simply corresponds to *conventional refraction at the Brewster angle.*[20]

The L-shaped PPWB structure shown in Fig 7.9 is used to numerically observe the SPs. The material labeled i is the material from which the TEM plane wave is launched, and the interface of interest is the interface between media 1 and 2. Since nonradiative SPs needs to be excited by evanescent waves,[21] we use a scheme similar to that of the attenuated total reflection (ATR) [35] used in optical SPs: The wave is incident on the interface with an angle θ_i larger than the critical angle, so that total internal reflection occurs in medium i and only evanescent waves penetrates in medium 1, as required to excite the nonradiative SP.

---

[20]Note that this Brewster angle curve is a continuous function of frequency, as a consequence of the fact that the constitutive parameters are dispersive.

[21]This should be clear from Eq. (7.18). For a nonradiative SP, $k_x > |k_i|$. Therefore, $k_{iz} = -j\alpha_i$ and the perpendicular dependence of the mode is the evanescent function $e^{-\alpha_{iz}}$. Consequently, the incident energy should have the same exponentially decaying form to couple to the SP if it is incident perpendicularly on the interface (z direction). Another possibility, which may be more practical for applications, would be to excite the SP by a coupling structure directed along the direction of the interface.

It is worth noting that, as a consequence of the phase matching condition of Eq. (2.51) at the interface, $k_{x1} = k_{x2}$, and of the fact that the phase and group velocities must be opposite in the LH medium, *the Poynting vectors have opposite directions on either side of the interface*, as suggested in Fig 7.9. This fact may be paralleled with the case of a coupled-line backward-wave coupler (Section 5.4.1), where the two lines would correspond to the two media building the interface here; phase propagates in the same direction on each line, but power is oppositely directed on either line.

Fig 7.10 shows the numerical results obtained by full-wave FEM simulations. In these simulations, the metamaterial (medium 2) is restricted to its LH range, which is seen in Fig 7.8(a) to extend from DC to around 5 GHz. The expected nonradiative SP can be clearly observed in Figs. 7.10(a), (b), and (c), corresponding to the points A, B, and C in Fig 7.8(b), and phase animations clearly showed propagation of this SP from left to right. We can observe that, as frequency is decreased from ω_{SP}, the guided SP wavelength is also decreased, as expected from Fig 7.8(b) with the excellent agreement reported in Table 7.2, and that the SP becomes less and less confined to the interface, as a consequence of the fact that a_i decreases in Eq. (7.18). Fig 7.10(c) corresponds to the limit case where $\omega = k_0 c / \sqrt{\varepsilon_{r1}}$; therefore, according to Eq. (7.18), $k_{z1} \to \infty$, which is a trend

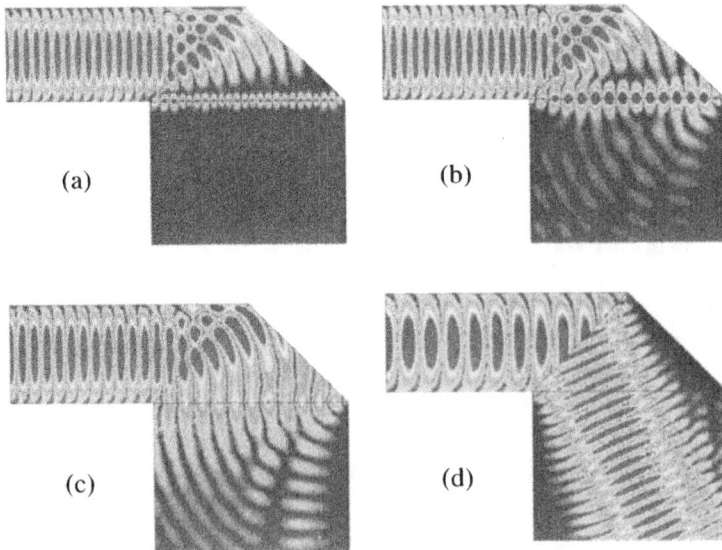

Fig. 7.10. Full-wave FEM simulated RH/LH-interface SPs ($\overline{E}$-field magnitude) at the points A (a), B (b), C (c) and D (d) of Fig 7.8(b) using the scheme of Fig 7.9, except for (d), where medium i has been replaced by air and medium 1 has been merged into medium 2 (both metamaterial). From [37]. © IEEE; reprinted with permission.

TABLE 7.2. SP Data at the Points A, B, C, and D of Fig 7.8.

	ε_{r2} (1)	μ_{r2} (1)	f (GHz)	λ_x^{th} (mm)	λ_x^{sim} (mm)
A	−0.5907	−1.0545	3.3006	43.8	44.6
B	−0.6652	−1.1719	3.2007	74.3	76.0
C	−0.7538	−1.3277	3.0810	97.3	97.4
D	−1.3783	−2.4280	2.5006	166.2	—

nicely observed in the field distribution. Finally, below point C in Fig 7.8(b), k_{z1} becomes real, and therefore only nonradiative SPs, coupling only to propagating waves, can be observed. In fact, this simply corresponds to conventional refraction and here to the well-established phenomenon of negative refraction, which can be observed in Fig 7.10(d).

It seems that the first investigations on MTM SPs were carried out by Ruppin [38, 39, 40], based on the plamonic type expressions of Eqs. (1.2) and (1.4) for ε and μ.[22] Ruppin described [38] the SP at the surface of a LH sphere and showed the following. When the real part of either the permittivity or the permeability, but not both, are negative, the RCS (or "extinction cross-section") of the sphere is relatively high, as a consequence of the Mie resonances associated with SP propagating at the surface of the sphere. In contrast, in the LH range of the sphere, a dip appears in the RCS because the sphere becomes transparent to electromagnetic radiation. Fields penetrate inside and through the structure, so that RCS strongly decreases. Ruppin [39] also solved the problem of a semi-infinite half-space (similar to that discussed above except for the fact that μ is resonant here) and showed that a LH/RH interface supports SP modes with possibly two SP resonances for each polarization. Ruppin [40] also treated the problem of SP at the interfaces of a LH slab. He observed that, in the case of the slab, the number of SP branches is twice the number of the corresponding branches of a semi-infinite medium, due to the interaction between the modes that are localized on either side of the slab, which cause a split up. This study is particularly relevant for a deeper understanding of the subwavelength slab of Pendry (Section 2.11).

Several further studies related to MTM SPs have been conducted and have demonstrated the enriched possibilities provided by MTMs in SP phenomena. For instance, Podolskiy et al. presented metal nanowire composites supporting SPs [23] and acting as a LH MTM [41]. Ishimaru et al. addressed the problem of the interface between a RH medium and an arbitrary MTM from the point of view of complex poles, branch points, and zeros in the $\varepsilon - \mu$ diagram, analyzed the location of the poles in the proper and improper Riemann sheets, and discussed forward/backward surface waves, Zenneck waves, and the relations between the Brewster angle and Sommerfeld poles [42]. Alù et al. investigated the problem of this RH/LH interface excited by a monochromatic thin sheet of currents and

---

[22]Because μ is resonant in these expressions, the results of Ruppin are different from those discussed above where both constitutive parameters are nonresonant, according to Eq. (7.20).

verified the antiparallelism existing between the Poynting vectors on either side of the interface [43].

7.8 ANTENNA RADOMES AND FREQUENCY SELECTIVE SURFACES

MTMs can also be used as antenna radomes or frequency selective surfaces (FSSs) to control the directive emission of radiating sources. The availability of MTMs with arbitrarily tailorable constitutive parameters creates opportunities for engineering radomes and FSSs with unique performances. One of the main challenges in this field will be the development of 3D isotropic MTMs (Section 7.3) since the radiation from small sources, the radiation patterns of which one may typically wish to control, is omnidirectional.

For example, the parabolic refractor described in Section 7.2 may be used as an interface or lens radome for a highly directive in-plane endfire antenna. Along the same line, 2D plano-concave NRI SRR-TW lens was demonstrated by Parazzoli et al. [44], where stronger focusing characteristics, lighter weight, and reduced aberration (due to larger radius of curvature for the same focal length) were pointed out as advantages of a curved NRI lens over a normal lens.

Ziolkowski et al. [45] demonstrated the possibility to increase the gain of an antenna by surrounding it with a LH spherical shell. In this case, the LH shell may be designed to act as a matching elements between the radiating part of the antenna and free space. This idea was inspired by the observation that, whereas a small dipole antenna within a RH medium exhibits a large capacitive reactance (i.e., not matchable to free space), the same antenna within a LH medium exhibits a large inductive reactance. The authors of the paper verified analytically and numerically transposition to achieve excellent matching of a small dipole with a shell surrounding the radiator. Increase in gain by orders of magnitude was reported. Although such a LH shell is not readily available in practice, novel 3D MTMs (Section 7.3) may provide a solution to engineer it in future.

Another idea to manipulate radiation of an antenna by a MTM was suggested by Enoch et al. [46]. Consider a radiating source embedded in a MTM slab of refractive index n_1 in contact with air ($n_2 = 1$), as depicted in Fig 7.11. If the source is very small, it will radiate within the MTM in a relatively omnidirectional manner and therefore rays will be incident on the MTM-air interface under various angles θ_1. By naive consideration of Snell's law of refraction [Eq. (2.55)], it may be anticipated that, if the MTM exhibits $n_1 \approx 0$, then the output angles θ_2 for all the rays, whatever the values of θ_1, will be close to zero, since then $\theta_2 \approx \sin^{-1}\left[n_1/n_2 \sin\theta_1\right]$. This corresponds to the situation illustrated in Fig 7.11, where all the incident rays are refracted in air in a quasi-parallel manner to form a quasi-plane wave. Consequently, very directive radiation is achieved. Because $n = \sqrt{\varepsilon_r\mu_r}$, quasi-zero refractive index may be obtained either from $\varepsilon_r = 0$ or from $\mu_r = 0$. The authors of [46] considered the former case, and used a TW lattice [16] to generate $\varepsilon_{r1} \approx 0$, by operating the medium slightly above the plasma

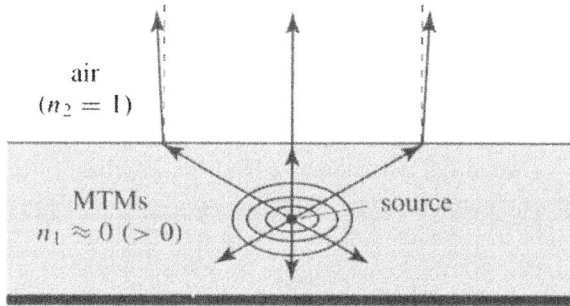

Fig. 7.11. Directive emission obtained by a MTM with refractive index close to zero [46].

frequency (in order to have $\varepsilon_r > 0$; otherwise, if $\varepsilon_r < 0$, the medium is opaque). The structure has been demonstrated experimental [46].

It may be anticipated that various other kinds of radomes and FSSs will be developed by using novel MTMs concepts and structures.

7.9 NONLINEAR MTMs

It seems that investigations on nonlinear MTMs were first reported by Zharov et al. [47]. This paper theoretically considers two different types of nonlinearities in the SRR/TW structure: a nonlinearity due to the presence of a nonlinear dielectric (Kerr effect, $\varepsilon_r \propto |\overline{E}|^2$ [48]) embedding the TWs and SRRs and a nonlinearity due to dependence of μ_r on the local field intensity in the gap of the rings. Moreover, the paper points out the *hysteresis*-type dependence of the nonlinear magnetic permeability and shows that this hysteresis allows second-order *phase transitions* from LH to RH properties and back. A similar analysis for the SRR array alone and with parameters appropriate to optical frequencies (Section 7.4) was performed by O'Brien et al. [24], where negative permittivity at telecommunication wavelengths and enhanced nonlinear effects (Kerr type) associated with bistable behavior are discussed. Another report in extension to [47] was presented by Zharova et al. [49]. In this paper, the nonlinear capacitance of the SRR (Fig 1.3), introduced in [47] to account for the fact that the microscopic electric field can become much higher than the macroscopic electric field carried by the propagating electromagnetic wave, is expressed in the form

$$C(V) = \frac{r^2\varepsilon}{4d_g}\left(1 + \alpha\frac{|V|^2}{V_c^2}\right), \qquad (7.22)$$

where r is the average radius of the SRR, ε is the linear part of the permittivity of the dielectric material inside the SRR gap, d_g is the size of the gap, V is the voltage across the gap, V_c is the characteristic nonlinear voltage, and

$\alpha = \pm 1$ is a *focusing factor*. The value $\alpha = +1$ corresponds to a *focusing nonlinearity*, where the eigenfrequency decreases with increasing intensity of the electromagnetic field because of the growth of the SRR capacitance, whereas the opposite effect obtained with the value $\alpha = -1$ is referred to as *defocusing nonlinearity*. Two nonlinear phenomena are described in [47] in conjunction with the incidence of a wave on a nonlinear LH slab: nonlinearity-induced suppression of the wave transmission (by switching of $\mu_{r,eff}$ from positive to positive values) and nonlinearity-induced transmission and generation of spatiotemporal solitons.

In fact, many novel *solitary waves* or *solitons* and *shock-wave* phenomena [50, 51] may occur in nonlinear MTMs to their potential for quasi-unlimited diversity in terms of electromagnetic parameters and nonlinearity manipulations. Solitary waves and solitons[23] are pulse-like stationary waves, which may propagate without ever altering in shape as they propagate in a *dispersive and nonlinear* medium as a consequence of an exact balancing between the phenomena of *group velocity dispersion (GVD)* and *self-phase modulation (SPM)* [51]. GVD, also simply referred to as dispersion (as in this book), corresponds to frequency-dependent nature of the group velocity v_g and henceforth of the group delay t_g, which leads to signal distortion in general. If $dt_g(\omega)/d\omega > 0$, dispersion is called *normal* and higher spectral components experience larger delays than lower spectral components; the opposite case of $dt_g(\omega)/d\omega < 0$ is called *anomalous* dispersion. SPM is associated with dependence of the refractive index on some field intensity quantity ι (e.g., electric field for Kerr effect), $n(t) = n_{lin} + n[\iota(t)]$, which induces a *chirp* $\Delta\omega(t)$ in the instantaneous frequency, $\omega(t) = d\phi(t)/dt = d[\omega_0 t - n(t)k_0\ell]/dt = d\{\omega_0[t - n(t)\ell/c]\}/dt = \omega_0 + \Delta\omega(t)$, where $\Delta\omega(t) = -(\ell\omega_0/c)dn(t)/dt$; if $dn(t)/dt > 0$ a pulse propagating through the medium is red-shifted ($\Delta\omega < 0$), whereas it blue-shifted ($\Delta\omega > 0$) if $dn(t)/dt > 0$. Shock waves are waves that exhibit extremely sharp rising or tailing or both edges at the output of a medium with SPM or/and GVD [51].

The interest of solitons and shock-waves is evident for applications requiring pulse width conservation, compression, or expansion. An example of this the new ultra wideband (UWB) technology, where information is directly coded by electromagnetic pulses and data rates are directly proportional to the narrowness of the transmission pulses [52]; thus, novel pulse compression and manipulation (for modulation) techniques are strongly needed. It was suggested by Caloz et al. [53] that LH TLs and MTM TLs could provide new avenues to meet these new challenges. While optical solitons (bright solitons) are generally characterized by anomalous GVD and negative SPM ($dn(t)/dt < 0$ for a signal of increasing intensity), the LH-TL based on nonlinear series varactors exhibits also anomalous GVD, but positive SPM, which may lead to efficient pulse compression and manipulation by exploiting the tailorable properties of LH or CRLH TLs. RH nonlinear TLs have been known for many years and shown by Lonngren et al.

---

[23]Solitons are a particular type of solitary waves, which have the specific property of being orthogonal, in the sense that when two of these waves cross one another in the medium their intensity profile is not altered (only phase shifts are imparted as a result of the interaction).

to support both solitons [54] and shock waves [55]; in this case, because a RH TL is naturally nondispersive, dispersion has to be generated artificially, typically by using a PRH series-L shunt-C network operated in the vicinity of the lowpass cutoff. The fact that GVD is an intrinsic property of LH and CRLH MTMs suggests that stronger and richer nonlinear effects will be obtainable in LH nonlinear TLs than in RH nonlinear TLs.

Various types of natural materials can be used in conjunction with MTMs to develop novel nonlinear engineering applications, such as for instance ferrimagnetic [56, 57] or ferroelectric films [58].

7.10 ACTIVE MTMs

Active MTMs represent entirely unexplored fields to date. It may be predicted that novel active applications will soon emerge from the new paradigm of MTMs and MTM TL structures and extend the range of available active microwave and optoelectronic devices [59].

An immediate direction for integration of active components into MTMs is compensation of losses, which has been a critical issue in many MTMs and in particular in SRR/TW structures. In the same manner as (passive) varactors were distributed along MTM TL structures (Section 6.3.1), transistors may be implemented, for instance to equalize and modulate the profile of the fields along a LW structure to control its radiation characteristics, or to develop enhanced gain-bandwidth distributed power amplifiers [60]. In the future, MTMs may become part of complex radio-frequency integrated circuits (RFICs) and monolithic microwave integrated circuits (MMICs) technology implementations [61].

7.11 OTHER TOPICS OF INTEREST

In this last section, we will list a number of topics of interest that did not fit in the previous sections, either because they have not been really demonstrated yet or because they are of purely prospective nature.

Although its has been the objective of most researchers to develop isotropic MTMs, *anisotropic* MTMs may be useful in specific applications, such as quasi-optical beam-forming (Section 7.2) or reflector systems. An *orthogonally anisotropic* RH/LH TL MTMs has been presented by Caloz et al. [62]. Balmain et al. [63] presented a *resonance-cone* "spatial spectrum analyzer" MTM constituted by a planar wire-grid network with inductors along one direction and capacitors along the other orthogonal direction. The idea of anisotropy can be pushed further to *nonuniform* or *graded-index* MTMs, which have been used in the beam-tuning CRLH antenna, demonstrated in Section 6.3.2 and in the focusing LW interface presented in [64].

Another interesting topic is the *subwavelength cavities* proposed by Engheta [65]. This cavity consists of the juxtaposition of RH and LH slabs within a PPWG and exploits the phenomenon of *phase compensation/conjugation* between the RH and the LH media to achieve a resonance not depending on the sum of the RH and LH thicknesses, but only on their ratio.[24] Although the practical implementation of such a cavity is not readily possible, because reduction of actual size of the cavity is limited by the size of the implants in current MTMs, cavities of sizes far less than the conventional $\lambda_g/2$ are theoretically possible.

Another exciting idea, also proposed by Engheta, is the futuristic concept of NRI nano-circuits and nano-TLs at optical frequencies [67] (Section 7.4). Optical lumped nano-inductors and nano-capacitors would be conceivable from plasmonic nano-layers exploiting displacement loop currents (Section 7.5) produced in an array of plasmonic nano-spheres arranged in a specific pattern to form a single wavelength "loop macro-inclusion".

Even more futuristic—or audacious!—is the exploration by Lakhtakia et al. of MTMs in the context of *special and general relativity*, and their speculations on the possible occurrence of the NRI phenomenon in "scenarios encompassing space exploration, navigation, and communications; remotely guided extraterrestrial manufacturing industries; and the unraveling of the evolution of our universe after the last Big Bang" [68, 69].

Finally, it may be safely predicted that the combination of currently exploding nanotechnologies [70] with novel MTMs will give birth to "intelligent" novel innovative MTMs and structures with unique performances, such as programmable or automatic reconfigurability for instance. MTMs may be implemented in semi-conductor technologies and under various new forms, including ferroelectric, ferrimagnetic/ferromagnetic, chemical, organic, or biological substances. The wealth of potential concepts and applications of MTMs is limited only by our imagination.

REFERENCES

1. D. R. Smith, W. J. Padilla, D. C. Vier, S. C. Nemat-Nasser, and S. Schultz. "Composite medium with simultaneously negative permeability and permittivity," *Phys. Rev. Lett.,* vol. 84, no. 18, pp. 4184–4187, May 2000.

2. R. A. Shelby, D. R. Smith, and S. Schultz. "Experimental verification of a negative index of refraction," *Science,* vol. 292, pp. 77–79, April 2001.

3. C. A. Balanis. *Antenna Theory, Analysis and Design,* Second Edition, John Wiley & Sons, 1997.

4. J. D. Kraus and R. J. Marhefka. *Antennas,* Third Edition, McGraw Hill, 2001.

---

[24]This concept bears some resemblance with the zeroth-order resonator presented in Section 6.6, in the sense that both resonators can be independent on their overall physical length. Their are two fundamental differences between these two concepts. First, whereas the zeroth-order resonator is made one single MTM (CRLH), Engheta's resonator includes two media (one RH and one LH). Second, whereas the field distributions (magnitude and phase) along the zeroth-order are constant (infinite wavelength), the fields distributions in the subwavelength resonator strongly vary across the structure and exhibit surface plasmons at the RH/LH interface (Sections 2.11 and 4.4.4) [66].

5. R. E. Collin. "Artificial Dielectrics," (Chapter 12), in *Field Theory of Guided Waves,* Second Edition, Wiley-Interscience, 1991.

6. D. Schurig and D. R. Smith. "Universal description of spherical aberrationfree lenses composed of positive or negative index media," *ArXiv Physics e-Prints,* 0307088, July 2003.

7. C. Caloz, A. Lai, and T. Itoh. "Wave interactions in a LH mushroom structure," in Proc. *IEEE-AP-S USNC/URSI National Radio Science Meeting,* Monterey, CA, June 2004.

8. A. Lai, W. -Y. Wu, K. M. K. H. Leong, T. Itoh, and C. Caloz. "Quasi-optical manipulation of microwaves using metamaterial interfaces," *in Proc. IEEE-AP-S USNC/ URSI National Radio Science Meeting,* Washington, DC, July 2005.

9. B. E. A. Saleh, and M. C. Teich. *Fundamentals of Photonics,* Wiley-Interscience, 1991.

10. L. Rayleigh. "On the influence of obstacles arranged in rectangular order," *Phil. Mag.,* ser. 5, vol. 34, pp. 481–502, 1892.

11. N. A. Khizhniak. "Artificial anisotropic dielectrics: I-II-III," *Zh. Tekh. Fiz.,* vol. 27, pp. 2006–2037, 1957.

12. W. Rotman. "Plasma simulation by artificial dielectrics and parallelplate media," *IRE Trans. Antennas Propagat.,* pp. 82–84, 1962.

13. L. Lewin. "The electrical constants of a material loaded with spherical particles," *in Proc. Inst. Elec. Eng.,* vol. 94, pp. 65–68, 1947.

14. C. L. Holloway, E. F. Kuester, J. Baker-Jarvis, and P. Kabos. "A double negative (DNG) composite medium composed of magnetodielectric spherical particles embedded in a matrix," *IEEE Trans. Antennas Propagat.,* vol. 51, no. 10, pp. 2596–2603, Oct. 2003.

15. O. G. Vendik and M. S. Gashinova. "Artificial double negative (DNG) media composed by two different dielectric sphere lattices embedded in a dielectric matrix," *in Proc. European Microwave Conf.,* Amsterdam, The Netherlands, pp. 1209–1212, Oct. 2004.

16. J. B. Pendry, A. J. Holden, D. J. Robbins, and W. J. Stewart. "Low frequency plasmons in thin-wire structures," *J. Phys. Condens. Matter,* vol. 10, pp. 4785–4809, 1998.

17. J. B. Pendry, A. J. Holden, D. J. Robbins, and W. J. Stewart. "Magnetism from conductors and enhanced nonlinear phenomena," *IEEE Trans. Micr. Theory. Tech.,* vol. 47, no. 11, pp. 2075–1084, Nov. 1999.

18. M. Born and E. Wolf. *Diffraction by a Conducting Sphere; Theory of Mie,* Section 14.5 in *Principle of Optics: Electromagnetic Theory of Propagation, Interference, and Diffraction of Light,* Seventh Edition, Cambridge University Press, 1999.

19. A. Ishimaru. *Wave Propagation and Scattering in Random Media,* Wiley-IEEE Press, 1999.

20. H. Liu, S. N. Zhu, Y. Y. Zhu, Y. F. Chen, and N. B. Ming, and X. Zhang. "Piezoelectric-piezomagnetic multilayer with simultaneously negative permeability and permittivity," *App. Phys. Lett.,* vol. 86, pp. 102904:1–3, 2005.

21. I. Vendik. *Private communication.*

22. E. M. Lifshitz, L. D. Landau, and L. P. Pitaevskii. *Electrodynamics of Continuous Media: Volume 8,* Second Edition, Butterworth-Heinemann, 1984.

23. V. A. Podolskiy, A. K. Sarychev, and V. M. Shalaev. "Plasmon modes in metal nano-wires and left-handed materials," *J. Nonlin. Opt. Phys. Mat.*, vol. 11, no. 1, pp. 65–74, 2002.

24. S. O'Brien, D. McPeake, S. A. Ramakrishna, and J. B. Pendry. "Near-infrared photonic band gaps and nonlinear effects in negative magnetic metamaterials," *Phys. Rev. B,* vol. 69, pp. 241101:1–4, 2004.

25. G. Shvets and Y. A. Urzhumov. "Engineering the electromagnetic properties of periodic nanostructures using electrostatic resonances," *Phys. Rev. Lett.,* vol. 93, pp. 2439021–4, 2004.

26. M. L. Povinelli, S. G. Johnson, J. D. Joannopoulos, and J. B. Pendry. "Toward photonic-crystal metamaterials: creating magnetic emitters in photonic crystals," *Appl. Phys. Lett.,* vol. 82, pp. 1069–1071, 2003.

27. N. Engheta. "Ideas for double-negative metamaterials, negative-index nano-layers, nano-circuit elements and transmission lines at optical frequencies," *in Proc. Latsis Symposium,* Lausanne, Switzerland, pp. 54–56, March 2005.

28. H. J. Schneider and P. Dullenkopf. "Slotted tube resonator: a new NMR probe head at high observing frequencies," *Rev. Sci. Instrum.,* vol. 48, no. 1, pp. 68–73, 1977.

29. W. H. Hardy and L. A. Whitehead. "Split-ring resonator for use in magnetic resonance from 200–2000 MHz," *Rev. Sci. Instrum.,* vol. 52, no. 2, pp. 213–216, 1981.

30. M. C. K. Wiltshire, J. B. Pendry, I. R. Young, D. J. Larkman, D. J. Gilderdale, and J. V. Hajnal. "Microstructured magnetic materials for RF flux guides in magnetic resonance imaging," *Science,* vol. 291, pp. 849–851, Feb. 2001.

31. M. C. K. Wiltshire, J. V. Hajnal, J. B. Pendry, D. J. Edwards, and C. J. Stevens. "Metamaterial endoscope for magnetic field transfer: near field imaging with magnetic wires," *Optics Express,* vol. 11, no. 7, pp. 709–715, April 2003.

32. T. J. Yen, W. J. Padilla, N. Fang, D. C. Vier, D. R. Smith, J. B. Pendry, D. N. Basov, and X. Zhang. "Terahertz magnetic response from artificial materials," *Science,* vol. 303, pp. 1494–1496, March 2004.

33. H. O. Moser, B. D. F. Casse, O. Wilhelmi, and B. T. Saw. "Terahertz response of a microfabricated rod-split-ring-resonator electromagnetic metamaterial," *Phys. Rev. Lett.,* vol. 94, pp. 063901:1–4, Feb. 2005.

34. D. L. Woolard, W. R. Loerop, and M. S. Shur (Editors). *Terahertz Sensing Technology,* (2 volumes) World Scientific, Selected Topics in Electronics and Systems, vol. 32, 2003.

35. H. Raether. *Surface Plasmons,* Springer-Verlag, 1988.

36. J. A. Stratton. *Electromagnetic Theory,* McGraw-Hill, 1941.

37. C. Caloz, C.-J. Lee, D. R. Smith, J. B. Pendry, and T. Itoh. "Existence and properties of microwave surface plasmons at the interface between a right-handed and a left-handed media," *IEEE AP-S USNC/URSI National Radio Science Meeting,* Monterey, CA, June 2004.

38. R. Ruppin. "Surface polaritons of a left-handed material slab," *Solid State Communications,* vol. 116, pp. 411–415, 2000.

39. R. Ruppin. "Surface polaritons of a left-handed medium," *Phys. Lett. A,* vol. 277, pp. 61–64, 2000.

40. R. Ruppin. "Surface polaritons of a left-handed material slab," *J. Phys. Condens. Matter,* vol. 13, pp. 1811–1818, 2001.

41. V. A. Podolskiy, A. K. Sarychev, and V. M. Shalaev. "Plasmon modes and negative refraction in metal nanowire composites," *Optics Express,* vol. 11, no. 1, pp. 65–74, 2002.

42. A. Ishimaru, J. R. Thomas, and S. Jaruwatanadilok. "Electromagnetic waves over half-space metamaterials of arbitrary permittivity and permeability," *IEEE Trans. Antennas Propagat.,* vol. 53, pp. 915–921, March 2005.

43. A. Alù and N. Engheta. "Radiation from a traveling-wave current sheet at the interface between a conventional material and a metamaterial with negative permittivity and permeability," *Microwave Opt. Technol. Lett.,* vol. 35, no. 6, pp. 460–463, Dec. 2002.

44. C. G. Parazzoli, R. B. Greegor, J. A. Nielsen, M. A. Thompson, K. Li, A. M. Vetter, and M. H. Tanielian. "Performance of a negative index of refraction lens," *App. Phys. Lett.,* vol. 84, no. 17, pp. 3232–3234, April 2004.

45. R. W. Ziolkowski and D. Kipple. "Application of double negative materials to increase the power radiated by electrically small antennas," *IEEE Trans. Antennas Propagat.,* vol. 51, no. 10, pp. 2626–2640, Oct. 2003.

46. S. Enoch, G. Tayeb, P. Sabouroux, N. Guérin, and P. Vincent. "A metamaterial for directive emission," *Phys. Rev. Lett.,* vol. 89, no. 21, pp. 213902:1–4, Nov. 2002.

47. A. A. Zharov, I. V. Shadrivov, and Y. S. Kivshar. "Nonlinear properties of left-handed metamaterials," *Phys. Rev. Lett.,* vol. 91, no. 3, pp. 037401:1–4, July 2003.

48. R. W. Boyd. *Nonlinear Optics,* Second Edition, Academic Press, 2003.

49. N. A. Zharova, I. V. Shadrivov, A. A. Zharov, and Y. S. Kivshar. "Nonlinear transmission and spatiotemporal solitons in metamaterials with negative refraction," *Optics Express,* vol. 13, no. 4, pp. 1291–1298, 2005.

50. A. C. Scott, F. Y. F. Chu, and D. W. McLaughlin. "The soliton: a new concept in applied science," *Proc. IEEE,* vol. 61, pp. 1443–1483, 1973.

51. G. P. Agrawal. *Nonlinear Fiber Optics,* Third Edition, Academic Press, 2001.

52. S. Roy, J. R. Foerster, V. S. Somayazulu, and D. G. Leeper. "Ultrawideband radio design: the promise of high-speed, short-range wireless connectivity," *Proc. IEEE,* vol. 92, no. 2, pp. 295–311, Feb. 2004.

53. C. Caloz, I.-H. Lin and T. Itoh. "Characteristics and potential applications of nonlinear left-handed transmission lines," *Microwave Opt. Technol. Lett.,* vol. 40, no. 6, pp. 471–473, March 2004.

54. J. A. Kolosick, D. L. Landt, H. C. Hsuan, and K. E. Lonngren. "Experimental study of solitary waves in a nonlinear transmission line," *App. Phys.,* vol. 2, pp. 129–131, Sept 1973.

55. K. E. Lonngren, D. L. Landt, C. M. Burde, and J. A. Kolosick. "Observation of shocks on a nonlinear dispersive transmission line," *IEEE Trans Circ. Syst.,* pp. 376–378, April 1975.

56. M. Chen, M. A. Tsankov, J. M. Nash, and C. E. Patton. "Backward-volume-wave microwave-envelope solitons in yttrium iron garnet films," *Phys. Rev. B,* vol. 49, no. 18, pp. 12773–12790, May 1994.

57. C. E. Zaspel, Y. G. Rapoport, K. Yashiro, and S. Ohkawa. "Shock formation in magnetic thin films," *IEEE Trans. Magnetics,* vol. 37, no. 4, pp. 2460–2462, July 2001.

58. M. P. Brown, S. J. Newnham, and P. W. Smith. "The application of SbSi and other ferroelectric materials to the generation of solitons and shock wave on nonlinear high

voltage transmission lines" *in Proc. Seventh Int. Conf. Diel. Mat. Meas. App.*, no. 430, pp. 76-81, Sept. 1996.

59. M. J. W. Rodwell, S. T. Allen, R. Y. Yu, M. G. Case, U. Bhattacharya, M. Reddy, E. Carman, M. Kamegawa, Y. Konishi, J. Pusl, and R. Pullela. "Active and nonlinear wave propagation devices in ultrafast electronics and optoelectronics," *Proc. IEEE*, vol. 82, no. 7, pp. 1037–1059, July 1994.

60. I. Bahl and P. Bhartia. *Microwave Solid State Circuit Design*, Wiley Interscience, 1988.

61. I. D. Robertson and S. Lucyszyn. *RFIC and MMIC design and technology*, IEEE Circuits, Devices and Systems Series 13, 2001.

62. C. Caloz and T. Itoh. "Positive / negative refractive index anisotropic 2D meta-materials," *IEEE Microwave Wireless Compon. Lett.*, vol. 13, no. 12, pp. 547–439, Dec. 2003.

63. K. G. Balmain, A. A. E. Lüttgen, and P. C. Kremer. "Power flow for resonance cone phenomenon in planar anisotropic metamaterials," *IEEE Trans. Antennas Propagat.*, vol. 51, no. 10, pp. 2612–2618, Oct. 2003.

64. I.-H. Lin, C. Caloz, and T. Itoh. "Near-field focusing by a nonuniform leaky-wave interface," *Microwave Opt. Technol. Lett.*, vol. 44, no. 5, pp. 416–418, March 2005.

65. N. Engheta. "An idea for thin subwavelength cavity resonators using metamaterials with negative permittivity and permeability," *IEEE Antennas Wireless Propagat. Lett.*, vol. 1, pp. 10–13, 2002.

66. A. Alù and N. Engheta. "Pairing and epsilon-negative slab with a mu-negative slab: resonance, tunneling and transparency," *IEEE Trans. Antennas Propagat.*, vol. 51, no. 10, pp. 2558–2571, Oct. 2003.

67. N. Engheta. "Ideas for double-negative metamaterials, negative-index nano-layers, nano-circuit elements and transmission lines at optical frequencies," in *Proc. Latsis Symposium*, Lausanne, Switzerland, pp. 54–56, March 2005.

68. A. Lakhtakia, T. G. Mackay, and S. Setiawan. "Global and local perspectives of grav-itationally assisted negative-phase-velocity propagation of electromagnetic waves in vacuum," *Phys. Lett. A*, vol. 336, pp. 89–96, Jan. 2005.

69. T. G. Mackay, A. Lakhtakia, and S. Setiawan. "Gravitation and electromagnetic wave propagation with negative phase velocity," *New J. Phys.*, vol. 7, no. 75, pp. 1–14, March 2005.

70. R. Waser (Editor). *Nanoelectronics and Information Technology*, Wiley-VCH, 2003.

INDEX

Electromagnetic Metamaterials: Transmission Line Theory and Microwave Applications,
By Christophe Caloz and Tatsuo Itoh
Copyright © 2006 John Wiley & Sons, Inc.

Printed and bound by CPI Group (UK) Ltd, Croydon, CR0 4YY

19/02/2023

03192937-0001